Desktop Encyclopedia of Telecommunications

Other McGraw-Hill Communications Books of Interest

Desktop Encyclopedia of Telecommunications

Nathan J. Muller

McGraw-Hill

New York San Francisco Washington, D.C. Auckland Bogotá
Caracas Lisbon London Madrid Mexico City Milan
Montreal New Delhi San Juan Singapore
Sydney Tokyo Toronto

Library of Congress Cataloging-in-Publication Data

Muller, Nathan J.
Desktop encyclopedia of telecommunications / Nathan J. Muller.
p. cm.
Includes index.
ISBN 0-07-044457-9
1. Telecommunication—Dictionaries. I. Title.
TK5102.M85 1998
384'.03—dc21 97-41307
CIP

McGraw-Hill

A Division of The ***McGraw-Hill*** *Companies*

1 2 3 4 5 6 7 8 9 0 DOC/DOC 9 0 3 2 1 0 9 8

ISBN 0-07-044457-9

Illustrated by Linda L. Tyke.

The sponsoring editor for this book was Steve Chapman, the editing supervisor was Stephen Moore, and the production supervisor was Pamela Pelton. It was set in Souvenir Light by Jan Fisher through the services of Barry E. Brown (Broker—Editing, Design and Production).

Printed and bound by R. R. Donnelley & Sons Company.

This book is printed on recycled, acid-free paper containing a minimum of 50% recycled, de-inked fiber.

To the Kostura girls . . .
Mary, Ann, Helen

Contents

 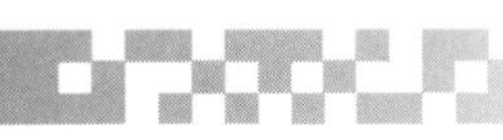

Introduction

Telecommunications encompasses many technologies and is one of the most rapidly advancing fields in the world today. Worldwide, the revenue from telecom products and services in 1996 stood at US$700 billion, a figure which represented 3 percent of the global economy. With continued deregulation and trade liberalization, telecommunications could mean global income gains of some US$1 trillion over the next decade or so.

Other benefits flow from the resulting competition in telecom markets. People and nations can communicate more easily and understand one another better, unencumbered by physical distance. Consumers enjoy more choice, better quality, and lower prices. Modernization and investment are enhanced worldwide, and more jobs are created.

This book provides nontechnical professionals with the essential knowledge required to succeed in this dynamic, fast-growing industry. This comprehensive volume offers readers a painless way to fill any knowledge gaps they might have, while providing new insights about the operational aspects of today's increasingly complex networks. Of course, this book also makes an excellent reference for those outside of the industry who want to better understand how telecommunications technologies are advancing and changing our everyday lives.

The topics described in the following pages span local and wide area networking environments and include coverage of both voice and data. In addition to explanations of technologies, equipment and services, and network applications, there also are discussions of key regulations and standards, and the various organizations that have contributed to the evolutionary growth of the telecommunications industry. Since technology by itself is of little benefit unless properly implemented, this book also contains management concepts that help put many of the other topics into perspective.

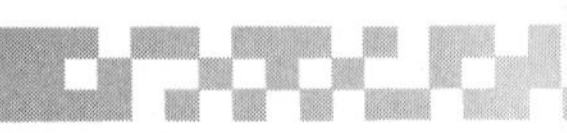

The information contained in this book, especially as it relates to specific vendors and products, is believed to be accurate at the time it was written and is, of course, subject to change with continued advancements in technology and shifts in market forces. Mention of specific products and services is for illustrative purposes only and does not constitute an endorsement of any kind by either the author or the publisher.

Desktop Encyclopedia of Telecommunications

Advanced Intelligent Network

The *Advanced Intelligent Network* (AIN) provides carriers with the means to create and uniformly support telecommunications services and features via a common architectural platform. New services are created and supported through processors, software, and databases distributed throughout the public network. These intelligent nodes are linked via Signaling System 7 (SS7) to support a variety of services and advanced call-handling features across multiple-vendor domains. By accessing these intelligent nodes, users are able to design and control their own services and customize features without telephone company involvement. The advanced intelligent network is a natural extension of the flexibility provided by the voice-oriented networks of AT&T, MCI, and Sprint that have been in operation since the early 1980s.

Advantages for carriers. The advanced intelligent network enables carriers to offer new services to subscribers and at the same time reduce both their capital spending and operating costs. They also have the flexibility to design and implement new services without having to rely on traditional switching vendors to support them. Most important, with impending competition in local exchange markets, the advanced intelligent network will provide the tools required by carriers to remain viable in a competitive environment.

Carriers traditionally purchased software and updates from the switch vendor and then loaded it into each switching system that provided the service. If the carrier had several different types of switches on its network, the process of introducing a new service was more cumbersome because the software upgrades had to be coordinated among the various switch manufacturers to ensure service continuity. This process delayed the availability of new services to some locations.

Because the services and features in an advanced intelligent network are defined in software programs distributed among fewer locations (intelligent nodes), telephone companies can develop and enhance the network software and eliminate switch manufacturer involvement altogether. Once the carrier develops new services, it can offer them to customers immediately via intelligent nodes distributed throughout the network. By accessing these nodes, customers can implement instantly a uniform set of services for maximum efficiency and economy, even across multiple locations.

By allowing users to design, add, or change their own networks and services from a management terminal, carriers are relieved of much of the administrative burden associated with customer service. In turn, this reduces the carrier's personnel requirements and, consequently, the cost of network operation.

Carriers also can control costs by aggregating customer demand to ensure full use of the network. This entails centralizing infrequently used or highly specialized capabilities and distributing frequently used and highly shared capabilities over a wider area. The result is more efficient utilization of network resources and lower operating costs. The distributed architecture of the advanced intelligent network makes all this possible.

Advantages for users. Rather than investing heavily in premises-based equipment to obtain a high level of performance and functionality via private networks, users are able to tap the service logic of intelligent nodes embedded in the public network for advanced calling features such as interactive voice response (or speech recognition) and bandwidth on demand to support multimedia applications. They also are able to design their own sophisticated hybrid networks without carrier involvement, and to manage them from an on-premises terminal as if they were private networks.

AIN allows users assemble the required resources, in the form of functional components, in accordance with their design specifications. Users also can test the integrity of network models by simulation prior to implementation. Service provisioning is virtually instantaneous. Because telephone companies can activate new services quickly and easily at all network locations, users will not have to wait months or years for new services to become available to all their locations. Users even can receive additional bandwidth within minutes of a request. As AIN technology and fiber-optic networks continue to evolve, large corporate users will be able to activate gigabit-per-second (Gbps) channels on demand to send data a thousand times faster than today's megabit-per-second (Mbps) channels.

These benefits portend substantial cost savings in up-front hardware investments and ongoing network operating costs. Ultimately, users no longer will have to settle for off-the-shelf networking solutions from carriers and hardware vendors. Nor will they have to rely on high-priced private networks for the desired levels of performance and functionality.

Building blocks of AIN. The intelligent network architecture (see Fig. A1) is composed of the following discrete elements that interact with each other to support the creation and delivery of services:

- *Service Switching Points* (SSPs) are distributed switching nodes that process calls by interacting with SCPs.
- *Service Control Points* (SCPs) are centralized nodes that contain service control logic.
- *Signal Transfer Points* (STPs) are tandem packet switches that route SS7 messages among SCPs and SSPs.
- *Service Management System* (SMS) is a centralized operations system for creating and introducing services.

- *Intelligent Peripheral Nodes* (IPNs) are intelligent nodes that provide a range of capabilities that are used in conjunction with many types of services.
- *Vendor Feature Nodes* (VFNs) are network nodes that provide a variety of value-added services that can compete with those offered by the local telephone company.
- *Signaling System 7* (SS7) is a separate, high-speed, packet-switched network that provides call control signaling between intelligent network nodes.

AIN applications. The following discussion highlights some of the new business services that AIN makes possible.

Area-wide networking. Designed for organizations and businesses with multiple locations, Area-Wide Networking enables subscribers to link their locations into a single communications network economically. Although the service is a fundamentally different solution because it is software-defined and simply overlays the existing network, it can be integrated with existing equipment and systems such as Centrex, PBXs, or key systems. It not only simplifies dialing between locations, it

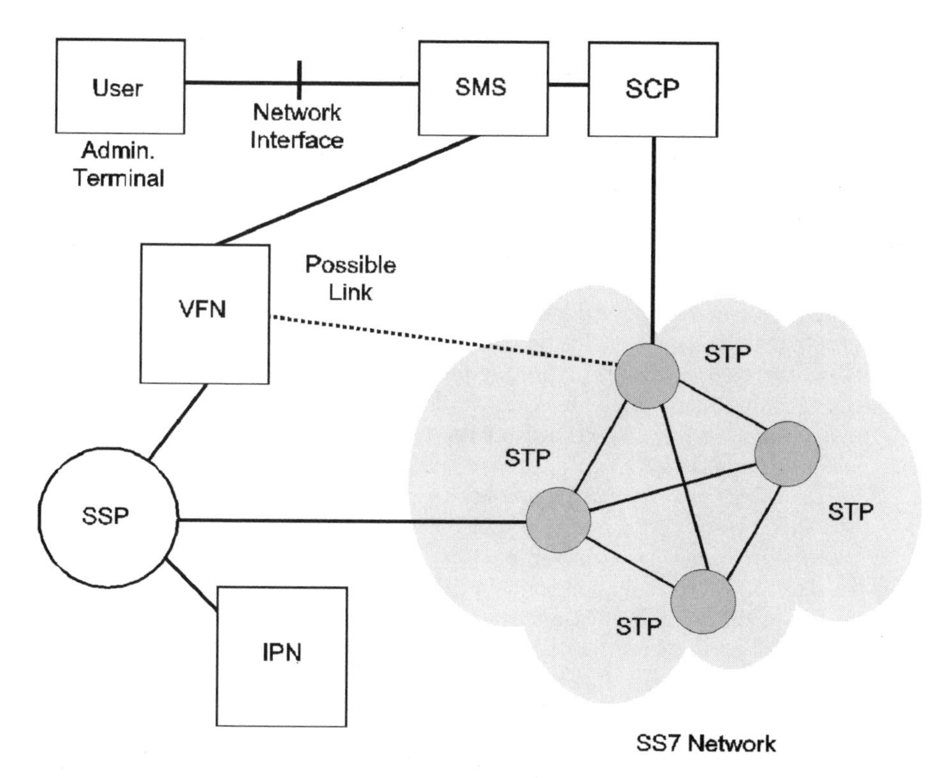

Figure A1 *Intelligent network elements, tied together with SS7, provide the infrastructure that makes it possible to deliver a variety of new telecommunications services with numerous advantages to both carriers and users.*

simplifies billing as well. Employees who work from home, for example, can be tied into the corporate network without extra dialing, enabling them to use office resources and make and receive business calls as if they were using the office phone system. Work-related calls are automatically billed to the company, saving the time and money associated with expense recording. The service offers several dialing plans, access-control options, and security features that can be customized to the customer's needs. Future add-ons are easily accommodated without the need for expensive equipment or associated maintenance costs.

Intelligent caller data. This service is designed to appeal to businesses that depend on incoming phone calls to sell products or services. It goes far beyond caller ID to give companies a report with aggregate information on incoming calls. The report summarizes quantity of calls by date, hour, area code and prefix, ZIP+4 code, and demographic code. This information can be used by businesses to establish a profile of their customers, test the effectiveness of various forms of advertising, and target their sales efforts for maximum effect.

Disaster routing. An organization's need to respond to its incoming calls and keep communications flowing must continue even if it is unable to maintain operations due to a flood, fire, storm, or power outage, or because a cable to the premises is severed. With Disaster Routing, companies have the control they need to ensure that critical calls will get through regardless of the disaster confronting them. Incoming calls to multiple telephone numbers are redirected to specified locations—even if the company has a PBX system with Direct Inward Dialing. Calls can be routed to another department, branch location emergency hot site, long-distance site, or any other specified location, enabling the company to continue business operations.

Intelligent call redirection. This enables customers to have incoming calls to multiple telephone numbers forwarded based on time of day, day of week, percentage of calls, or other variables. For example, a business can use this service to reroute calls immediately in the event of a natural disaster such as a fire or storm. It also can be used to give priority service to key customers by forwarding their calls to designated salespeople. In addition, it helps businesses manage resources during peak calling hours, or sends calls to alternate locations after regular business hours. The service is designed to appeal to businesses that depend on incoming phone calls, such as brokerage firms, banks, government agencies, mortgage companies, insurance claims centers, catalog companies, answering services, and hospitals.

Intelligent one-number calling. This is a service for businesses with several sites. It allows multiple locations to be linked by a single telephone number, ensuring that incoming calls are picked up at the right location. It also allows a business to advertise one phone number, cutting overhead costs. The basic service works by identifying the ZIP codes of incoming calls and automatically routing the calls to the location specified for trade areas established by the business. Several options are available that offer a business even greater control of its calls: time-of-day/day-of-week routing, specific date routing, and allocation routing. Used in combination with Intelligent Caller Data (described previously), this service allows businesses to get monthly reports and analyze incoming call volumes based on variables such as time-of-day, ZIP code, and demographic code. This information enables them to market specifically to certain customers and to target advertising dollars more efficiently.

Intelligent call screening. This service checks calls before their completion to prohibit access to corporate systems, restricting access to computers, PBXs, and faxes. Calls from phone numbers that have been authorized, or from callers who have been given an access code, gain access. All other calls are denied. Authorization is based on the calling party's number. The list of authorized callers can be changed, expanded, or updated via a password-protected PC interface at any time, or on an emergency basis, via Touch-Tone input into an Interactive Voice Response (IVR) system.

Reports detailing all calling activity are available. These reports track which authorized users connected and when, and track unauthorized usage for rejected calls. Reports are distributed either weekly or monthly and are available on either diskette or paper. If immediate review of either authorized telephone number list or access code list is required, the administrator can call the IVR.

Automatic callback. With this service, users can automatically place a call to the last number dialed without having to redial the full number—regardless of whether the call was answered, unanswered, or busy. This service is used to contact parties a caller has been unable to reach or to continue an interrupted conversation. When the feature is activated, the number that the user dialed last is rung again. If the line is idle, the call goes through. If the line is busy, the caller hears a special announcement, and the switch continues to monitor the number. When the line becomes idle again, the caller hears a special ring on his or her phone or a confirmation tone. When the caller picks up the phone, the telephone he or she was dialing rings.

Summary. The intelligent network provides the means for carriers to create and uniformly introduce and support new services. This intelligence is derived from sophisticated software and processors embedded within the public network. With the means to access this intelligence, users are able to engineer their own services, customize features, and exercise more control over their communications without telephone company involvement—no matter how many different carriers are involved. As the intelligent network expands, companies both large and small will be able to view the entire network as a customizable extension of everything they do in the office.

See also

HYBRID NETWORK

SIGNALING SYSTEM 7

VIRTUAL PRIVATE NETWORK

Advanced Peer-to-Peer Networking

APPN is IBM's next-generation SNA technology for linking devices without requiring the use of a mainframe. Specifically, it is IBM's proprietary SNA routing scheme for client-server computing in multiprotocol environments. As such, it is part of IBM's LU 6.2 architecture, also known as *Advanced Program-to-Program Communications*, which facilitates communications among programs running on different platforms.

SNA routing. Generally, APPN is used when SNA traffic must be prioritized by class of service, which routes traffic directly to end nodes, or when SNA traffic must be routed from peer to peer without going through a mainframe.

Included in the APPN architecture are *Automatic Network Routing* (ANR) and *Rapid Transport Protocol* (RTP) features. These features route data around network failures and provide performance advantages, closing the gap with TCP/IP. ANR provides end-to-end routing over APPN networks, eliminating the intermediate routing functions of early APPN implementations; RTP provides flow control and error recovery. To these features, HPR (described later) adds a more advanced feature called *Adaptive Rate Based* (ARB) congestion prevention.

ARB uses three inputs to determine the sending rate for data. As data is sent into the network, the rate at which it is sent is monitored. At the destination node, that rate is also monitored and reported back to the originating node. The third input is the allowed sending rate. Together these inputs determine the optimal throughput rate, which alleviates congestion by minimizing the potential for packet discards.

By enabling peer-to-peer communications among all network devices, APPN helps SNA users connect to local area networks and more effectively create and use client-server applications. APPN supports multiple protocols, including TCP/IP, and allows applications to be independent of the transport protocols that deliver them.

APPN's other benefits include allowing information routing without a host, tracking network topology, and simplifying network configuration and changes. For users still supporting 3270 applications, APPN can address dependent LU protocols as well as the newer LU 6.2 sessions, which feature protects a site's investment in applications relying on older LU protocols.

High Performance Routing (HPR) can be added to streamline SNA traffic so that routers can move the data around link failures or outages. HPR is used when traffic must be sent through the distributed network without disruptions. HPR provides link-utilization features that are important when moving SNA traffic over the wide area network and provides congestion control for optimizing bandwidth. HPR's performance gain comes from its end-to-end flow controls, which are an improvement over APPN's hop-by-hop flow controls.

Summary. There are other SNA routing techniques available. *Data Link Switching* (DLSw), for example, is used in environments consisting of a large installed base of mainframes and TCP/IP backbones. DLSw assumes the characteristics of APPN and HPR routing and combines them with TCP/IP and other LAN protocols. DLSw encapsulates TCP/IP and supports SDLC and high-level data link control (HDLC) applications (Fig. A2). It prevents session time-outs and protects SNA traffic from becoming susceptible to link failures during heavy congestion periods.

SNA traffic also can be routed over frame relay. Like DLSw, SNA and APPN protocols are encapsulated—in this case within frame relay frames. Frame relay provides SNA traffic with guaranteed bandwidth through *permanent virtual circuits* (PVCs) and, compared to DLSw, uses very little overhead in the process.

See also

ADVANCED PROGRAM-TO-PROGRAM COMMUNICATIONS (APPC)

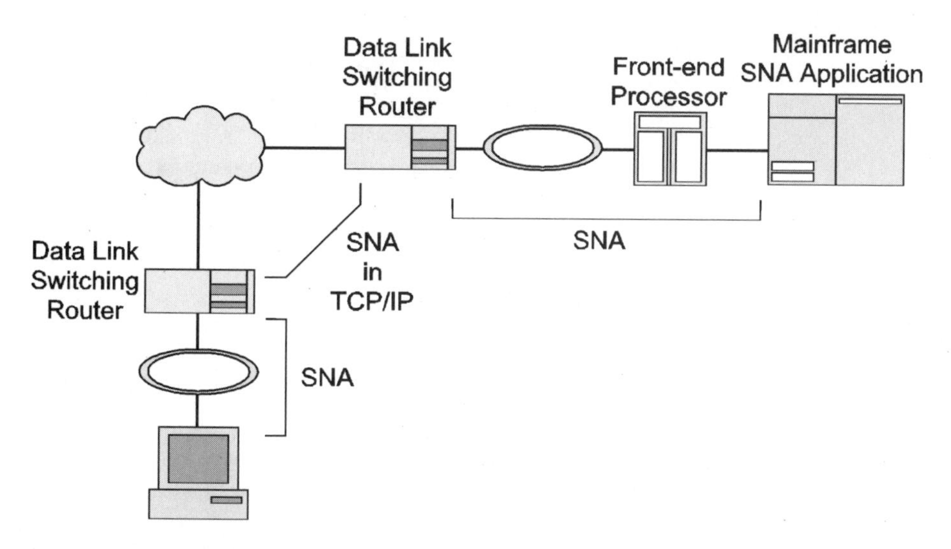

Figure A2 *A local DLSw router encapsulates SNA in TCP/IP. A remote DLSw router de-encapsulates TCP/IP so SNA traffic can be accepted by the target FEP or mainframe.*

Advanced Program-to-Program Communications

Unlike the traditional master-slave relationship that is implied in most micro-to-mainframe link products, peer protocols support direct communication between programs. One of these protocols is the *Advanced Program-to-Program Communication* facility of IBM's Systems Network Architecture (SNA). APPC is a set of functions that govern the way various categories of devices communicate as peers over the network, providing complete interoperability among them. In support of this objective, LU 6.2 emerged as a single, product-independent LU type, which marked a departure from previous LU types.

LU stands for *Logical Unit,* which is a software-defined access point through which users interact over the SNA network, rather than a physical access point such as a port. Logical units allow communication between users without each user having to know detailed information about the other's device type and characteristics. LU 6.2 software is used to implement a set of functions collectively known as Advanced Program-to-Program Communication.

The SNA network uses *Physical Units* (PUs) to indicate categories of devices and the resources they present to the network. The resources associated with a particular device category include the communications links. A combination of hardware and software in the device implements the physical unit, of which there are four types. Type 1 (PU 1) devices are "dumb" terminals, whereas Type 2 (PU 2) devices are user-programmable and have processing capabilities. Type 4 (PU 4) refers to the host node, and Type 5 (PU 5) refers to a communications controller. There is no Type 3 physical unit (PU 3).

The physical unit that implements the most comprehensive set of functions is Physical Unit Type 2, version 1 (PU 2.1). PU 2.1 supports connections to other PU 2.1 nodes, as well as conventional hierarchical connections to the mainframe, efficiently and economically. PU 2.1 also supports simultaneous multiple links and parallel sessions over a given link. PU 2.1 is used in conjunction with LU 6.2 in implementing APPC.

The SNA network also includes *System Service Control Points* (SSCPs), which provide the services required to manage the network, as well as establish and control the interconnections that allow users to communicate with one another. The SSCP provides broader functionality than an LU, which represents a single user, or a PU, which represents a device and its associated resources. The relationship of all three SNA components is illustrated in Fig. A3.

Summary. Among other things, APPC overcomes the inefficiencies that result when a PC is forced to emulate a 3270 terminal to access data on a mainframe. In the terminal emulation mode, the PC and mainframe must devote processing resources to servicing screen-by-screen data transfers. The PC can be appropriately equipped with an emulation board that uses its own processor to handle the increased load. But the mainframe can get bogged down when it is forced to handle requests from many PCs in the emulation mode. A popular solution to this problem

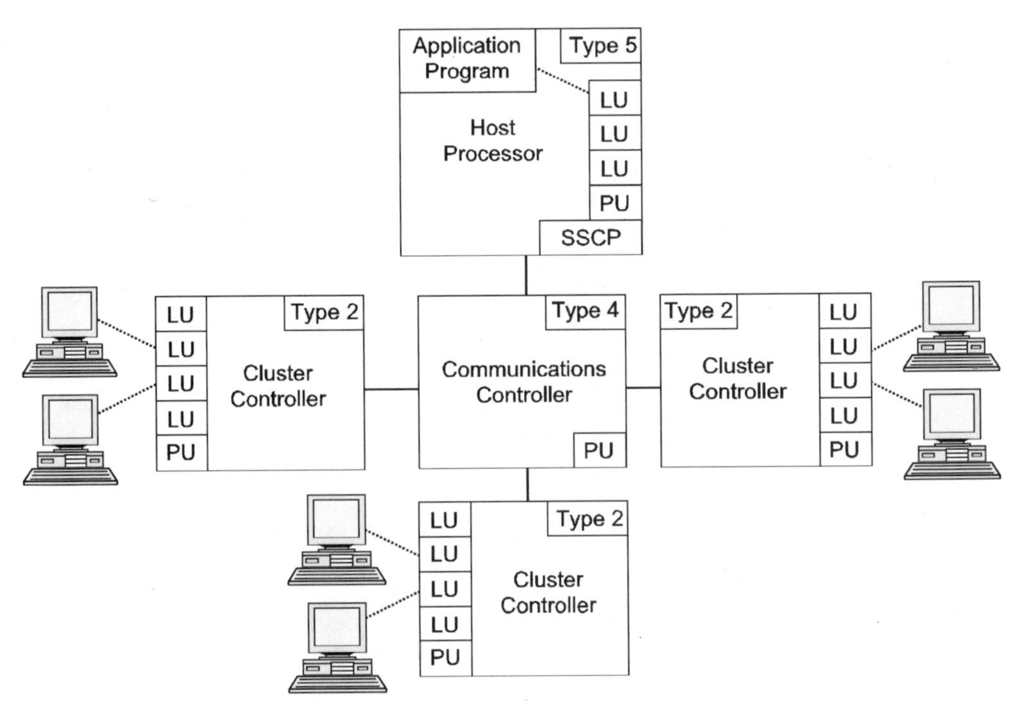

Figure A3 *Terminal-to-host communications in the SNA environment.*

is to use departmental or workgroup systems to service local PC users, thus offloading work from the mainframe. Adding APPC capabilities in the distributed environment makes computing and networking even more efficient and economical.

See also

ADVANCED PEER-TO-PEER NETWORKING (APPN)

Air-Ground Radiotelephone Service

The *Air-Ground Radiotelephone Service* is a radio service wherein common carriers can use wireless technology to provide telephone service to subscribers in aircraft in flight or on the ground. There are two versions of this service, one for general aviation and one for commercial aviation.

General aviation air-ground stations. General aviation air-ground stations are a network of independently licensed stations that employ a standardized duplex analog technology, called *Air-Ground Radiotelephone Automated Service* (AGRAS), to provide telephone service to subscribers flying over the United States or Canada in private aircraft, from small, single-engine craft to corporate jets. Because there are only 12 channels available for this service, it is not available to passengers on commercial airline flights. This service has been available to general aviation for more than 30 years.

Commercial aviation air-ground systems. Commercial aviation air-ground systems are nationwide systems that employ various analog or digital wireless technologies to provide telephone service to passengers flying in commercial aircraft over the United States, Canada, and Mexico. Passengers use credit cards or prearranged accounts to make telephone calls from bulkhead-mounted telephones, or, in larger jets, from seatback-mounted telephones. This service was available from one company on an experimental basis during the 1980s, and began regular competitive operations in the early 1990s. Currently there are three operating systems, one of which is GTE AirFone. As of December, 1996, over 1500 aircraft were equipped with AirFone service or equipment, which handled over 69 million calls.

When an AirFone call is placed over North America, information is sent from the phone handset to a transceiver in the plane's belly and then down to one of the 130 strategically placed ground radio base stations. From there it is sent to one of three main ground switching stations and then over to the public telephone network to the receiving party's location. When an AirFone call is placed over water, information is sent first to an orbiting satellite. From there the call transmission path is similar to the North American system, except that calls are sent to a satellite earth station instead of a radio base station. Calls can be placed to any domestic or international location.

To receive calls aboard aircraft, a passenger must have an activation number. In the case of AirFone, an activation number can be obtained by dialing 0 toll-free on board the aircraft, or 1-800-AIRFONE from the ground. For each flight segment the activation number will be the same, but the passenger must activate the phone for each flight segment and include his or her seat number. The person placing the call from the ground dials 1-800-AIRFONE and follows the voice prompts to enter the passenger's activation number. The passenger is billed for the call on a calling card or credit card, but gets to choose whether or not to accept the calls.

The following steps are involved when receiving a call:

- ➢ The phone will ring on the plane and the screen will indicate a call for the seat location.
- ➢ The passenger enters a personal identification number (PIN) to ensure that no one else can answer the call.
- ➢ The phone number of the calling party will be displayed on the screen.
- ➢ If the call is accepted, the passenger is prompted to slide a calling card or credit card to pay for the call.
- ➢ Once the call has been accepted, the passenger is connected automatically to the party on the ground.
- ➢ If the passenger chooses not to accept the call, he or she follows the screen prompts and no billing will occur.

Air-to-ground calls are very expensive. The cost to place domestic calls on the AirFone Service, for example, is US$2.99 to connect and US$3.28 per minute or partial minute, plus applicable tax. These rates apply to all data/fax and voice calls. Even calls to 800 and 888 numbers—which normally are toll-free on the ground—are charged at the same rate as regular AirFone calls. No billing ever occurs for the ground party.

See also
CELLULAR COMMUNICATIONS

American National Standards Institute

Founded in 1918, the *American National Standards Institute* is a federation of standards-developing organizations in the United States. ANSI represents the interests of about 1400 corporate, organizational, governmental, institutional, and international members through its headquarters in New York City and its satellite office in Washington, D.C.

ANSI does not itself develop standards, but facilitates their development by establishing consensus among qualified groups. ANSI ensures that its guiding principles—consensus, due process, and openness—are followed by the more than 175 distinct entities currently accredited by ANSI. These accredited organizations are committed to supporting the development of standards that address technological innovation, marketplace globalization and regulatory reform issues.

ANSI is the sole U.S. representative and dues-paying member of the two major nontreaty international standards organizations: the International Organization for Standardization, and, via the U.S. National Committee, the International Electrotechnical Commission.

ANSI was a founding member of the International Organization for Standardization and plays an active role in the governance thereof. ANSI is one of five permanent members of the governing ISO Council, and one of four permanent members of ISO's Technical Management Board. American participation, through the U.S. National Committee, is equally strong in the International Electrotechnical Commission. The USNC is one of 12 members on the IEC's governing Committee of Action.

Through ANSI, the U.S. has immediate access to the ISO and IEC standards development processes. ANSI participates in almost the entire technical program of both the ISO (78 percent of all ISO technical committees) and the IEC (91 percent of all IEC technical committees), and administers many key committees and subgroups (16 percent in the ISO and 17 percent in the IEC). As part of its responsibilities to the ISO and the IEC as the United States member body, ANSI accredits U.S. Technical Advisory Groups or USNC Technical Advisors, the primary purpose of which is to develop and transmit, via ANSI, the United States positions on activities and ballots of the international technical committee. U.S. standards in many cases are taken forward, through ANSI or its USNC, to the ISO or IEC, where they are adopted in whole or in part as international standards.

Summary. ANSI's program for accrediting third-party product certification has experienced significant growth in recent years. It continues to work for worldwide acceptance of product certifications performed in the United States, and to promote reciprocal agreements among U.S. accreditors and certifiers.

See also

INSTITUTE OF ELECTRICAL AND ELECTRONICS ENGINEERS (IEEE)
INTERNATIONAL ELECTROTECHNICAL COMMISSION (IEC)
INTERNATIONAL ORGANIZATION FOR STANDARDIZATION (ISO)
INTERNATIONAL TELECOMMUNICATIONS UNION (ITU)

Analog line impairment testing

Despite the trend toward digital lines, there are still many analog lines in place. Analog lines are used for voice and low-speed data via modems. Whether analog circuits are switched or dedicated, they must meet some very basic performance guidelines, which are published by Bellcore. By taking impairment measurements, a technician can determine whether carriers are complying with their stated levels of performance and, if not, get the problem solved.

The ability to test for impairments is especially important when an organization is using conditioned leased lines. A *conditioned line* is one that has been selected for its desirable characteristics—signal-to-noise ratio, intermodulation distortion, phase jitter, attenuation distortion, and envelope delay distortion—or treated with equalizers in order to improve the user's ability to transmit data at higher speeds than those normally possible over ordinary analog private lines. Since conditioning is provided by the carrier at extra cost, testing these facilities periodically allows users to verify that they are indeed getting the level of performance for which they are paying. The test equipment used for this purpose often is referred to as the *Transmission Impairment Measurement Set* (TIMS).

Analog TIMS can make a few very basic measurements by passively bridging into a circuit. These measurements require sending reference tones down the line and receiving them back. By analyzing the difference between what was sent and

what is received, the TIMS calculates the level of impairment. A TIMS can measure a variety of voice frequency (VF) impairments, including:

- Overall signal quality
- VF transmit level
- VF receive level
- Data carrier detect loss
- Dropouts
- Signal-to-noise ratio
- Gain hits
- Phase hits
- Impulse hits
- Frequency offset
- Phase jitter
- Nonlinear distortion

The measurements for selected parameters (such as those listed) then can be compared against the performance thresholds set by the network manager. If these thresholds are exceeded, data traffic might have to be rerouted to another facility until the primary line can be brought back into specification, or downspeeded to avoid the corrupting effects of the line impairment.

See also
PROTOCOL ANALYZERS

ARCnet

The Attached Resource Computer Network (ARCnet), introduced by Datapoint Corp. in 1977, was the first LAN to employ the wiring hub. Although more recent Ethernet (i.e., 10Base-T) employs a hub, and Token-Ring uses a hub-like device called a Media Access Unit (MAU), the wiring hub is an integral part of the original ARCnet's design, making it more reliable, flexible, and economical than Token-Ring and the earlier Ethernet topologies.

ARCnet operates at 2.5 Mbps and uses a *token-passing protocol.* When an ARCnet node receives the special packet called the *token,* it is permitted to send packets of data to other stations. While the Token-Ring protocol passes its token around a physical cable ring, ARCnet passes its token from node to node in order of the nodes' addresses, regardless of the physical arrangement.

Nodes can be located up to 2000 ft from an ARCnet hub through the use of RG-62 coaxial cabling. The total end-to-end length of the network can be 20,000 ft—nearly 4 miles. With twisted-pair wiring, each node can be located up to 400 ft from an ARCnet hub. As many as 254 connections are supported on a single ARCnet LAN via interconnected active hubs.

ARCnet was originally designed to be configured as a distributed star, which entails each node being directly connected to a hub (Fig. A4), with several hubs connected to each other. This design suits organizations that terminate cables in centrally located wiring closets. Later, an Ethernet-like bus topology was introduced for ARCnet, allowing nodes to be interconnected via a single run of cable. The two topologies can even be combined for maximum configuration flexibility. For example, instead of connecting a single ARCnet node to a port on an ARCnet hub, a bus cable with a maximum of eight nodes attached can be connected to the hub.

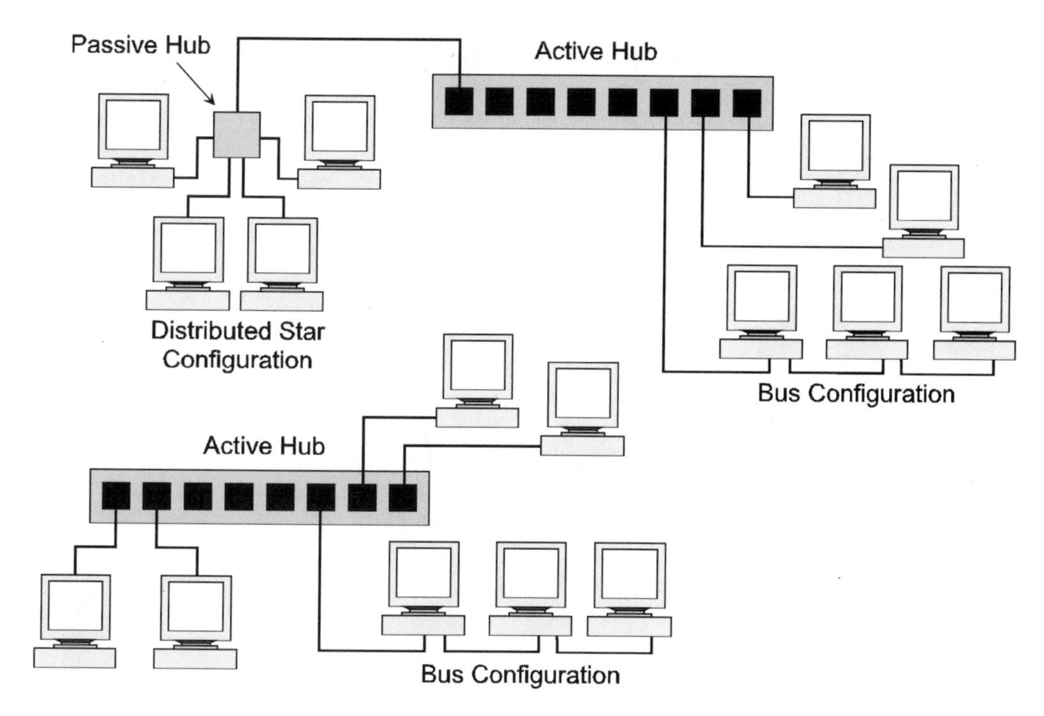

Figure A4 *ARCnet configuration incorporating both passive and active hubs.*

ARCnet makes use of two types of hubs: passive and active. *Passive hubs* are small, 4-port, nonpowered devices that support workstations at distances of up to 100 ft using coaxial cabling. *Active hubs* are 8- or 16-port powered units that support workstations at distances of up to 200 ft using coaxial cabling and up to 400 ft with twisted-pair wiring. By attaching passive hubs to each of an active hub's ports, a single active hub can support 24 or 48 workstations.

The primary advantage of this distributed-star arrangement include cost savings on cable installation and hub ports. A central active hub that uses twisted-pair wiring offers the best protection against network failure, since each station has its own dedicated connection to the active hub. Furthermore, the central-hub approach makes all the wiring accessible at one point, which simplifies troubleshooting, fault isolation, and network expansion.

From the beginning, ARCnet used internal transceivers in its hubs. Under Datapoint's concept of "conjoint networks," there was no need for bridges and routers to interconnect multiple LANs. Selected workstations and/or file servers could be configured to participate directly in up to six LANs at the same time. Access to each LAN or group of LANs was effectively controlled through hardware.

An improved version of ARCnet became available in 1992. ARCnet Plus offers 20 Mbps data transmission and is backward compatible with ARCnet, allowing

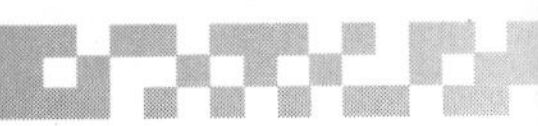

a mix of ARCnet and ARCnet Plus equipment to share the same network. When the network starts up, each ARCnet node broadcasts its capabilities, letting the others know that it can handle higher-speed transmissions. This permits two ARCnet Plus nodes to communicate at the maximum rate. If an ARCnet Plus node encounters a slower ARCnet node, the network steps down to 2.5 Mbps. ARCnet Plus uses packet lengths of up to 4096 bytes, versus 508 bytes under ARCnet. ARCnet Plus also supports more nodes: 2047 compared to ARCnet's 254.

See also

ETHERNET
STARLAN
TOKEN-RING

Asset management

The proper accounting of various hardware and software assets continues to be a neglected area of communications management. Without a thorough understanding of what assets an organization has, network managers cannot accurately plan departmental budgets or allocate costs. The failure to account for and manage such assets as desktop hardware and software, in-house cabling, and network lines has other ramifications as well. It can lead to cost overruns on projects, leave the door open to employee theft ("asset shrinkage"), and lead to misuse or abuse of the network. In some cases, such as when employees copy software, the lack of controls can expose the company to financial penalties for copyright infringement.

Types of assets. There are several types of assets that organizations must track. These fall into the general categories of hardware, software, network, and cable assets.

Hardware assets. Hardware inventory starts with identification of the major kinds of systems that are in use in the distributed computing environment—from the servers all the way down to the desktop computers—as well as their various components, including the CPU, memory, boards, and disk drives. The utilities that come with servers generally provide this kind of information, along with various performance metrics.

Most asset management products provide the following basic hardware configuration information:

- CPU: model and vendor
- Memory: type (extended or expanded) and amount (in kilobytes or megabytes)
- Hard Disk: amount and percentage of disk space used and available, volume number, and directories
- Ports: in use and available

Most hardware identification is based on the premise that if a driver is loaded, then the hardware must be present. Many of these drivers go unused but are not removed, however, resulting in inaccurate inventory. Fortunately, this situation is being addressed by industry standards, such as the Desktop Management Interface (DMI) and Plug and Play (PnP).

Hardware inventories can be updated automatically on a scheduled basis, such as daily, weekly or monthly. Typically included as part of the hardware inventory is the physical location of the unit, owner (workgroup or department), and name of the user. Other information may include vendor contact information and the unit's maintenance history. All of this information is manually entered and updated.

In addition to providing inventory and maintenance management, some products provide procurement management as well. They maintain a catalog of authorized products from preferred suppliers, as well as list and discount prices. They track all purchase requests, purchase orders, and deliveries. With some products, even the receipt of new equipment can be automated, with the system collecting information from scans of asset tags and bar codes. Warranty information also can be added.

Still other asset management packages accommodate additional information for financial reporting, such as:

- Cost: purchase price of the unit and add-in components
- Payment schedule: principal and interest
- Depreciation: one-time expense or multiyear schedule
- Taxes: local, state, federal (as applicable)
- Lease: terms and conditions
- Charge-back: cost charged against the budgets of departments, workgroups, users, or projects

This kind of information is manually entered and updated. Depending on the product, this information can be shared with spreadsheets and other financial applications and used for budget monitoring, expense planning, and tax preparation.

Software assets. Another technology asset that must be tracked is client software. Not only can software tracking (also called *applications metering)* reduce support costs, it can protect the company from litigation resulting from claims of copyright infringement, as when users copy and distribute software on the network in violation of the vendor's license agreement.

Asset management products that support software tracking automatically discover what software is being used on each system on the network by scanning local hard drives and file servers for all installed software. They do this by looking for the names of all executable files and arranging them in alphabetical order. They determine how many copies of the executable files are installed and look into them to provide the product name and the publisher. Files that cannot be identified absolutely are listed as found but flagged as unidentified. Once the file is eventually identified, the administrator can fill in the missing information.

The accumulated asset information can be used to build a software distribution list. The administrator can then automatically install future upgrades on each workstation appearing on the distribution list.

The administrator also can monitor the usage status of all software on the network to enforce license compliance. If several copies of an application are not being used, they can be made available to other users. If all copies of an application are in use, a queue is started and the application is made available on a first-come, first-served basis.

Network assets. The kinds of network assets that must be monitored, controlled and accounted for in inventory include repeaters, bridges, routers, gateways, hubs, and switches. These types of equipment are designed to be centrally managed and controlled.

Via the network management systems, the entire chassis of a hub, for example, can be viewed, showing the types of cards that are inserted into each slot. With a zoom feature, any card can be isolated and a representation of the ports, LEDs, and configuration switches can be displayed at the management console. A list of devices attached to any given port also can be displayed or printed out. Some of the other views available from a hub's management system include:

- *Configuration view* organizes the configuration values for a device and its model, including model name, IP address, security string and device name, type, location, and firmware version.
- *Resource view* is a special-purpose view for endpoint devices, showing where the endpoint device accesses its application resources, such as primary and secondary print servers, electronic mail server, and file server.
- *Cablewalk view* illustrates the connections that exist along a segment of cable (Ethernet, Token-Ring, or FDDI) and the devices connected to the segment.
- *Diagnostic view* organizes diagnostic and troubleshooting information for a device, including errors, collisions, events, and alarms.
- *Performance view* displays performance statistics, including load, hard and soft errors, and frame traffic.
- *Port performance views* summarize port-specific or board-specific performance statistics for each individual port or board.
- *Application view* organizes device application information, including device IP and Internet Control Message Protocol (ICMP) statistics.
- *Assigns view* allows a specific technician to be assigned to devices owned by network users.

The hub's management system also might provide a method for documenting the equipment and cable plant inventory through a third-party cable asset management system. This, and other third-party applications, typically are integrated with application programming interfaces (APIs).

Cable assets. Among the assets that must be managed is the cabling that connects all of the devices on the network, including coaxial cable (thick and thin), twisted-pair (shielded and unshielded), and optical fiber (single-mode and multimode). There are a number of specialized applications available that keep track of the wiring associated with connectors, patch panels, cross-connects, and wiring hubs. They use a graphical library of system components to display a network. Clicking on any system component brings up the entire data path, with all its connection points. These cable management products offer color maps and floor plans that are used to illustrate the cabling infrastructure in one or more buildings. A zoom feature can isolate backbone cables within a building, on a floor, or within an office.

Some cable asset management products can generate work orders for moving equipment or rewiring. Managers can create both logical and physical views of their facilities, and even view a complete data path simply by clicking on a connection. Some cable asset management products automatically validate the cabling architecture by checking the continuity of the data paths and the type of network for every wire. In addition, a complete picture of the connections can be generated and printed out. With this information, network administrators and technicians know where new equipment should go, what needs to be disconnected, and what should be reconnected. Some cable asset management products can even calculate network load statistics to facilitate proactive management and troubleshooting.

Like other types of asset management applications, cable management applications can be run as stand-alone systems or may be integrated with help desk products, hub management systems, and major enterprise management platforms such as IBM's NetView, Hewlett-Packard's OpenView, and SunSoft's Solstice SunNet Manager.

Summary. There are several approaches to asset management. Organizations can buy one or more software packages, use an integrated approach available with some help desk or network management systems, or outsource the asset management task to a systems integrator or computer vendor. In addition to containing the cost of technology acquisitions and reining in hidden costs, such asset management can improve help desk operations, enhance network management, assist with technology migrations, minimize asset shrinkage, and provide essential information for planning a corporate reengineering strategy.

See also

ELECTRONIC SOFTWARE DISTRIBUTION
NETWORK DESIGN
NETWORK MANAGEMENT

Asynchronous Transfer Mode

Asynchronous Transfer Mode (ATM) is a protocol-independent, cell-switching technology that offers high speed and low latency for the support of data, voice, and video traffic. ATM provides for the automatic and guaranteed assignment of bandwidth to meet the specific needs of applications, making it ideally suited to supporting multimedia. ATM also lends itself to upward and downward scaling, making it equally suited for interconnecting local area networks and building wide area networks. ATM-based networks may be accessed through a variety of standard interfaces, including frame relay.

Applications. ATM serves a broad range of applications very efficiently by allowing an appropriate Quality of Service (QoS) to be specified for each application. Various categories have been developed to help characterize network traffic, each of which has its own QoS requirements. These categories and QoS requirements are summarized in Table A1.

Table A-1. Quality of Service Requirements for Network Traffic.

Category	Application	Bandwidth Guarantee	Delay Variation Guarantee	Throughput Guarantee	Congestion Feedback
Constant Bit Rate (CBR)	Provides a fixed circuit for applications that require a steady supply of bandwidth, such as voice, video and multimedia traffic.	Yes	Yes	Yes	No
Variable Bit Rate (VBR)	Provides enough bandwidth for bursty traffic such as transaction processing and LAN interconnection, as long as rates do not exceed a specified average.	Yes	Yes	Yes	No
Unspecified Bit Rate (UBR)	Makes use of any available bandwidth for routine communications between computers, but does not guarantee when or if data will arrive at its destination.	No	No	No	No
Available Bit Rate (ABR)	Makes use of available bandwidth and minimizes data loss through congestion notification. Applicants include e-mail and file transfers.	Yes	No	Yes	Yes

Operation. QoS enables ATM to admit a constant bit rate (CBR) voice connection, while protecting a variable bite rate (VBR) connection for a transaction processing application and allowing an available bit rate (ABR) or unspecified bit rate (UBR) data transfer to proceed over the same network. Each virtual circuit will have its own QoS contract, which is established at the time of connection setup at the user-to-network interface (UNI). The network will not allow any new QoS contracts to be established if they will adversely affect its ability to meet existing contracts. In such cases, the application will not be able to get on the network until the network is fully capable of meeting the new contract.

When the QoS is negotiated with the network, there are performance guarantees that go along with it: maximum cell rate, available cell rate, cell transfer delay, and cell loss ratio. The network reserves the resources needed to meet the performance guarantees and the user is required to honor the contract by not exceeding the negotiated parameters. Several methods are available to enforce the contract; among them are traffic policing and traffic shaping.

Traffic policing is a management function performed by switches or routers on the ATM network. To police traffic, the switches or routers use a buffering technique referred to as a "leaky bucket." This technique entails traffic flowing (leaking) out of the buffer (bucket) at a constant rate (the negotiated rate), regardless of how fast it flows into the buffer. If the traffic flows into the buffer too fast, the cells will be allowed onto the network only if enough capacity is available. If there is not enough capacity, the cells are discarded and must be retransmitted by the sending device.

Traffic shaping is a management function performed at the UNI of the ATM network. It ensures that traffic matches the contract negotiated between the user and network during connection setup. Traffic shaping helps guard against cell loss in the network. If too many cells are sent at once, cell discards can result, which will disrupt time-sensitive applications. Because traffic shaping regulates the data transfer rate by spacing the cells evenly, discards are prevented.

Cell structure. Voice, video, and data traffic usually are comprised of bytes, packets, or frames. When the traffic reaches an ATM switch, it is segmented into small, fixed-length units called *cells.* The size of ATM cells is fixed at 53 octets; the cell consists of a 5-octet header and 48-octet payload (Fig. A5).

The cell header contains the information needed to route the information field through the ATM network. The header supports five functions:

- *Generic Flow Control* (GFC): This 4-bit field has only local significance; it enables customer premises equipment at the UNI to regulate the flow of traffic for different grades of service.
- *Routing Field* (RF): This 24-bit field contains a virtual path identifier/virtual channel identifier (VPI/VCI) combination used to route the cell through the network. The number of bits available for VPI and VCI subfields is negotiated at subscription time to the network.
- *Payload Type* (PT): This 3-bit field is used to indicate whether the cell contains user information or connection management information. This field also provides for network congestion notification.

- *Cell Loss Priority* (CLP): This 1-bit field, when set to a 1, indicates that the cell may be discarded in the event of congestion.
- *Header Error Check* (HEC): This 8-bit field is used by the Physical layer for detection and correction of bit errors in the cell header. The header carries its own error check to validate the VPIs and VCIs and prevent misdelivery of cells to the wrong UNI at the remote end. Cells received with header errors are discarded. Higher-layer protocols are responsible for initiating lost cell recovery procedures.

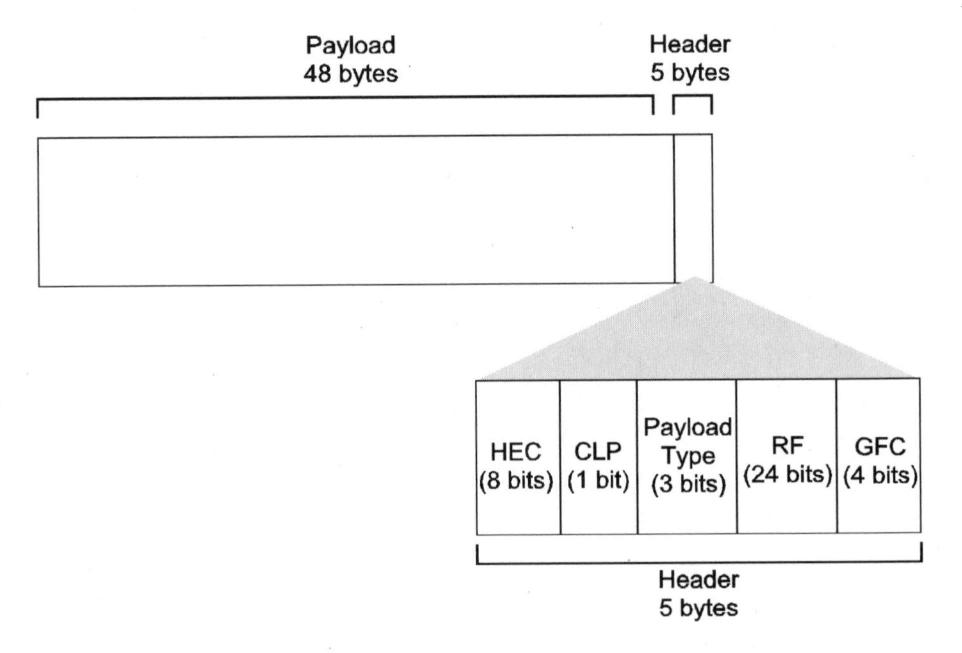

Figure A5 *ATM cell structure.*

There has been some concern about the high overhead of cell relay, with its ratio of 5 header octets to 48 data octets. With recent innovations in Wave Division Multiplexing (WDM) to increase fiber's already high capacity, however, ATM's overhead is no longer a serious issue. Instead, the focus is on ATM's unique ability to provide a quality of service in support of all applications on the network.

ATM layers. Like other technologies, ATM uses a layered protocol model. ATM has only three layers: the Physical layer, ATM layer, and Adaptation layer (Fig. A6):

- *Physical Medium-Dependent:* The ATM Physical layer currently defines several transport systems, including SONET, T3, optical fiber, and twisted pair. The Synchronous Optical Network provides the primary transmission infrastructure for implementing public ATM networks, offering service at

OC-1 (51.84 Mbps) to OC-12 (622.08 Mbps). Current definitions of SONET go up to OC-192 (9.952 Gbps). SONET facilities have only limited availability to many users, however. Therefore, the UNI outlines the use of DS3 and a Physical layer definition similar to Fiber Distributed Data Interface (FDDI) to provide a 100 Mbps private ATM network interface.

- *ATM Layer:* This layer provides segmentation and reassembly operations for data services that may use protocol data units (PDUs) different from those of an ATM cell. This layer then is responsible for relaying and routing, as well as multiplexing, the traffic through an ATM network.
- *ATM Adaptation Layer* (AAL): Residing between the ATM Layer and the higher-layer protocols, this layer provides the necessary services that are not part of the ATM Layer, in order to support the higher-layer protocols.

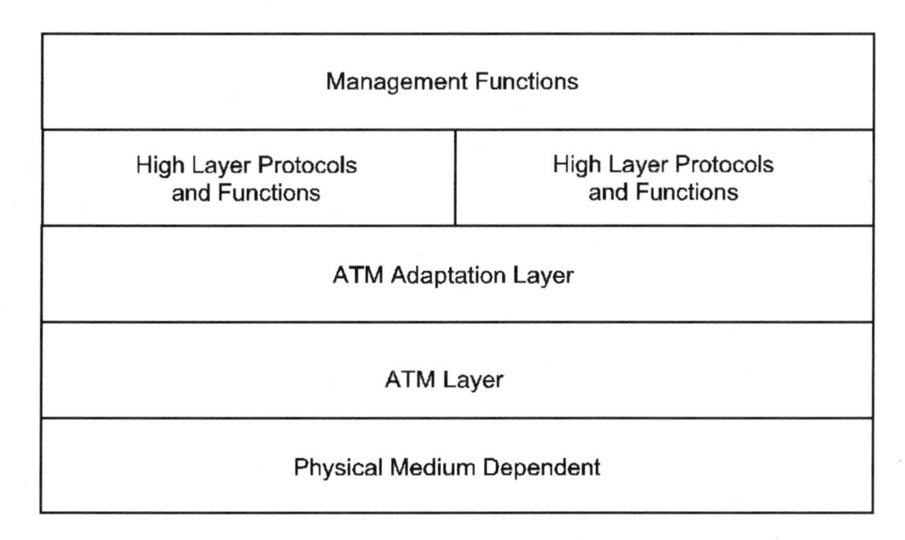

Figure A6 *ATM protocol model.*

Summary. A solid base of standards now exists to allow equipment vendors, service providers, and end users to implement a wide range of applications via ATM. The standards will continue to evolve as new applications emerge. The rapid growth of the Internet is one area where ATM can have a significant impact. With the Internet forced to handle a growing number of multimedia applications—telephony, videoconferencing, faxes, and collaborative computing, to name a few—congestion and delays are becoming ever more frequent and prolonged. ATM backbones will play a key role in alleviating these conditions, enabling the Internet to be used to its full potential.

See also
SYNCHRONOUS OPTICAL NETWORK (SONET)
WAVELENGTH DIVISION MULTIPLEXING (WDM)

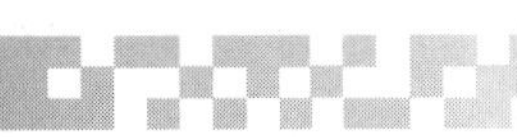

Attenuation

Attenuation is the decrease or reduction in the power of a signal; in other words, attenuation is signal loss. In fiber-optic links, for example, the glass fibers are not perfectly clear. The amount of light coming out of the distant end of a cable is always somewhat less than the amount transmitted into the fiber. Fiber loss may be due to several causes, among them:

- Absorption by impurities
- Scattering by impurities or by the defects at the core-cladding interface
- Rayleigh scattering by the molecules of the medium (i.e., silicon)
- Fiber bends and microbends
- Scattering and reflection at splices

All these phenomena contribute to the degradation of the fiber transmission. Excessive attenuation results in a received signal that is too weak to be useful. Attenuation usually is a linear quantity, increasing in direct proportion to the length of a fiber-optic cable run. Attenuation is measured in decibels (dB) and is the ratio between the output power of the fiber and the input power. The minimum attenuation of a standard telecommunication fiber occurs around the 1550 nm wavelength; it is on the order of 0.2 dB/km.

See also
DECIBEL

Automatic Call Distributors

Automatic Call Distributors (ACDs) provide an efficient method of queuing and processing heavy volumes of incoming voice calls. These systems greet callers with recorded messages, provide a menu of dial options, and route calls to appropriate individuals or the voice mail system. ACDs also can provide music while callers are on hold and make periodic announcements regarding queue status or holding time. ACDs can even be linked via private lines to form distributed call processing systems that can route incoming calls to corporate locations in other time zones.

By directing calls rapidly to agents who are available and have the information or technical expertise required by the caller, companies make efficient use of their communications systems and human resources. Many systems also provide a range of management capabilities, including reporting functions that incorporate such statistics as the number of incoming calls handled, the number of calls abandoned, and the system's peak calling capacities. These reports can be used to determine such things as agent productivity, as well as the need for more lines or staff.

ACD operation. A primary goal of the ACD is to maintain the productivity of agents through the efficient distribution of incoming calls. Supervisor, or master, positions are points within the service center that are staffed by managerial personnel who monitor individual calls, agents, and overall system activity.

ACDs typically answer a telephone call on the first ring or after a fixed number of rings, then examine preprogrammed processing tables for routing instructions while callers on hold listen to recorded announcements or music. ACDs also can answer calls dynamically by sensing the incoming call and searching through routing schemes before answering. After the call is answered, other systems (such as a voice response unit) may gather additional information and compare that information with customer databases before passing the call to an agent position.This basic structure is common to all ACDs. Systems differ primarily by the method of call allocation, the types of system management reports, and the various control features.

Applications. ACDs generally are used by companies that have call centers with high incoming call volumes and which have five or more agents whose responsibilities are almost entirely restricted to handling incoming calls. Typically, ACDs are used in the following call center environments:

- Customer service
- Help desk
- Order entry
- Credit authorization
- Reservations
- Insurance claims
- Catalog sales

A relatively new call center environment is the Web page on corporate IP-based intranets or the greater Internet. A company can set up a hot link on its Web page so that when a user clicks on the link from a multimedia-equipped PC, a voice connection is established over the IP network. When the call hits an IP-to-PSTN gateway, it is transferred to the ACD, where it gets routed to the next available agent. When used in this manner, the ACD can be used to support electronic commerce applications.

System components. ACD systems consist of incoming lines, agent positions, supervisor positions, and the switch itself. Often toll-free 800 lines are connected to ACD systems, but any type of line can be connected. ACD lines can be routed through a PBX that provides general carrier interface support for the business. This arrangement is most common in systems that use digital T1 trunks to carry both ACD-related calls and other general call traffic.

Types of ACDs. Various types of equipment and services are available to provide ACD features and functions, including:

- *Stand-alone systems* are used mainly in environments where the service center is separate from the rest of the business and ACD functions do not require integration with the corporate telephone system.
- *Integrated systems* are those in which ACD functionality is added to a PBX key telephone system with ACD software, providing call allocation and service supervision within the telephone system environment.
- *PC-based systems* use software, added to a multimedia-equipped PC, that includes functions such as voice mail, interactive voice response, intelligent queue announcements, and computer integration, along with the traditional ACD distribution and routing functions. The software collects call statistics and generates management reports.

- *Automatic call sequencers* are independent devices that perform the same type of call-to-agent station allocation as an ACD, but without complex load and time calculations. The systems rely on the PBX for routing calls; they have no switching matrix of their own.
- *Centrex-based systems* provide ACD functions and features as part of the telephone companies' Centrex services.
- *Central office–based systems* are those in which the telephone company provides ACD functions and features as a service, apart from Centrex.
- *Third-party services* are those in which third-party firms provide ACD service to other companies, as well as handle their call overflow. Operation is completely transparent to the caller.

Summary. Advances in call processing technology, improvements in the public switched telephone network, developments in computer telephone integration (CTI), the growing popularity of the Internet, and advancements in PC-related technologies (especially in the area of multimedia) have all combined to make it possible to use ACDs in almost any business that has a requirement for this capability.

See also

CENTREX
PRIVATE BRANCH EXCHANGE (PBX)

Automatic Number Identification

Historically, signaling information that enabled call routing and billing was transmitted through multifrequency in-band signaling on the same circuits used to connect the calling and called parties. Beginning in the 1980s, carriers worldwide began abandoning multifrequency in-band signaling in favor of out-of-band signaling. With out-of-band signaling, a packet network transmits signaling information on circuits separate from those used to connect the calling and called parties.

Compared to multifrequency in-band signaling, out-of-band signaling allows carriers to use their networks more efficiently; efficiency gains are realized because the carriers can set up and release calls more quickly. This increases the ratio of the time interoffice circuits are in use carrying a conversation to the time interoffice circuits are in the process of being connected or disconnected. Additionally, out-of-band signaling enhances flexibility in call handling and processing. It also avoids certain kinds of fraud that could be used with multifrequency signaling to defeat billing systems. The latest version of out-of-band signaling, deployed in carrier networks worldwide, is Signaling System 7 (SS7).

Privacy issue. The original purpose of Automatic Number Identification (ANI) was to enable carriers to bill customers for calls they made. ANI, however, also can facilitate many of the services made possible by the delivery of the calling party's number. As the interexchange carriers (IXCs) recognized the value of ANI to businesses for identifying customers, they began offering it as a service to those customers they served directly who were paying for the call, such as subscribers to 800 services. This raised the privacy issue among consumer groups.

After considering the issue, the FCC decided not to prohibit carriers from sharing ANI information with 800 service customers, even when a calling party requests privacy. For consumers, the major concern about ANI services was actually the reuse of the calling party information by 800 service customers. Accordingly, the FCC ruled that customer information gained from ANI services could not be reused or sold without the affirmative consent of the calling party. The only exception is that the information could be used to offer products or services to established customers where the products or services are directly related to products or services previously provided. Carriers providing ANI services are required to include these restrictions in contracts that offer the service.

Summary. ANI enables service providers and consumers to conduct transactions more efficiently. Computer services could recognize the calling party's number and either permit or deny access. Stockbrokers, travel agents, parts and equipment providers, and booksellers could route a call to a preprogrammed location closer to the calling party to expedite deliveries or services. Retailers could verify credit and billing information instantaneously. Customized services that depend on the caller's individualized preferences could be developed. In its decision to allow ANI to be used for nonbilling applications, the FCC observed that even small efficiencies on individual transactions become significant in an economy that averages more than one billion interstate calling minutes a day. These savings could lower service costs for suppliers, leading to lower prices for consumers.

See also

CALLER IDENTIFICATION
CUSTOM LOCAL AREA SIGNALING SERVICES
SIGNALING SYSTEM 7
TELEPHONE FRAUD

Bellcore

As one of the results of the breakup of AT&T in 1984, Bell Communications Research, Inc. (Bellcore) was established to provide engineering, administrative, and other services to the telecommunications companies of Ameritech, Bell Atlantic, BellSouth, NYNEX, Pacific Telesis, SBC Communications, and US West.

Since 1984 the organization has evolved to become a leading provider of communications software, and engineering and consulting services. Bellcore serves over 800 customers worldwide in 55 countries and in a variety of industries. Major United States telecommunications customers include AT&T, Cincinnati Bell, GTE, The Southern New England Telephone Company, Rochester Telephone, Sprint, Stentor, and various government entities.

Eighty percent of the U.S. public telecommunications network depends on software invented, developed, implemented, or maintained by Bellcore. Its employees are recognized leaders in the creation and development of such technologies as ADSL, AIN, ATM, ISDN, frame relay, PCS, SMDS, SONET, and video-on-demand. Bellcore issues standards recommendations in these and other areas.

More than 600 domestic and foreign patents for technical innovation are available for licensing from Bellcore. Bellcore-developed network systems handle every single 800 and 888 call placed in the United States each day. Consulting services include systems integration, local number portability, unbundling and interconnection, network integrity and reliability, fraud management, and pricing and costing analyses.

Headquartered in Morristown, N.J., Bellcore has sales offices in the United States, Canada, Mexico, Australia, and the Philippines, and in the cities of London, Hong Kong, Tokyo, and São Paulo. Bellcore has 5800 employees and annual revenues exceeding US$1 billion.

As a result of changing developments in the telecommunications industry and the owners' diverging strategies and business plans, Bellcore was purchased by Science Applications International Corporation (SAIC). The transaction is expected to be finalized in late 1997 after Bellcore's owners obtain the requisite regulatory approvals. Combined, SAIC and Bellcore will consist of more than 28,000 science, engineering, software, and administrative professionals.

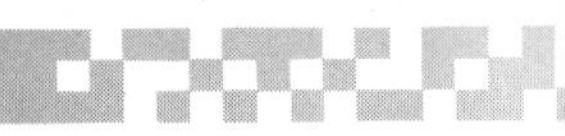

Summary. Bellcore is the nation's largest research consortium, performing fundamental research in computer science, phototonics, materials science, network architecture, and services. With the combined assets of SAIC, the two companies plan to serve markets where their combined skills can be leveraged. These include large software projects, advanced network designs, secure networks, Internet technologies, wireless communications, and other advances in telecommunications software systems and technology.

See also

BELL LABS

Branch office routing

As companies become more decentralized, there is an increasing need to tie remote branch offices into the corporate backbone network. With the wider availability of circuit-switched digital services, including ISDN, however, network planners now have an array of choices available for meeting the diverse data communications and LAN interconnectivity needs of branch offices.

Connection alternatives. In many cases, dial-up connections over ordinary phone lines are the most economical way to remotely access the corporate network. Since modem speed is limited to 33.6 kbps, this method of access is best suited for occasional use, such as retrieving electronic mail or downloading small files. Although 56 kbps modems (with competing technologies) are available, the higher speed is available only in the downstream direction, toward the user. If branch offices frequently download large files, this type of modem would be the better choice.

Although private lines such as T1 can be used for the connection, remote branch offices usually do not have enough traffic to justify the cost of installation and maintenance, and the high monthly charges. For many companies, more economical digital services like switched 56 kbps, frame relay, or ISDN are the answer, especially when mission-critical applications are involved. How economical these services are depends on the amount of traffic each remote node has. Instead of modems, small inexpensive routers are used at the branch office location (Fig. B1). These devices are connected to a larger backbone router, which is equipped with appropriate WAN interfaces and provides legacy integration and bandwidth management tools.

Branch router features. Like backbone routers, branch office routers can transmit data to other sites using such transmission methods such as ISDN, frame relay, and X.25. They can route TCP/IP, IPX, and AppleTalk. Some vendors' products support SMDS and ATM via their respective data exchange interfaces (DXI). They also can bridge nonroutable protocols transparently using such methods as source route and translation bridging. An increasing number of branch office routers support IBM's Data Link Switching (DLSw) architecture for routing SNA and NetBIOS traffic over TCP/IP.

The difference between branch office and backbone routers is mainly in the hardware, specifically the number of ports and types of interfaces available. The larger routers are also highly scalable, whereas branch routers offer limited scala-

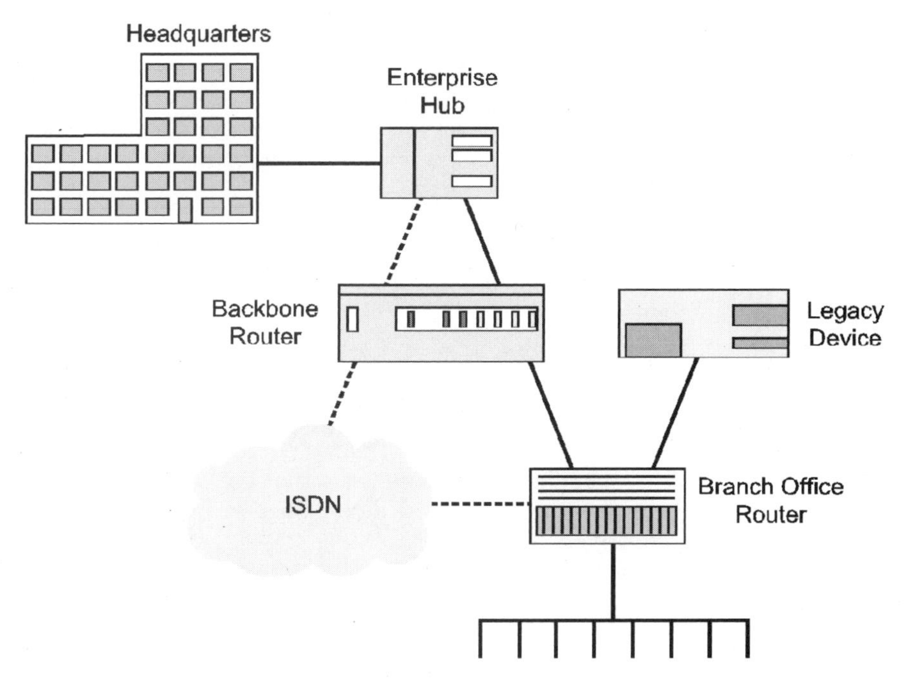

Figure B1 *The use of branch office routers saves money without sacrificing connectivity by delivering primary service via leased lines, with backup or additional bandwidth obtained through switched services like ISDN. Connecting a legacy device such as a cluster controller or X.25 switch to the branch router's second serial port provides another data link to headquarters.*

bility. Scalability can be achieved by purchasing stackable routers for branch offices. The main advantage of the stackable approach is that it lets network planners start small with their router installations and add capacity in affordable increments when growth warrants. When more connectivity is needed, additional routers are stacked. Each router in the stack can support from 2 to 24 interfaces, providing connectivity for Ethernet, Token-Ring, synchronous, ISDN BRI, and other types of networks.

Management. Because branch routers are often used at remote locations where technical expertise is not available, vendors offer Windows-based tools to make configuring the devices relatively simple. Using these tools, users can configure multiple routers simultaneously from the same computer screen, as well as create so-called *snap-ins,* that is, predefined configuration building blocks, to reduce configuration time and the chances of making errors. With these tools, interrelated configuration parameters can be viewed, checked, and changed. Alternatively, the LAN manager at the central site can dial into the branch office routers and configure them. In some cases the administrative activities of branch routers are offloaded to a central-site router, leaving remote routers to make only basic traffic-forwarding decisions.

Some router vendors offer a platform-independent, SNMP-based application designed expressly for simplifying router node management. They feature an intuitive, point-and-click graphical interface for simplifying network setup and expansion, real-time operations and monitoring, and real-time event and fault monitoring for efficient problem identification and isolation. Traffic priorities, which speed application response times and optimize WAN efficiency, also can be configured centrally through the management application. The same application can control all routers—from the smallest branch office routers to the largest backbone routers—allowing a single system for monitoring and controlling the backbone internetwork and the remote office links.

Cost containment features. Some branch routers offer cost containment features for minimizing usage charges over costly switched WAN services. Depending on vendor, one or more of the following features may be supported:

- *Bandwidth-on-demand* optimizes the use of WAN services availability by ensuring that a service is being used only when required and not being paid for when there are no data to be sent.
- *Bandwidth augmentation* optimizes performance by combining additional channels when extra bandwidth is required.
- *Data compression* minimizes information sent across the WAN by implementing 2-to-1 or 4-to-1 data compression, enabling up to two to four times the amount of data to be sent over the available dial-up bandwidth.
- *Spoofing* is the ability to keep “chatty” protocols such as TCP and Novell's SPX from running up service charges by filtering out overhead frames at one end of the link and emulating them at the other.
- *Time-based tariff management* is the ability to specify which WAN service should be used at a particular time of day, because WAN services have different telephone charges at different times of the day.
- *Transparent backup* activates a secondary link if the primary link fails. When the primary link comes back up, traffic is transparently transferred back to it without any loss of data.
- *Connection prioritization* is the ability to assign priority to applications that have an urgent need to access data remotely. For example, SNA traffic may be given a higher priority than IPX traffic because too much delay may cause the SNA host session to time out. Other time-sensitive protocols include DEC LAT and IP Telnet.

Some vendors offer an ISDN budgeting system, allowing network managers to set time-of-day and day-of-week operation over ISDN services. When the number of hours of usage has been exceeded, further ISDN calls are denied. Other cost containment features include reducing SNA polling across the wide area and filtering NetBIOS broadcasts across those links to minimize WAN costs.

Summary. Connecting remote offices to the backbone network is becoming a critical business requirement. An effective internetworking strategy can facilitate data communication and leverage centralized information resources to improve productivity at remote locations. Branch office routers connected to backbone routers equipped with appropriate WAN interfaces offer companies more flexibility than modems for interconnecting geographically dispersed offices, workgroups, and telecommuters.

See also

REMOTE CONTROL
REMOTE NODE

Bridges

Bridges are used to extend or interconnect LAN segments. At one level they are used to create an extended network that greatly expands the number of devices and services available to each user. At a higher level, bridges can be used for segmenting LANs into smaller subnets to improve performance, control access, and facilitate fault isolation and testing without impacting the overall user population.

The bridge does this by monitoring all traffic on the subnets that it links. It reads both the source and destination addresses of all the packets sent through it. If the bridge encounters a source address that is not already contained in its address table, it assumes that a new device has been added to the local network. The bridge then adds the new address to its table.

In examining all packets for their source and destination addresses, bridges build a table containing all local addresses. The table is updated as new packets are encountered and as addresses that have not been used for a specified period of time are deleted. This self-updating capability permits bridges to keep up with changes on the network without requiring that their tables be manually revised.

The bridge isolates traffic by examining the destination address of each packet. If the destination address matches any of the source addresses in its table, the packet is not allowed to pass over the bridge because the traffic is local. If the destination address does not match any of the source addresses in the table, the packet is allowed to pass onto the adjacent network. This filtering process is repeated at each bridge on the internetwork until the packet eventually reaches its destination. Not only does this process prevent unnecessary traffic from leaking onto the internetwork, it acts as a simple security mechanism that can screen unauthorized packets from accessing various corporate resources.

Bridges also can be used to interconnect LANs that use different media, such as twisted-pair, coaxial, and fiber-optic cabling, and various types of wireless links. In office environments that use wireless communications technologies such as spread spectrum and infrared, bridges can function as an access point to wired LANs. On the WAN, bridges even switch traffic to a secondary port if the primary port fails. For example, a full-time wireless bridging system can establish a modem connection on the public network if the primary wireline or wireless link is lost due to environmental interference.

In reference to the OSI model, a bridge connects LANs at the Media Access Control (MAC) sublayer of the Data Link layer. It routes by means of the Logical Link Control (LLC), the upper sublayer of the Data Link layer (Fig. B2).

Because the bridge connects LANs at a relatively low level, throughput often exceeds 30,000 packets per second (pps). Multiprotocol routers and gateways,

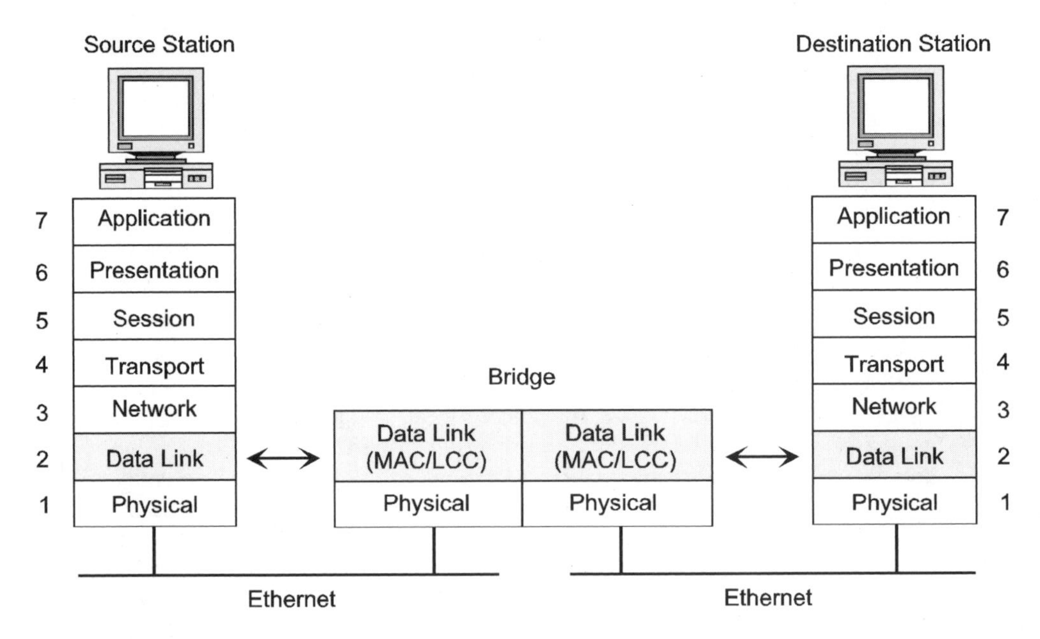

Figure B2 *Bridge functionality in reference to the OSI model.*

which also can be used for LAN interconnection, operate at higher levels of the OSI model. In performing more protocol conversions, routers and gateways usually are slower than bridges.

See also

GATEWAYS
OPEN SYSTEMS INTERCONNECTION (OSI)
REPEATERS
ROUTERS

Business process reengineering

Business process reengineering is a methodology that involves the restructuring of business operations and the application of appropriate technology to solving specific problems, as opposed to simply looking at technology from the traditional price-performance perspective.A business process reengineering plan usually is developed before the organization makes any commitments to vendors or carriers for new equipment or services. Often the plan is developed with the aid of a systems or network integrator or a third-party consulting firm.

When organizations begin planning for new systems or networks, or technology migrations, they first should assess and reorganize business processes with the idea of filtering out unnecessary procedures, eliminating redundancy, and streamlining workflow to enhance productivity, trim overhead costs, and improve customer response. Failure to engage in business process reengineering could result in the organization wasting enormous amounts of money in new equipment, lines, and services—because these alone do not improve organizational performance.

Preliminary analysis. A business process reengineering project starts with a preliminary analysis of the organization so it can be determined where to apply changes for maximum impact. There are several components to this type of analysis: a focus review, requirements definition, and strategic impact analysis.

Focus review. The focus review addresses an organization's strategic business objectives, identifying potential targets for improvement and providing a high-level cost/benefit analysis. The objective of the review is to develop a preliminary plan to improve the existing work environment. The focus review also helps management zero in on which workgroup or department would benefit most from the application of new systems or network technology, based on such parameters as volume of transactions, frequency of database access, distribution of work, and contribution to the core business.

Requirements definition. Once the decision has been made to improve a specific business process, a requirements definition is formulated. This builds on the preliminary input from the focus review to further define, analyze, and document the specific needs of the target workgroup or department. It lays the groundwork for all subsequent design and implementation activities. This step identifies the proposed project's major inputs, outputs, and volumes; defines the current workflow processes, identifying those that should be automated and those that should not; and defines all hardware, software, and network services that will be required for implementation. Often it is necessary to interview staff at various levels in the organization to obtain a clear picture of how work is performed.

With the current manual processing methods identified and understood, appropriate steps can be recommended to eliminate inefficient or outmoded practices and thereby streamline operations. This analysis ensures that the workgroup or department's total requirements have been explored and understood prior to the company's commitment to purchase new systems or network services. In addition, the requirements definition describes the types of third-party services that are required, such as consultants or integration services, which will be brought in under the direction of a project management team. The requirements definition also provides an initial installation schedule and cost estimates for implementing the solution.

The completion of these services results in a detailed design specification, which describes the actual solution and how it will be implemented. The information included in this document includes the workflow analysis, system or network configuration, end-user training, acceptance test procedure, and complete project schedules and timetables.

Strategic impact analysis. The strategic impact analysis takes these processes a step further, evaluating the potential effect of the proposed solution on the entire enterprise. The resulting report recommends an appropriate architecture, method of enterprise integration, and a specific implementation approach.

Preinstallation planning. Once management has signed off on the proposal, the systems or network integrator uses the information gathered for preinstallation planning. This starts with the creation of a detailed design document, which defines all hardware and software up front so that the actual installation will be performed with minimal business interruption in the shortest possible timeframe. Eventually this document also will be used to facilitate the implementation of the acceptance test plan, which enables the customer to verify that all requirements are being addressed prior to the acceptance of the new system or network.

Installation and implementation. The system or network integrator should provide a site analysis so that equipment and software can be installed immediately upon arrival with minimal impact on the workforce and daily business operations. At this phase of the project's life cycle, the integrator installs the hardware and software components and brings the system or network to an operational state. The integrator then initiates a verification process to confirm that these components are running properly. All aspects of the new system or network are documented. Upon acceptance by the customer, the system or network is put into service in the production environment.

Postinstallation activity. The goal of the integrator's postinstallation activities is to provide for the ongoing support of the installed system or network for a predetermined time period. This ensures that unanticipated problems are resolved quickly. The mechanisms for doing this can include any number of support services such as remote monitoring, on-site hardware repair service, overnight shipping of replacement components and modular subsystems, and access to technical staff via a toll-free number, a dial-up BBS, or World Wide Web (WWW) site on the Internet.

Management services. The system or network integrator should provide management services spanning the entire project life cycle, including training. An integrator typically offers specialized courses that are conducted at regional training centers or on location at customer sites. Several training paths should be available, including courses for system administrators, network managers, help desk personnel, and technicians. The training should range from basic principles of operation to administrative and technical functions.

Once the new system or network is up and running and has been accepted by the customer, the integrator's project management team may no longer be required. Technical assistance retainer plans are usually available, however, that includes the on- site services of the project manager on either an ongoing or periodic basis.

Summary. Business process reengineering seeks to analyze the current functions of a workgroup, department, or enterprise for the purpose of planning and implementing appropriate technology that will improve the efficiency of operations, the productivity of staff, and the response to customers. Business process reengineering precedes any capital investment in technology or commitment to vendors or carriers. Often, business process reengineering is an ongoing activity. The overriding goal of business process reengineering is to make the organization more agile with respect to addressing the needs of a dynamic, global marketplace and eventually to secure competitive advantage.

See also

DOWNSIZING
NETWORK INTEGRATION
SYSTEMS INTEGRATION
WORKFLOW AUTOMATION

Cable television networks

Cable television began around 1949 in Lansford, Pa. A local TV shop owner noticed a decrease in television sales and wanted to find out the reason. After talking to town residents, he discovered that low sales were due in large part to the poor reception in the area. The closest station was in Philadelphia, about 65 miles away, and there was a mountain that overlooked Lansford.

After helping some residents in the outlying areas set up antennas for better reception, the shop owner came up with an idea to help with the town's reception problem. He built an antenna on top of the mountain and, using an amplifier, he was able to boost the signal back to full strength. Next he ran coaxial cable down the mountain and into the town, charging people a fee to connect to the cable. The first community antenna television (CATV) system was born, consisting of three channels and a few hundred subscribers.

At around the same time, other towns claimed to have put the first CATV system in operation. Regardless of which town actually was first, the intent was the same—to provide clear reception to remote areas that otherwise had trouble receiving a signal. CATV began moving almost immediately into metropolitan areas such as New York City, where reception was difficult because the tall buildings caused multiple-signal interference (multipath) or blocked the signal entirely.

By 1952 there were 70 CATV systems with 14,000 subscribers in operation nationwide. By the 1960s, however, growth in the CATV market had all but stopped. Cable service had been installed in most of the major market areas. Technology also limited cable's growth. Until the mid-1970s, most cable systems only had enough capacity for 12 channels.

Major growth in cable market began to take off again after 1975. The availability of satellite receivers allowed cable operators to take specific signals and insert them in their channel lineup. This led to cable-only programming. Cable system operators began adding programs such as movie (HBO), sports (ESPN), and shopping (HSN) channels, as well as the so-called superstations (TBS). The technology also allowed cable companies to give subscribers pay-per-view programming. With this new service, a subscriber pays a one-time fee to view a special event, such as a concert or sporting event, or watch a first-run movie.

Program delivery. The actual video signals delivered to the cable system can be generated from three basic sources:

- *Satellite or microwave receivers:* Program sources include national networks such as CNN, HBO, and ESPN, and local sources such as commercial and public television. Usually these program sources run 24 hours per day, but may be interrupted by inserting locally originated programming or commercials.
- *Videotape:* Videotape recorders are used to deliver prerecorded material such as commercials, infomercials, public-service programs, and movies. The use of videotape recorders is undesirable due to the labor involved in getting the tapes made, moved to the playback site, and played. Instead, multimedia servers, which automate program delivery, are increasingly being used.
- *Multimedia servers:* Servers store and play multimedia programming that includes graphics, animation, sound, text, and digital MPEG video. These computers may accept real-time data from weather services, Internet information sources, computer databases, and satellite data networks for automated delivery on a scheduled or demand basis.

CATV is today the primary method of program distribution in the United States, where approximately 60 million subscribers access programming from a cable TV network. There are 11,000 networks nationwide and a million miles of cable plant. These networks reach 95 percent of all households, making information, entertainment, and education available to almost everyone who chooses to subscribe.

Subscribers pay a monthly fee for a set of basic services and may select optional packages of premium services for an additional monthly fee. In addition, subscribers may choose pay-per-view programs by calling the cable operator to request a specific program from a menu of choices that changes daily. Usually there is a nominal extra charge for each additional television set that is set up to receive cable programming. All services are itemized on the monthly bill from the cable TV operator.

This market generates about US$25 billion in revenue per year from subscribers. The funds are generally split two ways: financing the operating costs of existing networks and constructing new systems, and providing payment for programming like HBO, MTV, and the Disney Channel.

Summary. To survive in the new competitive climate ushered in by the Telecommunications Act of 1996, cable companies are investing billions of dollars to upgrade their networks for full-duplex operation. Deregulation lets the telecommunications companies deliver local data services and even TV programming. It also lets the cable companies deliver local voice telephony, which (like broadband data) requires two-way transmission and switching capabilities. Other services CATV operators can offer are Internet access and video-on-demand. Among the technology choices for upgrading CATV networks for these advanced services are *hybrid fiber/coax* (HFC) and *fiber in the loop* (FITL).

See also

HYBRID FIBER/COAX
FIBER-IN-THE-LOOP
MODEMS

Call centers

Call centers are specialized environments that are equipped, staffed, and managed to handle a large volume of incoming telephone calls. A call center typically has an automatic call distributor (ACD) to connect calls to an order taker, customer service representative, help desk operator, or some other type of agent. Calls that cannot be answered immediately are put in a queue until the next agent becomes available. While on hold, callers might listen to music or advertising and get periodic barge-in messages informing them of their queue status. They also might get a menu of dialing choices so their calls can be routed in the most appropriate way.

When the call is answered, the agent addresses the caller's immediate needs and takes down relevant information about the caller, which is entered into a computer database. This information can be referred to the next time the person calls. In addition, this information can be used for a variety of other purposes, including the preparation of shipping labels for ordered merchandise, follow-up sales calls, direct mail advertising, and consumer surveys.

The elements of a typical call center (Fig. C1) include the following:

- ➢ *Telephone lines and services:* Call centers usually use digital lines such as 56 kbps, fractional T1 or T1, and ISDN; services may consist of 800 toll-free numbers to take orders, or 900 numbers to provide a service, the cost of which is charged to the caller's phone bill.
- ➢ *Switching system:* Most often a stand-alone ACD or integrated ACD/PBX system is used to support call center operations. ACD functions can be provided by the telephone company's central office, however, or as part of the carrier's Centrex services.
- ➢ *Telephone instruments:* Telephones equipped with a headset permits keyboard entry of customer information by the agent.
- ➢ *Workstations:* Agents enter customer information at their workstations, usually by filling out standard forms that appear on the display.
- ➢ *Host computer and database:* Customer records are stored on a central database. The host computer can sort the records in any number of ways and generate appropriate reports.
- ➢ *Management information tools:* Managers can query the database and retrieve information that can reveal such things as new sales opportunities, levels of customer satisfaction, and call center performance.

Summary. A call center can consist of only two or three agents, or as many as several thousand. The agents might operate from a single location, or they might be distributed around the world. Sometimes calls will be automatically transferred across time zones so the organization can provide customers with 24-hour service. The size of the call center and the distribution of its personnel will determine what kind of ACD system and lines are required:

See also

AUTOMATIC CALL DISTRIBUTORS

COMPUTER-TELEPHONY INTEGRATION

HELP DESKS

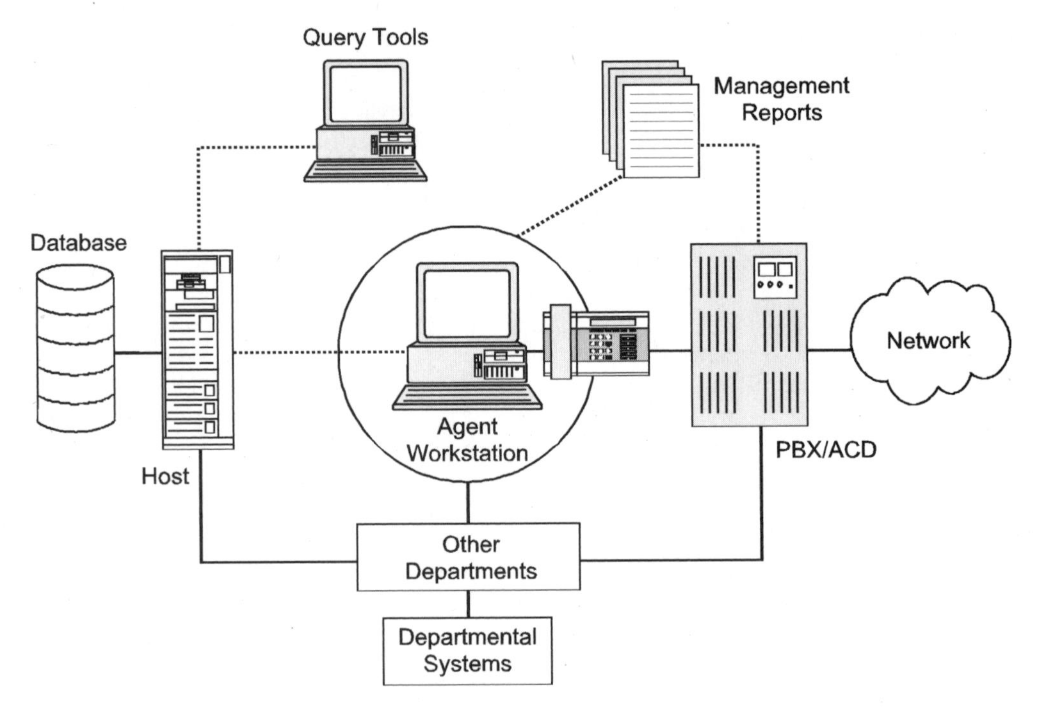

Figure C1 *Typical call center elements.*

Caller identification

Caller identification, or *caller ID,* is an optional service offered by telephone companies for an additional fee. It lets the user determine who is calling before answering the phone. The name and number of the calling party is displayed on the LCD screen of a telephone or a separate device connected to the phone.

While caller ID has been available in local calling areas for many years, only since 1995 has it been available on out-of-state calls. That year the FCC issued rules governing national caller ID services. These rules give callers the choice of delivering or blocking their telephone number for any interstate call they make.

The FCC's national caller ID rules protect the privacy of the called and calling party by requiring telephone carriers to make available free, simple, and consistent per-call blocking and unblocking arrangements. Each time a user picks up the phone to make a call, he or she can make a choice as to whether or not to block the number from being displayed to the called party. In addition, if per-line blocking is available, the caller can select that option. Per-line blocking is a service that automatically blocks the caller's telephone number from being delivered on all calls made from the line.

With per-call blocking, callers can block delivery of their phone numbers on a call-by-call basis by dialing *67 before dialing the target number. Many states allow consumers to select per-line blocking. Under this option, the telephone number will be blocked for every call without the requirement to dial extra digits. If the caller subscribes to the per-line blocking service but wants to allow the number to be transmitted to the called party, he or she must dial *82 before the telephone number each time a call is made.

Some services also transmit the name of the calling party. The FCC's caller ID rules require that, when a caller requests that his or her number be concealed, a carrier may not reveal the name of the subscriber to that line as well. Calls to emergency, 911-type lines are exempted from the FCC's caller ID rules. State rules and policies govern the obligation of carriers to honor caller privacy requests to emergency numbers.

Requesting privacy on calls to 800 and 888 numbers might or might not prevent the display of one's telephone number. When a toll-free number is dialed, the called party pays for the call. Typically the called party is able to identify the telephone number of incoming calls using a technology called *Automatic Number Identification* (ANI). When used with computer-telephony integrated (CTI) applications, the phone number of the incoming call can be matched against a database record, enabling a customer service representative, for example, to have information about the caller displayed on a computer terminal. This allows the representative to have all relevant information about the customer immediately available so the call can be handled in the most expeditious manner possible.

Summary. Nationwide caller ID offers many benefits for consumers and for the economy as a whole. Nationwide caller ID brings consumers rapid and efficient service, encourages the introduction of new technologies and new services to the public, and enables service providers and consumers to conduct transactions more efficiently.

See also

COMPUTER-TELEPHONY INTEGRATION
CUSTOM LOCAL AREA SIGNALING SERVICES (CLASS)

Calling card

Mobile professionals often need telecommunications services while away from the office and find calling cards helpful for making calls or leaving voice messages. Most long-distance carriers offer calling cards at no extra charge, and apply calls made with these cards to the corporate discount plan.

The caller simply dials the carrier's 800 number, followed by the telephone number, and finally the card number. The 800 prefix ensures that a caller reaches the carrier's network from any telephone and that all eligible calls are captured and aggregated for discount purposes. Charges are automatically posted to the account; an operator does not need to verify the call.

Calls placed with these cards cost less than third-number billed or collect calls and are connected without operator assistance. If a rotary telephone is used, however, the operator places the call without an additional charge for operator assistance.

Convenience features. Depending on the carrier, there are extra features available to call card users. For example, when using a Touch-Tone telephone, a caller can place sequential calls without reentering the card number. A speed-dial feature enables a cardholder to preprogram the calling card with frequently called numbers.

When the called party does not answer or the number is busy, the caller simply dials a few digits to leave a recorded message. Some cards have a magnetic strip for use at phones equipped with a strip reader, so callers can place calls without manually entering the card number.

Callers can use their calling cards to place local toll, long-distance, international, AirFone1, Inflight Phone, and RailFone2 calls. Callers can place cellular calls while out of their immediate regional serving areas if their cellular companies use the same long-distance carrier. The card also can be used to set up teleconferences. A cardholder can even dial a call directly and charge it to the calling card. And when the company migrates to another calling plan or the company changes telephone numbers, the same calling cards can continue to be used.

Calling cards also can simplify record-keeping and travel expense reporting because all card calls, domestic and international, are itemized and appear on a caller's monthly telephone bill. Billing codes can be used to identify certain calls placed with the calling card for charge-back purposes.

The carrier offers a toll-free number where customer service representatives can be reached to handle most card-related issues, including reporting lost cards, requesting additional cards, or reporting a problem with a card.

The cards themselves can be customized with the corporate name and logo or a special design. This involves a setup fee and per-card charge; depending on the design, a minimum number of cards might be required.

Security. Calling cards can be protected against fraudulent use by designating the numbers, area codes, or countries that can be called with a calling card and specifying a dollar amount that a cardholder can charge. The card can even be limited to calling the corporate headquarters or branch office.

An optional personal identification number (PIN) embedded into the card's magnetic strip discourages fraud if the card is lost, stolen, or examined by outside parties. This strip contains pertinent information, such as PIN and account number, that is read by most card-reader telephones. Cardholders usually can select their own PINs and choose whether or not to have them printed on the fronts of the cards.

In addition, an enhanced fraud protection process enables the carrier to identify potential fraudulent use on a real-time basis and quickly notify cardholders so they can take immediate corrective action. For example, the AT&T Fraud Analysis and Surveillance Center (FASC) monitors card calls around the clock for unusual calling activity and attempts to contact the cardholder for authorization. If the calling activity continues, prior to cardholder notification, the FASC deactivates the card.

Summary. Calling cards offer mobile professionals the means to access telecommunications services while away from the office. In addition to being less expensive than collect calls, calls placed with these cards can apply to the corporate discount plan. Numerous convenience features make today's calling cards easy to use in a variety of circumstances. Security is enhanced through the use of PINs and the carrier's own fraud protection systems.

See also

LOCAL TELECOMMUNICATIONS CHARGES
LONG-DISTANCE TELECOMMUNICATIONS CHARGES

Cellular communications

Cellular telephony provides communications service to automobiles and handheld portable phones, and interconnects with the public telephone network using radio transmissions based on a system of cells and antennas.

The cellular concept was developed by AT&T's Bell Laboratories in 1947, but it was not until 1974 that the FCC set aside radio spectrum between 800 and 900 MHz for cellular radio systems. The first cellular demonstration system was installed in Chicago in 1978; three years later the FCC formally authorized 666 channels for cellular radio signals and established Cellular Geographic Servicing Areas (CGSAs) to cover the nation's major metropolitan centers.

At the same time, the FCC created a regulatory scheme for cellular service, which specified that two competing cellular companies would be licensed in each market. For each city the commission ruled that one license would be reserved for the local telephone company (a *wireline* company), and the other license would be granted to another qualified applicant. When the number of applicants became prohibitively large, the Commission amended its licensing rule and specified the use of lotteries to select applicants for all but the top 30 markets. Cellular service is now available virtually everywhere in the United States.

Applications. Cellular telephones allow users to optimize their schedules by turning nonproductive driving and out-of-the-office time into productive—and often profitable—work time. Cellular solutions not only facilitate routine telephone communications, they also increase revenue potential for people in professions that have high-return opportunities as a direct result of being able to respond promptly to important calls. Developing countries that do not have a well- developed communications infrastructure are increasingly turning to cellular technology so they can take part in the global economy without having to go through the resource-intensive step of installing wire.

Technology. Cellular networks rely on relatively short-range transmitter/receiver (transceiver) base stations that serve small sections (or *cells)* of a larger service area. Mobile telephone users communicate by acquiring a frequency or time slot in the cell in which they are located. while a master switching center called the Mobile Transport Serving Office (MTSO) links calls together using traditional copper technology. Figure C2 illustrates the link from the MTSO to the base stations in each cell. The MTSO also has links to local telco central offices, so cellular users can communicate with users of conventional telephones.

Cell sites. Cell boundaries are neither uniform nor constant. The usage density in the area, as well as the landscape, the presence of major sources of interference (e.g., power lines, buildings), and the location of competing carrier cells all contribute to the definition of cell size. Cellular boundaries change continually, with no limit to the number of frequencies available for transmission of cellular calls in an area. As the density of cellular usage increases, individual cells are split to expand capacity. By dividing a service area into small cells with limited-range transceivers, each cellular system can reuse the same frequencies many times. Emerging technologies such as Code Division Multiple Access (CDMA) and Expanded Time Division Multiple Access (E-TDMA) promise further capacity gains.

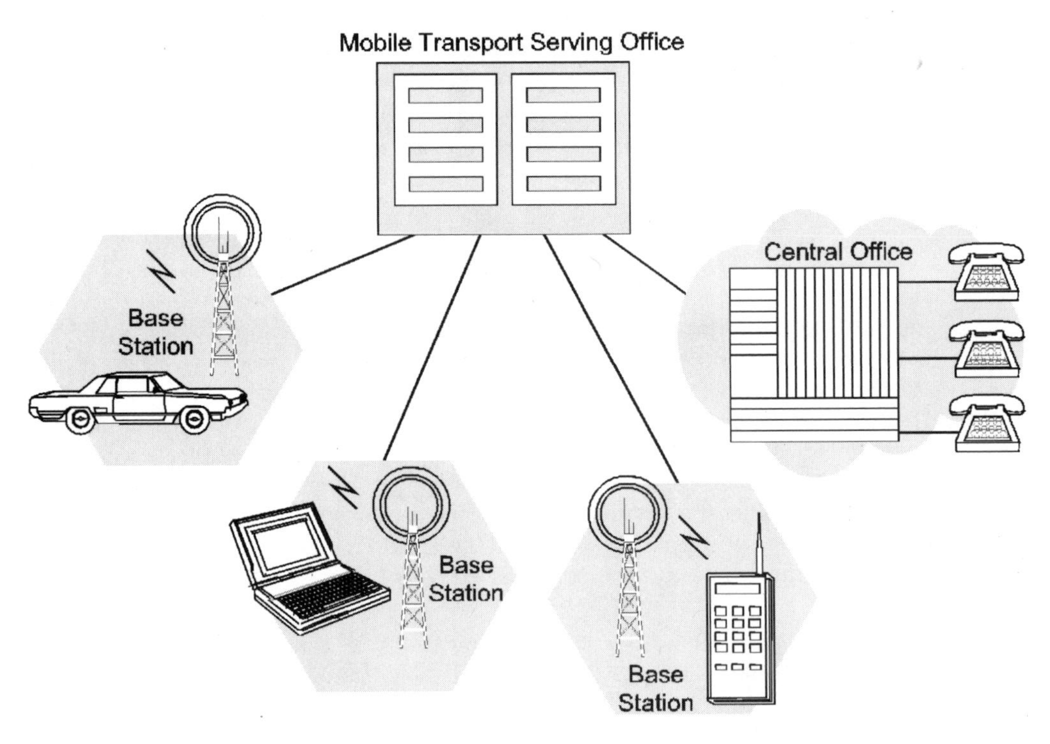

Figure C2 *A typical cellular network configuration.*

Master control center. In a typical cellular network, the master switching center (master control center) operates similarly to a telephone central office, and links with other offices. The cellular switching center supports trunk lines to the base stations that establish the cells in the service area. Each base station supports a specific number of simultaneous calls, usually from 3 to 15, depending on the underlying technology (i.e., CDMA, TDMA, or some derivative).

Transmission channels. Most cellular systems provide two types of channels: a control channel and a traffic channel. The base station and mobile station use the control channel to support incoming and outgoing calls, monitor signal quality, and register when a user moves into a new zone. The traffic channel is used only when the station is off-hook and actually involved in a call.

The control and traffic channels are divided into time slots. When the user initiates access to the control channel to place a call, the mobile station randomly selects a subslot in a general-use time slot to reach the system; the system then assigns a time slot to the traffic channel. For an incoming call to a mobile station, the base station initiates conversations on the control channel by addressing the mobile station in a time slot, which at the same time reserves that time slot for the station's reply. If a user's call attempt collides with another user's call attempt, both instruments automatically reselect a subslot and try again. After repeated collisions, if no time slots are available within a predetermined time, the system rejects service requests for incoming and outgoing calls.

When a mobile telephone user places a call, the cell in which the user is traveling allocates a slot for the call. The call slot allows the user access through the base station to the master switching center, essentially providing an extension on which the call can be placed. The master switching center, through an element of the user-to-base station connection, continuously monitors the quality of the call signal, and transfers the call to another base station when the signal quality reaches an unacceptable level due to the distance traveled by the user, obstructions, or interference. If the user travels outside of the system altogether, the master switching center terminates the call as soon as the signal quality deteriorates to an unacceptable level.

Roaming. Roaming occurs when a user moves out of the home area and into the serving area of another cellular carrier. In most cases, however, the cellular carrier belongs to an extension service such as MobiLink, which provides a service hand- off between cellular carriers. Instead of calls being dropped because the user strayed beyond a service boundary, the call is handed off to the next cellular carrier. The caller does not need to preregister with another service provider when traveling outside the home area. With MobiLink, for example, the subscriber merely dials a standard code to handle incoming and outgoing calls when traveling outside the home area: dialing *18 activates roaming and *19 deactivates roaming.

Cellular telephones. The cellular telephone is the most visible part of the cellular system. Cellular telephones incorporate a combination of multiaccess digital communications technology and traditional telephone technology, and are designed to appear to the user as familiar residential or business telephone equipment. Manufacturers use miniaturization and digital signal processing technology to make cellular phones feature-rich yet economical.

Cellular instruments consist of a transceiver operating in the 900 MHz band, an analog/digital converter, and a supervisory/control system that manages calls and coordinates service with both the base station and the master switching center. Cellular telephones can be powered from a variety of sources, including vehicle batteries, ac adapters, and rechargeable battery sets.

Traditional cellular instrument types include handheld, transportable, and car telephones. Advances in cellular technology are creating additional types of telephones, however, including modular and pocket phones. The trend in cellular instruments is toward multipurpose transportable telephones.

There are dual-mode cellular phones that can be used with in-building wireless PBXs as well as with the outside cellular service. The handset registers itself with an in-building base station and takes its commands from the wireless PBX. For out-of-building calling, the handset registers with the nearest cell site transceiver. Aside from convenience, an added benefit of the dual-mode phone is that calls made off the corporate premises can be aggregated with business calls made at home or on the road for the purpose of achieving a discounted rate on all calls.The future all-digital cellular system holds the promise of a national personal communications system with even more sophisticated applications, such as automatic security and alarm reporting, vehicle locating service, integrated paging service, ambulance patient monitoring, and navigation assistance. Digital cellular will also offer several advantages for data communications, including high-speed, error-free information transport, and data encryption to safeguard privacy.

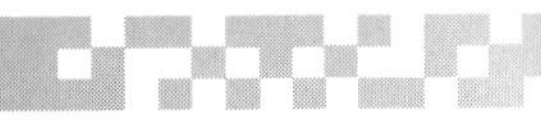

Summary. No other communications technology has taken the world by storm quite like cellular. Many factors influence the cellular industry, including deregulation and competitive pricing among the various cellular telephone companies, not to mention competition from other technologies such as CDPD and PCS. All of these factors have combined to reduce the price of equipment and service in recent years to the point where cellular is no longer a service for professionals on the go, but for the average consumer as well.

See also

CELLULAR TELEPHONES
PERSONAL COMMUNICATIONS SERVICES

Cellular telephones

In recent years, cellular telephones have emerged as a "must have" item among mobile professionals and consumers alike, growing in popularity every year since they were introduced in 1983. Their widespread use for both voice and data communications is a product of significant progress that has been made in portability, availability of network services, and the declining cost for equipment and network services.

System components. There are several categories of cellular telephone. Mobile units are mounted in a vehicle. Transportable units can be moved easily from one vehicle to another. Pocket phones, weighing in at less than 4 ounces, can be conveniently carried in a jacket pocket or purse. There are even cellular telephones that users can wear as a fashion accessory. Regardless of how they are packaged, cellular telephones consist of the same basic elements.

Handset/keypad. The handset and keypad provide the interface between the user and the system. This is the only component of the system with which, under normal operation, the user needs to be concerned. Any basic or enhanced system features are accessible via the keypad, and once a connection is established, this component provides similar handset functionality to that of any telephone. Until a connection is established, however, the operation of the handset differs greatly from that of a conventional telephone.

Rather than initiating a call by first obtaining a dial tone from the network switching system, the user enters the dialed number into the unit and presses the SEND function. This conserves the resources of the cellular system since only a limited number of talk paths are available. Once the network has processed the call request, the user will hear conventional call progress signals such as a busy signal or ringing. From this point on, the handset operates in a customary manner. To end a call, the END function key is pressed on the keypad. In addition to these functions, the handset typically contains a display that shows dialed digits as well as status of other features, a CLEAR key that enables the user to correct misdialed digits, functions that enable storage of numbers for future use, and other enhanced features that can vary greatly from one phone to the next.

Logic/control. The logic/control functions of the phone include the numeric assignment module, or NAM, for programmable assignment of the unit's telephone number by the service provider, and the electronic serial number of the unit, which is a fixed number unique to each instrument. When signing up for service, the carrier makes a record of both numbers. When the unit is in service, the cellular network interrogates the phone for both of these numbers in order to validate that the calling/called cellular telephone is that of an authentic subscriber.

This component of the phone also serves to interact with the cellular network protocols that determine what control channel the unit should monitor for paging signals, which indicate the network's effort to connect a call coming into the phone, determine and select the voice channels that the unit should utilize for a specific connection, and monitor the received control signals of cell sites when the phone is in either standby or in-use mode, so that the phone and network can coordinate transitions to adjacent cells as conditions warrant.

Transmitter/receiver. The transmitter/receiver unit of the telephone is under the command of the logic/control unit. Powerful 3 W telephones typically are of the vehicle-mounted or transportable type; their transmitters are understandably larger and heavier than those contained within lighter-weight handheld cellular units. These more powerful transmitters require significantly more electricity than handheld units that transmit at power levels of only a fraction of a watt; they normally use the main battery within a vehicle or a relatively heavy rechargeable battery to do so. A diplexer unit within the phone enables the transmitter and receiver to utilize a single antenna while simultaneously transmitting and receiving.

Antenna system. The antenna for a cellular telephone can consist of a flexible rubber antenna mounted on a handheld phone, an extendable antenna on a pocket phone, or the familiar curly stub seen attached to the rear window of many automobiles. Antennas and the cables used to connect them to radio transmitters must have electrical performance characteristics that are matched to the transmitting circuitry, frequency, and power levels. Use of antennas and cables that are not optimized for use by these phones can result in poor performance. Improper cable, damaged cable, or faulty connections can render the telephone completely inoperative.

Power sources. Cellular phones typically are powered by a rechargeable battery. Nickel cadmium (NiCad) batteries are the oldest and cheapest power source available for cellular phones. Newer nickel–metal hydride (NiMH) batteries provide extended talk time compared to lower-cost conventional nickel-cadmium units. They provide the same voltage as NiCad batteries, but offer at least 30 percent more talk time than NiCad batteries and take approximately 20 percent longer to charge.

Lithium ion batteries offer increased capacity and are lighter in weight than similar-sized NiCad and NiMH batteries. These batteries are optimized for the particular model of cellular phone, which helps ensure maximum charging capability and long life for the battery.

Newer cellular phones may operate with optional high-energy AA alkaline batteries that can provide up to 3 hours of talk time or 30 hours of standby time. These batteries take advantage of the new lithium–iron disulfide technology, which results in 34 percent less weight than standard AA 1.5 V batteries (15 vs. 23 g per battery) and 10-year storage life, double that of standard AA alkaline batteries.

Vehicle-mounted cell phones can be optionally powered via the vehicle's 12 V dc battery by using a battery eliminator that plugs into the dashboard cigarette lighter socket. This saves useful battery life by drawing power from the vehicle's electrical system and comes in handy when the phone battery has run down. A battery eliminator will not recharge the phone battery, however; recharging the battery can only be done with a special charger.

Lead-acid batteries are used to power transportable cellular phones when the user wishes to operate the phone away from the vehicle. The phone and battery are usually carried in a vinyl pouch.

Options and features. Cellular telephones offer many features and options, including:

- *Voice activation:* Sometimes called "hands-free operation," this feature allows the user to establish and answer calls by issuing verbal commands. This feature allows a driver to control the unit without becoming visually distracted by the telephone.
- *Memory functions:* These allow storage of frequently called numbers to simplify dialing. Units may offer as few as 10 memory locations or more than 100.
- *Multiple numeric assignment module:* This feature allows a single phone to be used with multiple carriers. The phone then can be used to access the best carrier for a specific location in areas where two local service providers might have different coverage gaps. This feature also is useful in cutting the cost of roaming into other service areas where surcharges might apply.
- *Visual status display:* This conveys information on numbers dialed, state of battery charge, call timers, roaming indication, and signal strength. Cellular phones differ widely in the number of characters and lines of alphanumeric information they can display. Among the recent innovations in display technology is the use of dedicated icons, which enhance ease-of-use by visually identifying the phone's features.
- *Programmable ringer tones:* Some cellular phones allow the user to select the phone's ringer tone.
- *Silent call alerts:* Features include visual or vibrating notification in lieu of an audible ring signal. This can be particularly useful in locations where the sound of a ringing phone would constitute an annoyance.
- *Security features:* These include password access via the keypad to prevent unauthorized use of the telephone, as well as features to help prevent access to the phone's telephone number in the event of theft.
- *Voice messaging:* This allows the phone to act as an answering machine. A limited amount of recording time, about 4 minutes, is available on some telephones. However, carriers also offer voice messaging services that are not dependent on the phone's memory capacity. When the phone is left in standby mode, callers reach the answering device, which functions exactly as most voice-mail systems. Further, air time is not required for users to retrieve the messages that were left while they were in meetings or otherwise occupied.
- *Call restriction:* This feature allows the user to permit use of the phone by others to call selected numbers, local numbers, or emergency numbers without permitting them to dial the world at large.
- *Call timers:* These provide the user with information as to the length of the current call; some telephones also can maintain a running total of air time for all calls. These features make it easier for users to keep track of call charges.

With the increased use of cellular telephones for personal use, choice of color and styling is playing a greater role in the phone selection process. Cellular phones come in such diverse colors as Sunstreak (yellow), Dark Spruce, Eggplant, Regatta Blue, and Temptation Teal.

Cellular data communications. As much as cellular telephones are a useful tool for mobile voice communications, they also are becoming indispensable for users who require portable data communications capabilities. Cellular networks are used for the transmission of fax traffic, electronic mail, remote order entry and inquiries, file transfers, and most data communications applications for which the wired telephone network is used. Remote metering locations for pipelines, electrical substations, and other unattended locations that may be far from the nearest telephone line rely on cellular equipment to provide a connection.

Cellular phones and the networks through which they communicate originally were not designed for data communications purposes and, until recently, adapters were the only means by which a conventional data modem could interconnect with the cellular network. Even then, not all cellular phones were capable of connecting to a modem. To provide cellular connectivity, a cellular phone must have an outlet into which the user can insert a cable that connects the phone to a modem, the other end of which is plugged into a portable computer. The phone also must support special signaling features that allow it to communicate with the modem. In a properly designed cellular modem, the modem automatically reconfigures itself for cellular operation, allowing the user to send e-mail and faxes, or access on-line service providers such as CompuServe or America Online, or "surf the Internet" with the portable computer.

Fortunately, a great many telephones now incorporate data communications interface capabilities as part of the cell phone unit. Cellular phones do not present to a modem the typical dial tone and electrical characteristics of a standard telephone line. And while the process of a network hand-off from one cell site to another can be quite acceptable for carrying voice traffic, it can effectively terminate any data communications session in progress.

Adapter units compensate for this phenomenon when used in conjunction with telephones that are not inherently data-capable; data-ready phones do not require these adapters. The adapter units or special cellular-capable modems require that the remote end of the link between the phone and the Mobile Transport Serving Office (MTSO) also have a device that can communicate using the same cellular-capable protocol. These devices can be provided by the carrier within the MTSO in a pooled configuration available to all users, or the user can ensure that a proper unit is installed at the remote computer location to which the cellular phone is attempting to communicate.

Modems are available that allow the remote user to utilize them for both cellular and land lines (wired phone lines). The popular units for users of current generation of laptop computers are the PCMCIA card modems; these take far less space than conventional modems or the early cellular modems, and they also allow the computer to utilize standard phone lines when available. The cellular network is not as capable of carrying high-speed data communications as is the wired network, but speeds near 9.6 to 14.4 kbps are possible for data and fax traffic. Cellular digital packet data (CDPD) transmission, which currently is being implemented in a number of areas, promises to provide more reliable data communications via existing cellular networks, and at slightly higher throughput rates (to 19.2 kbps), although not at speeds equivalent to that of land lines. CDPD is appropriate for most applications that might also use conventional packet networks, such as bursty transaction processing or routine short-duration use by individuals.

Cellular data calls are subject to the same impairments as cellular voice calls, specifically multipath distortion, signal fading, fluctuating power levels, poor frequency response, and external noise. A variety of factors, such as tall buildings, electronic equipment, and street traffic, affect the quality of the connection. Although cellular modems contain advanced, cellular-specific error correction protocols (e.g., MNP-10, MNP-10EC, HST, ETC, EC2, and TX-CEL) to compensate for the external factors that impact cellular transmission, data calls should occur from a stationary position, away from power lines or electrical equipment, to ensure the highest transmission speed.

Summary. Cellular phones are growing more intelligent, as evidenced by the availability of units that are part cellular phone and part palmtop computer. These devices not only support data communications, but voice messaging, e-mail, fax, and Internet access as well. Third-party software provides the operating system and such applications as calendaring, card file, and to-do lists. With more cellular phones supporting data communications, cellular phones are becoming available that provide connectivity to PC desktop and databases via infrared or serial RS-232 connections. Information even can be synchronized between cell phones and desktop computers to ensure that the user is always accessing the most up-to-date information.

See also

CELLULAR COMMUNICATIONS

PERSONAL COMMUNICATIONS SERVICES

Central office switches

Today's central office switches are available from several manufacturers worldwide, differing in line capacities, services and features, and network environments supported. They not only switch ordinary telephone calls, but support digital voice, text, image, and data communications via ISDN. The environments supported can include a stand-alone office, a distributed network, colocation with an analog office, or remote configurations that allow services and features to be extended to isolated areas via remote switch units.

The switches themselves are modular; with various software upgrades, card additions, and adjunct system connections, they can be equipped to provide access to Centrex/business group features. They can function as a database gateway, connecting telephony functions with online databases, or interactive services and packet networks. They also can be equipped to support LAN interconnection over the wide area network (WAN). In recent years, support for a variety of broadband technologies has been added, including frame relay, SMDS, ATM, and SONET.

A switch can handle incoming traffic at a rate of about 25,000 Erlangs. (An Erlang is defined as the equivalent of one line continuously operating at 100 percent capacity.) It is capable of processing as many as 1.5 million busy-hour call attempts (BHCAs). Each digital line unit can support hundreds of individual subscriber lines. A mobile exchange subsystem can be added to the central office switch to accommodate as many as 40,000 radio subscribers per exchange. The switch's service management subsystem accommodates operator and administrative terminals. The switch also hosts data polling systems for traffic analyses and automated billing.Today's digital central office switches are very versatile, particularly through their support of advanced intelligent network (AIN) services. The switch's ISDN and AIN capabilities complement each other in that ISDN supports service access, while the AIN supports service control and execution. Both facilities are based on Signaling System 7 (SS7), and both enable users to quickly obtain and utilize new services.

Features. The central office switch offers subscribers a wealth of basic and advanced features. Basic features are those provided to subscribers whose facilities are not connected to ISDN lines; advanced features are those provided to ISDN users.

Basic features include abbreviated dialing, alarm call, call rerouting–busy, call waiting, call rerouting–no answer, call charge indicator, toll-free calling, conference calling, direct dialing to extension, emergency call areas, hot line, call trace, incoming call block, individual call record, outgoing traffic limitations, override block, subscriber with special services, subscriber priority, and three-way calling.

Advanced features are provided to subscribers whose systems are equipped with ISDN Basic Rate (2B+D) or Primary Rate (23B+D) Interfaces. Advanced features include automatic callback, call forwarding, call hold, call pickup, call rerouting when busy, call waiting, charge handling, data transmission, dedicated connection, display information, incoming call block, multiline hunt group, subscriber-programmed features, user groups, and closed user groups.

System components. The central office switch accommodates several types of hardware modules, which may be called by different names by different manufacturers:

- *Digital Line Units* (DLU) interface subscriber traffic, including analog or ISDN signals, to the switch. These units provides analog-to-digital conversion and initial signal processing, and they concentrate traffic over T1 links to the Line Trunk Groups (LTGs). DLUs can accommodate any mix of single-party, dual-party, ground-start, coin-operated, direct inward dial, and ISDN lines.
- *Remote Control Unit* (RCU) houses multiple DLUs in a remote deployment configuration. Whatever service or feature is available from the central office usually is available to users connected at the RCU. The RCU can provide emergency local switching for users connected to the same RCU, however, without carrying traffic to the LTGs and the Switching Network (SN).
- *Integrated Packet Handler* (IPH) performs the ISDN packet handler functions for ISDN packet subscribers. This allows ISDN-equipped customers to both originate and receive packet transmissions at their desktop terminals.
- *Line Trunk Group* (LTG) connects subscriber and trunk lines with the CCSNC. LTGs also function as expansion elements to the switch.

- *Switching Network* (SN) interfaces signals from the DLU, controls input and output switching, and provides switching control functions.
- *Coordination Processor* (CP) controls and coordinates the system through various operation, administration, and maintenance functions.
- *Common Channel Signaling Network Controller* (CCSNC) handles the transfer of SS7 messages in the distributed environment, specifically between the Service Switching Point, the network Signal Transfer Points, and the Service Control Point. The switch's CCSNC application module enables the switch to separate call handling from network management.

Software components. The operating system and applications software controls switching, administration, and maintenance of the switch and its required databases. The operating system provides organization programs for system managers (e.g., time administration, memory administration, input and output elements, and safeguarding). Application software can be configured to provide local, transit, and long-distance exchange, as well as radio relay. Through software, various ISDN, intelligent network, and mobile radio system features can be incorporated in the switch on a modular basis. The switch functions typically provided by software include:

- *Local Number Portability* provides the means for subscribers to keep the same telephone number if they decide to switch local or long-distance service providers.
- *Advanced Centrex Services* include improved remote access for telecommuters to Centrex groups at the central office switch.
- *Centrex Attendant Console* supplies additional call-handling and business group capabilities to support Key Telephone System (KTS) environments.
- *Enhanced KTS Business Group Applications* provide advanced ISDN features to analog instruments, including caller ID and visual message-waiting indicator.
- *Centrex Command and Control Workstation* supplies a graphical user interface (GUI) for Centrex users for access and control of their facilities. Capabilities include traffic collection, maintenance, message detail data recording, trunk testing, and authorization code control, outgoing facility routing, time-of-day redirection, alternate facility restriction, and automatic flexible routing.

Summary. Today's modular central office switches allow carriers to build different types of switching centers using various hardware combinations. The switch can function as an end office, an access tandem, or a remote unit capable of serving rural communities. The switch's multiple processors and modular components simplify system modification: Upgrades or fixes are confined to a subsystem rather than to the central switching processor, thereby reducing the likelihood that problems will occur in other parts of the network. The modular design also allows software to be updated and new processors and subsystems to be added with relative ease, creating a ready migration path for future technologies and services.

See also
ADVANCED INTELLIGENT NETWORK

CENTREX
INTEGRATED SERVICES DIGITAL NETWORK
SIGNALING SYSTEM 7

Centrex

Centrex (short for *central office exchange)* is a switching service that directs calls for business and residential customers through a telephone company's switch, rather than through a customer-owned, premises-based switch. Centrex provides a full complement of station features, remote switching, and network interfaces to the customer premises. Centrex offers remote options for businesses with multiple locations, providing features that appear to users and the outside world as if the remote sites and the host switch were one system.

Individual Centrex users have access to direct inward dialing features, as well as station identification on outgoing calls. Each station has a unique line appearance in the central office, in a manner similar to residential telecommunications subscriber connections. A Centrex call to an outside line exits the switch in the same manner as a toll call exits a local exchange. Users dial a 4- or 5-digit number without a prefix to call internal extensions, and dial a prefix (usually 9) to access outside numbers.

The telephone companies operate, administer, and maintain all Centrex switching equipment for their customers. They also supply the necessary operating power for the switching equipment, including backup power to ensure uninterrupted service during commercial power failures.

Centrex also is offered through resellers that buy Centrex lines in bulk from the local exchange carrier. Using its own or commercially purchased software, the reseller packages an offering of Centrex, and perhaps other basic and enhanced telecommunications services, to meet the needs of a particular business. The customer gets a single bill for all the local, long-distance, 800, 900, and calling card services at a fee that is less than the customer would pay otherwise.

Centrex features. Centrex service offerings typically include direct inward dialing (DID), direct outward dialing (DOD), and automatic identification of outward dialed calls (AIOD). Advanced digital Centrex service provides all of the basic and enhanced features of a modern PBX in the areas of voice communications, data communications, networking, and ISDN access. Commonly available features include voice mail, electronic mail, message center support, and modem pooling.

Centrex, as part of the public telephone network switching operation, provides a full complement of network interfaces to the customer premises. Centrex offers remote options for businesses with multiple locations, providing features and system operations that appear to users and the outside world as if the remote sites and the host switch were one system. For large networks, the Centrex switch can act as a tandem switch, linking many PBXs through an electronic tandem network. Centrex also is compatible with most private switched network applications, including the Federal Telecommunications System (FTS) and the Defense Switched Network (DSN).

Many users subscribe to Centrex service primarily because of its networking capabilities, particularly the opportunity to set up a virtual city-wide network without major cost or management concerns. With City-Wide Centrex, a business can

set up a network of business locations with a uniform dialing plan, a single published telephone number, centralized attendant service, and full feature transparency for a few dollars more per month than a single Centrex site would cost.

Customer premises equipment. Centrex customer premises equipment is available for lease or purchase from a number of vendors, including the telephone company. Centrex CPE combines the advanced features of a PBX with the convenience and flexibility of Centrex. Popular Centrex CPE products include multiline phones, PC-based attendant workstations, attendant consoles, telemanagement packages, voice processing equipment, line monitors, and tip-and-ring scanners.

Centrex telephones. Phones made specifically for Centrex service allow users to access a wealth of Centrex features with the touch of a button, without having to memorize codes. Many even provide single-stroke transfer capability to remote sites. When users have the proper equipment, the service is used more efficiently and the features are used more frequently, resulting in better value for the company's Centrex investment dollars.

PC-based attendant workstations. Screen-based attendant consoles combine Centrex access and improved calling options with Windows databases. This allows an attendant to work in a Windows word processing application, for example, and hot-key over to answer an incoming call. Some PC attendant workstations come equipped with database directories that can support multiple telephone, fax, and paging numbers for each entry and enable single-keystroke dialing. Those that support Automatic Number Identification (ANI) and directory name look-up can do "screen pops" of caller information to give truly personalized call handling service.

LAN access workstations. Centrex workstations that interface with LANs also allow attendants to do more than just handle calls. Hot-keying lets them answer and transfer calls, and still complete computer-based tasks over the local area network.

Centrex answering consoles. Centrex answering consoles can take the form of multibutton telephones, conventional-looking attendant consoles, or PC-based platform systems. Centrex attendant consoles allow call handlers to perform single-button call transfers. They also have line status displays that let the attendant know when a line is in use or idle. Many consoles also allow the attendant to reprogram extensions, access features, and make other system rearrangements without the help of the telephone company.

Message desk. As an option, the telephone company can provide a data line, called a station message desk interface (SMDI), to the customer. Centrex CPE is available that interfaces with the SMDI link to give full voice mail and Centrex integration. It allows a call to an unanswered station to go directly to the person's voice mailbox without the caller having to reenter the extension number.

Call accounting system. Many options are available for businesses that use Centrex to obtain station message detail recording (SMDR) data in order to increase system efficiency. The telcos can provide call detail information to their customers, or, alternatively, customers can use CPE line scanners and PC telemanagement products to obtain the same functionality at a lower price. Some systems record SMDR right from the switch and store the information until it can be transferred to a PC for processing.

Administration systems. In the past, Centrex customers had to go through the telephone company to change an extension on their Centrex service. Now Centrex administration systems let users reconfigure their own Centrex service. Platform-based consoles allow users to turn enhanced CO services (e.g., CO voice mail) on and off; to activate or deactivate lines; to add or reassign trunks; and to rearrange long distance dialing patterns to route calls over the least expensive lines.

Summary. Centrex offers high-quality, dependable, feature-rich telephone service that supports a variety of applications. For many users, Centrex offers distinct advantages over on-premises PBX or key/hybrid systems. Centrex users save money over the short term because there is no outlay of cash for an on-premises system, and Centrex installations charges are usually low. If the service is leased on a month-to-month basis, there is little commitment and no penalty for discontinuing the service. A company can pick up and move without worrying about moving and reinstalling a telecommunication system that may not be right for a new location.

Centrex systems can be expanded almost indefinitely by adding communication paths, memory, intercom lines, tie lines, and CO lines as needed. PBX and key systems usually have maximum capacities. If there is a Centrex problem, repair is immediate and inexpensive. There is no need to invest in spare parts inventory, test equipment, or technical staff. As the CO switching equipment is updated, the Centrex services are also updated.

See also

PRIVATE BRANCH EXCHANGE (PBX)

Channel banks

A channel bank is a low-end multiplexer that consolidates up to 24 voice and data channels of 64 kbps each onto a higher-speed digital facility. Pulse Code Modulation (PCM) is used to convert analog voice to digital signals suitable for transmission over T1 lines, which support the DS1 rate of 1.544 Mbps. Some channel banks can multiplex as many as 96 voice channels in support of the DS2 rate of 6.312 Mbps.

A channel bank might be used on the front end of an analog PBX, for example, to transport a bundle of voice conversations via a T1 line to another analog PBX across town. There, another channel bank receives the higher-speed digital signal and converts it back to the individual analog channels for acceptance by the PBX. Most digital PBXs now support T1, however, eliminating the need for a channel bank and allowing the PBX to connect to the span with only a Channel Service Unit (CSU). The CSU is required at the front end of a circuit to equalize the received signal, filter both the transmitted and received waveforms, and interact with the carrier's test facilities.

Channel banks are ideally suited for handling voice because their basic rate and bit pattern are matched to the management of voice signals, both for multiplexing at the DS1 rate and in preparation for additional downstream multiplexing at higher rates.

Channel banks vs. multiplexers. Whereas channel banks originally were designed to accept analog inputs, T1 multiplexers are designed to accept digital inputs. Although T1 multiplexers can be equipped to accept analog inputs via

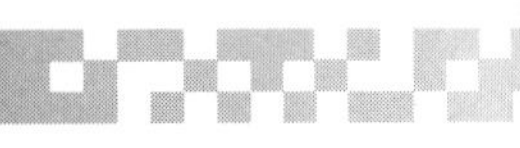

optional plug-in cards, they are very versatile devices that are especially adept at handling data streams. For example, multiplexers can compress data to increase the number of channels available for voice, prioritize traffic to avoid congestion on the network, and downspeed data to avoid corruption at higher rates when links experience service degradation.

Even though channel bank manufacturers have equipped their products with more data features and intelligence, resulting in entirely new devices called *intelligent channel banks,* T1 multiplexers still offer more functionality, especially in the area of network management. Network management information even can be inserted into each digital channel for end-to-end supervision and control.

When data channels experience high error rates, the multiplexer can reroute them to other links and leave unaffected voice channels on the primary link. T1 multiplexers also are more flexible in managing the available bandwidth, implementing software-based reconfigurations that change automatically by time of day, or by events such as link failure and bit error rate thresholds. And when disaster threatens to bring down the entire network, the multiplexer can implement preplanned disaster recovery scenarios under stored program control, calling into service any combination of available private leased lines and public network services.

Summary. Channel banks offer an economical way to connect analog communications equipment to digital T1 service. They are ideal for static network environments where low cost is the primary concern. Today's channel banks also offer data ports for direct digital connectivity. Password-protected remote management via a modem connection enables system performance to be monitored and tests to be performed from any centralized PC or workstation, thus reducing down time.

See also
CHANNEL SERVICE UNIT (CSU)
MULTIPLEXERS
T-CARRIER FACILITIES

Channel Service Unit

Since digital transmission links are capable of transporting signals between data terminal equipment (DTE) in a form nearer to the original, there is no need for complex modulation/demodulation (modem) techniques, as is the case with sending data over analog connections. Instead, Channel Service Units (CSUs) are used at the front end of the digital circuit (Fig. C3) to equalize the received signal, filter the transmitted and received waveforms, and interact with both the user's and carrier's test facilities via a supervisory terminal or network management system. The FCC's Part 68 registration rules require that every T-carrier circuit be terminated by a CSU. These devices can be used to set up a T1 line with a PBX, channel bank, T1 multiplexer, or any other DSX1-compliant DTE.

Key CSU functions. Line build-out (LBO) is a functional requirement of all Part 68–registered T1 CSUs. An LBO is an electronic simulation of a length of wireline that adjusts the signal power so that it falls within a certain decibel range at both ends of the circuit. This is determined by looping a test signal back over the receive pair and measuring for signal loss. This procedure also helps reduce the potential

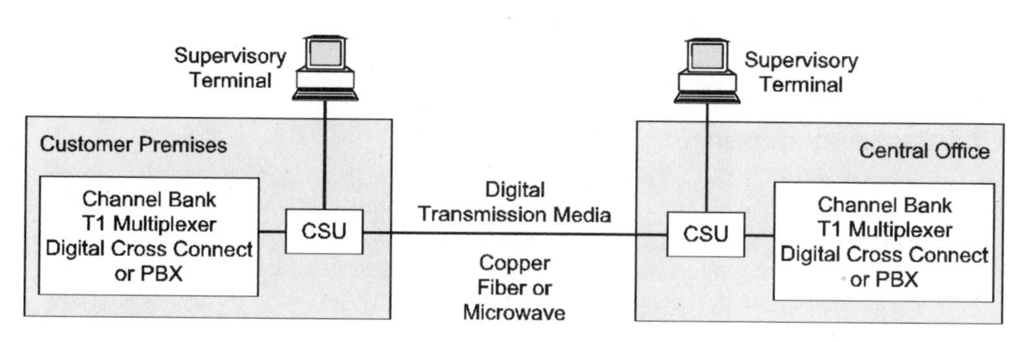

Figure C3 *A Channel Service Unit (CSU) terminates each end of the T-carrier facility.*

for one T1 transmitter to generate crosstalk in the receiver of other services within the same cable binder. Once line loss is determined, the telephone company can tell the customer what setting to use on the local CSU.

All T1 CSUs provide a repeater to reconstitute signals that have been attenuated and distorted by the T1 span line or the customer's in-house cabling. This function is also part of the FCC Part 68 equipment registration requirements for CSUs.

In addition to equalizing the transmitted signal through the LBO and regenerating the received line signal, the CSU ensures that the user's DTE does not send signals that could disrupt the carrier's network. For example, very long strings of 0s do not provide timing pulses for the span line repeaters to maintain synchronization. The CSU monitors the data stream from the attached DTE so that the *1s density rule* (e.g., the customer's data must have at least a 12.5 percent pulse density) is not violated. This rule ensures that there are sufficient pulse transitions for the span line repeaters to maintain timing synchronization. The CSU will inject pulses if excessive 0s are being transmitted by the attached DTE.

The CSU also provides isolation between the DTE and the network to protect equipment and telephone company technicians from the potentially harmful line voltages and lightning surges which may propagate beyond the DTE.

The CSU provides the functionality to troubleshoot circuit and transmission problems. For proactive network management, these include LEDs that indicate both the status of the network and equipment connections, and whether or not any alarm thresholds or error conditions have been detected. This lets technicians at either end of the circuit isolate problems in minutes instead of hours. The CSU also contains a buffer that stores collected performance information, which can be accessed by both the carrier and user for diagnostic purposes.

Summary. Because the CSU interfaces user equipment to carrier facilities, it provides a window on the network, allowing both the carrier and the customer to perform testing up to the same point. CSU access to the network has prompted vendors to equip it with increasingly sophisticated diagnostic and network management features. These go a long way toward enhancing user control and ensuring the integrity of T-carrier facilities.

See also
DATA SERVICE UNIT (DSU)

Client-server networks

For much of the 1990s, the client-server architecture has dominated corporations' efforts to downsize, restructure, and otherwise reengineer themselves for survival in an increasingly global economy. Frustrated with the restrictive access policies of traditional MIS managers and the slow pace of centralized, mainframe-based applications development, the client-server approach grew out of the need to bring computing power and decision-making down to the user, so businesses could respond faster to customer needs, competitive pressures, and market dynamics.

Architectural model. The client-server architecture is not new. A more familiar manifestation of the architecture is the decades-old corporate telephone system, with the PBX acting as a server and the telephones acting as the clients. All the telephones derive their features and user access privileges from the PBX, which also processes incoming and outgoing calls. What is relatively new is the application of this model to the LAN environment, which is data-oriented. Here, an application program is broken out into two parts, client and server, that exchange information over the network (Fig. C4).

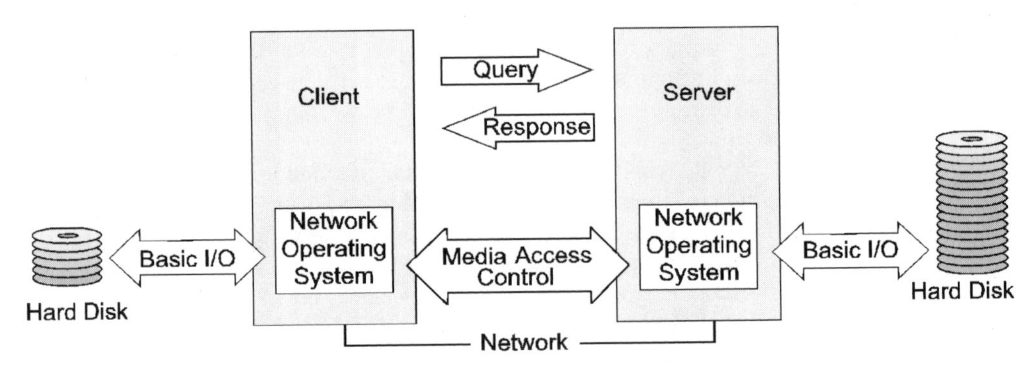

Figure C4 *Simplified model of the client-server architecture.*

Client. The client portion of the program, or *front end,* is run by individual users at their desktops and performs such tasks as querying a database, producing a printed report, or entering a new record. These functions are carried out through a database specification and access language that operates in conjunction with existing applications; the most widely used of these is Structured Query Language (SQL). The front end part of the program executes on the user's workstation, drawing upon its random access memory (RAM) and central processing unit (CPU).

Server. The server portion of the program, or *back end,* is resident on a computer that is configured to support multiple clients, offering them shared access to numerous application programs as well as to printers, file storage, database management, communications, and other resources. The server not only must handle simultaneous requests from multiple clients, but also perform such administrative tasks as transaction management, security, logging, database creation and updating, concurrency management, and maintaining the data dictionary. The *data dictionary* standardizes terminology so that database records can be maintained across a broad base of users.

Network. The network consists of the transmission facility, usually a LAN. Among the commonly used facilities for LANs is coaxial cable (thick and thin), twisted-pair wiring (shielded and unshielded), and optical fiber (single- and multimode). In some cases, wireless facilities such as infrared and spread-spectrum RF are used to link clients and servers.

A medium access control protocol is used to regulate access to the transmission facility. Ethernet and Token-Ring are the two most popular medium access control protocols. When linking client-server computing environments over the wide area network (WAN), other communications protocols come into play, such as private T1 links, which provide a transmission rate of up to 1.544 Mbps. Although the Internet Protocol (IP) remains the most common transport protocol used over the WAN, frame relay and other fast packet technologies are gaining in popularity.

Summary. To date, the promise of client-server is somewhat disappointing. With client-server networks, administration and management difficulties tend to multiply, while costs become nearly impossible to track. According to some industry estimates, the total cost of owning a client-server system is about 3 to 6 times greater than it is for a centralized mainframe system, while the software tools for managing and administering client-server cost 2½ times more than comparable mainframe tools. Regardless of these factors, many organizations have successfully implemented client-server networks and have achieved efficiency and productivity gains.

See also

LOCAL AREA NETWORK
SYSTEMS NETWORK ARCHITECTURE (SNA)

Code Division Multiple Access

Code division multiple access (CDMA) is based on spread-spectrum technology, a family of digital communication techniques originally developed for military communications and control applications in World War II. Spread-spectrum uses carrier waves that consume a much wider bandwidth than that required for simple point-to-point communication at the same data rate. This results in the carrier wave looking more like random noise than real communication between a sender and receiver. Originally there were two motivations for implementing spread-spectrum: to resist enemy efforts to jam vital communications, and to hide the fact that communication was even taking place.

The use of CDMA for civilian mobile radio applications was articulated as a theory in the late 1940s, but its practical application for cellular communications did not take place until the 1980s. For cellular telephony, CDMA is a digital multiple access technique specified by the Telecommunications Industry Association (TIA) as IS-95.

Commercial applications of CDMA became possible because of two key developments. One was the availability of low-cost, high-density digital integrated circuits, which reduce the size, weight, and cost of the mobile phones. The other was the realization that optimal multiple-access communication depends on the ability of all mobile phones to regulate their transmitter power to the lowest level that will achieve adequate signal quality.

CDMA changes the nature of the mobile phone from a predominately analog device to a predominately digital device. CDMA receivers do not eliminate analog processing entirely, but they separate communication channels by means of a pseudorandom modulation that is applied and removed in the digital domain, not on the basis of frequency. This allows multiple users to occupy the same frequency band; this frequency reuse results in high spectral efficiency.

TDMA systems commonly start with a slice of spectrum, referred to as one *carrier.* Each carrier is then divided into time slots. Only one subscriber at a time is assigned to each time slot, or channel. No other conversations can access this channel until the subscriber's call is finished, or until that original call is handed off to a different channel by the system. For example, TDMA systems, designed to coexist with AMPS systems, divide 30 kHz of spectrum into three channels. PDC divides 25 kHz slices of spectrum into three channels. GSM systems create eight time-division channels in carriers 200 kHz wide.

CDMA systems divide the radio spectrum into carriers 1250 kHz (1.25 MHz) wide. Unique digital codes, rather than separate RF frequencies or channels, are used to differentiate subscribers. The codes, called *pseudorandom code sequences,* are shared by both the mobile station (i.e., cellular phone) and the base station. All users share the same range of radio spectrum.

One of the unique aspects of CDMA is that while there are ultimate limits to the number of phone calls that can be handled by a carrier, this is not a fixed number. Rather, the capacity of the system depends on how coverage, quality, and capacity are balanced to arrive at the desired level of system performance. Since these parameters are tightly intertwined, operators cannot have the best of all worlds: three times wider coverage, 40 times capacity, and high quality sound. For example, the 13 kbps vocoder provides better sound quality, but reduces system capacity compared to an 8 kbps vocoder. Higher capacity might be achieved through some degree of degradation in coverage or quality.

System features. CDMA has been adapted for use in cellular communications with the addition of several system features that enhance efficiency and lower costs.

Mobile station sign-on. Upon powering up, the mobile station already knows the assigned frequency for CDMA service in the local area and will tune to that frequency and search for pilot signals. Multiple pilot signals typically will be found, each with a different time offset. This time offset distinguishes one base station from another. The mobile station will pick the strongest pilot, and establish a frequency reference and a time reference from that signal. Once the mobile station becomes synchronized with the base station's system time, it can then register. *Registration* is the process by which the mobile station tells the system that it is available for calls and notifies the system of its location.

Call processing. The user makes a call by entering the digits on the mobile station keypad and hitting the SEND button. If multiple mobile stations attempt a link on the access channel at precisely the same moment, a collision occurs. If the base station does not acknowledge the access attempt, the mobile station will wait a random time interval and try again. Upon making contact, the base station assigns a traffic channel, whereupon basic information is exchanged, including the mobile station's serial number. At that point, the conversation mode is started.

As a mobile station moves from one cell to the next, another cell's pilot signal, which is strong enough for the station to use, will be detected. The mobile station will then request a soft hand-off, during which it is actually receiving both signals via different correlative elements in the receiver circuitry. Eventually the signal from the first cell will diminish and the mobile station will request from the second cell that the soft hand-off be terminated. A base station does not hand off the call to another base station until it detects acceptable signal strength.

This soft hand-off technique is touted by CDMA advocates as a marked improvement over the hand-off procedure used in analog FM cellular systems, where the communication link with the old cell site is momentarily disconnected before the link to the new site is established. For a short time, the mobile station is not connected to either cell site, during which the subscriber hears background noise or nothing at all. Sometimes the mobile stations ping-pong between two cell sites as the links are handed back and forth between the approaching and the retreating cell sites. Other times, the calls are simply dropped. Because a mobile station in the CDMA system has more than one modulator, it can communicate with multiple cells simultaneously to implement the soft hand-off.

At the end of a call placed over the CDMA system, the channel will be freed and may be reused. When the mobile station is turned off, it will generate a power-down registration signal that tells the system that it is no longer available for incoming calls.

Voice detection and encoding. With voice activity detection, the transmitter is activated only when the user is speaking. This reduces interference levels—and, consequently, the amount of bandwidth consumed—when the user is not speaking. Through interference averaging, the capacity of the system is increased. This allows systems to be designed for the average rather than the worst interference case. According to the IS-95 CDMA standard, however, no interfering signal that is significantly stronger than the desired signal can be received, because it would then jam the weaker signal. This has been called *the near-far problem* and means that high cell capacity does not necessarily translate into high overall system capacity.

The speech coder used in CDMA operates at a variable rate. When the subscriber is talking, the speech coder operates at the full rate; when the subscriber is not talking, the speech coder operates at only one-eighth of the full rate. Two intermediate rates are also defined to capture the transitions and eliminate the effect of sudden rate changes. Since the variable rate operation of the speech coder reduces the average bit rate of the conversations, system capacity is increased.

Privacy. Increased privacy is inherent in CDMA technology. CDMA phone calls will be secure from the casual eavesdropper because, unlike a conversation carried over an analog system, a simple radio receiver will not be able to pick out individual digital conversations from the overall RF radiation in a frequency band.

A CDMA call starts with a standard rate of 9.6 kbps. This is then spread to a transmitted rate of about 1.25 Mbps. Spreading means that digital codes are ap-

plied to the data bits associated with users in a cell. These data bits are transmitted along with the signals of all the other users in that cell. When the signal is received, the codes are removed from the desired signal, separating the users and returning the call to the original rate of 9.6 kbps.

Because of the wide bandwidth of a spread-spectrum signal, it is very difficult to identify individual conversations for eavesdropping. Since a wideband spread-spectrum signal is very hard to detect, it appears as nothing more than a slight rise in the "noise floor" or interference level. With analog technologies, the power of the signal is concentrated in a narrower band, which makes it easier to detect with a radio receiver tuned to that set of frequencies.

The use of wideband spread-spectrum signals also offers more protection against cloning, an illegal practice whereby a mobile phone's electronic serial number is taken over the air and programmed into another phone. All calls made from a cloned phone are "free" because they are billed to the original subscriber.

Power control. CDMA systems rely on strict control of power at the mobile station to overcome the near-far problem. If the signal from a near mobile station were to be received at the cell site receiver with too much power, it would be overloaded by that particular mobile station. This would overwhelm the signals from the other mobile stations located farther away. The goal of CDMA is to have the signals of all mobile stations arrive at the base station with exactly the same power: the closer the mobile station to the cell site receiver, the lower the power necessary for transmission; the farther away the mobile station, the greater the power necessary for transmission.

Two forms of adaptive power control are employed in CDMA systems: open-loop and closed-loop.

- *Open-loop power control* is based on the similarity of loss in the forward and reverse paths. The received power at the mobile station is used as a reference. If it is low, the mobile station is assumed to be far from the base station and transmits with high power. If it is high, the mobile station is assumed to be near the base station and transmits with lower power. The sum of the two power levels is a constant.
- *Closed-loop power control* is used to force the power from the mobile station to deviate from the open-loop setting. This is achieved by an active feedback system from the base station to the mobile station. Power control bits are sent every 1.25 ms to direct the mobile station to increase or decrease its transmitted power by 1 dB. Lack of power control to at least this accuracy greatly reduces the capacity of CDMA systems.

With these adaptive power control techniques, the mobile station transmits only enough power to maintain a link. This results in an average power requirement that is much lower than that for analog systems, which usually do not employ such techniques. CDMA's lower power requirement translates into smaller, lightweight, longer-life batteries (approximately 5 hours of talk time and over 2 days of standby time) and makes possible smaller, lower-cost handheld computers and hybrid computer- communications devices. CDMA phones can easily weigh in at less than 8 ounces.

Spatial diversity. Among the various forms of diversity is that of spatial diversity, which is employed in CDMA and in other multiple access techniques, including FDMA and TDMA. Spatial diversity helps to maintain the signal during the call hand-off process when a user moves from one cell to the next. This process entails antennas in two different cell sites maintaining links with one mobile station. The mobile station has multiple correlative receiver elements that are assigned to each incoming signal and can add these.

CDMA uses at least four of these correlators, three that can be assigned to the link and one that searches for alternate paths. The cell sites send the received data, along with a quality index, to the MTSO (mobile telephone switching office), where a choice is made regarding which is the better of the two signals.

Not all of these features are unique to CDMA; some can be exploited by TDMA-based systems as well. Spatial diversity against fading and power control, for example, already exists in all TDMA standards today, while soft hand-off is implemented in the European DECT cordless telecommunications standard, which is based on TDMA.

Summary. There are conflicting performance claims for TDMA and CDMA. Since both TDMA and CDMA have become TIA standards (IS-54 and IS-95, respectively), vendors are now aiming their full marketing efforts toward the cellular carriers. Proponents of each technology have the research to back up their claims of superior performance. Of the two, CDMA suffered a credibility problem early on because its advocates made grandiose performance claims for CDMA that could not be verified in the real-world operating environment. In some circles, this credibility problem lingers today. Of note, however, is that both technologies have been successful in the marketplace, each having been selected by many cellular carriers around the world. Both are capable of supporting emerging PCS networks and providing such services as wireless Internet access, short message service, voice mail, facsimile, and paging.

See also

DIGITAL ENHANCED CORDLESS TELECOMMUNICATIONS
SPREAD-SPECTRUM RADIO
TIME DIVISION MULTIPLE ACCESS

Communications Services Management

Communications Services Management (CSM) entails the provision of telecommunications facilities and services within a multiple-tenant office building, campus, or office park. The CSM concept began in the early 1980s, when it was known as Shared Tenant Service (STS), intelligent buildings, smart buildings, or multiple-tenant telecommunications.

Under this arrangement, a provider installs a high-capacity central PBX in a multitenant office building and offers a suite of telecommunications services to tenants on a shared basis. The PBX usually is owned by the provider or jointly by the provider and the property owner. To ensure success, a large anchor tenant usually must be persuaded to subscribe to the service.

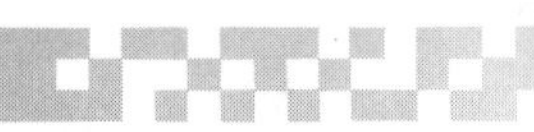

The costs of equipment, lines, services, and maintenance are distributed among the tenants. The tenants benefit by obtaining a level of service and support they normally would not be able to afford by themselves. Organizations find CSM appealing for a variety of other reasons, including:

- The convenience of one-stop shopping for all of their telecommunications needs
- Cost savings from bulk-rate, discounted services,
- Time savings from having the CSM provider do all the service and support planning, and
- Reduced operating costs, since the need for technical expertise and spares inventory is eliminated.

Types of services. CSM firms provide a wide variety of telecommunications services through a complex web of partnerships that include local and long-distance carriers, computer companies, integration firms, and consultants. Thus, CSM companies can provide a variety of standard and custom services that meet the needs of many different constituencies. These services typically include:

- Basic local telephone service,
- Discounted intrastate, interstate, and international telephone service,
- Telecommunications system rental and maintenance,
- Management of system moves, adds, and changes,
- Custom billing and management reports,
- Help desk and message center services,
- Cable and wiring installation and maintenance,
- Consulting, customization, and systems integration,
- Directory publishing,
- Network design and consulting,
- LAN/WAN service and support,
- Disaster planning, hot-site management, and recovery,
- Data communications and computer facilities,
- Audio conferencng,
- Videoconferencing,
- Public facsimile (FAX) and telex transmission,
- FAX-On-Demand storage and retrieval services,
- Cable and satellite television, both public and private,
- Voice-activated (voice recognition) functions,
- Satellite, microwave, and wireless services (public, shared, and private), and
- Custom call processing, including caller ID–enabled applications.

Summary. Many companies are finding that managing their own telecommunications services is too great a drain on resources and diverts their attention from core business concerns, especially considering the fast-changing nature of the technology. Businesses are drawn to CSM arrangements because of the simplicity they offer, as well as the cost savings. The availability of on-site service, less capital expense, and a wider range of features and capabilities are also reasons for signing with a CSM provider.

See also
OUTSOURCING

Computer-telephony integration

In recent years, a new generation of applications has emerged that relies on the integration of voice and data technologies to increase productivity and customer response. Computer-telephony integration (CTI) makes use of the advanced call processing capabilities of digital telephone systems and the open application environment of PCs and LANs via intelligent computer-to-telephone system interfaces. The resulting benefits are especially suited to telephone-intensive environments such as customer service and telemarketing.

In rudimentary form, CTI has been around for more than a decade. The original CTI approach was to link the PBX with a host, such as a mainframe or minicomputer. With today's LAN capabilities, the LAN server can act as the host, bringing integrated voice-data applications to every desktop. It is not just LANs that make this possible, but advanced PBXs as well. Just about all of today's PBXs and digital phones are based on the use of the computer as the controlling device for implementing a wide variety of call handling features.

Many of the attributes of CTI trace their lineage from the call center environment; specifically:

- *Automated Call Distribution* (ACD) manages incoming calls in a variety of ways, including holding them in queue and spreading them among available operators.
- *Automatic Number Identification* (ANI) provides the system with the telephone number of the incoming call to identify the caller.
- *Database Matching* provides the means to look up data about a customer based on ANI before the call is answered.
- *Call Accounting* entails collecting call-related information for cost containment, internal billing, and trend analysis.

Today's concept of CTI takes this integration further, treating voice as simply another data type that can be manipulated by the user. In this integrated environment, voice is a messaging format on a par with electronic mail, facsimile, and even paper. Once a visual link to a voice-processing system or PBX is in place, it becomes possible to work with voice mail and other telephone functions as desktop applications, tie them into e-mail messaging systems, and create entirely new categories of applications that are telephone-enabled.

Applications. CTI is an architecture that allows one or more computer applications to communicate with the telephone switch. Among the possible applications of CTI are:

- *Inbound call information:* Information passed from the telephone network to the telephone switch, such as the caller's telephone number and the number dialed, is passed to computer applications. Applications then can identify the caller (by the calling number) and the purpose of the call (from the number dialed). This allows the application to automatically deliver to a workstation, as the telephone rings, caller information and data specific to the purpose of the call.

- *Computerized call processing:* Commands passed from computer applications instruct the telephone system to perform call processing functions such as make a call, answer a call, or other call functions. This allows for application-controlled call routing based on inbound call information and numbers in a computer database.
- *Outbound calling:* CTI increases productivity in outbound calling environments. With automated dialing applications, agents proceed from one active call to the next. No time is wasted listening to busy signals or unanswered ringing, or manually dialing numbers.

In these and other cases, computer applications use a call processing server and APIs to originate, answer, and manipulate calls. The call processing server interfaces to the telephone switch and invokes the required function, as requested by the client applications. The server keeps track of call status information on the telephone switch side and session status on the application side, making the logical association between the two.

The role of APIs. The challenge for CTI vendors has been to come up with a standardized way of allowing developers to build and implement integrated voice-data solutions that work across vendor domains. Their strategy has been to establish standardized application programming interfaces (APIs) that work across vendor boundaries.

Intel and Microsoft have developed the Telephony Application Programming Interface (TAPI), which is intended to create a single specification for Windows application developers to use in connecting their products to the telephone network. At the LAN server level, AT&T and Novell offer their jointly developed NetWare Telephony Services API, which is intended to make NetWare the platform for this kind of integration. Other vendors, such as Sun Microsystems and Hewlett-Packard, also offer APIs for developing CTI applications.

Telephone Application Programming Interface (TAPI). TAPI allows custom applications to be built around inexpensive personal computers; specifically, the Windows Telephony API provides a standard development interface between PCs and the myriad telephone network APIs. TAPI is intended to insulate software developers from the underlying complexity of the telephone network. TAPI allows developers to focus entirely on the application without having to take into account the type of telephone connection: PBX, ISDN, Centrex, cellular, or plain old telephone service (POTS). They can specify the features they want to use without worrying how the hardware is ultimately linked.

Application classes. TAPI facilitates the development of three classes of Windows applications. The first class of applications will be telephone-enabled versions of existing applications, such as word processors. TAPI creates standard access to telephone functions such as call initiation, call answering, call hold, and call transfer for Windows applications. TAPI addresses only the control of the call, not its content. the specification can be applied to any type of call, however, whether voice, data, fax, or even video.

The second class of telephone-centric applications might embrace visual call control or telephone-based conferencing and collaborative computing. Although such applications currently are available, they have been limited by incompatible APIs. The third class of applications will enable the telephone to act as an input/output device for sound data, including voice across data networks.

Application components. An actual TAPI product implementation will comprise three distinct components:

- The TAPI-aware application
- A TAPI dynamic link library (DLL)
- One or more Windows drivers to interface to the telephone hardware

A TAPI application is any piece of software that makes use of the telephone system. An obvious example might be a personal information manager (PIM), which could dial phone numbers automatically. An application becomes TAPI-compliant by writing to the applications programming interfaces defined in the TAPI specification.

The TAPI DLL is the next major component. The application talks to the DLL using the standard APIs. The DLL translates those API calls and controls the telephone system using the device driver.

The final component is the Service Provider Interface (SPI), which is a driver that is unique to each TAPI hardware product. It is analogous to a each sound card having its own driver. The TAPI specification supports more than one type of telephone adapter. In turn, the adapters can support more than one line.

TAPI-enabled features. TAPI facilitates the development of applications which allow the user to control the telephone from a Windows PC. A number of possible control features may become available in future applications, including:

- *Visual call control* provides a Windows interface to such common PBX functions as call hold, call transfer, and call conferencing. Replacing difficult-to-remember dialing codes with Windows icons will make even the most complicated telephone system functions easy to implement.
- *Call filtering,* in conjunction with ANI, allows the user to specify the telephone numbers allowed to get through. All others will be routed to an attendant, message center, or voice mailbox. Or the call can be automatically forwarded to another extension while the user is out of the office.
- *Customized menuing systems* allow users to build menuing systems to help callers find the right information, agent, or department. Using the drag-and-drop technique, the menuing system can be revised daily to suit changing business needs. The menu system can be interactive, allowing the caller to respond to voice prompts by dialing different numbers. A different voice message can be associated with each response. Voice messages can be created instantly via the PC's microphone.

NetWare Telephony Services API. While TAPI defines the connection between a single phone and a PC, the NetWare Telephony Services API (TSAPI) defines the connection between a networked file server and a PBX. TSAPI is the result of a joint effort by Novell and AT&T to integrate computer and telephone functions at the desktop using a logical connection established over the LAN.

In connecting a NetWare server to the PBX, individual PCs are given control over telephone system functions. TSAPI is implemented with NetWare Loadable Modules (NLMs) that run on Novell servers, along with another NLM containing a PBX driver. No special hardware is required at the desktop; the PBX supports its own physical connection and uses its own software. The physical link is an ISDN Basic Rate Interface card in the server, which allows for the connection between the NetWare server and the PBX.

NetWare Telephony Services consists of a Telephony Server NLM, a set of dynamic link libraries (DLLs) for the client, and a sample server application (a simple point-and-click telephone listing that is integrated with directory services). Novell also offers a driver for every major PBX. Alternatively, users can obtain a driver from their PBX vendor.

The NLM's features include drag-and-drop conference calling; the ability to put voice, facsimile, and electronic mail messages in one mailbox; third-party call control; and integration between telephones and computer databases. Noteworthy among these is third-party call control, which provides the ability to control a call without being a part of it. This feature would be used for setting up a conference call, for example.

Unlike Microsoft's TAPI, which allows only first-party call control, third-party constructs are an integral part of NetWare Telephony Services. The command *Make Call,* for example, has two parameters, one for addressing the originating party and the other for addressing the destination party. An application using this command therefore would allow users to designate an address different from their own as the originating party and establish a connection without becoming a participant in the call. This third-party call control also lets users set up automatic routing schemes.

Summary. CTI removes the barriers between telephony and other information and productivity tools, providing users with substantial gains in efficiency and information management in an easy-to-use environment. Under the CTI concept, the most appropriate pieces of technology are combined in practical applications for a more productive workplace.

See also

PRIVATE BRANCH EXCHANGE (PBX)
LOCAL AREA NETWORK (LAN)

Cordless telecommunications

The familiar cordless telephone, introduced in the early 1980s, has become a key factor in reshaping voice communications. Because people cannot be tied to their desks, as many as **70 percent** of business calls do not reach the right person on the first attempt. This situation has seen dramatic improvement with cordless technology, which makes phones as mobile as their users. Now almost **70 percent** of business calls reach the right person on the first attempt.

Cordless vs. cellular. Although cellular phones and cordless phones are both wireless, the terms have come to have quite distinct and separate meanings, based on their areas of use and the differing technologies developed to meet user requirements.

Briefly, *cellular telephones* are intended for off-site use in cars or other forms of transport. The systems are designed for a relatively low density of users. Here, macrocellular technology provides wide-area coverage and the ability to make calls while traveling at high speeds. *Cordless telephones,* on the other hand, are designed for users whose movements are within a well-defined area. The cordless user makes calls from a portable handset linked by radio signals to a fixed base station (Fig. C5). The base station is connected either directly or indirectly to the public network. Cellular and cordless also have their own technology and standards.

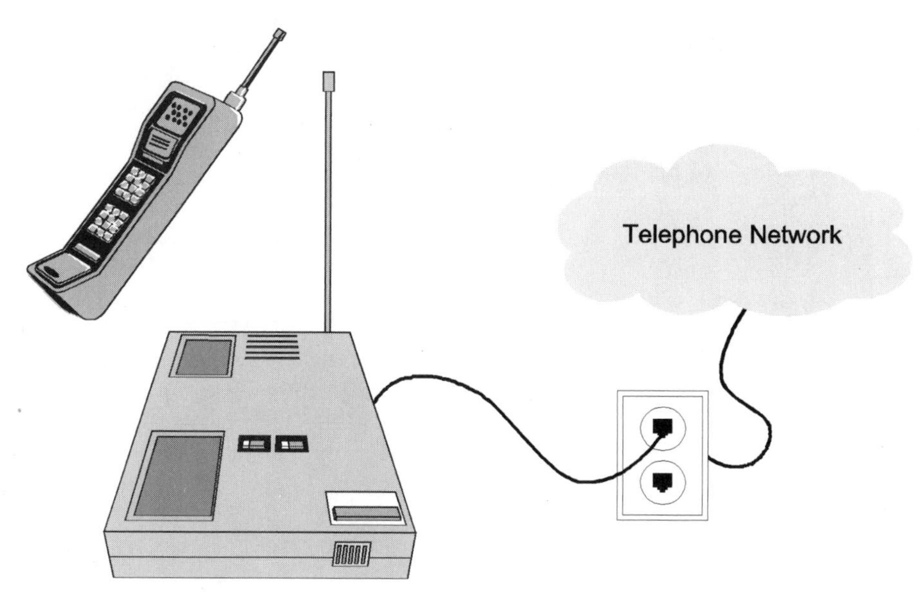

Figure C5 *The familiar cordless telephone found in many homes.*

Cordless standards. The cordless system standards are referred to as CT0, CT1, CT2, CT3, and DECT, with CT standing for *Cordless Telecommunications.* CT0 and CT1 were the technologies for first-generation analog cordless telephones. Comprising base station, charger, and handset, and primarily intended for residential use, they had a range of 100 to 200 m. They used analog radio transmission on two separate channels, one to transmit and one to receive. The potential disadvantage of CT0 and CT1 systems is that the limited number of frequencies can result in interference between handsets, even with the relatively low density of residential subscribers.

Also targeted at the residential user, CT2 represented an improved version of CT0 and CT1. Using Frequency Division Multiple Access (FDMA), the CT2 system creates capacity by splitting bandwidth into radio channels in the assigned frequency domain. In the initial call setup, the handset scans the available channels and locks onto an unoccupied channel for the duration of the call. Using Time Division Duplexing (TDD), the call is split into time blocks that alternate between transmitting and receiving.

The Digital Enhanced Cordless Telecommunications (DECT) standard started as a European standard for cordless communications, with applications that included residential telephones, telepoint, the cordless PBX, and cordless local-loop access to the public network. DECT was designed primarily to solve the problem of providing cordless telephones in high-density, high-traffic office and other business environments.

CT3, on the other hand, is a technology developed by Ericsson in advance of the final agreement on the DECT standard and is designed specifically for the cordless PBX application. Since DECT is essentially based on CT3 technology, the two standards are very similar. Both enable the user to make and receive calls when within the range of a base station. Depending on the specific conditions, this amounts to a radius of between 50 and 250 m from the station. To provide service throughout the site, multiple base stations are set up to create a picocellular network. Hand-off between cells is supported by one or more radio exchange units that are ultimately connected to the host PBX.

Both DECT and CT3 have been designed to cope with the highest-density telephone environments, such as city office districts, where user densities can reach 50,000 per km^2. A feature called Continuous Dynamic Channel Selection (CDCS) ensures seamless hand-off between cells, which is particularly important in a picocellular environment where several hand-offs may be necessary, even during a short call. The digital radio links are encrypted to provide absolute call privacy.

The two standards are based on a multicarrier Time Division Multiple Access/Time Division Duplexing (TMDA/TDD). They do not use the same operating frequencies, though, and consequently have different overall bit rates and call-carrying capacity.

It is the difference in frequencies that governs the commercial availability of DECT and CT3 around the world. Europe is committed to implementing the DECT standard with the frequency range of 1.8 to 1.9 GHz. Other countries, however, have made frequencies in the 800 to 1000 MHz band available for cordless PBXs, thereby paving the way for the introduction of CT3.

Summary. Many of the problems arising from the nonavailability of staff to a wired PBX can be avoided with cordless telephones. They are, for instance, ideal for people who by the very nature of their work can be difficult to locate (maintenance engineers, warehouse staff, messengers, etc.), and for those places on a company's premises that cannot be effectively covered by a wired PBX (warehouses, factories, refineries, exhibition halls, dispatch points, etc.).

A key advantage of cordless telecommunications is that it can be integrated simply into the corporate telecom system with add-on products and without the need to replace existing equipment. Another advantage of cordless telecommunications is that the amount of telephone wiring is dramatically reduced. Since companies typically spend between 10 and 20 percent of the original cost of their PBX on wiring the system, the use of cordless technology can have a significant impact on costs. There also is considerable benefit in terms of administration. When moving offices, for example, employees need not change their extension numbers, nor does the PBX need to be reprogrammed to reflect the change.

See also

CELLULAR COMMUNICATIONS
DIGITAL ENHANCED CORDLESS TELECOMMUNICATIONS

Custom Local Area Signaling Services

Custom Local Area Signaling Services (CLASS) is a collection of services offered to local area residential and business customers on a presubscribed basis. These features support both local and interoffice applications where Signaling System 7 (SS7) is deployed. CLASS features include:

- *Call Return* redials the last caller, whether the call was completed or not. The user does not have to know the number to return the call because the service provides it. The service also lets the user know if the number was blocked, private, or out of the area. If the line is busy, the service checks for a free line for up to 30 min.
- *Caller ID* displays the 10-digit telephone number of an incoming call. This service requires the subscriber to have a CLASS-compatible display phone or display unit.
- *Calling Number Delivery Blocking* allows users to block their numbers from being displayed on a display telephone or display unit.
- *Call Trace* allows the user to trace harassing or life-threatening telephone calls on demand. The telephone company releases traced information to appropriate law enforcement officials. The user must file a complaint with the proper authorities and fill out a complaint form at the telco's office. This feature usually works only with calls within the local service area.
- *Repeat Dialing* automatically calls the last number the user dialed, whether it was answered, unanswered, or busy. The service can redial a busy number for up to 30 min. When the line is free, the subscriber hears a distinctive ring, indicating the call is established.
- Preferred Call Forwarding allows the user to forward only selected calls to a special number. It allows the user to store up to 6 numbers. When incoming calls originate from one of those numbers, those calls are forwarded to the desired number.
- *Call Block* allows the user to reject calls from selected numbers. It allows the user to store up to 6 numbers. When incoming calls originate from one of those numbers, the user's phone will not even ring. Callers hear a recording indicating the user is not accepting calls.
- *Call Selector* allows the user to choose numbers that will ring distinctively. The user can store up to 6 numbers. When incoming calls originate from one of those numbers, the user's phone rings distinctively.
- *Wake Up Service* allows the user to program the phone to ring at a certain time.

Summary. New custom calling features are continually being added. In most cases, CLASS services entail an extra charge of between US$2.50 and US$5.00 per month and are itemized on the monthly phone bill. Periodically, many telephone companies run promotions on new calling features, giving their customers 30 days to try out the services at no charge. Unless the customer specifically cancels the service, however, they will be billed automatically after the trial period.

See also

ADVANCED INTELLIGENT NETWORK
SIGNALING SYSTEM 7

Data compression

Data compression has become a standard feature not only of modems, but of most bridges and routers as well. In its simplest implementation, compression capitalizes on redundancies found in the data. The algorithm detects repeating characters or strings of characters and represents them as a symbol or token. At the receiving end, the process works in reverse to restore the original data.

The compression ratio tends to differ by application. The compression ratio can be as high as 6 to 1 when the traffic consists of heavy-duty file transfers. The compression ratio is less that 4 to 1 when the traffic is mostly database queries. When there are only "keep-alive" signals or sporadic query traffic on a T1 line, the compression ratio can dip below 2 to 1. Encrypted data exhibits little or no compression because the encryption process expands the data and uses more bandwidth. If data expansion is detected and compression is withheld until the encrypted data is completely transmitted, however, the need for more bandwidth can be avoided.

The use of data compression is particularly advantageous in the following situations:

- When data traffic is increasing due to the addition or expansion of LANs and associated data-intensive, bursty traffic,
- When LAN and legacy traffic are contending for the same limited bandwidth,
- When reducing or limiting the number of 56/64 kbps lines is desirable to reduce operational costs, and
- When lowering the Committed Information Rate (CIR) for frame relay services or sending fewer packets over an X.25 network can result in substantial cost savings.

The greatest cost savings from data compression most often occurs at remote sites, where bandwidth typically is in short supply. Data compression can extend the life of 56/64 kbps leased lines, thus avoiding the need for more expensive fractional T1 lines or N×64 services. Depending on the application, a 56/64 kbps leased line can deliver 112 to 256 kbps or higher throughput when data compression is applied.

Types of data compression. Today there are several different data compression methods in use over wide area networks; among them are TCP/IP header compression, link compression, and multichannel payload compression. Depending on the method used, there can be a significant trade-off between lower bandwidth consumption and increased packet delay.

TCP/IP header compression. With TCP/IP header compression, the packet headers are compressed but the data payload remains unchanged. Since the TCP/IP header must be replaced at each node for IP routing to be possible, this compression method requires hop-by- hop compression and decompression processing. This adds delay to each compressed/decompressed packet and puts an added burden on the router's CPU.

Link compression. With link compression, the entire frame—both protocol header and payload—are compressed. This form of compression typically is used in LAN-only or legacy-only environments. This method requires error correction and packet sequencing software, however, which adds to the processing overhead already introduced by link compression and results in increased packet delays. Like TCP/IP header compression, link compression also requires hop-to-hop compression and decompression, so processor loading and packet delays occur at each router node the data traverses.

With link compression, a single data compression vocabulary dictionary or history buffer is maintained for all virtual circuits compressed over the WAN link. This buffer holds a running history about what data has been transmitted to help make future transmissions more efficient. To obtain optimal compression ratios, the history buffer must be large, requiring a significant amount of memory. The vocabulary dictionary resets at the end of each frame. This technique offers lower compression ratios than multichannel, multihistory buffer (vocabularies) data compression methods. This is particularly true when transmitting mixed LAN and serial protocol traffic over the WAN link and frame sizes are 2048 bytes or smaller. This translates into higher costs, but if more memory is added to get better ratios, this increases the up-front cost of the solution.

Mixed-channel payload data compression. By using separate history buffers or vocabularies for each virtual circuit, multichannel payload data compression can yield higher compression ratios that require much less memory than other data compression methods. This is particularly true in cases where mixed LAN and serial protocol traffic traverses the network. Higher compression ratios translate into lower WAN bandwidth requirements and greater cost savings.

Performance varies because vendors define payload data compression differently, however; some consider it to be compression of everything that follows the IP header. But, because the IP header can be a significant number of bytes, header compression must be applied for overall compression to be effective. This adds to the processing burden of the CPU and increases packet delays.

External data compression solutions. Although bridges and routers can perform data compression, external compression devices are often required to connect to higher-speed links. The reason is that data compression is extremely processor-intensive, with multichannel payload data compression being the most burdensome. The faster the packets must move through the router, the more difficult it is for the router's processor to keep up.

The advantages of internal data compression engines are that they can provide multichannel compression, lower cost, and simplified management. By using a separate internal digital signal processor (DSP) for data compression, however, instead of the software-only approach, all of the other basic functions within the router can continue to be processed simultaneously. This parallel processing approach minimizes the packet delay that can occur when the router's CPU is forced to handle all these tasks by itself.

Summary. Data compression will become increasingly important to most organizations as the volume of data traffic at branch locations begins to exceed the capacity of the wide area links. Multichannel payload solutions provide the highest compression ratios and reduce the number of packets transmitted across the network. Reducing packet latency can be effectively achieved via a dedicated processor like a DSP, and by employing end-to-end compression techniques rather than node-to-node compression/decompression. All of these factors contribute to reducing WAN circuit and equipment costs as well as improving the network response time and availability for user applications.

See also

VOICE COMPRESSION

Data Service Unit

The Data Service Unit (DSU) is a device that connects various data terminal equipment via RS-232 or V.35 interfaces with widely available digital services that offer 56/64 kbps access, including Digital Data Service (DDS) and frame- based services such as SMDS, frame relay, and ATM. Typical applications for the DSU include LAN interconnection, Internet access, and remote PC access to local hosts (Fig. D1).

Digital Data Service (DDS) operates at speeds of 2.4 to 56/64 kbps in support of point-to-point or multipoint applications. Most DSUs have a built-in asynchronous-to-synchronous converter, accommodating asynchronous input devices that operate at speeds of 2.4, 4.8, 9.6, 19.2, and 38.4 kbps, as well as synchronous input devices that operate at 2.4, 4.8, 9.6, 19.2, and 56/64 kbps. When packaged with Channel Service Unit (CSU) functions, the DSU/CSU device interfaces with T1 services at 64 kbps and N×64 kbps up to 1536 kbps (1.536 Mbps).The DSU converts the binary data pulse it receives from the DTE to the bipolar format required by the network. The DSU also supplies the transmit and receive logic, as well as timing. Any device that connects directly to the digital line (via an external or internal CSU) must perform these functions, or it needs a DSU. Any piece of

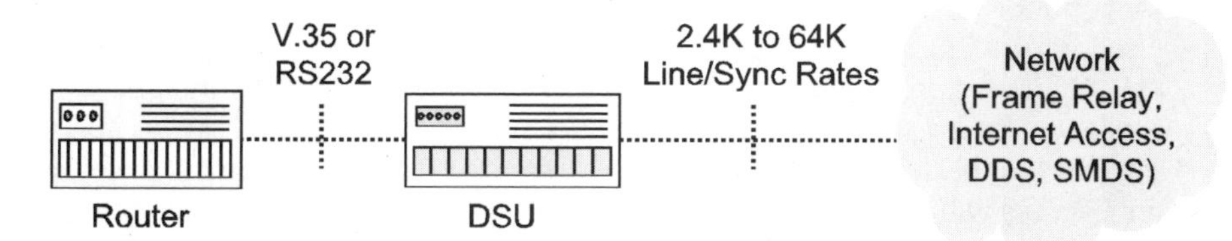

Figure D1 *Role of DSU in providing low-speed access to digital services.*

equipment that does not have a bipolar port needs a DSU to connect to a CSU. The most common type of access device is the combination unit, which offers DSU and CSU functionality.

Like the CSU, the DSU also provides the means to perform diagnostics. The front panel of the DSU provides a set of LED indicators that show the status of the V.35 DTE interface, various test modes, and loop status. The device responds to standard loopback commands from the service provider. Included with the remote loopback is the capability to select a bit error rate test (BERT) pattern and have the results displayed on the front panel. Like CSUs, DSUs can be managed either by the vendor's proprietary network management system, or by the user's own Simple Network Management Protocol (SNMP) tools.

Network testing services. In addition to the diagnostic and loopback test capabilities built into many of today's DSUs, users can also rely on telephone company–provided test systems like the Automated Bit Access Test System (ABATS) for AT&T's Dataphone Digital Service (DDS). For nationwide AT&T Accunet DDS customers, ABATS actually can command each and every CSU/DSU termination into a Line Loopback (LL) and Remote Terminal Loopback (RT) test to isolate network problems.

ABATS is activated when a user places a call to the Serving Test Center (STC) to report a problem. The STC proceeds to interact with the ABATS Test Center, where technicians troubleshoot the customer's entire network, right down to the CSU/DSUs. Many smaller organizations, with neither the diagnostic test equipment nor the technical expertise required to run sophisticated diagnostic systems, have come to rely on this method for maintaining their networks.

Users also can subscribe to AT&T's Customer Test Service (CTS), which is available over interLATA DDS circuits only. By positioning an asynchronous terminal and a leased or dial-up data set on the customer's premises, the user can access the network test system. From this terminal, the user can draw upon the service provider's resources to identify a problem on the network and isolate its cause by initiating the appropriate loopback tests.

Tests initiated through ABATS and CTS are disruptive, however, because they are conducted on an in-band basis, that is, the test signals replace the user's production data. To lessen the impact, users may schedule circuit testing for off-peak hours.

User-controlled diagnostics. Carrier-provided diagnostic services, like ABATS and CTS, provide only the most rudimentary test capabilities. Because these methods are based strictly on loopback testing, which is always interfering, other network control and diagnostic systems are frequently chosen instead. Of special importance for DDS users is the optional Secondary Channel (DDS/SC), which is used for nonintrusive testing.

By providing a completely independent, low-speed data channel, network surveillance can be performed on a continuous and nondisruptive basis. Also, nondisruptive loopback testing may be performed over the secondary channel, without interfering with the production data, which continues to flow over the primary channel.

Summary. Data Service Units, as well as Channel Service Units, not only provide an interface between DTE and the carrier's network, such devices also help network managers fine -tune their networks for performance and cost savings. For example, when using the diagnostic capabilities of a DSU/CSU connected to a frame relay network, the network manager can monitor traffic on each permanent virtual connection (PVC) to set an appropriate committed information rate (CIR) and allowable burst rate on each circuit. In addition, the delay between network nodes can be measured, as well as the performance of a line between the user and local carrier, to see if the carrier is actually delivering the level of service promised.

See also
CHANNEL SERVICE UNIT
DIGITAL DATA SERVICES

Data switches

Data switches have been in mainframe computer environments since their introduction in 1972. A data switch is a port-selection or port-contention device that permits a larger number of users to share a limited number of host ports. Today's data switches also perform the necessary protocol conversions that allow PCs to communicate with the mainframe. They have evolved to become a very economical means of controlling access to the mainframe via partitioning (Fig. D2) and other security features. Data switches typically support data transmission speeds of up to 19.2 kbps for asynchronous data and 64 kbps for synchronous data.

Connectivity. The modular architecture of data switches permits incremental growth from as few as two computer lines locally to thousands of computer ports worldwide. The data switch also can serve as a LAN server, or gateway to packet and T1 networks for remote host access.

Microcomputers may be connected to the data switch in a variety of ways, including direct connection with EIA-232C cabling for distances of up to 50 ft. For longer distances of 1 to 2 miles (between buildings, for example), PCs and peripherals may be connected to the data switch via line drivers or local multiplexers. Some data switch vendors have integrated these functions into their data switches in the form of optional plug-in cards. Connections between PCs and the data switch usually are accomplished via extra twisted pairs in telephone wire that already is in place in most office environments. Remote data switches may be accessed via dedicated or dial-up phone lines.

When a user enters the connection command and personal password, the data switch will attempt to complete the requested connection. If the requested port is busy, the user is put into queue and notified of changes in position with a screen message. Some users may be assigned a higher priority in the queue than others. When a high-priority user attempts to access a busy port, he or she bumps other waiting users by automatically assuming first place in the queue. All other contenders for that port are then notified of their new status.

Features. Not only do some data switches permit different configurations to be loaded, but they can implement one or the other automatically on a scheduled ba-

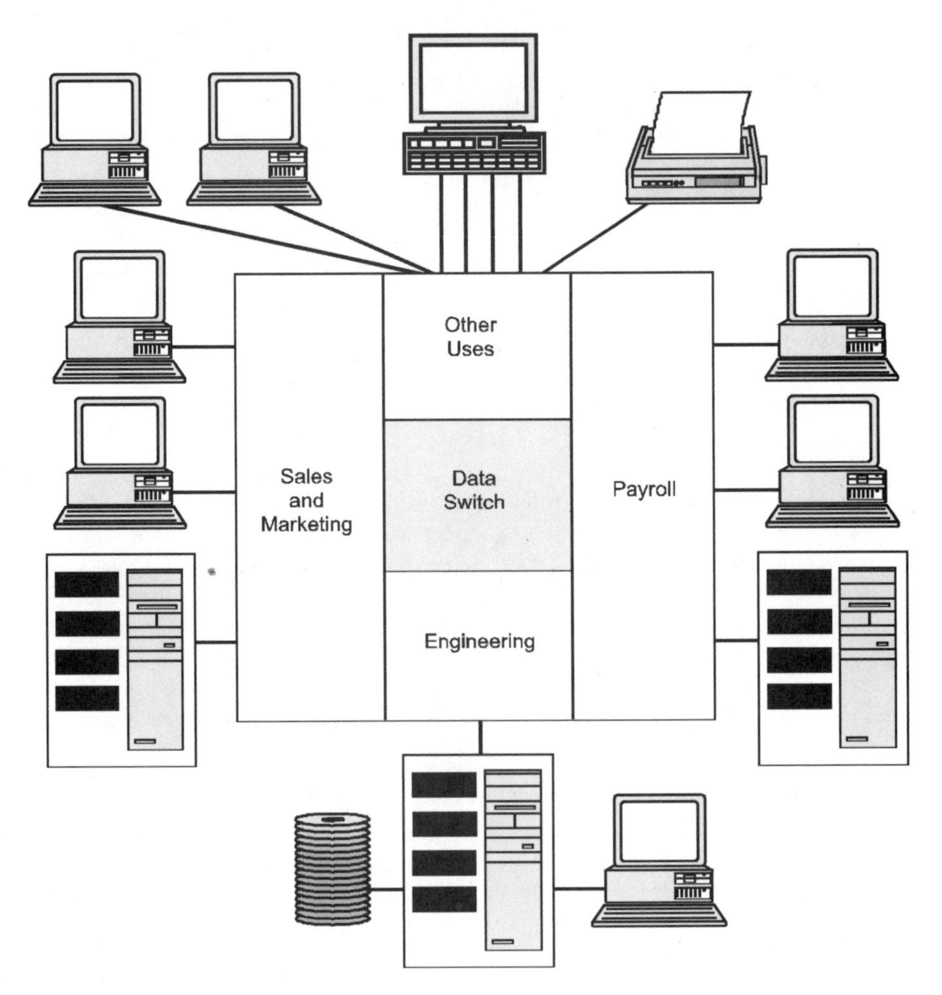

Figure D2 *Micro-to-mainframe connectivity, with partitioning, implemented by a data switch.*

sis. A late-night crew of CAD/CAM people can plod along with cumbersome design work on high-performance workstations, for example, and be restricted to accessing the mainframe from 6 P.M. to 8 A.M. The class-of-service feature denies these users the ability to access sensitive financial information that also might be stored in the mainframe. When the workday begins at 8 A.M., the time-of-day clock changes the data switch back to the primary configuration for general access, but without interrupting existing connections. This enhances overall network performance in that available resources may be reallocated during the day for optimal usage based on the varying needs of different classes of users. Various alternative configurations may be stored on diskette and implemented by an authorized PC with only a few keystrokes.

A session-toggling feature enhances operator productivity by permitting two connections, one designated primary and the other designated secondary. The operator can toggle back and forth between two host sessions on different links to perform multiple tasks simultaneously. For example, a batch file transfer can be in progress over the secondary link while a real-time database search is being performed over the primary link. When the batch file transfer is completed, the primary link can be put on hold while another file transfer is initiated over the secondary link.

The data switch automatically adapts to different transmission rates. The network administrator does not have to match terminals with computer ports; each computer port may be set at its highest rate. The data switch uses a buffer to perform the rate conversion for any device that communicates with another device at a faster or slower rate. This means that users do not have to be concerned about speed and network administrators do not have to waste time changing the transmission speeds of computer ports to accommodate lower-speed devices. Thus, a computer port set at 19.2 kbps may send data to a much slower printer. For reliable data rate conversion, however, the connecting devices must be capable of flow control; if not, there is a risk of losing data.

When (software-based) XON/XOFF is used for flow control, the switch buffer will be prevented from overflowing. When the buffer is in danger of overflowing, an XOFF signal is sent to the computer, telling it to suspend transmission. When the buffer clears, an XON signal is sent to the computer, telling it to resume transmission. These settings also are used for reformatting character structures, enabling devices from different manufacturers to communicate with one another through the data switch.

Administration. Instead of being confined to one terminal, many data switches allow the network administrator to log into the computer from any terminal. Once connected, there are a variety of functions that can be invoked to enhance operating efficiency.

- A *broadcast feature* allows the network administrator to transmit messages to individual users, with delivery controlled by the time-of-day clock. The same message may be sent every hour, or a different message may be sent to different users simultaneously.
- A *special link feature* lets the network administrator make permanent connections between a terminal and a port. Sometimes called "nail-up," this feature allows the continuous access to certain devices like printers.
- With a *force disconnect feature,* the network administrator can disconnect any port, any time, for any reason. Open files are even closed automatically with the proper disconnect sequences, including any required control characters. This capability also is found in a *time-out feature,* which enhances system efficiency by automatically disconnecting idle ports after a predefined period of inactivity.
- When the data switch is equipped with a *logging port,* the network administrator can obtain a complete record of port connects and disconnects to aid in maintaining security. When used with the optional *security callback,* this feature provides a precise audit trail of connections, as well as the users who made them. In addition, all alarms may be logged for output to a PC or designated port on the host to aid in network analysis and management.

The network administrator may decide to group ports with similar characteristics and capabilities under a class name to permit a more efficient utilization of shared resources. Password-protected classes also can be used to restrict access and enhance system security. The class of service may designate the use of particular host ports for general access, while other classes may designate a high-speed printer, a modem pool, or a gateway to a packet network.

Summary. Data switches are an economical alternative to LANs when host connectivity needs are relatively simple. They also make a suitable LAN server when networking needs grow more complex. Gateway functions may be added to data switches through plug-in cards that provide the appropriate protocol converters. This permits asynchronous or synchronous terminals from one vendor to communicate with the mainframes and network nodes provided by other vendors. It also permits the data switch to connect to various types of LANs.

See also
ETHERNET
TOKEN-RING

Data warehousing

A data warehouse is an extension of the database management system (DBMS), which consolidates information from various sources into a high-level, integrated form used to identify trends and make business decisions. For a large company, the amount of information in a data warehouse could be several trillion bytes, or terabytes (TB). The technologies used to build data warehouses include relational databases, powerful and scalable processors, and sophisticated tools to manipulate and analyze large volumes of data and identify previously undetectable patterns and relationships. Benefits include increased revenue and decreased costs due to the more effective handling of data.

Data warehousing applications are driven by such economic needs as cost reduction or containment, revenue enhancement, and response to market conditions. In addition, the more effective handling of corporate data—identifying patterns of repeat purchases, the frequency and manner with which customers use a company's products, and by their propensity to switch vendors when they are offered better prices or more targeted features—can improve customer satisfaction and cement customer loyalty. A change of only a few percentage points of customer retention can equate to hundreds of millions of dollars to a company.

Data warehousing frameworks typically comprise four features: data residing in one or more database systems, software to translate data, connectivity software to transfer data between databases and platforms, and end-user query tools. Key system components include an information store of historical events (the data warehouse), warehouse administration tools, data manipulation tools, and decision support systems (DSS) that allow strategic analysis of the information.

The effectiveness of a data warehouse architecture depends on how well it addresses the issues related to each of the following major components:

- *Warehouse population:* A central repository of warehouse data is obtained by downloading and consolidating data from various operational systems. The data consist of historical (transaction-based) events and related infor-

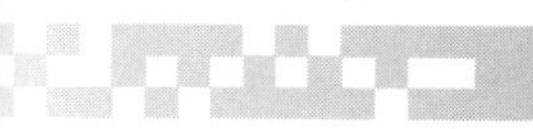

mation needed to isolate and aggregate those events. Volume tends to be high, so performance and cost are key considerations for warehouse and operational data sources.

- *Warehouse volume:* The data in a voluminous warehouse can be made more accessible by arranging them into *data marts,* or specialized subsets of the data warehouse. It might take days for a query to run through a multiterabyte data warehouse. Data marts emerged with the aim of improving system and network performance, since it is not always necessary for everyone in the organization to have direct access to all the data in the warehouse.
- *Warehouse administration:* This component focuses on maintaining the metadata—the data about data—that provides analytical derivation, exception recognition, integrity, controls, and security. Metadata, which resides above the warehouse data, defines the rules and content of the views provided from the entire domain of available information. It maps user queries to the operational data sources needed to satisfy requested parameters.
- *Operational data store:* The ODS draws its data from the various operational systems in the corporation, but it also can add information derived from data keys in a data mart created by a data warehouse. As such, it not only can add the value of consolidating a common view of enterprise data, but it also can add data derived from trend analyses done on the data marts.

A comprehensive warehouse contains most or all of the following data types:

- Historical events downloaded from operational systems, such as invoice transactions, claims, and payments. The level of granularity maintained is determined during the design phase of the data warehouse architecture.
- Master entities referenced by event, such as customer, product, vendor, and patient.
- Master entity hierarchical roll-ups, such as customer, store, district, region, and country.
- Summarization of historical events, such as monthly sales by store, district, and region. These constitute preemptive queries. Data is not real-time; it aggregates when periodically added to the warehouse. In this way, data requested by a user are always accurate. In many systems, critical data are maintained in both a granular state and with one or more levels of summarization for rapid access.
- Miscellaneous domain data, such as codes, flags, validation, and translation data. Domain data are important for assigning the correct attributes to data fields and for correlating different codes that are used on separate operational systems.

Another data type is *metadata.* Metadata contains information needed to map the warehouse to DSS views, operational sources, and related data in the warehouse. It determines the analytical strength of the warehouse in terms of overall flexibility, extensibility, and adaptability. Examples of metadata include the source of warehouse information, data aggregation methods and rules, purge and retention periods, replication and distribution rules, the history of extracts, and event data from outside sources such as subscription databases.

Decision support system. The DSS supports managerial decisions. It provides the user interface and tools for the heuristic analysis of large amounts of data. The DSS works with metadata to offer a flexible, responsive, interactive, intuitive, and easy-to-use method of constructing and executing warehouse queries.

Decision support systems let users develop a hypothesis, test its validity, and create queries based on the validated hypothesis. Among the key characteristics of a DSS are: an automatic monitoring capability to control runaway queries; flexibility to transparently request data from the central, local, or desktop database; data-staging capabilities for temporary data stores for simplification, performance, or conversational access; a drill-down capability to access exception data at lower levels; import capabilities to translators, filters, and other desktop tools; a scrubber to merge redundant data, resolve conflicting data, and integrate data from incompatible systems; and usage statistics, including response times. A key capability of the DSS is *data mining,* which uses sophisticated tools to detect trends, patterns, and correlations hidden in vast amounts of data. Information discoveries are presented to the user and provide the basis for strategic decisions and action plans that can improve corporate financial performance.

Multidimensional analysis. *On-line analytical processing* (OLAP) is a sophisticated form of DSS that provides business intelligence through multidimensional analysis of information stored in relational and other tabular (two-dimensional) databases. Users can interrogate data warehouses dynamically and intuitively by slicing, dicing, and otherwise carving up the cube to answer complex, real-world business questions, such as, "Which products sell best in each region?" or "How can inventory be reduced to free working capital?" The cubes themselves are optimized, via built-in algorithms and programmable business rules, for efficient calculation.

Summary. Data warehouses are reaching the terabyte level because businesses and government agencies are not only collecting more data, but are keeping it longer for the purpose of analyzing trends. Raw data pulled from transaction-processing systems and other sources generally accounts for a fraction of the overall size of warehouses, however. Indexes, temporary files, backup capacity, mirrored data, and workspaces make up the rest. Furthermore, as databases grow, modeling the database, loading it with data, scrubbing the data, and creating indexes becomes more complicated and time-consuming. Performance can bog down as ad hoc queries plow through the data warehouse.

Moving to a distributed architecture by deploying data marts can make large data warehouses more manageable, but the disadvantage of this approach is that becomes harder to build a central repository of corporate information if this is important to the business. A distributed *logical warehouse* is another option. The database can be segmented to achieve greater efficiency and, in the process, provide redundancy and load balancing to guard against data loss and improve overall performance.

See also

STORAGE MEDIA

Decibel

The decibel (dB) is a unit of measurement expressing ratios using logarithmic scales to give results related to human audio or visual perception. Many different attributes are given to the base reference point, which is 0 dB, and subsequent measurements are relative to that reference point. Many performance levels are quoted in dB, such as a transmission link's signal-to-noise ratio (S/N). High S/N facilitates error-free transmission, whereas low S/N degrades transmission.

See also

HERTZ

Dialing parity

A local exchange carrier (LEC) must provide dialing parity to competing providers for all local toll and long-distance calls. This means that a subscriber should not have to dial an access code (e.g., 10-ATT) to place a local toll call. Access codes may discourage customers from choosing alternative carriers and give an advantage to the LECs, which require no access code.

Instead of using the local telephone company for local toll calls, customers may presubscribe to an alternative carrier to handle them. With presubscription, there is no access code to dial, nor special equipment to buy. Phone numbers are dialed in the ordinary manner. Presubscription is a result of deregulation at the state level. In states where local toll presubscription has not been ordered by state authorities, it will occur simultaneously with the LECs' eventual entry into the long-distance market.

Digital Crossconnect Systems

In 1981, AT&T developed a T-carrier device called the Digital Crossconnect System (DCS) to automate the entire process of circuit provisioning. Instead of having a technician manually patch access lines to long-haul transport facilities, for example, the DCS allowed customer-ordered circuits to be set up between two points from a remote location via keyboard command, thereby expediting circuit setup.

AT&T started offering these capabilities to its Accunet T1.5 subscribers in 1985. Instead of waiting for service orders to be processed before they could rearrange their networks, users were given the ability to do it themselves via a tariffed service called *Customer Controlled Reconfiguration* (CCR). With a terminal connected to AT&T's central control system, customers can reconfigure their networks without carrier involvement. Since the reconfigurations are software-defined, they can be implemented in a matter of minutes. Most long-distance and local telephone companies now offer this capability as a service. Corporations also can implement this capability via crossconnect systems installed on their private networks.

Central control system. The reconfiguration capabilities of crossconnect systems are useful for disaster recovery, bypassing critical systems so that scheduled preventive maintenance may be performed, meeting peak traffic demand, and implementing temporary applications such as videoconferencing.

The network manager issues reconfiguration instructions via a dial-up or dedicated connection from an on-premises terminal to the carrier's central control system. The central control facility holds network routing maps for each subscriber. With the maps held in memory, each subscriber may invoke predefined alternate configurations with only a few keystrokes from the on-premises control terminal. At periodic intervals, the carrier's control system distributes these configurations to the various crossconnect systems that will be involved in implementing the routing changes.

Several customers can share the crossconnect system, allowing the telephone company to offer virtual private networks to small companies, many of whom appreciate the benefits of private networks but still cannot afford to implement multinode configurations of their own. Hubbing off the central office crossconnect system makes a viable alternative to setting up a separate network because the arrangement can provide all of the control and flexibility of a private network, but without the huge capital investment in equipment and the risk of obsolescence. Access to management features is password-protected to prevent one customer interfering with the network of another customer. The telephone company is responsible for circuit testing and system maintenance.

Drop and insert. Aside from customer-controlled reconfiguration, the key feature of digital crossconnect systems is *drop and insert.* This refers to the capability of the DCS to exchange channels from one facility to another, either to implement appropriate routing of the traffic, reroute traffic around failed facilities, or to increase the efficiency of all the available digital facilities. This capability also is offered by some T1 and T3 multiplexers.

The most sophisticated crossconnects provide three levels of switching: DS3, DS1, and DS0. At the DS3 level, a battery of 28 T1 facilities can be switched. At the DS1 level, the entire composite of 24 channels (DS0s) can be switched from one T1 facility to another. At the DS0 level, individual 64 kbps channels can be switched from one DS1 stream to another, and other channels can be inserted in their place. In other words, while one or more DS0s can be dropped at an intermediate location, others can be inserted into the bit stream at that time for transmission to another location (Fig. D3).

Some crossconnect systems can even perform switching at the sub-DS0 level, allowing individual subchannels operating at 2.4, 4.8, or 9.6 kbps to be bundled into one 56/64 kbps channel and then separated at another crossconnect for individual routing to their respective destinations.

Summary. Although the DCS is more accurate and flexible than a manual patch panel, it is not as well suited for setting up calls or real-time disaster recovery because of the time it takes to establish the desired pathways. For these functions a PBX and T1 multiplexer, respectively, are required. And unlike the central office switch, which sets up, supervises, and tears down communication paths every time a call is placed, a crossconnect system keeps communication paths in place for continuous use over a period of months or even years. But for networks with constantly changing needs, circuits may be added, deleted, or rearranged as demand warrants. This capability can result in tremendous cost savings because it eliminates the tendency to over-order facilities to meet any contingency and eliminates the delay in going through the carrier for ordering circuit reconfigurations.

See also
CENTRAL OFFICE SWITCHES
MULTIPLEXERS
PRIVATE BRANCH EXCHANGE

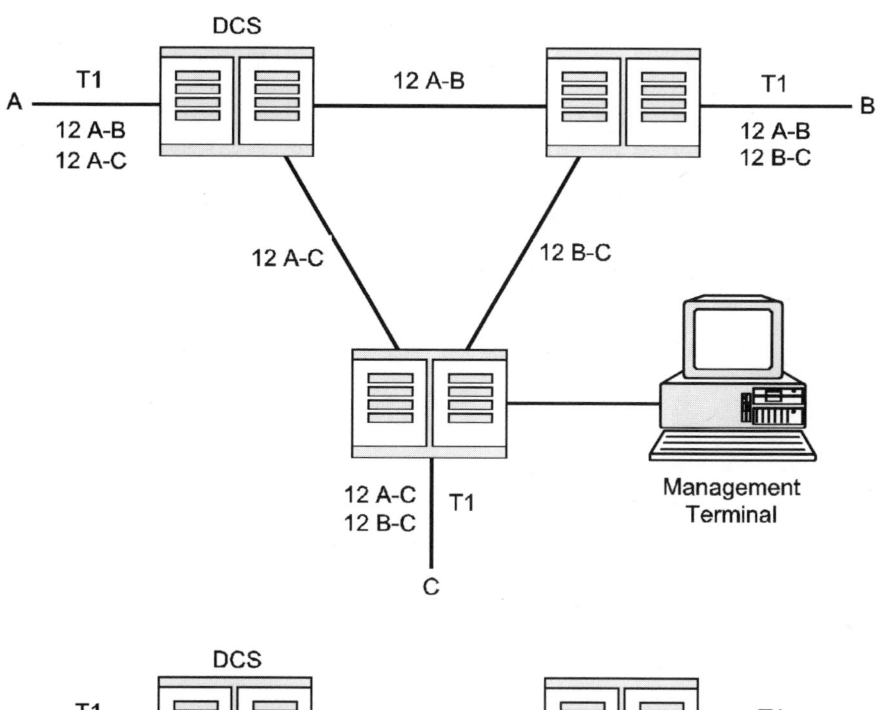

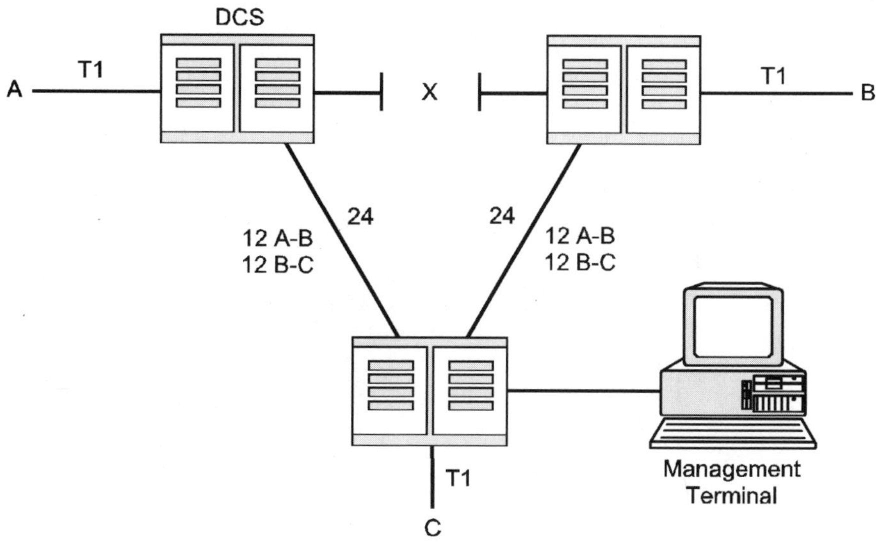

Figure D3 *Rerouting channels in a DSC network via a management terminal under customer-controlled reconfiguration.*

Digital data services

AT&T pioneered digital communications with the introduction of its Dataphone Digital Service (DDS) in the early 1970s. The acronym DDS has come to refer either to the digital data service transport method, or to the brand name of the AT&T service itself. Until 1984, when T1 facilities were tariffed, 56 kbps DDS facilities were the fastest digital systems commercially available. DDS facilities typically include rates of 2.4, 4.8, 9.6, and 56 kbps. Later, 19.2 and 64 kbps speeds were added to DDS, along with an optional secondary channel for low-speed data applications such as diagnostics. A 56 kbps DDS circuit normally is referred to as *56K service,* while 19.2, 9.6, 4.8, and 2.4 kbps lines are referred to as *subrate services.*

Lower-speed (2.4 to 56 kbps) data terminal equipment interconnected via DDS by means of standard RS-232C or V.35 interfaces requires both the CSU and DSU functions. Most new DDS circuits are terminated in an integrated CSU/DSU device.

DDS is a hub-based service and is available in point-to-point and multipoint synchronous configurations. With hub-oriented services, the user's traffic must be routed through a series of special hub offices. Because there are fewer DDS hubs nationwide (AT&T has about 100 hubs), longer circuit mileage is typical, which inflates the cost of the service. The average back-haul distance with DDS is 60 miles, compared with more economical generic digital services (GDS) implemented from serving wire centers (SWCs). Since there are about 20,000 SWCs among all carriers nationwide, the average back-haul distance for GDS is only 6 miles. Of course, the advantage of DDS is that it offers a higher-quality transmission and less down time than GDS.

Secondary channel. Optionally, DDS with a secondary channel (DDS/SC) may be provided to the customer. Since this secondary channel operates at a relatively low bit rate, it typically is used by customers to measure the end-to-end error-rate performance of the primary channel. Vendors of DDS equipment often use the secondary channel to provide remote unit configuration and monitoring capabilities. These include host control over remote DSU optioning, surveillance of the remote DSU/terminal interface, remote alarming, performance testing of the network and equipment, and reporting of reference information resident in the firmware of each remote CSU/DSU. Table D1 summarizes the primary and secondary channel bit rates.

User-controlled diagnostics. In providing a completely independent, low-speed (auxiliary) data channel, DDS/SC allows network surveillance to be performed on a

Table D.1 DDS Primary and Secondary Channel Bit Rates

Primary Channel	Secondary Channel
2.4 kbps	133 bps
4.8 kbps	266 bps
9.6 kbps	533 bps
19.2 kbps	1066 bps
56 kbps	2132 bps
64 kbps	not available

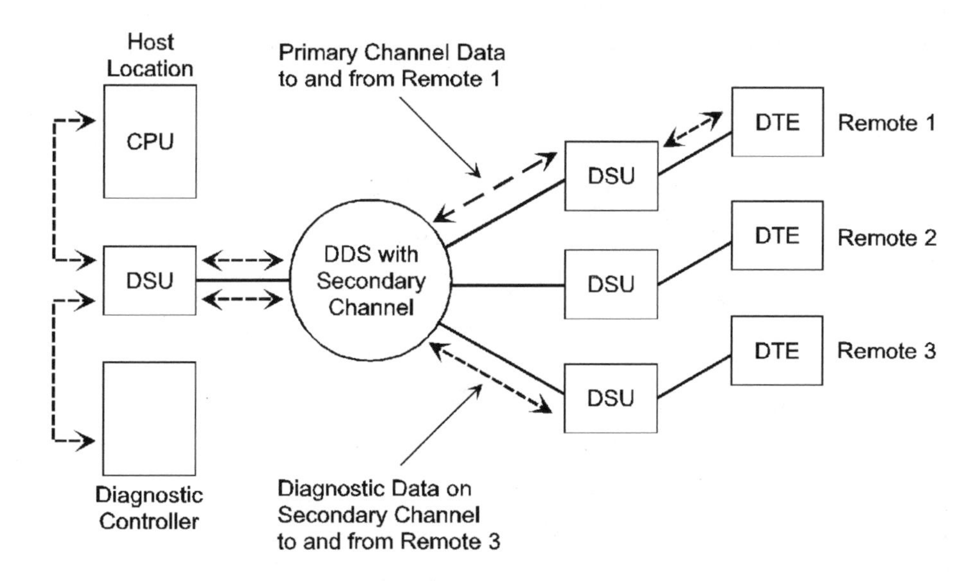

Figure D4 *An application of DDS with Secondary Channel (DDS/SC). The secondary channel provides the means through which the diagnostic controller can poll location 3. At the same time, the host, using the primary channel, may poll location 1 for normal production data.*

continuous and nondisruptive basis over the secondary channel without interfering with the production data (Fig. D4), which continues to flow over the primary channel.

While there are CSU/DSUs that provide a virtual secondary channel on conventional DDS circuits (by multiplexing test data along with user data), these devices do not provide the testing flexibility of DDS/SC. Because the primary and secondary channels derived with this approach are not truly independent, continuous real-time tests cannot be performed without impacting production data.

Summary. Initially, DDS met the emerging need for a high-speed, high-quality private-line service. The addition of a secondary channel gave users the means to control and diagnose problems without interfering with normal production data on the primary channel. Generic digital services (GDS) made their debut in the late 1980s as a more economical alternative to DDS. The lower cost of GDS is mainly attributable to its implementation through carrier-serving wire centers, which eliminates the problem of traffic back-hauling. Because traffic does not have to travel as far with GDS, it has become a popular alternative to DDS.

See also

GENERIC DIGITAL SERVICES

Digital Enhanced Cordless Telecommunications

The Digital Enhanced Cordless Telecommunication (DECT) standard defines a protocol for secure digital telecommunications that offers an economical alternative to existing cordless and wireless solutions. DECT uses Time Division Multiple Access (TDMA) technology to provide ten 1.75 MHz channels in the frequency band between 1.88 and 1.90 GHz. Each channel can carry up to 12 simultaneous two-way conversations. Speech quality is comparable to conventional land-based phone lines. Whereas conventional analog cordless phones have a range of about 100 m, a DECT version can operate reliably up to 300 m. What started out as a European standard for replacing analog cordless phones has been continually refined by the European Telecommunication Standards Institute (ETSI) to become a worldwide standard that provides a platform for wireless local loops (WLL) and wireless LANs as well.

Advantages. A key advantage of DECT is *dynamic reconfiguration,* which means that implementation does not require advance load, frequency, or cell planning. Other wireless architectures require a predetermined frequency allocation plan. Conventional analog cellular networks, for example, are organized in honeycomb fashion. To avoid conflict from adjacent cells, each base station is allotted only a fraction of the allowable frequencies. Changing a particular station's frequency band to accommodate attempts to increase network capacity by adding more base stations entails an often difficult and expensive hardware upgrade. However sparsely the base stations are constructed at the start of an installation, all possible base stations must be assigned frequencies before any physical systems are put into place.

In a DECT system, planning for uncertain future growth is unnecessary. This is because a DECT base station can dynamically assign a call to any available frequency channel in its band. The 12 conversations occurring at any one time can take place on any of the 10 channels in any combination. The handset initiating a call identifies an open frequency and timeslot on the nearest base station and grabs it. DECT systems also can reconfigure themselves on the fly to cope with changing traffic patterns. Therefore adding a base station requires no modification of existing base stations and no prior planning of channel allocations.

Compared to conventional analog systems, DECT systems do not suffer from interference or crosstalk. Neither different mobile units nor adjacent DECT cells can pose interference problems because DECT manages the availability of frequencies and time slots dynamically. This dynamic reconfiguration capability makes DECT useful also as a platform for wireless local loops. DECT allows the deployment of a few base stations to meet initial service demand, with the easy addition of more stations as traffic levels grow.

Voice compression (i.e., ADPCM) and the higher levels of the DECT protocol are not implemented at the base stations, but are handled separately by a concentrator. The concentrator routes calls between the WLL network and the public-switch telephone network (PSTN). This distributed architecture frees up base station processing power so it can better handle the up to 12 concurrent transmission and reception activities.

For high-end residential and small-business users, DECT permits wireless versions of conventional PBX equipment, supporting standard functions such as incoming and outgoing calls, call hold, call forwarding, and voice mail without having to install new wiring. In this application, DECT dynamic reconfiguration means that implementation does not require advance load, frequency, or cell planning. Users can begin with a small system, then simply add to it as needs change.

When implemented, the DECT/GSM Interworking Profile will allow a single handset to address both DECT systems and conventional cellular networks. This will allow users to take advantage of the virtually free wireless PBX service within a corporate facility and then seamlessly switch over to GSM when the handset passes out of range of the PBX base station. When the call is handled by GSM, appropriate cellular charges accrue to the user.

Wireless local loops. Although residential cordless communication represents the largest current market for DECT-based products, other applications look promising for the future. In developing countries, where lack of a universal wired telecommunications infrastructure can limit economic growth, DECT permits the creation of a wireless local loop, thereby avoiding the considerable time and expense required to lay wirelines. Wireless local loops can be implemented in several ways, which are summarized in Fig. D5.

In a small cell installation in densely populated urban or downtown areas, the existing telephone network can be used as a backbone that connects the base stations for each DECT cell. These DECT base stations may be installed on lampposts or other facilities. Customer boxes (i.e., transceivers) installed on the outsides of houses and office buildings connect common phone, fax, and modem jacks inside. Through the transceivers, customers use their telephone, fax, and modem equipment to communicate with the base stations outside. In addition, customers can use DECT-compliant mobile phones, which can receive and transmit calls to the same base station.

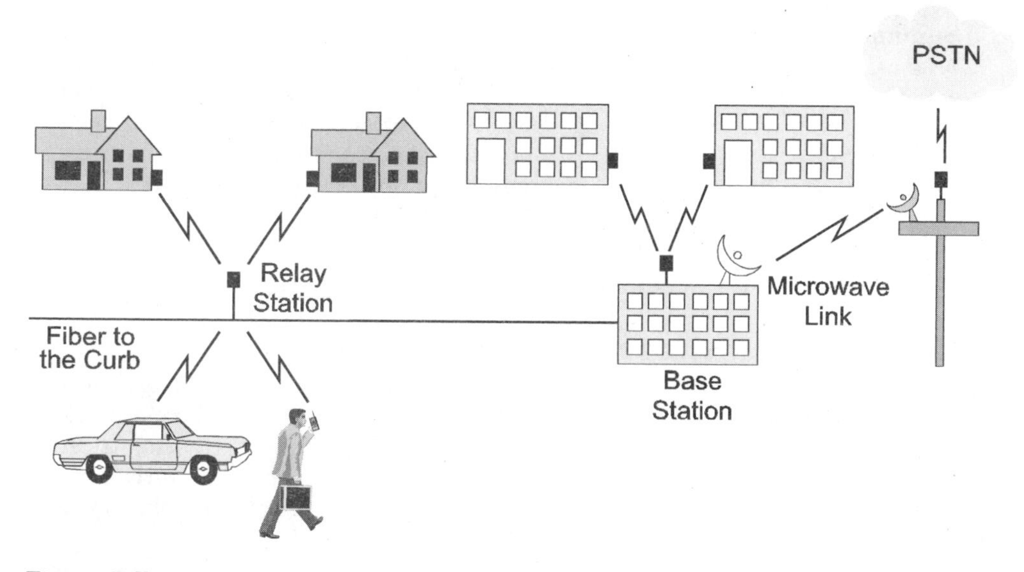

Figure D5 *DECT supports the deployment of wireless local loops that offer a high degree of configuration flexibility and cost savings over conventional wireless and wireline solutions.*

In larger cell installations, such as suburban or rural areas, fiber-optic lines may provide the backbone which connects local relay stations to the nearest base station. These relay stations transmit and receive data to and from customer boxes. In these installations, the customer box must have a direct line of sight to the relay station.

Network feeds over long distances may be accomplished via microwave links, which is more economical than having to install new copper or fiber lines. Large cells can be converted easily into smaller cells by installing additional base units or relay stations. Since DECT has a self-organizing air interface, no top-down frequency planning is necessary, as is the case with other wireless connection techniques such as GSM or DCS 1800.

While the first WLL installations will focus on regular telephone and fax services, DECT paves the way for enhanced services. Multiple channels can be bundled to provide wider bandwidth, which can be tailored for each customer and billed accordingly. This allows the mapping of ISDN services all the way through the network to the mobile unit.

Wireless LANs. In many data applications with low bit rate requirements, DECT can be a cost-effective solution. One example is remote wireless access to corporate LANs. By bundling channels, full duplex transmission of up to 480 kbps per frequency carrier is theoretically possible. For multiple data links, a DECT base station can be complemented by additional DECT base stations controlled by a DECT server. This forms a multicell system for higher traffic requirements. With a transparent interface to ISDN, data access and video conferencing through wireless links can be realized. Such installations may also include such services as voice mailboxes, automatic callback, answering and messaging services, data on demand, and Internet access.

Summary. DECT was originally intended as a replacement for analog cordless systems, but has become a platform for wireless local loops and wireless LANs as well. Once thought of as a European standard, DECT is spreading worldwide. It can be directly connected to such fixed-network technologies as ISDN, Ethernet, and ATM. At the same time, it has the potential to be interconnected with existing mobile communication systems, such as those based on GSM, DCS 1800, and PCS 1900 standards, thereby offering the widest possible connectivity options.

See also

GLOBAL SYSTEM FOR MOBILE (GSM) TELECOMMUNICATIONS

Digital loop carrier systems

Digital loop carrier systems (DLCS) enable telephone companies to expand their networks in an economical manner to accommodate additional subscribers and efficiently handle the higher traffic load. Expanding a network with such systems has provided telcos with the flexibility to accommodate growth, expanded efficient use of digital or analog central office facilities, and provided a cost-effective alternative to replacing or upgrading older offices or wire centers.

Operating environment. Within the Local Access and Transport Area (LATA) are numerous wire centers, which serve specific communities of interest. A wire center may be a 50,000-line central office, or it may be a Community Dial Office (CDO) serving only a few hundred subscribers. Service also may be provided by a Remote Terminal (RT), which, for example, gives tenants of a distant office park access to the telephone network.

The central office typically serves subscribers via copper pairs (24–26 AWG) within a 2.5-mile (12,000-ft) radius. To reach subscribers farther away, two conventional strategies have been employed. One is to set up a CDO with trunks linking it to the central office. Another is to set up a remote terminal to link a fairly limited number of subscribers to the network. The remote terminal may be linked to a CDO via a secondary feeder, or a series of secondary feeders may link remote terminals directly to the central office via the primary feeder. Subscribers connected through the CDO or the remote terminals have access to all the features of the central office to which they are ultimately linked.

Territories served by remote terminals are called *carrier serving areas* (CSA). This is simply a planning entity consisting of a distinct geographical area served by a carrier remote terminal site. The CSA concept (Fig. D6) allows the telco to plan its network to accommodate incremental growth; the telco can provide service to an ever-increasing number of subscribers in any number of remote locations.

The primary feeder from the central office to the CDO may consist of as many as 3600 copper wire pairs. A secondary feeder from a CDO to a remote terminal may consist of as many as 1800 copper wire pairs. From the secondary feeder, the pairs branch off to individual subscriber locations. A problem occurs when all the wire pairs in a primary or secondary feeder are used up. The telco must find a way to continue accommodating additional subscribers and efficiently handle the higher traffic load.

Before digital loop carrier systems were introduced in the mid-1970s, the only viable solution was to segment each CSA and add more wire and equipment. That was an extreme solution, however, involving increased installation and maintenance costs, particularly in metropolitan areas where costs for conduit/duct expansion were prohibitive. Today, with demand often outpacing feeder capacity, telcos require a simpler, more cost-effective way of coming to grips with the growth problem.

One very attractive solution is to use T1 lines to free up copper wire pairs and increase the traffic capacity of the network—and, in the process, retain a high degree of flexibility in network planning. The following scenario illustrates the three possible applications for digital loop carrier systems: CDO replacement, feeder relief (which also eliminates the need for conduit/duct expansion in major metropolitan areas), and efficient distribution.

Applications. The XYZ Company, a nationwide electronics firm, has just announced plans for a new manufacturing plant in an undeveloped location outside of city limits. While county officials extolled the virtues of free enterprise, not to mention the 200 new jobs it would create for their constituents, city officials were already drawing up plans to annex the site of the planned facility to broaden the tax base. The XYZ Company, which cannot afford the additional tax burden, abruptly changes its location plans to a small town across the county border, well beyond the reach of the city. Unfortunately, the new location is also well beyond the capability of the local telephone company to serve with present facilities.

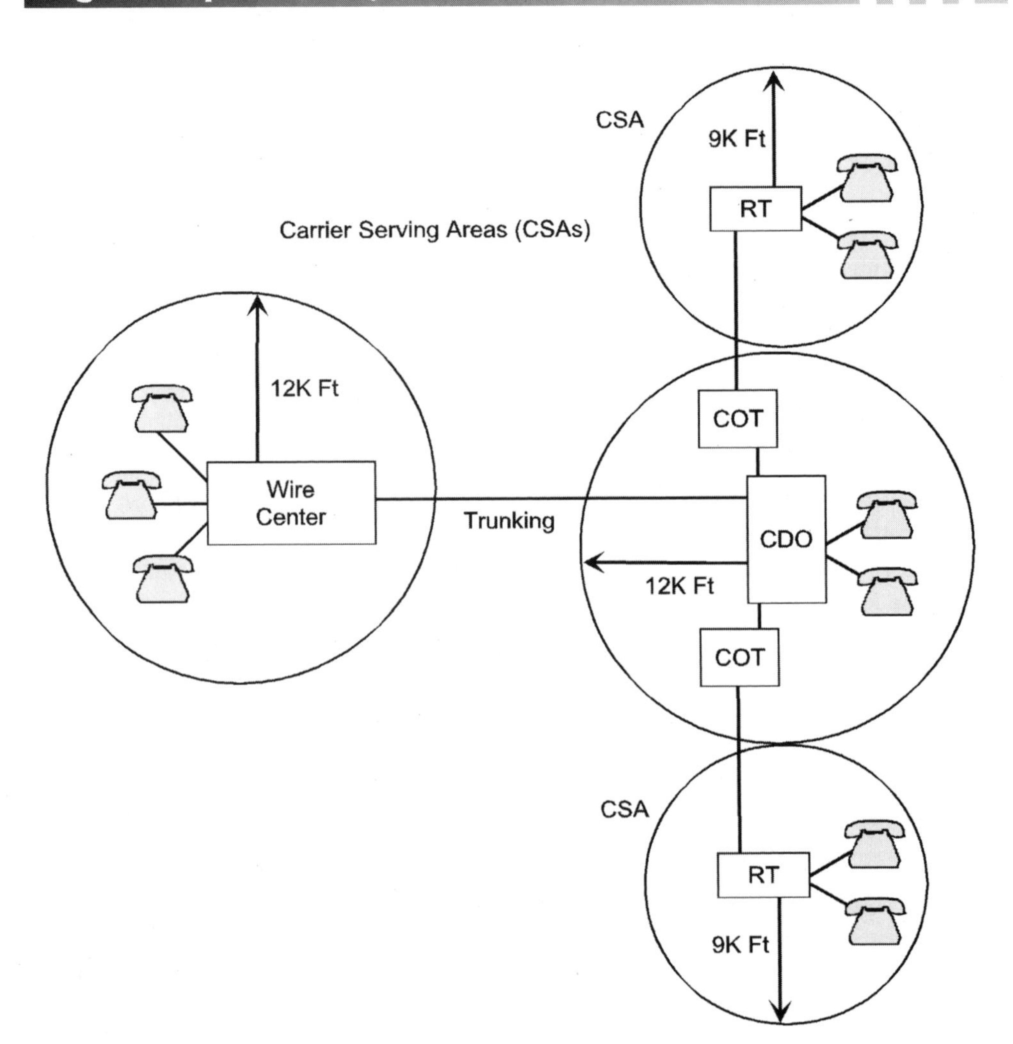

Figure D6 *Carrier serving areas are territories served by remote terminals.*

Anticipating growth around the city, the telco installed community dial offices. It had planned to serve the XYZ Company's new plant with an underutilized CDO nearby, but it was caught completely off guard by XYZ's last-minute decision to build the facility in a small town 10 miles away, where 50 residents were served by a remote terminal.

With potentially hundreds of new subscribers to serve, the telco would quickly run out of wire pairs at its remote terminal site, yet the increase in subscribers was not enough to justify adding a CDO. In this case, the telco chose as the best solution the replacement of an existing CDO with a DLCS (Fig. D7). With a DLCS, the telco achieved several objectives.

First, it retired wire pairs from the remote terminal to the CDO through the use of T1 span lines, which in turn handled the higher concentration of traffic from the

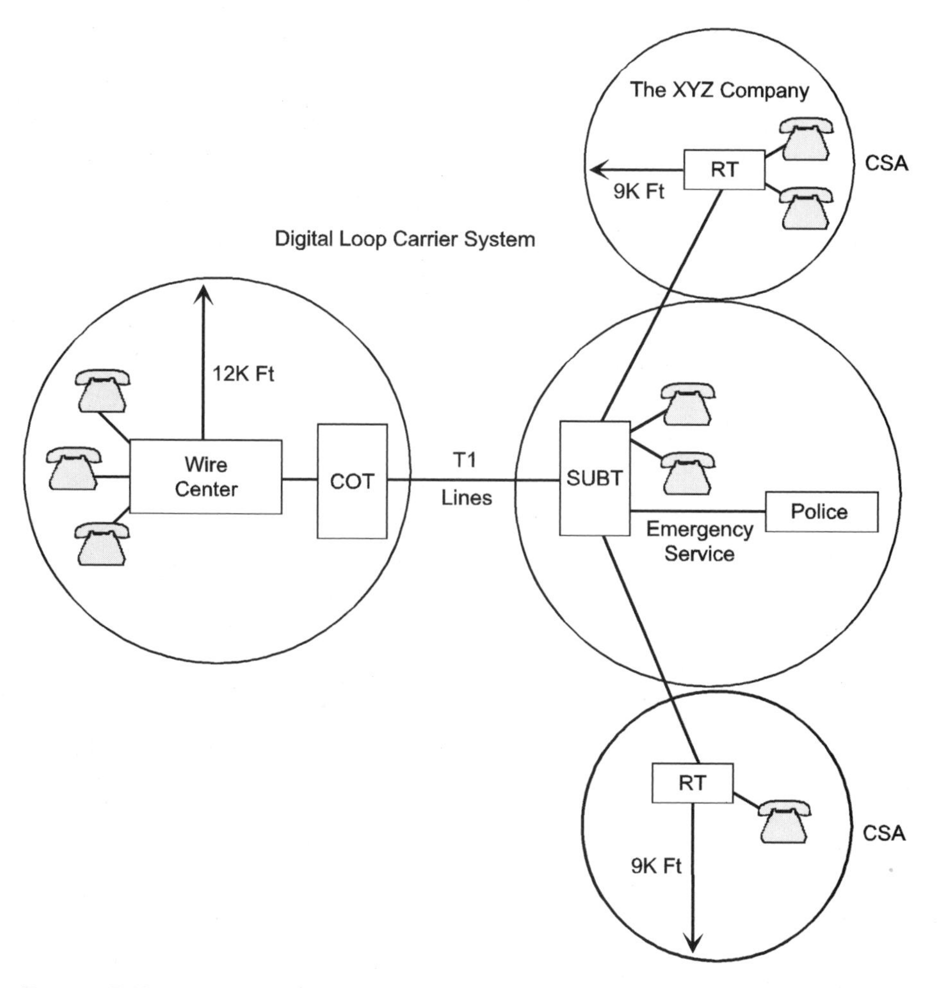

Figure D7 *Digital loop carrier system.*

small town. Since the T1 span lines carry traffic in digital format, the DLCS can be readily integrated into the telco's digital switch. Second, the telco could offer emergency and other special services, such as data, coin, and FX (foreign exchange) lines, simply by adding plug-in modules to the DLCS's subscriber terminal (SUBT) and linking the subscriber terminal to the central office terminal (COT) via T1 lines. Also, the architecture of the DLCS permits deferral of a significant portion of the terminal costs (plug-ins) until forecasts are clearer or until specific services are actually needed.

Third, with the T1 span lines between the DLCS subsystems (COT and SUBT), enough wire pairs in the feeder plant could be retired or used for distribution to accommodate future subscriber growth. The telco also achieved a degree of planning flexibility it did not have before, which will save on future installation and maintenance.

An additional but less apparent benefit of the DLCS is that it eliminates the need for "load coils." Present wire pair media are limited to the voice frequency

bandwidth of approximately 4 kHz, which is sufficient for POTS (plain old telephone service), but not for the high-speed data services. Load coils are used to improve baseband (4 kHz) transmission for distances longer than 18,000 ft. Wire pairs without load coils (nonloaded) are capable of supporting wider bandwidths than the 4 kHz baseband, providing 144 kbps data services and supporting short-range video.

Summary. Digital loop carrier systems were introduced in the mid-1970s as solutions for economical telco network expansion. Savings have accrued not only from reducing analog facilities by up to 80 percent, but from building and real estate liquidation, and maintenance efficiencies. Today DLCSs have evolved into economic, efficient vehicles for generating new revenues for telephone companies. These systems will continue to evolve to meet increasingly sophisticated network needs.

See also
CENTRAL OFFICE SWITCHES
LOCAL ACCESS AND TRANSPORT AREAS

Digital subscriber line technologies

Local loops have been badly neglected from a technology upgrade perspective. This is true not only for telco copper pairs, but also for end user–owned loops in campus environments and high-rise buildings. Existing local loops were designed for voice telephony, and the amount of bandwidth can be further limited by bridged taps and wire gauge changes. If the telcos are going to offer advanced services, such as high-speed Internet access, they must have an economical means to increase the bandwidth of existing twisted pairs. The answer is being provided by new digital subscriber line (DSL) technologies. In fact, there is a whole lineup of technologies dealing with digital subscriber lines that offer high-speed services over existing copper facilities.

ADSL. One of these new technologies is *Asymmetrical Digital Subscriber Line* (ADSL), which is an affordable local-loop upgrade technology that allows for the transmission of about 6 Mbps over existing twisted-pair copper wiring: specifically, 1.544 to 6.144 Mbps downstream (CO to customer) and 16 to 640 kbps upstream (customer to CO). This is enough bandwidth to support multimedia (video, audio, graphics, and text) to the customer premises. ADSL carves up the local loop bandwidth into several independent channels suitable for any combination of services, including voice, ISDN, video-on-demand programming, and interactive gaming. The VLSI electronics at both ends compensate for line impairments, increasing the reliability of high-speed transmissions and reducing trouble calls.

Two competing modulation schemes underlie ADSL: discrete multitone (DMT) and carrierless amplitude/phase modulation (CAP). CAP is a proprietary solution that delivers higher bandwidth (7 Mbps downstream), but the ANSI ADSL standard supports DMT.

Given that copper cabling encompasses more than 95 percent of the nation's local phone lines, ADSL is quite compatible with what is in the ground today. As a result, ADSL is very attractive to the telcos, which are collectively seeking a viable, quick, and effective weapon against TV cable companies.

HDSL. High-bit-rate Digital Subscriber Line (HDSL) is slower than ADSL, with data moving between 384 kbps and 1.5 Mbps. Because HDSL is symmetric, it is an ideal low-cost alternative to leased T1 lines, the application for which it was originally intended. HDSL also could be used by telecommuters seeking remote LAN access. One big benefit of HDSL is wider reach for local loops. ISDN lines are only specified for local loops up to 12,000 ft, but HDSL lines can reach as far as 16,500 ft without sacrificing speed.

ISDL. ISDN DSL (IDSL) operates at speeds up to 128 kbps, considerably less than the speeds offered by other SDL technologies. Like other forms of DSL, it operates over the same copper wires used for single-line voice telephone service. IDSL differs from standard ISDN in one important way, however: it is a dedicated service. This means users pay a flat fee rather than the per-minute charges typical of switched ISDN, something that would be attractive to users who spend a lot of time online.

Unlike ADSL, IDSL is a data-only service, lacking the analog voice line that ADSL offers. IDSL could be attractive to users because it can use standard ISDN customer premises equipment. From the telco's perspective, IDSL pulls traffic that would otherwise tie up a voice switch and drops it onto a frame relay network that is designed to handle nailed-up permanent virtual circuits. Off-loading Internet traffic from voice switches is important because the telco community has been complaining to the Federal Communications Commission (FCC) that the time customers spend on the Internet is causing major bottlenecks in the local loop.

RADSL. Rate Adaptive Digital Subscriber Line (RADSL) technology offers data throughput up to 400 times faster than a 14.4 kbps modem over existing copper phone lines, with downstream rates of 600 kbps to 7 Mbps, and upstream return rates of 128 kbps to 1 Mbps.

A standard ADSL connection delivers up to 6 Mbps downstream and 384 kbps upstream over a distance of 12,000 ft, leaving out users who live farther from the provisioning central office. With RADSL, however, the carrier could offer less bandwidth over a longer distance. On the downstream path, modems based on the RADSL chips could deliver as little as 600 kbps at distances of up to 21,300 ft and 128 kbps at distances of over 25,000 ft.

RADSL technology is a significant advance over other DSL implementations because its transmission speed is rate-adaptive based on the length and signal quality of an existing telephone line. RADSL products will have the option to select the highest practical operating speed automatically or as specified by the telecom service provider.

SDSL. Symmetric Digital Subscriber Line (SDSL) technology extends the high bandwidth capacity of the telco network into a home or business using only one single-pair copper wire phone line. Single-Pair SDSL will enable network access transmission rates from 144 kbps (the upper limit of ISDN BRI) up to 2 Mbps, and will enable the option for voice POTS to operate simultaneously and independently of the data transmission on the same phone line.

SDSL supports a variety of applications, including extension of T1/E1 services from the PSTN to connect customer premise equipment, such as routers, multiplexers, and PBXs. Other applications include cross-campus interconnection of devices requiring T1/E1 or fractional T1/E1 speeds, and interconnection of cellular service cell towers to the service provider's mobile telephone switching office. The high-speed technology also can be used to deliver videoconferencing, digital imaging, remote LAN access, and distance learning.

VDSL. Another digital subscriber line technology is *very high-speed DSL* (VDSL), which is capable of transmitting almost 52 Mbps over a single pair of copper wires. Data can travel at the SONET OC-1 rate of 51.84 Mbps from 100 to 300 m, with rates of up to 1.6 Mbps on the upstream return path.

SDL vs. ISDN. ADSL outshines ISDN for sheer bandwidth, offering 6.144 Mbps downstream and 640 kbps upstream. Basic rate ISDN supports only two 64 kbps bearer channels and a 16 kbps signaling channel. Another differentiator is that ISDN is a dial- up service intended for periodic use, whereas ADSL is a dedicated line that handles more constant demand, which is characteristic of Internet access.

ISDN will continue to have a niche for telecommuters because of its flexibility. With two 64 kbps channels, users can add or drop one channel to get more bandwidth for data, or to clear up a channel to take a voice call. One channel also can be used for fax. ISDN also supports multiple line appearances, which is very useful for telecommuters and home-based businesses who must be able to differentiate personal from business calls before picking up the phone. ISDN is also a more secure service.

ISDN is a switched service, meaning that anyone can call anyone and receive calls from anyone. The traffic is switched through a standard voice telephone switch that has been enhanced to handle ISDN. ADSL, on the other hand, relies on the service provider to establish a short list of others to whom a user can connect. An ADSL line carries an analog voice channel that operates on the same line at lower frequencies than the broadband channels, which are not switched. As a result, the broadband channel is essentially permanent and available 24 hours a day, which is attractive to users who routinely spend several hours a day online. Whereas ADSL is billed at a flat rate per month, ISDN typically is billed by usage, making it a very expensive way to surf the Web.

From the telco point of view, ADSL will be the less expensive service for them to provision in areas where demand for ISDN is low, and will be less complicated for users to configure.

Role of ATM. The telcos eventually want to push ATM toward the edge of their networks. This fits in with the trend toward telecommuting, where, ideally, users would want to access the corporate LAN without experiencing diminished performance, as is currently the case with remote access solutions over dial-up lines. ATM over ADSL offers speeds up to 5 Mbps, which is slower than some other DSL technologies but fast enough for LAN access and connectivity to services such as frame relay. Provisioning ATM to the home costs more than alternatives, such as Ethernet, but it removes the need for large cell-conversion devices. Traffic can be dumped directly onto the telco's ATM switches, where virtual circuits can be established to a variety of networks and services.

The ATM-over-ADSL architecture consists of an ADSL network terminator (ANT) at the user's site and an access adapter in the carrier network. The ANT comes with an ATM network interface card for PCs and a 10 Mbps Ethernet interface to connect with corporate LANs.

Summary. DSL technology can go a long way toward realizing the potential of the Internet and enhancing business models that leverage the power of multimedia networks. The different flavors of DSL will coalesce around specific market niches and applications. For example, ADSL may become favored for home uses such as video-on-demand, telecommuting, and high-speed Internet access. Already there are 56 kbps modems with integral ADSL capabilities and plans to support ATM-over-ADSL via PC card and desktop modems. On the other hand, HDSL could be favored by small businesses and power users, where very high performance at the T1 level is required. It could also be used to extend backbones across campus networks. VDSL might be the choice to deliver Internet traffic or limited multimedia traffic to large businesses.

See also

Asynchronous Transfer Mode (ATM)
Integrated Services Digital Network (ISDN)

Direct broadcast via satellite (DBS)

The availability of economical digital satellite broadcasting (DBS) technologies is a direct response to consumer demand for entertainment programming, Internet connectivity, and multimedia applications. DBS offers a currently viable product with more program choices for consumers and a platform for the development of future services. Much of the growing popularity of DBS services is attributable to the picture quality provided by digital technology.

One of the most popular DBS services is DirecTV. Introduced in the United States in 1994 by Hughes Electronics and Thomson Consumer Electronics (now Thomson MultiMedia), DirecTV is now marketed worldwide. DirecTV was the first DBS service to deliver up to 175 channels of digital-quality programming. The satellite service is provided by DSS and the standard system consists of an 18-in dish, a digital set-top decoder box, and a remote control. The DSS system features an on-screen guide that lets users scan and select programming choices using the remote. Customers also can use the remote control to instantly order pay-per-view movies, as well as set parental controls and spending limits.

The DirecTV installation includes an access card, which provides security and encryption information, and allows customers to control the use of the DSS system. The access card also enables DirecTV to capture billing information. A standard telephone connection is used to download billing information from the DSS decoder box to the DirecTV billing center. This telephone line link allows DirecTV subscribers to order pay-per-view transmission as desired.

DSS allows users to integrate local broadcast channels with satellite-based transmissions. In markets where broadcast or cable systems are in place, users can maintain a basic cable subscription, or connect a broadcast antenna to the DSS digital receiver to receive local and network broadcasts. A switch built into the remote control allows consumers to switch instantly between DirecTV and local stations.

Internet access is provided through DirecPC, a product that uses DirecTV technology in conjunction with a PC to deliver high-bandwidth, satellite-based access to the Internet. The DirecPC package includes a satellite dish and an expansion card designed for a PC's I/O bus. The "receiver card" transmits data from the Internet to the computer at 400 kbps, a rate 14 times faster than that of a 28.8 kbps modem connection.

Users connect to the Internet service provider (ISP) through a modem connection, but the ISP is responsible for routing data through the satellite uplink and transmitting the data to the receiver card and into the computer. The service also provides users with the option to "narrowcast" software from the head end of a network to branch users during off-peak hours. Additionally, DirecPC transmits television broadcasts from major networks, such as CNN and ESPN, to the user's computer system.

Operation. DBS operates in the Ku band, the group of frequencies from 12 to 18 GHz. TV shows and movies are stored on tape or in digital form at a video server, while live events are broadcast directly to a satellite (Fig. D8). Stored programs are sent to the uplink (ground-to-satellite) center manually via tape or electronically from the video server over fiber-optic cable. Live events also pass through the uplink center. There all programs, whether live or stored, are digitized (or redigitized) and compressed before they are uplinked to the satellites. All DBS systems use the MPEG-2 compression scheme because it delivers a clean, high-resolution video signal and CD-quality sound. The satellites broadcast up to 200 channels simultaneously via the downlink. The home satellite dish picks up all the channels and sends them via a cable to a set-top decoder. The set-top decoder tunes one channel, decodes the video, and sends an analog signal to the TV.

Service providers. More than one million U.S. residents have installed small TV satellite dishes to receive programming via satellite services. At this writing there are four DBS system in operation: PrimeStar, EchoStar, DSS (Digital Satellite Service), and AlphStar.

Ordering PrimeStar service is similar to receiving cable: After the order is placed, a technician installs the dish and activates programming. DSS, EchoStar, and AlphaStar services also give users the option of installing the dish themselves. The dish must be placed so it can capture a clean signal from the nearest satellite, usually on the roof and facing south. To activate service, the user calls the programming provider to obtain a unique satellite dish address. DSS lets users choose between two program sources, USSB and DirecTV.

Equipment. The key component of the DBS system is the <B> dish antenna, which comes in various sizes. Dish size depends on the strength of the satellite signal; the stronger the signal, the smaller the dish can be. Users select the dish based on their geographic proximity to the satellite source. This also explains why it is necessary to install the dish so that it points in a specific direction. If the satellite sits on the southern horizon, the dish must be pointed south.

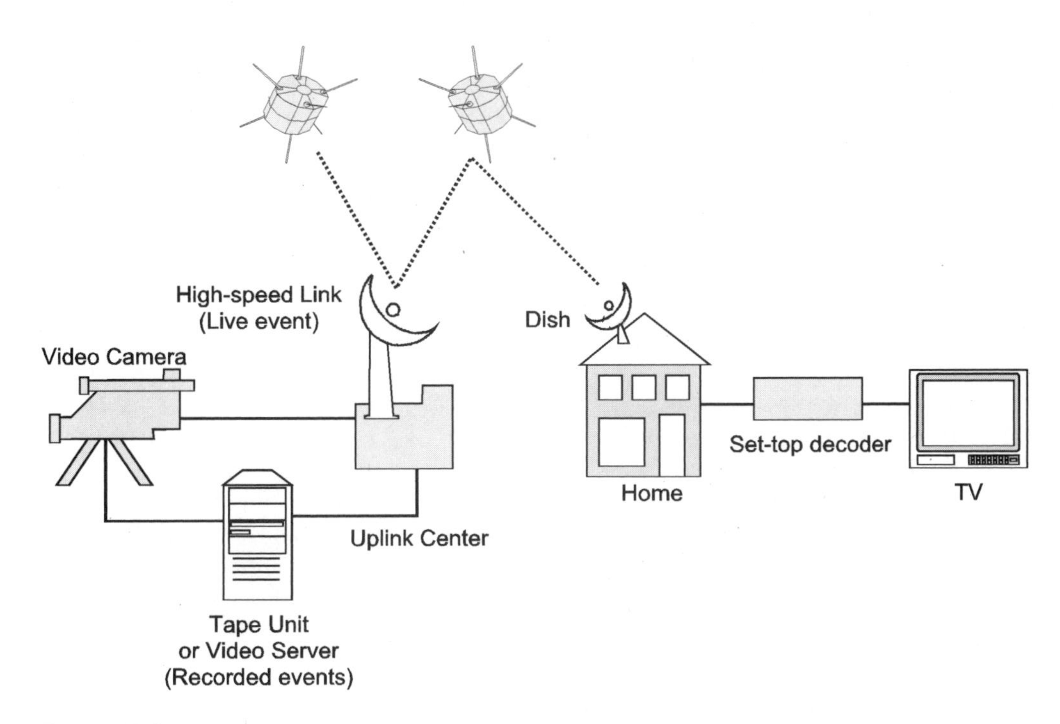

Figure D8 *Typical DBS configuration.*

The user also needs a receiver-decoder unit, which tunes in one channel from the multitude of channels it receives from the dish. The decoder then decompresses and decodes the video signal in real time so the programs can be watched on the television set. These set-top units may also include a phone line connection for pay-per-view ordering and Internet access. Taping DBS programs requires the set-top unit to be tuned to the correct channel. To make recording easier, some receiver-decoders include an event scheduler and an on-screen programming guide.

As with most audio-video components, DBS units come with a remote control. Some manufacturers offer a universal remote that also can be used to operate the TV and VCR.

The accessories available for DBS systems deal with secondary and tertiary installations. Users can buy additional receiver-decoder units or multiroom distribution kits, which use either cable or radio frequencies to transmit the signals from the original set-top unit to other rooms. Some kits allow the VCR to be plugged into the distributor.

Programming. Each of the four DBS systems currently available provide similar core services. The differences lie in the availability of premium movie channels, audio channels, and pay-per-view events.

With up to 200 channels from which to choose, the on-screen programming guide can become an important factor when selecting a service. Most guides allow users to sort the available programming based on content area—such as sports, movies, comedies—or list favorite channels at the top of the menu. Depending on the equipment selected, users can even store the favorite-channel profiles of several family members.

Parental lockout enables adults to block specific channels or programming with a specific content rating, or to set a maximum pay-per-view spending limit. Channel-blocking options are protected by passwords; with multiprofile units, parents can customize the system for each child.

Summary. Industry experts say there could be as many as 15 million DBS households by the turn of the century, giving DBS 20 percent of the market by 2000, up from its current 6 percent market share. This has attracted AT&T and MCI to the market. AT&T, for example, resells the DBS services of DirecTV, offering one-stop shopping, from ordering DSS equipment and service, to home installation, customer service, and financing.

The future of DBS service may include specialized markets for information services. It is already being used for Internet access and could be deployed in ways to offer sales force automation, interenterprise communications, home shopping, and demographically targeted programming. DBS services also are positioned to take advantage of emerging technologies such as interactive multimedia and high-definition television (HDTV).

See also

SATELLITE COMMUNICATIONS

Distance learning

Distance learning is the use of digital networks and video conferencing technology to deliver a variety of education programs to peripheral locations. It can be used to deliver courses to rural areas, for example, or to link together schools in a district to provide a common curriculum. A university can extend its reach to suburbs, offering the same quality of instruction that is available on campus. A corporation or government agency can use distance learning to educate and inform staff, while cutting the cost of travel and the amount of lost productivity time.

In the years ahead, this method of education is expected to have an increasingly significant impact on learning in the home, school, and workplace.There are numerous reasons for this optimistic outlook for distance learning. According to the United States Distance Learning Association (USDLA), the following factors will expand the importance of distance learning:

- Mounting frustration with school systems nationwide, driven by unacceptable student performance and underachievement.
- Parents' interest in additional learning services for their children outside of the traditional school environment.
- Business interest in the ability of information technology and networks to deliver training and learning services in a timely and cost-effective manner.
- The need of employees in both the public and private sector to continually update their job skills and knowledge.

All 50 states have distance learning projects underway. Most involve the cooperation of state and federal government agencies, businesses, and one or more telephone companies.

Applications. Although a variety of high-speed networks and videoconferencing technologies are integral components of distance learning programs, any successful program must focus on the instructional needs of the participants, rather than on the technology itself. An important factor for successful distance learning is a competent instructor who is experienced, at ease with the equipment, and uses the media creatively. The effectiveness of distance learning programs also often hinges on the level of interactivity between the instructor and students and among the students themselves.

Public schools. Another factor that will increase the popularity of distance learning programs has to do with the chronic budgetary problems of the nation's public school systems and the widespread anti-tax sentiment that prevents an easy remedy. Distance learning networks can go a long way toward making up budgetary shortfalls by permitting schools in the same town or city to make use of the same teacher to deliver live instruction to multiple classrooms simultaneously.

Tying local schools together via distance learning networks also can go a long way toward balancing the quality of education between rich and poor schools. For example, the best math, science, and language teachers at "rich" schools can deliver courses to students at "poor" schools at the same time using interactive videoconferencing systems. And courses that might not be available at one school can be accessed from another school without requiring students or teachers to change locations during school hours.

Colleges. Colleges have been offering distance learning programs since the 1980s. Now they are seeking to leverage their distance learning networks to enhance the educational opportunities of geographically dispersed students. One way to do this is by tying in high-speed automated library services. Via the distance learning network, remote users have access to the college's CD-ROM–based library services, including books, journals, videos, government documents, and databases.

Additionally, such networks permit users to share and retrieve information available at other academic and public libraries via the college's library consortium membership and—at the state, national and international level—via the Internet. Through Internet access, the distance learning network can deliver such information as bulletin board data and graphics, database and archival resources, and computer programs, as well as file transfer and retrieval, and electronic mail facilities.

Corporations. Increasingly, corporations are delivering training programs to remote locations via their private networks. A properly designed distance learning program can reduce training time by 30 to 50 percent and can eliminate costs associated with travel. Added benefits include consistent quality of delivery, consistent content, and the consistent application of professional standards at each location.

Associations. Associations also are offering distance learning programs for their members. Medical and hospital associations, for example, operate VSAT-based video conferencing networks that, among other things, offer instruction on new surgical procedures, new drugs, and new medical technologies. Associations representing the automotive, retail, travel, and hospitality industries are among the many other providers of distance learning programs for their members.

Military. The military is also among the largest providers of distance learning programs. In addition to providing instruction on a variety of military topics, these programs include college- and graduate-level courses to help members of all the armed forces obtain their degrees and upgrade their management and communications skills.

Government agencies. Government agencies are continually subject to public criticism and congressional budget cuts. Often significant amounts of money can be saved by cutting back on travel, without negatively impacting staff operations. For many agencies the answer is distance learning, which can be used to provide instruction on new regulations, compliance matters, administrative policies and procedures, and improve supervisory and management skills.

Summary. Distance learning has demonstrated its value in a variety of environments, including public schools and universities, corporations, and government agencies. It can make learning independent of time and distance constraints. It can afford learners in rural areas the same educational opportunities as those in urban areas. It can help spread the cost of technology, computer courseware, and multimedia references and instruction to a broader base of users.

See also

VIDEOCONFERENCING

Dominant carrier status

To encourage competition in the long-distance market, the FCC in 1980 devised the dominant/nondominant regulatory scheme for rate and entry regulation. The FCC defined a dominant carrier as one that "possesses market power" and noted that control of bottleneck facilities was "...evidence of market power requiring detailed regulatory scrutiny." The Commission also determined that, if a common carrier was determined to be "nondominant," regulatory requirements would be "streamlined." Specifically, tariffs filed by nondominant carriers would be presumed lawful and would be subject to reduced notice periods.

Because AT&T held 90 percent of the long-distance market in 1980 and could use its position to control prices, it was classified by the FCC as dominant. MCI and Sprint, on the other hand, were classified as nondominant.

The FCC also found that AT&T also was dominant in the provision of international service. AT&T controlled the overwhelming share of that market, had exclusive operating agreements with the carriers in most major foreign markets, and had few rivals in the provision of essential U.S. international submarine cable facilities. The Commission had ample reason to conclude that AT&T exercised market power and should be regulated as dominant for its provision of international calling services.

Applying dominant carrier regulatory safeguards to AT&T enabled the FCC to monitor changes in AT&T's circuit capacity that could indicate anticompetitive activity.

Over the past decade, competitive conditions have changed significantly. AT&T's competitors now hold operating agreements and international facilities for all major markets. They share ownership of all major international facilities with AT&T, and the new, state-of-the-art submarine cable facilities have reserve capacity available to all owners that exceeds AT&T's own capacity on the facilities. Moreover, there are three facilities-based networks for domestic long-distance services that compete with AT&T's network to link international facilities to U.S. customers. This domestic competition prevents AT&T from leveraging control over its domestic network to shut out competition on the international segment. In short, it is no longer plausible to view AT&T as controlling bottleneck facilities.

Changes in market share and consumer behavior also reflect significant shifts in the market structure. AT&T's share of the overall domestic market has declined to less than 60 percent, and it is now below 70 percent in all the top 50 international markets. Demand elasticity is substantial, as demonstrated by great volatility in household choice of long-distance and international carriers in response to specialized pricing and marketing plans. These developments collectively reveal a market in which the FCC believed AT&T cannot unilaterally exercise market power.

In 1995 the FCC found that AT&T was no longer a dominant carrier because it no longer possessed individual market power. Accordingly, AT&T was relieved of the regulatory burdens imposed by the FCC's dominance standard.

Summary. Continuing to apply the dominant carrier paradigm to AT&T does little to bolster competition because AT&T does not control bottleneck facilities (including operating agreements), and faces substantial rivalry by well-established competitors. In fact, aspects of dominant carrier regulation may hinder competition under current market conditions if applied to a carrier that no longer has market power. In particular, the longer tariff-filing notice periods applicable to AT&T as a dominant carrier subject to price cap regulation may have potential anticompetitive consequences once AT&T is no longer dominant. The longer notice periods for AT&T, still by far the largest carrier in this relatively concentrated market, can serve as a price signaling device that facilitates other carriers' ability to price just on par or slightly below the AT&T levels. It also means that AT&T cannot react as quickly and certainly as its competitors in making bids to business customers.

In essence, the disparity in notice periods slows rivalry in the market because the bidding for significant business customers is a major competitive stimulus in the market. Once AT&T's competitors have the facilities, operating agreements, and market credibility necessary to compete for large business customers, as they now do, then restricting the competitiveness of the largest carrier only reduces competitive performance in the market.

Downsizing

Downsizing refers to the process whereby companies ease the load of the central mainframe by distributing appropriate processing and information resources to local area networks, specifically the servers and microcomputers located throughout the organization.[1] This arrangement can yield optimal results to all users: It provides microcomputer users with ready access to the information they need in a friendly format, while permitting mission-critical applications and databases to remain on the mainframe where security and access privileges can best be applied.

1 The term *downsizing* also has been used to describe how organizations restructure themselves to become more competitive by trimming the number of employees, outsourcing specialized tasks, and spinning off operations unrelated to the core business.

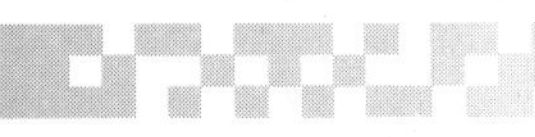

Benefits of downsizing. In moving applications and information closer to departments and individuals, knowledge workers at all levels in the organization are empowered, improving the quality and timeliness of decision-making and allowing a more effective corporate response to market dynamics. In addition:

- Downsized applications might run as fast or faster on microcomputers and workstations than on mainframes, at only a fraction of the cost.
- Even if downsizing does not improve response time, it can improve response-time consistency.
- In the process of rewriting applications to run in the downsized environment, there is the opportunity to realize greater functionality and efficiency from the software.
- With greater reliance on off-the-shelf applications, the number of programmers and analysts needed to support the organization is minimized.
- The lag time between applications development and business processes can be substantially shortened, enabling significant competitive advantage to be realized.
- The organization need not be locked into a particular vendor, as is typically the case with centralized host-based architectures.

The distributed environment. The distributed computing environment brought about by downsizing can take many forms.

Dedicated file servers. Downsizing has also been used to describe how organizations restructure themselves to become more competitive by trimming the number of employees, outsourcing specialized tasks, and spinning off operations unrelated to the core business.

Dedicated file servers can be used on the local area network to control access to application software and to prevent users from modifying or deleting certain types of files. Several file servers can be deployed throughout the network, each supporting a single application (e.g., electronic mail, facsimile, graphics, or specific types of databases). Metering tools can be included on the server to monitor usage and prevent unauthorized copying, which might violate various software licenses.

Minicomputers. When connected to a LAN, a minicomputer can act as a file server. The drawbacks of this solution are higher cost, amount of setup time needed, limited expansion possibilities, and the apparent need for continued administrative vigilance to keep things working properly.

Superservers. With so many organizations looking to downsize applications from mainframes to local area networks, the superserver concept is becoming a key strategic tool. Superservers are high-end microcomputers specifically equipped to act as network servers. These systems typically come with multiple high-speed processors and redundant subsystems, offer data storage in the gigabyte or terabyte range, and employ mainframe-like data restore and security techniques. Although a superserver may be used to support multiple applications, it is ideal for processing-intensive applications (e.g., CAD/CAM), relational database applications, and applications based on expert systems or neural networking.

Mainframes. The mainframe is still valued for its unequaled number-crunching capability. The mainframe can continue supporting traditional applications that require processing power and access to legacy applications, but also can act as a server in the distributed computing environment. In the case of IBM, for example, the addition of special software allows the mainframe to act as a server:

- *LAN Resource Extension and Services (LANCES):* A server-based software product that is used with NetWare LANs.
- *Data Facility Storage Management Subsystem (DFSMS):* A suite of software programs that automate storage management.
- *Network File Server (NFS):* Originally designed to operate on LANs, this software can now operate with MVS on the mainframe.
- *File Transfer Protocol (FTP):* The FTP server application can be used as part of the native TCP/IP stack under VTAM, or as a single third-party FTP server application also running as a VTAM application.

The mainframe also can play the role of master server in a hierarchical arrangement, backing up, restoring, and archiving data from multiple LAN servers.

Client-server. With the client-server approach, an application program is divided into two parts on the network. The client portion of the program, or *front end,* executes on the user's workstation, enabling such tasks as querying databases, producing printed reports, or entering new records. The server portion of the program, or *back end,* is resident on a computer (i.e., server) that is configured to support multiple clients. This setup offers users shared access to numerous application programs, as well as to printers, file storage, database management, communications, and other capabilities.

Cooperative processing. Cooperative processing is a variation of the client-server architecture in which one or more clients are used to off-load some of the work usually done by the central server. Among the most popular techniques for cooperative processing applications are Advanced Program-to-Program Communications (APPC) in the SNA environment and IP sockets in the Unix environment. Emulation is most often used in the Windows environment. The most basic cooperative processing approach converts a 3270 application to a more user-interactive application, making it mouse-based with colors, pull-down menus, and other Windows-like features. Other options include integrating different applications via front ends that merge the results from several unlike systems and display them on a single screen. The arrangement reduces overall processing costs.

There are tools available that bring together the power of APPC with the graphical user interface of Microsoft Windows. With such tools, Windows-based APPC transaction programs can be created which communicate with partner programs running on other computer systems. Such APPC conversations occur on a direct peer-to-peer basis in an efficient cooperative processing environment, eliminating the overhead associated with traditional terminal emulation communications.

Peer-to-peer data sharing. Perhaps the most economical method of distributed computing is peer-to-peer data sharing. Users publish parts of their hard disks to let others access them. Although this approach is economical in that it does not rely on a dedicated file server, it has several disadvantages that can outweigh cost savings. Unlike servers, this data sharing scheme does not have a central administration facility to enforce database integrity, perform backups, or provide security. In addition, the performance of each user's hard disk may degrade significantly when accessed by other users.

An example of a peer-to-peer technology is IBM's Advanced Peer-to-Peer Network (APPN), also known as LU6.2. As PCs obtained greater functionality via more powerful microprocessors, end users acquired greater data processing autonomy, creating the need for a decentralized networking strategy. Through LU6.2, IBM accomplished several objectives:

- More efficient handling of commands and information in a program-to-program context,
- Support for multiple sessions so that users could establish as many connections as needed,
- Connectivity with any application environment whose resources might be required, and
- Compatibility with a variety of hardware systems to circumvent the need for large software upgrades to accommodate added features.

IBM has defined a next-generation SNA, replacing its hierarchical network structure with a foundation on which users can build multiprotocol peer-to-peer enterprise networks. This is accomplished using its High Performance Routing (HPR) products for the mainframe and front-end processor (an extension to the existing APPN technology), and AnyNet, IBM's conversion technology that lets applications interact regardless of underlying protocols.

Summary. Determining the benefits of downsizing, the extent to which it will be implemented, who will do it, and the justification of its up-front costs are highly subjective activities. The answers will not appear in the form of cookie- cutter solutions from market-savvy vendors. In fact, the likelihood of success will be improved greatly if the downsizing effort is approached within the context of business process reengineering.

See also
BUSINESS PROCESS REENGINEERING
CLIENT-SERVER NETWORKS

Electronic commerce

The increasing popularity of the Internet has awakened companies to the numerous commercial opportunities that entail the sale of goods and services to a vast global marketplace. According to various industry estimates, about 15,000 merchants have already set up shop on the Web. When security ceases to be a problem and consumers are assured of risk-free transaction handling, the number of merchants could double every year for at least the next 5 years. There is everything to gain and little to lose: The cost of creating a virtual storefront is negligible—as low as US$2000 for a software package that includes customizable templates—and its market reach is global.

Payment systems. There are a number of tools available for creating and managing secure electronic payment systems. Such tools offer an Internet cash register and an electronic wallet that can be integrated into a Web browser. With these tools, retail merchants of any size can set up and manage virtual storefronts on the Internet with templates included in the software. With the storefront in place, companies can then process credit card payments over the Internet. From Internet clients, the cash register takes payment information protected with 128-bit Secure Sockets Layer (SSL) encryption. It also enables retailers to add product and related searching, sales tax and audit reports, discounts, and coupons.

The electronic wallet application organizes a user's credit card numbers, shipping address, digital IDs, coupons, vouchers, and any other information needed for buying goods or services over the Internet. It also stores receipts and provides a transaction ledger to track transactions and aid with future purchases.

Online banking. The Internet is suited not only for buying and selling, but for online banking as well. Via the Internet, bank customers can monitor account balances, transfer money from their savings to checking accounts, pay bills electronically, apply for car loans, and prequalify for mortgages. Hundreds of financial service organizations currently offer, or are implementing, remote access banking and/or bill payment services.

Among the growing number of virtual banks on the Internet is Atlanta Internet Bank, a service of Carolina First Bank. The FDIC-insured virtual bank offers a comprehensive portfolio of services that can be accessed online via its Web page.

Customers can choose any service a traditional bank has (Fig. E1). Clicking on the "Transfer Funds" icon, for example, allows the user to open a secure connection for transferring money from a money market fund to his or her checking account. Some virtual banks, such as Security First Network Bank (SFNB), even offer a check imaging feature that lets users view and print out a copy of each check that has been cleared in the last 90 days. Customers also can transfer funds back and forth between accounts through point-and-click commands.

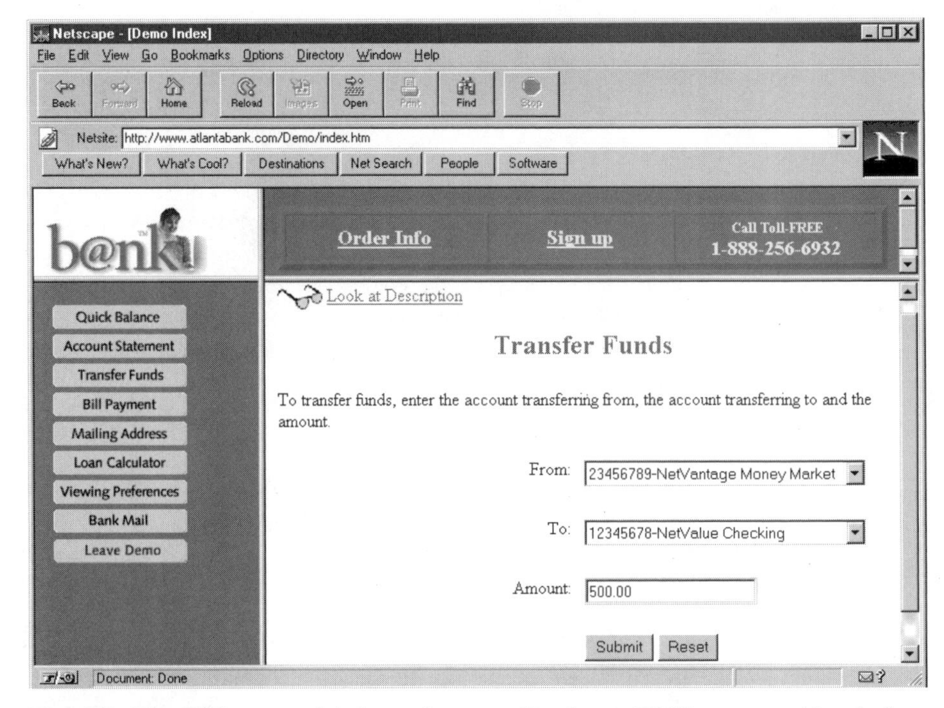

Fig. E1 *The Web page of Atlanta Internet Bank, an FDIC-protected bank that operates on the Internet.*

Regular Internet checking accounts have no monthly fees or minimum balance requirements. Customers get 20 free electronic payments each month, unlimited check-writing privileges, and a free ATM or debit card. SFNB is even insured by the FDIC, just like traditional banks. SFNB expects to offer customers online discount stock brokerage, insurance purchasing, and consumer loans. A single balance sheet will display all assets and liabilities, brokerage activity, and account information.

To extend these and other financial services over the Internet requires that retail companies, credit card issuers, and financial institutions implement secure transaction systems that guard against tampering by intruders and protect sensitive financial information from getting into the wrong hands.

Standards. Several major industry groups have developed methods that allow buyers and sellers to engage in secure electronic transactions over the Internet.

Secure Electronic Transaction (SET). SET is a technical specification for securing payment card transactions over open networks such as the Internet. SET will help the buyer and seller complete a transaction and have it authorized by a bank. The proposed Internet standard secures payment card transactions over the Internet by using a combination of data encryption, user authentication services, and digital certificates to safeguard a buyer's card number and ensure that the transaction data is immune from tampering.

The major advantage of SET over existing security systems is the addition of digital certificates that associate the cardholder and merchant with a financial institution and the Visa or MasterCard payment system. Digital certificates will prevent a level of fraud that the existing systems do not address. The certificates also will provide cardholders and merchants with a higher degree of confidence that the transaction will be processed in the same high-quality manner that conventional Visa and MasterCard transactions are being handled today.

Once the transaction is handed off to the payment card company's Web site, the SET protocol is shed and the data rides as it usually would across MasterCard, Visa, and other private transaction-processing networks.

Under SET, encryption is tamper-proof because of data partitioning: The credit card number is different than the payment information going across the Internet. Today's electronic funds transfer (EFT) systems are based on the same concept, where EDI information acts as a wrapper for the EFT payment message. Corporations have settled financial transactions this way for years; SET merely extends the technique to consumers.

The specification is open and free to anyone who wishes to use it to develop SET-compliant software. Electronic transaction systems based on SET are expected to start coming online in 1998.

Open Financial Exchange (OFX). Microsoft, financial software publisher Intuit, and electronic commerce service provider CheckFree have developed a single, unified technical specification that allows financial institutions to exchange financial data over the Internet with other customers. Open Financial Exchange (OFX) consolidates three proprietary standards: Microsoft's Open Financial Connectivity, Intuit's OpenExchange, and CheckFree's electronic banking and payment protocols. OFX is implemented using Internet security standards, specifically Secure Sockets Layer (SSL) and Private Communication Technology (PCT).

OFX lets financial institutions exchange financial data over the Internet with users of desktop and Web-based financial software by streamlining the process required for connecting to multiple customer interfaces and systems. By making it more compelling for financial institutions to implement online banking, OFX opens the door to online transaction services for a growing number of consumers, allowing them to manage finances online with the institution of their choice.

OFX supports a wide range of financial activities, including consumer and small-business banking; consumer and small-business bill payment; and investments, including stocks, bonds, and mutual funds. The companies plan to add other financial services that will include financial planning, insurance, and tax preparation and filing. Some 50 financial institutions, including Visa Interactive, Fidelity, Schwab, and Royal Bank of Scotland, have endorsed the OFX standard.

Joint Electronic Payments Initiative (JEPI). The Joint Electronic Payments Initiative (JEPI), which is sponsored by CommerceNet and the World Wide Web consortium (WWW3), provides a universal payment platform to allow merchants and consumers to transact business over the Internet using different forms of payment. JEPI will allow different payment instruments and protocols to exchange information. Since not all merchants accept all forms of payment and transport mechanisms, a payment negotiation protocol could provide a standard way through which applications can negotiate the appropriate payment methods.

JEPI allows clients and servers to negotiate payment instruments, protocol, and transport among them. JEPI consists of two parts: an extension layer that rides on the HyperText Transfer Protocol (HTTP), and a negotiations protocol that identifies the appropriate payment methodology. The protocols make payment negotiations automatic for end users, happening at the moment of purchase based on configurations within the browser.

The new protocol can be used to build a type of financial services middleware that will help merchants and buyers identify whether specific transactions will be handled by electronic mail or by file transfer, for example. It will also help users identify whether the transaction supports such protocols as the payment card industry's Secure Electronic Transaction specification.

Among the companies involved in JEPI are Microsoft, IBM, CyberCash, Xerox, British Telecom, and Digital Equipment Corporation. The companies are incorporating the payment negotiation protocol into their electronic commerce systems.

Summary. In a few years, most organizations will have an electronic commerce capability. By year-end 2000, at least 10 percent of European and North American retail trade will be attributable to online services. Visa estimates that by the turn of the century, annual purchases by credit card will amount to a US$1 trillion business for card issuers. Credit card companies anticipate that Internet-based buying will contribute a substantial portion of this amount, perhaps reaching the US$200 billion some research firms have predicted for the year 2000.

From the consumer's perspective, certain online transactions finally make economic sense. For example, on a transaction-by-transaction basis, online banking, actually can be cheaper than postage stamps. With point-and-click ease, users now can launch Quicken, Microsoft Money, or a proprietary banking software package and pay bills, open accounts, close accounts, transfer money between accounts, monitor stock portfolios, make adjustments to investments, and even create monthly ledgers for tax purposes. These online services offer automation, immediacy, and flexibility—key selling points, especially among harried urban consumers. And even the smallest businesses can now automatically execute monthly financial transactions and save money on bookkeeping—over the Internet through participating financial institutions.

See also
ELECTRONIC DATA INTERCHANGE

Electronic data interchange

Since the 1970s, electronic data interchange (EDI) has been promoted as the way to enhance business transactions with speed, accuracy, and cost savings. In essence, EDI is the electronic transfer of business documents between companies

in a structured, computer-processable data format. Because the information is transmitted over a value-added network (VAN) and does not have to be rekeyed at the other end, the speed and accuracy of business transactions is greatly improved. Since business documents are in electronic form, cost savings can range from US$3 to US$10 per transaction over manual, paper-based procedures.

The idea behind EDI is simple enough: Instead of processing a purchase order, for example, with multiple paper forms and mailing it to the supplier, the data are passed through an application link, where software maps the data into the standard machine-readable data format. The data are then transmitted to the supplier's computer, passing into an application link that maps it to the internal format expected by the supplier's order entry system.

While electronic mail also transfers business data electronically, it generally uses a free rather than a structured format. Because the sender might choose any format, it would be difficult to design an application program that would directly accept electronic mail input from many different sources without significant manual editing. This is the situation EDI standards are intended to overcome.

Benefits of EDI. EDI offers many other benefits to participants. Data flow within the organization is streamlined, making it easier to develop and maintain complete audit trails for all transactions. Having all of this information online also provides the means to track vendor performance, perform cost-benefit analyses, improve project management, and enhance overall corporate financial control.

A reduced order-to-pay cycle is the natural result of eliminating the use of regular mail and decreasing the time required to process paper at both ends of the transaction. This means that buyers can wait longer before replenishing stock ordered from suppliers, thereby reducing inventory and associated costs. The inherent efficiency of EDI means that buyers can pay for goods sooner, perhaps qualifying for further discounts. At the same time, sellers can improve their cash positions through timely payments from buyers.

With more accurate and timely data, planning and forecasting can be greatly improved. This means that companies can better plan the receipt of materials to coincide with assembly line schedules or continuous control processing operations, thus eliminating unnecessary downtime caused by material shortages.

Finally, dramatic cost savings associated with daily business transactions can accrue to users of EDI because the manual tasks of sorting, matching, filing, reconciling, and mailing paper documents are eliminated.

For many small- to mid-sized companies, the pressure for an EDI implementation is increasingly coming from their larger customers, who indicate that EDI is the preferred method of transacting business. In some cases, EDI implementation becoming a prerequisite for continuation of the business relationship. Some larger companies are even subsidizing EDI implementation for their smaller suppliers, recovering the cost by realizing even greater efficiencies and economies in transaction processing.

The role of VANs. The main service a VAN offers is a reliable, secure clearinghouse for EDI transmissions. Its operation is similar to an electronic mail system. The user sends the VAN a single transmission containing several interchanges destined for various trading partners. The VAN provides assurances that the data were received intact and distributed to each recipient's electronic mailbox. In turn, the service provider sends the contents of each mailbox to its subscriber upon request. The VAN also provides security to guarantee that no one but the addressee can access the data. The VAN performs other services, including compliance checking and translation of data from one protocol standard to another. VANs also provide software installation and troubleshooting assistance to their subscribers.

The VAN supplies EDI translation software to its subscribers, which maps the data from an application (order entry) and translates it into the EDI format. This software runs on a PC equipped with a modem and communications software to provide dial-up access to the VAN.

Another approach, which represents a more thorough adoption of EDI, entails a company integrating several business applications into the EDI system and using its own communication facilities and services to carry the transaction messages to an EDI gateway that connects to a VAN (Fig. E2). Each application is reengineered so that the output is already in the prescribed EDI format, making further translations unnecessary.

Interactive vs. batch processing. The problem with traditional batch processing for EDI transactions is that as a company adds trading partners and increases the volume of messages, batch jobs begin to contend with one another for access to computer and communications resources. EDI messages, such as ship notices, may be delayed because the EDI translator is busy processing a large batch job, or because all available communications channels are being used. Aside from rescheduling jobs, these problems can be solved through the purchase of additional equipment or the leasing of more lines.

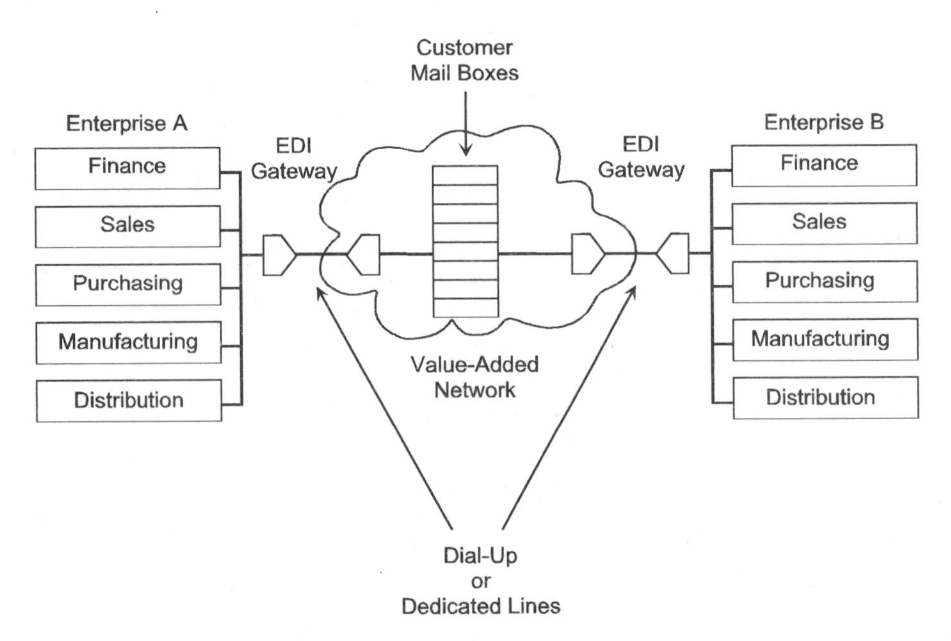

Fig. E2 *Typical network configuration for EDI.*

To improve this situation, companies are adopting fast-batch EDI and interactive EDI. *Fast-batch EDI* essentially speeds up the store-and-forward process by sending messages through electronic mailboxes on value-added networks directly into a recipient's computer system. *Interactive EDI* typically involves establishing a two-way link that allows trading partners to exchange records or fields rapidly within an EDI message, rather than the entire message.

Web-based EDI. The latest innovation in EDI entails use of the World Wide Web for sending business documents. This allows smaller firms that traditionally have not used EDI software and VANs to participate in the process at a very low entry cost. The common format for Web-based document exchanges is the HyperText Markup Language (HTML). This can work in several ways. In one scenario, a company's HTML-based forms are converted behind a VAN's Web site (i.e., gateway) into EDI-formatted messages before being passed on to the recipient. In another scenario, two trading partners might exchange documents over the Web directly, without going through a VAN. A back end program puts the information in the right format so it can be stored in a database. Security is maintained during transmission over the Web through the Secure HyperText Transfer Protocol (SHTTP) or Secure Sockets Layer (SSL).

Summary. In the future, businesses that have not implemented EDI (or will not) will be at a competitive disadvantage. Companies that derive considerable benefit from EDI will be reluctant to deal with those who force them to get bogged down in manually processing paper transactions. This goes against the trend toward flatter, decentralized organizational infrastructures, which many companies believe is necessary to compete more effectively.

See also

ELECTRONIC MAIL
VALUE-ADDED NETWORKS
WORLD WIDE WEB

Electronic mail

The popularity of electronic mail (e-mail) in recent years has paralleled that of the Internet. More mail is now delivered in electronic form over the Internet than is delivered by the U.S. Postal Service. That amounts to a few billion messages a month. Message delivery usually takes only minutes over the Internet, instead of days or weeks, as is typical with postal services in many countries.[1]

Advantages of e-mail. Automating information delivery and processing with electronic mail can dramatically reduce the cost of doing business because the manual tasks of sorting, matching, filing, reconciling, and mailing paper files are virtually eliminated. There also are attendant cost savings on overnight delivery services, supplies, file storage, and clerical personnel.

1 For more detailed information on e-mail over the Internet, including how to build mail-enabled forms for Web pages, see my book, *The Totally Wired Web Toolkit*, published by McGraw-Hill in 1997 (ISBN 0-07-044434-X).

Users can even send e-mail from within any application that supports Microsoft's Messaging Applications Programming Interface (MAPI) specification. Thus, a file done in a MAPI-compliant word processor or spreadsheet application, for example, can be e-mailed as an attachment without the user having to leave that application.

Through mail gateways, e-mail can be sent to people who may subscribe to some other type of service, such as CompuServe, America Online, or Microsoft Network (MSN). Some gateway services, such as RadioMail, can even transport messages from the Internet to wireless devices such as personal digital assistants (PDAs) and portable computers equipped with radio modems. And with more paging services supporting short-text messaging, e-mail even can be sent over the Internet to alphanumeric pagers.

The servers and gateways on the Internet take care of message routing and delivery. The e-mail address contains all the information necessary to route the message. If e-mail cannot be delivered because of some problem on the Internet, an error message is returned that not only explains the reason, but estimates the time of delivery.

Internet protocols. The dominant mail protocols used on the Internet currently are the Simple Mail Transfer Protocol (SMTP) and Post Office Protocol 3 (POP3). When installed on a server, SMTP gives it the ability to send and route messages over the Internet. POP3 is also installed on a server, giving it the ability to hold incoming e-mail until the recipient is ready to download it to his or her own computer. Once downloaded, the e-mail message can be opened, edited for reply, cut and pasted into another application, filed for future reference, or deleted (Fig. E3).

POP was designed to support offline message access. Messages must be downloaded before they can be read and then deleted from the mail server. This mode of access is not compatible with access from multiple computers because it tends to distribute messages across all of the computers used for mail access.

A new protocol is emerging, called the Internet Mail Access Protocol (IMAP4), which allows users to access messages on the server as if they were local. For example, e-mail stored on an IMAP server can be manipulated from a desktop computer at home, a workstation at the office, and a notebook computer while traveling, without the need to transfer messages or files back and forth between these computers.

IMAP's ability to access messages (both new and saved) from more than one computer has become extremely important as reliance on electronic messaging and use of multiple computers increase. Key goals for IMAP include:

- Full compatibility with other Internet messaging standards, such as MIME.[2]
- Message access and management from more than one computer.
- Message access without reliance on less efficient file access protocols.
- Support for online, offline, and disconnected access modes.
- Support for concurrent access to shared mailboxes.
- Client software needs no knowledge about the server's file storage format.

2 MIME stands for *Multipurpose Internet Mail Extensions*. It is a technique for encoding text, graphics, audio, and video files as attachments to SMTP-compatible Internet mail messages.

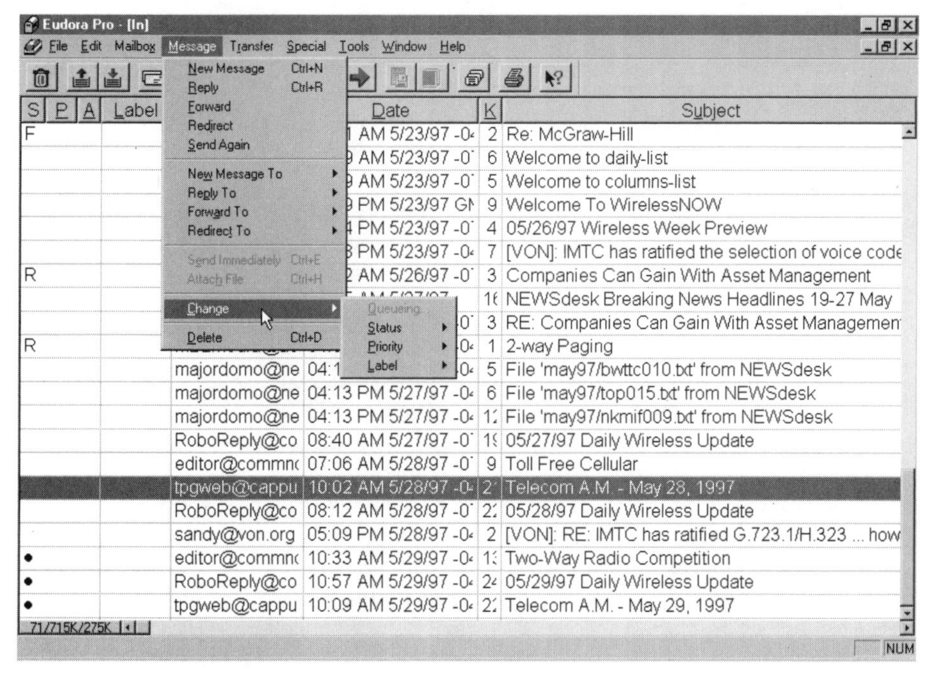

Fig. E3 *A typical e-mail interface, in this case QualComm's Eudora Pro.*

The IMAP protocol includes operations for creating, deleting, and renaming mailboxes; checking for new messages; permanently removing messages; setting and clearing flags; server-based MIME parsing (relieving clients of this burden) and searching; and selective fetching of message attributes and text.

Summary. Only 10 years ago, e-mail was considered a fad by most companies. Now there is a great appreciation for e-mail and its role in supporting daily business operations. In fact, the popularity of e-mail is now so great that many companies are having to rethink the capacity of their communications links and systems to accommodate the growing traffic load. Steps also are being taken to minimize unnecessary traffic, such as by storing only one copy of e-mail attachments on a server, rather than allowing the same attachment to be duplicated to all recipients of the message.

See also

FACSIMILE

PAGING

UNIVERSAL MESSAGING

Electronic software distribution

With a growing population of PC and workstation users deployed across widely dispersed geographical locations—each potentially using different combinations of operating systems, applications, data bases, and network protocols—software has become infinitely more complex and difficult to install, maintain, and meter. The ability to perform these tasks over a network from a central administration point

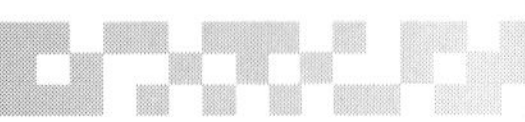

can leverage investments in software, enforce vendor license agreements, and greatly reduce network support costs.

Industry experts estimate that the average 5-year cost of managing a single desktop PC exceeds US$45,000, and that the 5-year cost of deployment and managing changes to new client-server applications will average nearly an additional US$45,000 to US$55,000 per user. These support costs can be cut by as much as 50 percent by automating the distribution and maintenance of software from a central administration facility. Electronic software distribution (ESD) tools also can provide useful reports that can aid in problem resolution.

Automating file distribution. The complexity of managing the distribution and implementation of software at the desktop requires that network administrators make use of automated file distribution tools. By assisting an network administrator with tasks like packaging applications, checking for dependencies, and offering links to event and fault management platforms, these tools reduce installation time, lower costs, and speed problem resolution.

One of these tools is a programmable file distribution agent. It is used to automate the process of distributing files to particular groups or workstations. A file distribution job can be defined as software installations and upgrades, startup file updates, or file deletions. Using a file distribution agent, these types of changes can be applied to each workstation or group automatically.

The agent can be set up to collect file distribution status information. The network administrator can view this information at the console to determine if files were distributed successfully. The console provides view formats and types that allow the administrator to review status data, such as which workstations are set up for file distributions, the stations to which files have been distributed, and the number of stations waiting for distributions.

Because users can be authorized to log in at one or more workstations, the file distribution agent determines where to distribute files based on the primary user (owner) of the workstation. The owner is established the first time a hardware/software inventory is taken of the workstation. Before automated file distributions are run, the hardware inventory agent is usually run to check for resource availability, including memory and hard disk space. Distributions are made only if the required resources are available to run the software.

Via scripts, the network administrator can define distribution criteria, including the group or station to receive files and the day or days on which the files are to be distributed. Scripts usually identify the files for distribution and the hardware requirements needed to run the file distribution job successfully. Many vendors provide templates to ease script creation. The templates are displayed as a preset list of common file tasks. Using the templates, the network administrator can outline a file distribution script, then use the outline to actually generate the script.

To help network administrators prepare for a major software distribution, some products offer routines called "wizards" that walk administrators through the steps required to assemble a "package," which is defined as a complete set of scripts, files, and recipients necessary to complete a distribution and install the software successfully. To reduce network traffic associated with software distributions, some products automatically compress packages before they are sent to another server or workstation. At the destination, the package is automatically decompressed when accessed.

When a file distribution job is about to run, target users receive a message indicating that files are about to be sent and requesting that they choose to either continue or cancel the job.

Managing installed software. Maintaining a software inventory allows the network administrator to determine quickly what operating systems and applications are installed on various servers and clients. In addition to knowing what software packages are installed and where they are located, the network administrator can track application usage to ensure compliance with vendor license agreements (Fig. E4).

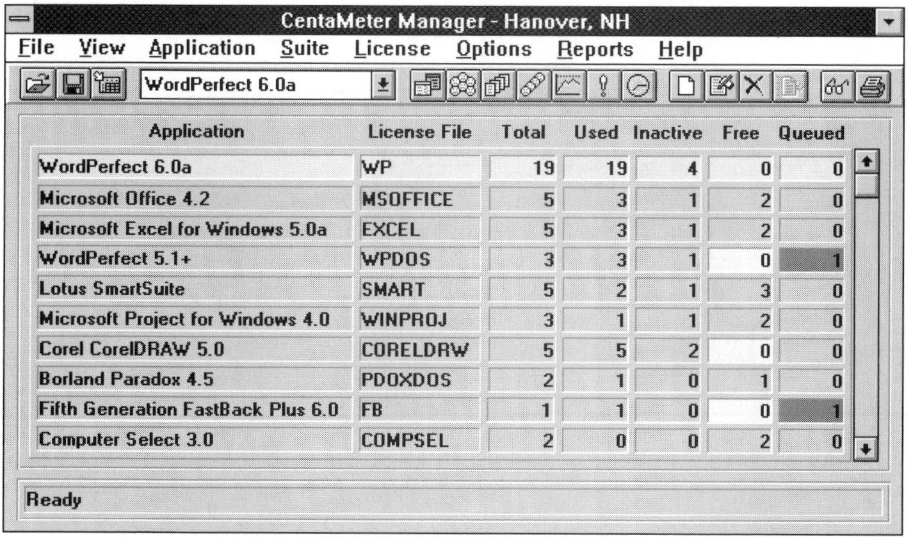

CentaMeter Manager - Hanover, NH

File View Application Suite License Options Reports Help

WordPerfect 6.0a

Application	License File	Total	Used	Inactive	Free	Queued
WordPerfect 6.0a	WP	19	19	4	0	0
Microsoft Office 4.2	MSOFFICE	5	3	1	2	0
Microsoft Excel for Windows 5.0a	EXCEL	5	3	1	2	0
WordPerfect 5.1+	WPDOS	3	3	1	0	1
Lotus SmartSuite	SMART	5	2	1	3	0
Microsoft Project for Windows 4.0	WINPROJ	3	1	1	2	0
Corel CorelDRAW 5.0	CORELDRW	5	5	2	0	0
Borland Paradox 4.5	PDOXDOS	2	1	0	1	0
Fifth Generation FastBack Plus 6.0	FB	1	1	0	0	1
Computer Select 3.0	COMPSEL	2	0	0	2	0

Ready

Fig. E4 *Software license management screen from Tally Systems's CentaMeter.*

The ESD tool creates and maintains a software inventory by scanning all the disk drives on the network. Usually software management tools come with preset lists of software packages they can identify during a scan of all the disk drives. Some tools can recognize several thousand software packages. Software that cannot be identified during a scan is tagged for further inquiry and manual data entry. The next time a scan is done, the added software packages will be properly identified.

Whether the initial software inventory is established by drive scanning or manual entry, the following information about the various software packages usually can be added or updated at any time:

- *Package information:* Number of available software licenses, product manufacturer, and project code to log application use to a particular project.
- *Software availability:* Information such as when the software can be accessed by users (open or closed). The software package can be closed dur-

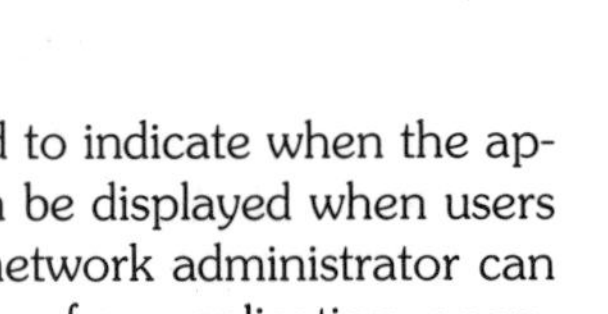

ing the upgrade process and a message displayed to indicate when the application will be available. A startup message can be displayed when users open particular applications. For example, the network administrator can post a message telling users that this is a beta copy of an application, or announce the date a new release of the software will be installed. Being able to relay such information to end users helps prevent unnecessary support calls to the help desk.

- *Files in package:* Executable files associated with the software and files for which to verify integrity, to indicate, for example, the presence of a computer virus or unauthorized file access.
- *Additional information about the application:* Items such as which departments in the organization have access to the application, or the vendor's technical support phone number.
- *Optional information:* Information such as whether the software is a Windows or OS/2 application, for example, and the directory to which it should be added.

In addition, access to applications can be restricted by user, workgroup, department, or project. Certain financial software, for example, can be restricted to accounting personnel. Personnel management software can be restricted to the human resources department. CAD/CAM software can be restricted to an engineering workgroup. Locking out unauthorized users prevents inadvertent data loss or malicious damage to files.

Metering software usage. The ability to track software usage helps the network administrator ensure that the organization complies with software license agreements, while making sure that users have access to required applications. Tracking software usage also helps reduce software acquisition costs; accurate usage information can be used to determine which applications are run most before deciding on upgrades and how many copies to buy.

Metering allows the network administrator to control the number of concurrent users of each application. The network administrator also can choose to be notified of the times when users are denied access to particular applications because all available copies are in use. This may identify the need to purchase additional copies of the software or pay an additional license charge to the vendor so more users can access the application.

Before users are granted access to metered software, the software inventory is checked to determine whether or not there are copies available. If no copies are available, a status message is issued, indicating that there are no copies available. The user waits in a queue until a copy becomes available.

Metering software can save money on software purchases and ensure compliance with software copyright laws. For example, if there are 100 users of Microsoft Word on an enterprise network and only half that number use it concurrently, the software metering tool's load balancing feature automatically handles the transfer of software licenses from one server to another on a temporary or permanent basis. Load balancing helps network administrators purchase licenses based on need rather than on the number of potential users at a given location.

License management capabilities are important to have because it is a felony under Federal law to copy and use (or sell) software. Companies found guilty of copyright infringements face financial penalties of up to US$100,000 per violation. The Software Publisher's Association (SPA) runs a toll-free hot line and receives about 40 calls a day from whistle-blowers. It sponsors an average of 250 lawsuits a year against companies suspected of software copyright violations. Since 1988, every case the SPA has been involved in has been settled successfully. Having a license management capability can help the network administrator track down illegal copies of software and eliminate a company's exposure to litigation and financial risk.

Some software metering packages allow application usage to be tracked by department, project, workgroup, and individual for charge-back purposes. Charges can be assigned on the basis of general network use, such as time spent logged on to the network or disk space consumed. Reports and graphs of user groups or department charges can be printed out or exported to other programs, such as an accounting application.

Although companies may not require departments or divisions to pay for application or network usage, charge-back capabilities can still be a valuable tool for breaking down operations costs and planning for budget increases.

Summary. ESD tools are essential for managing software assets. They can help trim support costs by permitting software to be distributed and installed from a central administration point; ensure compliance with vendor license agreements, thereby eliminating exposure to lawsuits for copyright infringement; and help companies manage software to minimize their investments while meeting the needs of all users.

See also

ASSET MANAGEMENT

Emergency Service 911

The number 911 is the designated universal emergency telephone number in North America. Dialing this number puts the caller in immediate contact with a Public Safety Answering Point operator who arranges for the dispatch of appropriate emergency service—ambulance, fire, police, rescue—based on the nature of the reported problem. Since its inception in 1968, this concept has amply demonstrated its value by saving countless lives in thousands of cities and towns across the United States and Canada.

Although 911 represents a major advance in the 7-digit, multiple-agency emergency service concept of years past, the systems that implemented the abbreviated dialing plan in the early years had serious limitations. The Public Safety Answering Point (PSAP) operators had to rely on their note-taking ability during crisis situations, which often involved near-hysterical callers who had to be calmed before a meaningful exchange of information could occur. This procedure wasted valuable time, risked omission of vital information, and left too much room for error.

If the caller hung up before providing location information, there was no way for the PSAP operator to know who called without involving the telephone company, whose call trace procedures might or might not have been successful in obtaining that information. Without the originating telephone number, the PSAP operator could not determine whether calls were really emergencies or just hoaxes. Even when the phone number is provided by the telephone company, without additional information from the caller, the PSAP operator often did not know what type of emergency service to dispatch to the location.

Advances in technology led to the introduction of "enhanced 911" (E911) service in the mid-1980s. The primary enhancement consisted of automating the entire sequence of events leading to the dispatch of appropriate emergency services. Automation eliminated human error and provided precise location information, which resulted in faster response times, less wasted effort on false alarms, and substantial cost savings on the delivery of emergency services. Automation also resulted in a dramatic increase in the number of lives saved.

System components and operation. While a number of sophisticated management tools are now available to increase the operating efficiency of E911 systems, there are only four standard capabilities (Fig. E5) required to provide E911 service: Automatic Number Identification (ANI), Automatic Location Information (ALI), selective routing, and call transfer.

The selective routing system located at the local central office identifies 911 calls and matches the caller's directory number with its assigned primary and secondary PSAPs. The ANI control system typically is located at the PSAP. The device answers incoming calls from the central office and then requests the ANI for decoding.

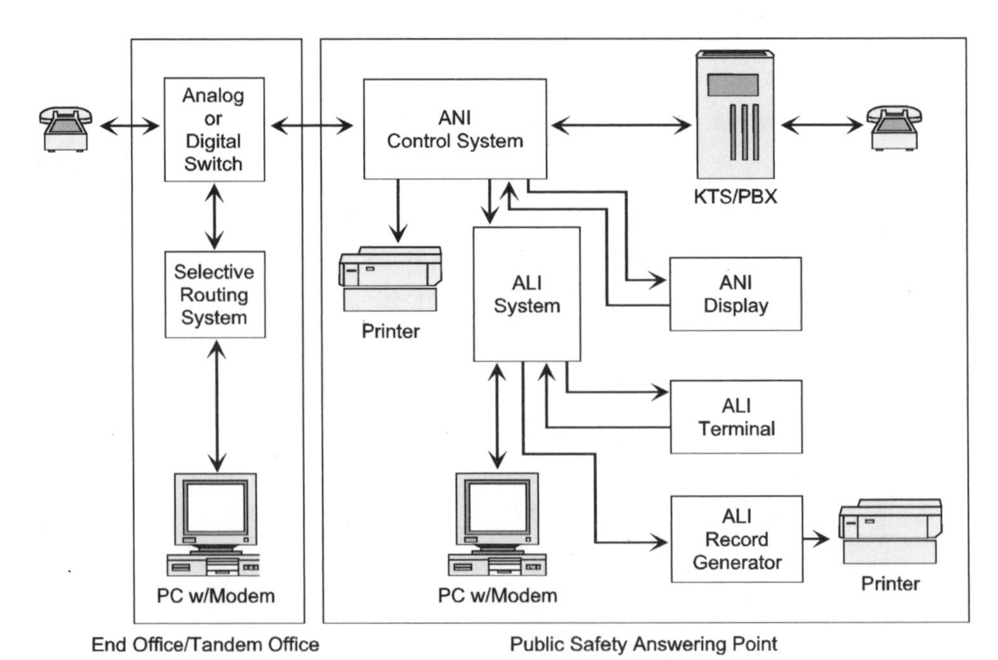

Fig. E5 *Schematic showing the main hardware components of an E911 system.*

Next, the ANI control system forwards call ringing to the PSAP's phone system and sends ringback to the caller via the central office. (The caller hears this as a distinct second ring; the first comes from central office ring circuits and the second comes from the PSAP's ANI control system.) The calling number goes to the ANI display at the PSAP attendant station. Metropolitan PSAPs may be equipped with Automatic Call Distributors (ACDs), which put calls in queue until an operator answers. After a timeout period, the ACD will divert the call to a recorded announcement or send it back to the tandem office for rerouting to the designated secondary PSAP. Some manufacturers offer ACD as an integral part of E911 systems.

At the same time that the calling number goes to the ANI display, it is forwarded to the ALI system, where it is cross-referenced with the Automatic Location Identification (ALI) database containing detailed location and identification information. This database includes a file for each directory number. Each file contains such standard information as street address, occupant name, and the nearest facility for police, fire, medical service, and poison control.

This database can be customized to include the unique attributes of commercial and residential structures, hazardous conditions (explosives, chemicals, and radioactive materials), and specific information about the occupants (disabled person, children, elderly, etc.). Expanded screen formats allow the input of even more detailed information, which can be accumulated on a continuous basis. Database maintenance is accomplished via a PC, with updates loaded to the main computer on demand, or automatically according to a predefined schedule. The data entry system includes several security features, including passwords, to prevent unauthorized access.

Some E911 systems also can call up maps that can zero in to any level of detail showing the nearest locations of such things as power lines, gas and water mains, hydrant positions, and buried cable. With this information, firefighters and rescue teams can arrive at the scene mentally prepared and appropriately equipped to deal with virtually any emergency.

The ALI system is located at the PSAP. In addition to interfacing with the ANI control system and matching ANI information with the appropriate database file on the caller, the ALI system automatically forwards location and identification information to the proper agency's attendant position terminal. These terminals may be located at the police or fire station, or at a centralized emergency response center.

An optional ALI record generator system interfaces with the telco's change order system to automatically assign the nearest PSAP, police and fire departments, emergency medical service, and poison control center to street addresses. When used with an ALI system, the record generator system automatically updates all onsite PSAP databases.

The selective routing system is a multifunction device that may be located at either a tandem or end office. As its name implies, it selectively routes 911 calls, by directory number, to primary/secondary PSAPs. If the primary PSAP is busy, or there is no response due to a failure in shared equipment, the selective routing system will reroute the call to a predefined secondary PSAP.

Most E911 systems include an automatic call-back capability, which is used to verify the legitimacy of incoming calls. Optional single-button call transfer allows PSAP operators to transfer calls on a discretionary basis to appropriate secondary

PSAPs or emergency service providers. And a call sharing feature allows two or more PSAPs or emergency services to share information, which is displayed simultaneously on operator terminals. Local police (sheriffs and state troopers, for example) can share information and immediately agree on appropriate jurisdiction before dispatching services and personnel.

Based on the processing options selected, the E911 system also can:

- Print a copy of the information displayed on the screen.
- Accept input from the PSAP operator regarding the type of response to the call and its outcome.
- Write a record to the daily call file, showing the incoming telephone number, date, and time for later report generation.

In addition, the system optionally can produce a variety of reports, such as an online log report, which shows all incoming calls and a variety of other information in chronological order. Also available are monthly and quarterly reports, showing the number of calls by operator, the number of calls by response, and other statistical information.

Summary. Emergency 911 service has become a valuable tool in rendering prompt and appropriate assistance to people in critical need. Most states have laws that mandate prompt action on all calls received by a PSAP operator. Unfortunately, the 911 system is so taken for granted that many calls are not for emergencies at all, and expensive resources end up being needlessly expended on trivial pursuits. PSAP operators now receive calls on such matters as garbage collection dates, late mail delivery, a leaky faucet or heater the landlord won't fix, directions to stores and restaurants, and whether or not to see a lawyer for this or that problem. The 911 systems in some communities are so bogged down with nonemergency calls that public awareness campaigns are launched periodically by telephone companies and emergency service providers with the aid of the local news media.

See also
AUTOMATIC NUMBER IDENTIFICATION

Ethernet

The Ethernet LAN, which operates at 10 Mbps, originated as a result of the experimental work done by Xerox Corporation at its Palo Alto Research Center (PARC) in the mid-1970s. Once developed, Ethernet quickly became a de facto standard with the backing of DEC and Intel. Xerox licensed Ethernet to other companies, who developed products based on the formal specification issued by Xerox, Intel, and DEC in 1980. Because Ethernet was an "open" system, it allowed interoperability among the equipment of different vendors. This prompted a number of independent vendors to embrace Ethernet as the industry standard for local area networks.

The "pure" Ethernet specification jointly developed by Xerox, DEC, and Intel defined the functions performed by the lowest levels of functionality: the Physical and Data Link layers of the Open Systems Interconnection (OSI) Reference Model. Essentially, Ethernet is a multiaccess, packet-switched network that uses a passive broadcast medium. Much of the original Ethernet design was incorporated into the 802.3 standard developed in 1980 by the Institute of Electrical and Electronic Engineers (IEEE).

Ethernet is based on a bus topology that is contention-based, meaning that stations vie with one another for access to the network, a process that is controlled by a statistical arbitration scheme. Each station "listens" to the network to determine if it is idle. Upon sensing that no traffic currently is on the line, the terminal is free to transmit. If the channel is already in use, the station backs off and tries again.

Frame format. The IEEE 802.3 standard defines a multifield frame format that differs only slightly from that of pure Ethernet (Fig. E6).

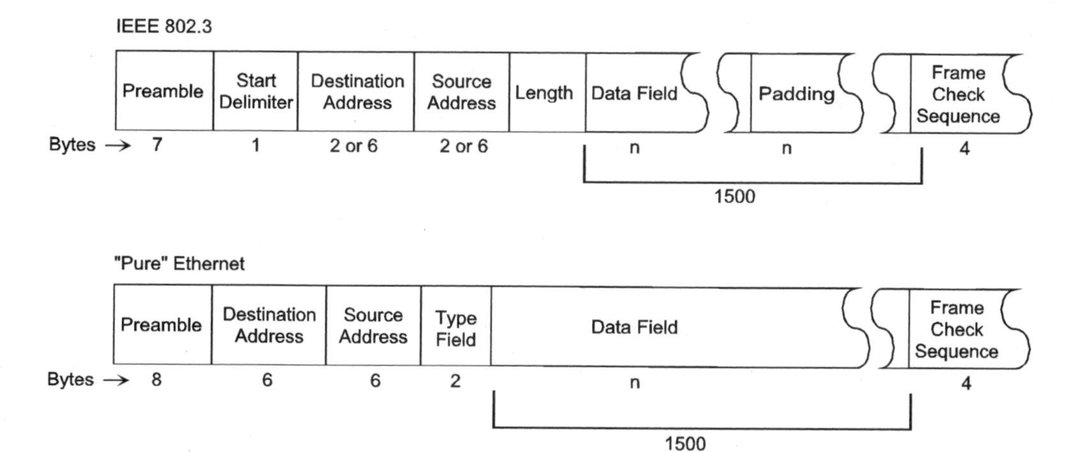

Fig. E6 *Comparison of Ethernet frame formats: IEEE 802.3 and "pure" Ethernet.*

Preamble. The frame begins with an 8-byte field called a *preamble,* which consists of 56 bits having alternating 1 and 0 values. These are used for synchronization and to mark the start of the frame. The same bit pattern is used for the pure Ethernet preamble as the IEEE 802.3 preamble, which includes the one-byte start frame delimiter field.

Start frame delimiter. The IEEE 802.3 standard specifies a start frame delimiter field, which is really a part of the preamble. This is used to indicate the start of a frame.

Address fields. The destination address field identifies the station(s) that are to receive the frame. The source address field identifies the station that sent the frame. If addresses are locally assigned, the address field can be either 2 bytes (16 bits) or 6 bytes (48 bits) in length. A destination address can refer to one station, a group of stations, or all stations. Ethernet specifies the use of 48-bit addresses, while IEEE 802.3 permits either 16- or 48-bit addresses.

Length count. The length of the data field that follows is indicated by the 2-byte length count field. This IEEE 802.3-specified field is used to determine the length of the information field when a pad field is included in the frame.

Pad field. To detect collisions properly, the frame that is transmitted must contain a certain number of bytes. The IEEE 802.3 standard specifies that if a frame being assembled for transmission does not meet this minimum length, a pad field must be added to bring it up to that length.

Type field. Pure Ethernet does not support length and pad fields, as does IEEE 802.3. Instead, two bytes are used for a type field. The value specified in the type field is only meaningful to the higher network layers and is not defined in the original Ethernet specification.

Data field. This portion of the frame is passed by the client layer to the Data Link layer in the form of octets (bytes). The minimum frame size is 72 bytes, while the maximum frame size is 1526 bytes, including the preamble. If the data to be sent uses a frame that is smaller than 72 bytes, the pad field is used to stuff the frame with extra bytes. In defining a minimum frame size, there are fewer problems to contend with in collision handling. If the data to be sent uses a frame that is larger than 1526 bytes, it is the responsibility of the higher layers to break it into individual packets in a procedure called *fragmentation*. The maximum frame size reflects practical considerations related to adapter card buffer sizes and the need to limit the length of time the medium is tied up in transmitting a single frame.

Frame check sequence. A properly formatted frame ends with a frame check sequence, which provides the means to check for errors. When the sending station assembles a frame, it performs a cyclical redundancy check (CRC) calculation on the bits in the frame. The sending station stores the result of this calculation in the 4-byte frame check sequence field before sending the frame. At the receiving station, an identical CRC calculation is performed and a comparison is made with the original value in the frame check sequence field. If the two values do not match, the receiving station assumes that a transmission error has occurred and requests that the frame be retransmitted. In pure Ethernet, there is no provision for error correction; if the two values do not match, notification that an error has occurred is simply passed to the client layer.

Media access control. Several key processes are involved in transmitting data across the network; among them are data encapsulation/decapsulation and media access management, which are performed by the MAC sublayer of OSI's Data Link layer.

Data encapsulation/decapsulation. Data encapsulation is performed at the sending station. This process entails adding information to the beginning and end of the data unit to be transmitted. The data unit is received by the MAC sublayer from the LLC sublayer. The added information is used to perform the following tasks:

- Synchronize the receiving station with the signal.
- Indicate the start and end of the frame.
- Identify the addresses of sending and receiving stations.
- Detect transmission errors.

The data encapsulation function is responsible for constructing a transmission frame in the proper format. The destination address, source address, type, and information fields are passed to the Data Link layer by the client layer in the form of a packet. Control information necessary for transmission is encapsulated into the offered packet. A CRC value for the frame check sequence field is calculated, and the frame is constructed.

When a frame is received, the data decapsulation function performed at the receiving station is responsible for recognizing the destination address, determining if it matches the station's address, performing error checking, and then removing the control information that was added by the data encapsulation function at the sending station. If no errors are detected, the frame is passed up to the LLC sublayer.

In the decapsulation process, frames are checked for errors. This can include a frame that is not a multiple of 8 bits or that exceeds the maximum packet length. It is also responsible for checking the address to see if the frame should be accepted and processed further. If it is, a CRC value is calculated and checked against the value in the frame check sequence field. If the values match, the destination address, source address, type, and data fields are passed to the client layer. What is passed to the station is the packet in its original form.

Media access management. The method used to control access to the transmission medium, known as *media access management* in IEEE terms, is called *link management* in Ethernet parlance. Link management is responsible for several functions, starting with collision avoidance and collision handling, which are defined by the IEEE 802.3 standard for contention networks.

Collision avoidance. This entails monitoring the line for the presence or absence of a signal (carrier). This is the "carrier sense" portion of CSMA/CD. The absence of a signal indicates that the channel is not being used and that it is safe to begin transmission. Detection of a signal indicates that the channel is already in use and that transmission must be withheld. If no collision is detected during the period of time known as the *collision window,* the station acquires the channel and can complete the transmission without risking a collision.

Collision handling. When two or more frames are offered for transmission at the same time, a collision occurs, which triggers the transmission of a sequence of bits called a *jam.* This is the means whereby all stations on the network recognize that a collision has occurred. At that point, all transmissions in progress are terminated. Retransmissions are attempted at calculated intervals. If there are repeated collisions, link management uses a process called *backing off,* which involves increasing the retransmission wait time following each successive collision.

On the receiving side, link management is responsible for recognizing and filtering out fragments of frames that resulted from a transmission that was interrupted by a collision. Any frame that is less than the minimum size is assumed to be a collision artifact and is not reported to the client layer as an error.

New methods have been developed to improve the performance of Ethernet by reducing or totally eliminating the chance for collisions without having to segment the LAN into smaller subnetworks. Special algorithms sense when frames are on a collision course and will temporarily block one frame while allowing the other to pass.

Summary. Ethernet is the most popular type of local area network. Its success has spawned some interesting innovations. 10Base-T Ethernet, for example, enables LAN operation over ubiquitous unshielded twisted-pair (UTP) wiring, instead

of thick or thin coaxial cable. There is also isochronous Ethernet, which brings ISDN to the desktop to support multimedia applications. For those who find 10 Mbps inadequate for supporting large file transfers and graphics-intensive applications, there are high-speed versions of Ethernet, including Fast Ethernet at 100 Mbps and gigabit Ethernet at 1 Gbps.

See also
- ETHERNET (10BASE-T)
- ETHERNET (100BASE-T)
- ETHERNET (GIGABIT)
- ETHERNET (ISOCHRONOUS)
- ETHERNET (100VG-ANYLAN)

Ethernet (10Base-T)

The 10Base-T specification is a relatively new IEEE standard for providing 10 Mbps Ethernet performance and functionality over ubiquitously available unshielded twisted-pair wiring. This standard is noteworthy in that it specifies a star topology, unlike traditional 10Base2 and 10Base5, which use coaxial cabling laid out in a bus or ring topology. The star topology permits central-site network monitoring, which enhances fault isolation and bandwidth management. For new installations, twisted-pair wire is substantially less expensive as well as easier to install and maintain than Ethernet's original thick coaxial cable (10Base5) or the thin coaxial alternative (10Base2).

Performance. Ethernet traditionally has relied upon coaxial cable with multidrop connections grouped onto a LAN segment. Ethernet can be extended by joining multiple cable segments together with repeaters, which regenerate the signals. The key disadvantage of this topology is that any disturbance to the continuity of the cable at any point along the bus will isolate the segments.

In contrast, the star topology inherent in twisted-pair hub systems dedicates a link to each user. Therefore, if an individual link were to exhibit an impairment, only the user on that link would be affected. This is precisely the benefit promised by 10Base-T: Because the network can tolerate a malfunction of any end-user device or its physical link, the rest of the network will not be affected, resulting in improved network availability. This is made possible by the link test and autopartition logic inherent in the port-level circuitry of a 10Base-T hub.

A network management system can make problems even easier to identify. When a predefined error threshold is exceeded, for example, the network management system will alert the LAN administrator, allowing a technician to be dispatched before the user even realizes that a problem exists. Although this is usually accomplished by a screen message or visual alarm indication, some systems can even notify the LAN administrator via pager.

Aside from the obvious difference in transmission media, traditional and 10Base-T Ethernet are distinguished by distance: 10Base-T operates reliably over cable segments not exceeding 100 m, whereas traditional Ethernet LANs using thick coaxial cabling operate reliably over cable segments of up to 500 m in length or 200 m for 10Base2 Ethernets using thin coax. The cable limitation of 10Base-

T aside, its bit error rate performance is at least as good as 10Base2 and 10Base5 systems. The 10Base-T specification allows for bit error rates of no more than 1 in 100 million bits. The data encoding scheme used for 10Base-T systems is the same as that used for coaxial-based Ethernets: self-clocking Manchester Encoding.

Applications. Ethernet 10Base-T LANs are designed to support the same applications as traditional Ethernet LANs. The relatively short distance over which 10Base-T operates is rarely a factor in its ability to support these applications. During the standardization process, a survey by AT&T revealed that over 99 percent of employee desktops are located within 100 m of a telephone wire closet where all the connections meet at a patch panel.

The media constraints for unshielded twisted-pair wiring are continually being shattered, as demonstrated by the recent standardization of Fast Ethernet, which provides a data rate of 100 Mbps, and efforts to standardize gigabit Ethernet. With Fast Ethernet supported by a number of hub and switch vendors, a whole new range of applications is opening up, including LAN-based videoconferencing, voice messaging, collaborative computer-aided design (CAD), and high-speed imaging.

System components. Because 10Base-T specifies a star configuration, installation entails the use of some equipment not always found on traditional Ethernets. The most obvious new ingredient in 10Base-T LANs is the type of medium used: twisted-pair wiring.

The media. Most installed telephone wiring is of the type known as 24 AWG (American Wire Gauge). Even if existing telephone wiring follows the required star configuration and is within the roughly 100-m (330-ft) distance limitation, it still might not be suitable to handle the 10 Mbps data rate. This is because no particular wire gauge or type is specified in the 10Base-T standard, although 24 AWG is what most equipment vendors have used in conformance tests to confirm that their products transmit reliably at up to 100 m.

More important than gauge, however, are the attenuation (signal loss), impedance (resistance), delay, and crosstalk characteristics of the wiring. Minimally acceptable levels for all of these are spelled out in the 10Base-T specifications. Generally the inside wiring installed in the last 20 years for telephone connections meets the 10Base-T specifications at cabling runs of up to 100 m. Older wiring may not meet the standards, in which case poor or erratic LAN performance could result and maximum transmission distances could be considerably less than 100 m.

Ethernet 10Base-T was designed to eliminate the requirement for shielded wiring. It relies on the twists in twisted-pair wire to hold down frequency loss, which in turn improves the integrity of the signal. The twists minimize the effects of this loss by preventing high-frequency signal energy from radiating to and corrupting the 10Base-T signal being carried over nearby twisted pairs (crosstalk). Even the individual wire pairs in 25-pair telephone cable are twisted. This cable is widely used in and around 10Base-T hubs; it allows as many as twelve 10Base-T segments to be neatly carried and patched using the common 50-pin connector.

Adapter cards. Adapter cards, also known as network interface cards (NICs), are boards that are inserted into a microcomputer or peripheral expansion slot. The adapter card connects the device's bus directly to the unshielded twisted-pair wiring of the 10Base-T LAN, typically eliminating the need for transceivers. Depending on the type of adapter, it permits connections to thick or thin coaxial cable, as well as to unshielded twisted pairs. An adapter with an Attachment Unit Interface (AUI) port will permit connection to a transceiver, allowing the card to be used with thick and thin coaxial cables and fiber optic cables.

10Base-T adapters vary considerably in their capabilities and features for reporting link status. Some adapters notify the user of a miswired connection to the LAN. Many have LEDs that indicate link status after the connection is wired and plugged in, such as link, collision, transmit, and receive. Others provide minimal information, such as if the link is correctly wired and working; still others light up only when something is wrong. Still other cards have no LEDs to display link information. Some cards feature a menu-driven diagnostic program to help the user isolate problems with the cards, such as finding out if the adapter is capable of responding to commands from the software.

Transceivers. Transceivers connect microcomputers and peripherals already equipped with Ethernet cards to 10Base-T wiring. Typically it consists of a small external box with an RJ-45 jack at one end for connecting to the unshielded twisted-pair wiring and an AUI port at the other end for connecting to the Ethernet adapter card.

Media access unit. MAUs terminate each end of the 10Base-T link. As such, it accommodates two wire pairs, one pair for transmitting the Ethernet signal and the other pair for receiving the signal. The 10Base-T standard describes seven basic functions performed by the MAU. The transmit, receive, collision detection, and loopback functions direct data transfer through the MAU. The jabber detect, signal quality error test, and link integrity functions define ancillary services provided by the MAU.

The jabber function removes equipment from the network whenever it continuously transmits for periods significantly longer than required for a maximum-length packet. The signal quality error test detects silent failures in the circuitry, while the link integrity signal detects breaks in the wire pairs. Both assist in fault isolation.

The hub. All 10Base-T stations are connected to the hub via two twisted pairs, one for transmitting and the other for receiving, over a point-to-point link. In essence, the hub acts as a multiport repeater. It contains the circuitry to retime and regenerate the signal received from any of the wire segments that connect at the hub to each of the other segments. A 10Base-T hub is more than a simple repeater, however; it serves as an active filter that rejects damaged packets.

The multiport repeater provides packet steering, fragment extension, and automatic partitioning. The packet steering function broadcasts copies of packets received at one repeater port to all of its other ports. Fragment extension ensures that partially filled packets are sent to their proper destination. Automatic partitioning isolates a faulty or misconnected 10Base-T link to prevent it from disrupting traffic on the rest of the network.

There are several types of 10Base-T hubs, chassis-based solutions and standalone, stackable 10Base-T hubs that can be cascaded with appropriate cable connections.

The chassis models are high-end devices that provide a variety of connections to the wide area network (WAN), support multiple LAN topologies and media, and offer sophisticated network management, with SNMP typically included as a functional subset. This type of 10Base-T hub is also the most expensive. Aside from enhancing their products with more interfaces and network management features to further differentiate them from competitive offerings, vendors are continually increasing the port capacity of these devices to bring down the price-per-port, making them more appealing to larger users.

Stand-alone, stackable hubs are used to link workgroups into a seamless departmental network. Configured in a cascading arrangement, these hubs are used to connect small clusters of users and provide connections to other clusters of users scattered throughout a department.

Management. The promise of 10Base-T networks was not only that Ethernet would run over economical unshielded twisted-pair wiring and offer unprecedented configuration flexibility, but that it would offer a superior approach to LAN management. Some of the most important management capabilities include:

- Support for the IEEE 802.3 Repeater Management standard and the Internet's Repeater Management Information Base (MIB).
- Remote site manageability from a central management station.
- Provision of performance statistics not only at the port level, where such information has been traditionally available, but at the module and hub level as well.
- Autopartitioning, which entails the ability to automatically remove disruptive ports from the network.
- The ability to set performance thresholds and notify managers of a problem, or automatically take action to address the problem.
- Port address association features that connect the Ethernet media access control address of a device with the port to which it is attached.
- Source/destination address information, which aids network redesign, traffic redistribution, troubleshooting, and security.

To broaden the management of multivendor 10Base-T networks, vendors have included support for SNMP in their hub management systems.

Summary. Overall, 10Base-T offers a sound technical solution for most applications that would normally make use of Ethernet. In the office environment, packet throughput and error rates over twisted-pair wiring are the same as with coaxial systems, but over shorter distances. The standard offers protection against equipment and media faults that can potentially disrupt the network, and the signaling method used is reasonably immune from sources of electromagnetic interference commonly found in the office environment.

See also

ETHERNET
ETHERNET (100BASE-T)

ETHERNET (GIGABIT)
ETHERNET (100VG-ANYLAN)
SIMPLE NETWORK MANAGEMENT PROTOCOL

Ethernet (100Base-T)

The increasing complexity of desktop computing applications is fueling the need for high-speed networks. Emerging data-intensive applications and technologies—such as multimedia, groupware, imaging, and a continuation of the explosive growth in the use of high-performance database software packages on PC platforms—all tax today's client-server environments and demand even greater bandwidth and improved client-server response times.

A number of high-speed LAN technologies are available to address the needs for more bandwidth and improved response times. Among them is Fast Ethernet, or 100Base-T, a technology designed to provide a smooth migration from current 10Base-T Ethernet, the dominant 10 Mbps network type in use today, to high-speed 100 Mbps performance.

Compatibility. Fast Ethernet uses the same contention-based media access control (MAC) method that is at the core of 10 Mbps Ethernet: Carrier Sense Multiple Access with Collision Detection, or CSMA/CD. The Fast Ethernet MAC specification simply reduces the *bit time,* that is, the time duration of each bit transmitted, by a factor of ten, allowing a 10× boost in speed. Fast Ethernet's scaled CSMA/CD MAC leaves the remainder of the 10Base-T MAC unchanged. The packet format, packet length, error control, and management information in 100Base-T are all identical to those in 10Base-T. All of the 100Base-T Fast Ethernet products as well as the 10Base-T products can be managed from a single SNMP management application.

Fast Ethernet allows the use of the same cabling technology already installed for 10Base-T networks. Implemented in the same star topology as a 10Base-T network, 100Base-T supports unshielded-twisted pair (UTP) and fiber-optic cabling, the media of choice for today's Ethernet networks.

Because no protocol translation is required, data can pass between 10Base-T and 100Base-T stations via a hub equipped with a 10/100 bridge module. Both technologies are capable of full-duplex operation. This compatibility allows existing LANs to be inexpensively upgraded to the higher speed as demand warrants.

Media choices. To ease the migration from 10Base-T to 100Base-T, Fast Ethernet can run over Category 3, 4, or 5 UTP cable, while preserving the critical 100-m segment length between hubs and end stations. The use of fiber allows more flexibility with regard to distance. For example, the maximum distance from a 100Base-T repeater to a fiber bridge, router, or switch using fiber-optic cable is 225 m (742 ft). The maximum fiber distance between bridges, routers, or switches is 450 m (1485 ft). The maximum fiber distance between a fiber bridge, router, or switch (when the network is configured for half-duplex) is 2 km (1.2 miles). By connecting together repeaters and internetworking devices, large, well-structured networks can be created easily with 100Base-T.

The types of media used to implement 100Base-T networks can be summarized as follows:

- *100Base-TX:* A two-pair system for data grade (EIA 568 Category 5) unshielded twisted-pair (UTP) and STP (shielded twisted-pair) cabling.
- *100Base-T4:* A four-pair system for both voice and data grade (Category 3, 4, or 5) UTP cabling.
- *100Base-FX:* A multimode two-strand fiber system.

Taken together, the 100Base-TX and 100Base-T4 media specifications cover all cable types currently in use in 10Base-T networks. Since 100Base-TX, 100Base-T4, and 100Base-FX systems can be mixed and interconnected through a hub, users can retain their existing cabling infrastructure while migrating to Fast Ethernet.

100Base-T also includes a media-independent interface (MII) specification that is similar to the 10 Mbps AUI. The MII provides a single interface that can support external transceivers for any of the 100Base-T media specifications.

Summary. Unlike other high-speed technologies, Ethernet has been installed for over 20 years in business, government, and educational networks. The migration to 100 Mbps Ethernet is made easier by the compatibility of 10Base-T and 100Base-T technologies, making it unnecessary to alter existing applications for transport at the higher speed. This compatibility allows 10Base-T and 100Base-T segments to be combined in both shared and switched architectures, allowing network administrators to apply the right amount of bandwidth easily, precisely, and cost-effectively. Fast Ethernet is managed with the same tools as 10Base-T networks, and no changes to current applications are required to run them over the higher-speed 100Base-T network.

See also

ETHERNET
ETHERNET (10BASE-T)
ETHERNET (GIGABIT)
ETHERNET (100VG-ANYLAN)

Ethernet (gigabit)

Ethernet is a highly scalable local area network technology, available in two versions: 10-Mbps Ethernet and 100-Mbps Fast Ethernet. A third version has been introduced that offers another order-of-magnitude increase in bandwidth. Offering a raw data rate of 1000 Mbps (1 Gbps), gigabit Ethernet maintains full compatibility with the huge installed base of Ethernet nodes through the use of LAN switches or routers. Because the frame format and size are the same for all Ethernet technologies, no other network changes are necessary.

Gigabit Ethernet supports new, full-duplex operating modes for switch-to-switch and switch-to-end-station connections and half-duplex operating modes for shared connections using repeaters and the CSMA/CD access method. Figure E7 illustrates the functional elements of gigabit Ethernet.

Current efforts in the IEEE 802.3z standards activity draw heavily on the use of Fibre Channel and other high-speed networking components. Encoding/decoding ICs and optical components for Fibre Channel are readily available and are specified and optimized for high performance at relatively low costs. Initial imple-

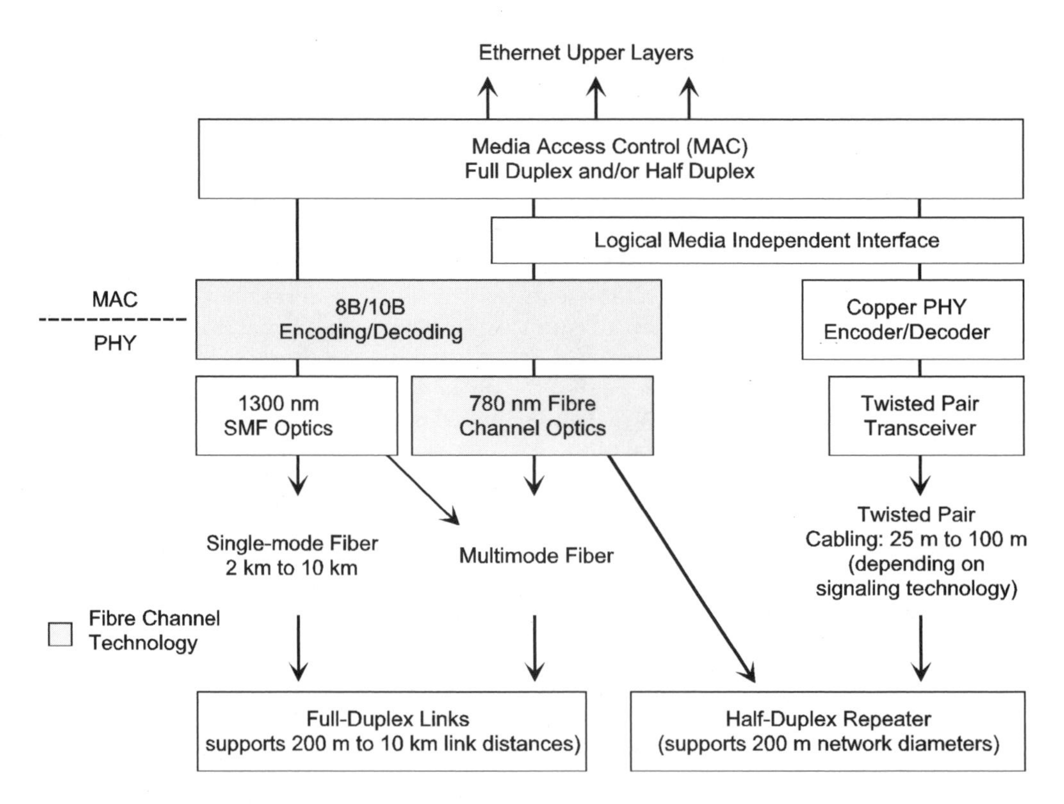

Fig. E7 *Functional elements of gigabit Ethernet.*

mentations of gigabit Ethernet will employ Fibre Channel's high-speed, 780-nm (short wavelength) optical components for signaling over optical fiber and 8B/10B encoding/decoding schemes for serialization and deserialization. Fibre Channel technology operating at 1.063 Gbps is being enhanced to run at 1.250 Gbps, thus providing the full 1000-Mbps data rate for gigabit Ethernet. For longer link distances, up to at least 2 km using single-mode fiber and up to at least 550 m on 62.5 μm multimode fiber, long-wavelength (1300-nm) optics also will be specified.

Beyond gigabit Ethernet's initial operation over optical fiber, the IEEE 802.3z standards committee is planning ahead for the expected advances in silicon technology and digital signal processing that will allow gigabit Ethernet to operate eventually over unshielded twisted-pair (UTP) cabling. To accommodate this, a logical interface will be specified between the MAC and Physical layers that will decouple Fibre Channel's 8B/10B encoding, thus allowing the use of other encoding schemes that more readily support cost-effective UTP cabling. In addition, there is investigation into mechanisms to support short link distances for use in computer room and wiring closet applications, as well as distances up to 100 m over Category 5 UTP cabling.

Table E1 summarizes the standardization efforts of gigabit Ethernet over various media.

Table E1. Standardizing Gigabit Ethernet.

Specification	Transmission Facility	Purpose
1000Base-LX	Long-wavelength laser transceivers	Support links of up to 550 m multi mode fier or 3000 m single-mode
1000Base-SX	Short-wavelength laser transceivers operating on multi-mode fiber	Support links of up to 300 m using 62.5 µm multimode or up to 550 m using 50 µm multimode
1000Base-CX	Shielded twisted-pair (STP) cable spanning no more than 25 m	Support links among devices located within a single room or equipment rack
1000Base-T	Unshielded twisted-pair (UTP) cable	Support links of up to 100 m using four-pair Category 5 UTP

Source: IEEE 802.3z Gigabit Task Force

Multimedia. In the past it was thought that video might require a different networking technology, one designed specifically for multimedia. But today it is possible to mix data and video over Ethernet through a combination of:

- Ethernet, enhanced by LAN switching.
- The emergence of new protocols, such as RSVP, that provide bandwidth reservation.
- The emergence of new standards such as 802.1Q or 802.1p, which will provide VLAN and explicit priority information for packets in the network.
- The widespread use of advanced video compression, such as MPEG2.

These technologies and protocols, summarized in Table E2, combine to make gigabit Ethernet a viable and economical solution for the delivery of video and multimedia traffic.

Network design. Network administrators today face a variety of internetworking choices and network design options. They are building intranets of increasing scale and combining routed and switched networks. Ethernet networks are shared (using repeaters) and switched, based on bandwidth and cost requirements. The choice of a high-speed network often restricts the choice of internetworking or network topology. Gigabit Ethernet overcomes these restrictions by its ability to work in switched, routed, and shared environments.

All of today's internetworking technologies, as well as emerging technologies such as IP-specific switching and layer 3 switching, are fully compatible with gigabit Ethernet, just as they are with Ethernet and Fast Ethernet. Gigabit Ethernet will be available in a shared, repeater-equipped hub—with the accompanying low cost per port—as well as on LAN switches and routers.

Summary. The seamless connectivity to the installed base of 10-Mbps and 100-Mbps equipment, combined with Ethernet's scalability and flexibility to handle new

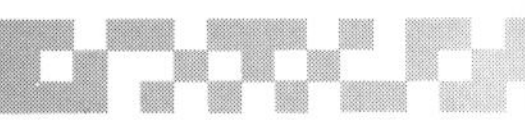

Table E2. Multimedia over Gigabit Ethernet.

Capabilities	Gigabit Ethernet	Fast Ethernet	ATM	FDDI
IP compatibility	Yes	Yes	Requires RFC 1557 and IP over LAN emulation today; I-PNNI and/or MPOA in future	Yes
Ethernet Packets	Yes	Yes	Requires LAN emulation or routing from cells to packets	No
Multimedia Support	Yes	Yes	Yes, but requires changes to applications	Yes
Quality of Service	Yes (via RSVP RSVP and 802.1 Q)	Yes (via RSVP RSVP and 802.1Q)	Yes (via SVCs)	Yes (via RSVP a 802.1Q)
VLANs with 802.1Q/p	Yes	Yes	Requires mapping SVCs to 802.1Q/p	Yes

Source: Gigabit Ethernet Alliance

applications and data types over a variety of media, makes gigabit Ethernet a practical choice for high-speed, high-bandwidth networking.

See also

ETHERNET
ETHERNET (10BASE-T)
ETHERNET (100BASE-T)
ETHERNET (ISOCHRONOUS)
ETHERNET (100VG-ANYLAN)
FIBRE CHANNEL

Ethernet (isochronous)

With billions of dollars already invested in Ethernet LANs, it is impractical to replace them or to install parallel networks to deliver high-bandwidth applications to individual desktops, especially multimedia applications that also might have to traverse the wide area network (WAN). Economical methods are available for improving and extending current LAN infrastructures so they can accommodate such applications. This is the idea behind a relatively new standard (IEEE 802.9a) called

ISLAN-16T, otherwise known as *isochronous Ethernet,* originally developed by National Semiconductor.

As its name implies, isochronous (literally, "same time") Ethernet simply adds the isochronous component, in the form of multirate ISDN, to standard Ethernet. The marriage of Ethernet and ISDN provides for the guaranteed quality of service (QoS) for voice in networked multimedia applications. In addition, isochronous Ethernet is interoperable with H.320, T.120, and MPEG standards for videoconferencing, document conferencing, and video distribution.

The premise behind isochronous Ethernet is that voice, not data or video, is the critical component in multimedia communications. During a conversation, even the slightest delay creates major disruption. Echoes caused by delays can confuse even the most articulate speakers. A split-second delay can disrupt a simple two-way conversation by causing one participant to ask a question again, only to interrupt the other's attempt to answer. Anyone who has experienced a telephone conversation via a round-trip satellite link is acquainted with the potentially disruptive effects of delay.

Guarding against this situation requires the use of isochronous protocols, in this case ISDN. A network is isochronous when it operates in real time, with *time* being defined by the worldwide standard 8 kHz clock for voice communications. Today's LANs are asynchronous and do not provide the 8-kHz clocking signal that the voice component of a multimedia application (delivered through an H.320 codec, for instance) requires. Because isochronous Ethernet inherently provides this synchronization, it offers the means to operate over the public switched telephone network in the WAN and deliver the same clocking to every desktop on the LAN. The IEEE 802.9a standard is the only LAN technology that extends 8-kHz clocking to the desktop. All others, including 25-Mbps ATM, require expensive buffering (to smooth out traffic bursts) and adaptation techniques to transmit and receive synchronous traffic such as H.320 voice and video. This makes real-time multimedia conversations possible with isoEthernet.

Isochronous Ethernet is a hybrid network that integrates standard 10 Mbps Ethernet (IEEE 802.3's 10Base-T) LAN technology with 6.144 Mbps of isochronous bandwidth, for a total of 16 Mbps available to any user, hence the IEEE designation of Integrated Service Local Area Network, or ISLAN 16-T (Fig. E8). Through an encoding scheme called 4B:5B, the total bandwidth is increased to 16 Mbps using the same 20-MHz clock that provides only 10 Mbps using the Manchester encoding scheme of traditional Ethernet.

Over the same ubiquitous Category 3 unshielded twisted-pair wire used for 10Base-T, isochronous Ethernet provides ninety-six 56/64 kbps ISDN B channels to bring packet plus wideband circuit-switched multimedia services to the desktop. This is at least 30 times more bandwidth than is needed to support a multipoint videoconference with six participants, each using 384 kbps, plus interactive chalkboarding and presentation graphics. In the background, the user can even receive a Group 4 facsimile, as well as e-mail and voice mail messages over separate channels.

Isochronous Ethernet can be integrated into an existing 10Base-T environment simply by putting an isochronous Ethernet hub into the wiring closet and con-

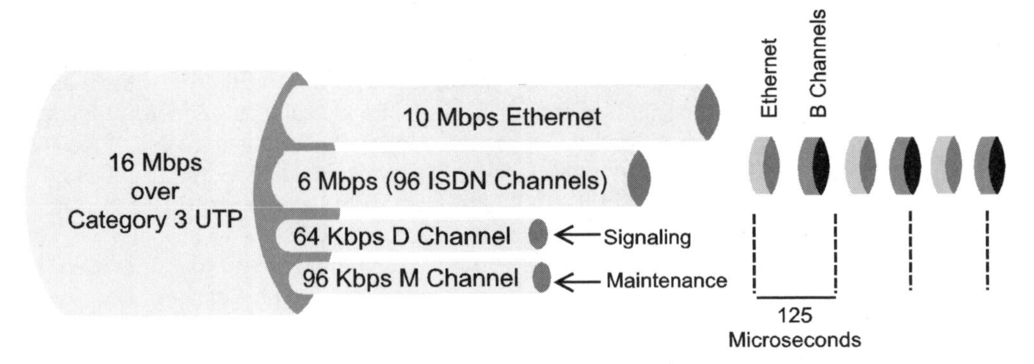

Fig. E8 *In essence, the IEEE 802.9a standard allows two networks to operate over the same 10Base-T wiring. Because Ethernet and ISDN transmissions do not share the same bandwidth, they do not affect one another's performance.*

necting the two hubs with an AUI. This linkage provide basic connectivity among all users.

Multimedia workstations, outfitted with isochronous Ethernet adapter cards, connect to the isochronous Ethernet hub. Because an isoEthernet hub and isoEthernet cards are needed only to connect a workgroup of multimedia users, existing investments in Ethernet 10Base-T infrastructure are protected, including bridges/routers, servers, applications, training, test equipment, and management tools. Furthermore, isoEthernet can be added incrementally on a user-by-user basis without changing the existing LAN applications, networking operating systems, or core network infrastructure.

If multimedia applications must traverse the WAN, ISDN links are used. The isoEthernet hub provides synchronization between WAN and LAN services. Because all the major local exchange carriers already support ISDN, they do not have to do any special circuit provisioning to accommodate multimedia applications. The same cannot be said of 25-Mbps ATM. With ATM, everything has to be changed and there is no interconnectivity between 25-Mbps ATM LANs and carrier-provided ATM services.

Summary. Isochronous Ethernet is one of the most attractive and affordable network solutions currently available for the delivery of multimedia. Although there are proprietary solutions from numerous vendors, isochronous Ethernet is the only solution that is capable of integrating LAN and WAN environments without carriers and companies having to make extensive changes to their networks.

See also

Asynchronous Transfer Mode
Ethernet
Integrated Services Digital Network

Ethernet (100VG-AnyLAN)

An alternative to 100Base-T Ethernet is 100VG-AnyLAN. Both operate at 100 Mbps. The 100VG-AnyLAN standard was developed by the IEEE 802.12 committee with the backing of AT&T and Hewlett-Packard. Like 100Base-T (the IEEE 802.3 Fast Ethernet standard), it can be deployed in both shared-media and switched implementations. Also like Fast Ethernet, 100VG supports current network topologies without significant reconfiguration, including 10-Mbps Ethernet, and 4- and 16-Mbps Token- Ring.

100VG-AnyLAN runs up to 100 m (node-to-node) over four-pair Category 3, 4, and 5 unshielded twisted-pair (UTP) wiring, 150 m over two-pair Category 5 shielded twisted-pair (STP) wiring, and up to 2000 m over fiber-optic cable (both single- and multimode).

Advantages. 100VG promises several advantages over 100Base-T. As its name implies, 100VG-AnyLAN can support both Ethernet and Token-Ring legacy applications. Each hub in a 100VG network can be configured to support either 802.3 10Base-T Ethernet frames or 802.5 Token-Ring frames. A single hub cannot support both frame formats at the same time, and all hubs on the network must be configured to use the same frame format. Because 100VG supports both Ethernet and Token-Ring, a router is used to move traffic between a 100VG network using Ethernet frames and one transporting Token-Ring frames. Going from either type of 100VG network and ATM (Asynchronous Transfer Mode), FDDI, or other topology also requires a router or a translating bridge (Fig. E9).

100VG eliminates packet collisions and permits more efficient use of network bandwidth. It does this by using a demand-priority access scheme instead of the traditional CSMA/CD scheme used in 10Base-T Ethernet and 100Base-T Fast Ethernet. Demand priority also permits rudimentary prioritization of time-sensitive traffic, such as real-time voice and video, making 100VG better suited for multimedia applications.

Media access control. 100VG-AnyLAN uses a unique media access control (MAC) layer in which a “collisionless” demand-priority access scheme (defined in the IEEE 802.12 standard) replaces pure Ethernet’s CSMA/CD scheme. Instead of collision detection or circulating tokens, demand priority uses a round-robin polling scheme implemented in a hub or switch. Round-robin polling makes it possible for every single-port node on a 100VG network to send one packet. If a multiport hub or switch serves as a node on the network, it must request access from the root hub at the next higher level. Multiport nodes can transmit one packet for each of their ports.

Using the demand-priority scheme, the 100VG hub or switch permits only one node to access the network segment at a time, thus eliminating the opportunity for packet collisions to occur. When a node needs to send data over the network, it places a request with the hub or switch. The hub or switch then services each node on the segment in sequence. If the node has data to send, the hub or switch per-

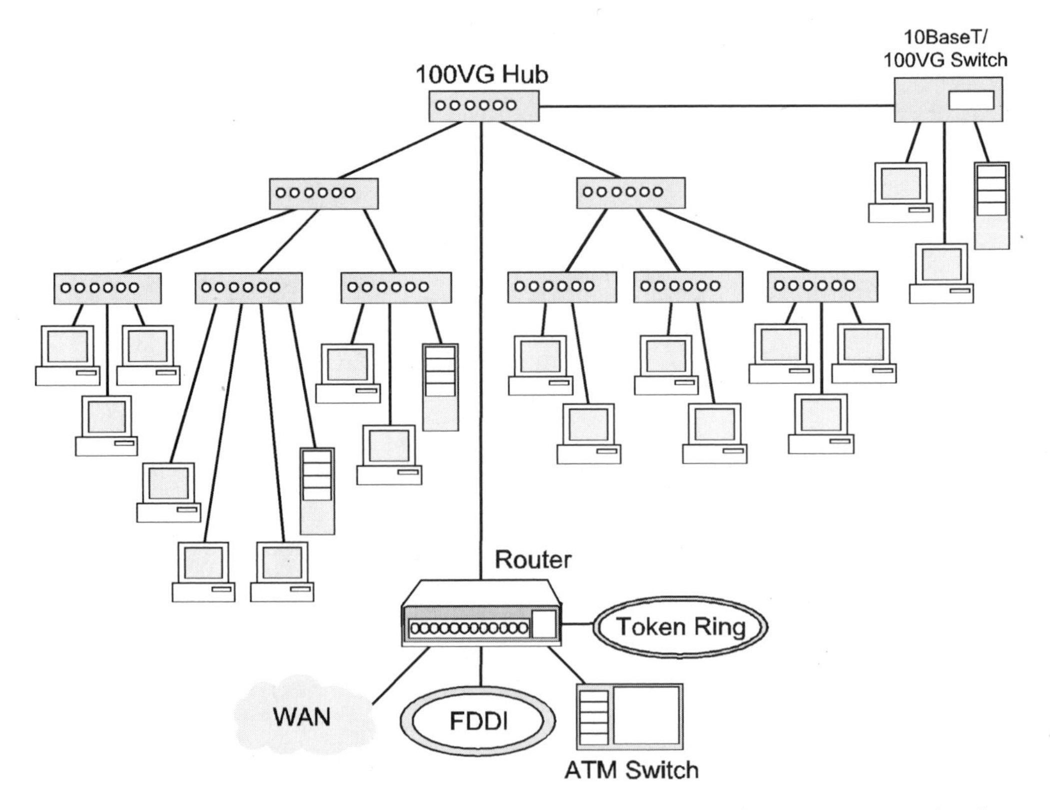

Fig. E9 *100VG-AnyLAN configuration supporting various other transport technologies.*

mits access to the segment. If it does not have anything to send, the hub moves on to the next node. Because all nodes requesting access are served during each polling round, all nodes are assured fair access to the network.

In 100VG networks that use Token-Ring frames, demand priority follows the same procedure, with the 100VG hub or switch essentially acting as the circulating token. Instead of waiting to catch the token before transmitting, a Token-Ring node waits for permission from the hub or switch. As in conventional Token-Ring, only one node is permitted to transmit across a network segment at any given time. Multiport 100VG nodes adhere to the same procedures when using Token-Ring packets as they do in Ethernet. When granted access to the root hub in a 100VG Token-Ring, the multiport 100VG hub or switch may transmit one packet per port.

Demand priority can overcome some inherent limitations of CSMA/CD, such as restricted network size. The more devices the network supports, the more collisions and contention for bandwidth. Although adding users can adversely affect network performance with 100VG, access to the network is always fair. The orderly operating procedure of demand priority makes it easier to add users to the network. Also, its ability to distinguish between high-priority and normal traffic ensures that bursty data applications in high-volume networks will not overwhelm time-sensitive multimedia traffic.

Multimedia. The most important requirement of multimedia transmission is that the receiver be able to receive a very long stream of information coherently. Ultimately, the application determines the degree to which this occurs. Because demand priority is a deterministic protocol, it allows multimedia application developers to model a worst-case scenario and use that information to determine when compression and timing routines should be invoked. By contrast, CSMA/CD is a first-come, first-served access method that offers no guarantee for data delivery and can result in poor performance.

With 100VG-AnyLAN's demand-priority access scheme, multimedia applications can request high-priority service from the 100VG hub. Ultimately, the combination of 100 Mbps, deterministic operation, and prioritization enhances the performance of multimedia applications.

Summary. At this writing, efforts are underway in the IEEE 802.12 working group to extend 100VG-AnyLAN to a 1-Gbps version that would compete with the new standard for gigabit Ethernet. The most likely use for the new technology will be to run multimedia over fiber-optic cable, interconnecting multiple buildings within a campus.

See also

ETHERNET
ETHERNET (10BASE-T)
ETHERNET (100BASE-T)
ETHERNET (GIGABIT)
ETHERNET (ISOCHRONOUS)

Facsimile

Facsimile transmissions have become an indispensable part of everyday business life. In the United States, faxing accounts for 30 percent of corporate telecommunications bills. Fortune 500 companies alone spend an average of US$15 million annually on fax-related transmission charges. In other countries, faxing accounts for 40 to 65 percent of the bill due to the higher cost of telecommunications.

The reason for the high cost of faxing is that each document must be scanned into an image before it can be sent electronically. Because charges are based on the duration and distance of the call, the cost of faxing adds up quickly. Even with compression, it takes a significant amount of time to transfer a document through the network, depending on the speed of the modem/fax and the quality of the connection. Before transmission even begins, the devices at each end must negotiate a common transmission rate, which also takes time.

Various industry estimates peg the number of fax machines worldwide at about 100 million at year-end 1996. The devices themselves come in a variety of configurations: stand-alone units, combination modem/fax cards for personal computers and workstations, and servers for network-based faxing. Network fax servers offer several benefits, including one-time installation, centralized management capabilities, and sophisticated routing features. They also can be more economical than equipping every office PC with its own modem/fax card.

Fax servers. Fax servers consist of hardware and software components. The hardware is a server equipped with multiple modem/fax boards, communications interfaces, an appropriate amount of memory and hard disk storage, and a battery backup subsystem. The software handles the formatting and conversion of faxes as well as the sending and receiving of faxes. The software also offers various inbound/outbound routing features.

There are several automated inbound routing techniques in use today. With direct inward dialing (DID), for example, each user on the network is assigned a personal fax number. Faxes are sent over a dedicated trunk line, received by the fax server, and then routed according to the fax number used. Users are notified about an incoming fax in the same way they are notified about e-mail, and they can then retrieve their fax from the server.

With dual-tone multifrequency dialing (DTMF), the sender dials the recipient's special extension number after the fax connection is made. The receiving machine

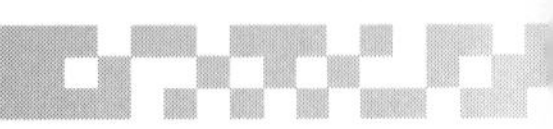

then identifies the recipient by the tones and routes the fax accordingly. Tones sometimes signal the sender to enter the extension number, or an automated voice prompt may request the extension number.

With channel- or line-based routing, a separate fax line is assigned to each recipient or department within a company. These lines are connected to the fax server, which receives all incoming faxes and routes them to the appropriate person or group.

Source ID routing routes incoming faxes according to where they originated. A list of fax numbers is maintained at the fax server so incoming faxes can be forwarded to the recipient who has been designated to get all faxes from a particular fax number.

Fax servers equipped with optical character recognition (OCR) can read the recipient's name or identification number on the fax cover sheet and then route the fax accordingly. This technology only works if the information is typed on the cover sheet; handwriting cannot yet be read reliably.

Once an incoming alert has been delivered a fax recipient can view the incoming fax onscreen, store it, print it, or forward it. Voice notification of incoming faxes is an emerging trend that allows the fax server to place a call to a telephone, pager, or e-mail address to alert a user to an incoming fax. This is a key feature of universal messaging.

Among the outbound routing features, fax servers offer broadcasting and delayed transmission capabilities. Broadcasting allows the same document to be faxed automatically to multiple recipients. Users can choose to send a fax to an entire group or only to certain individuals listed in the directory. Delayed transmission permits outgoing faxes to be collected for transmission after normal business hours when the telephone rates are lower. This capability is especially useful for fax transmissions to international locations.

Many fax servers also provide image processing capabilities. Logos, signatures, and other graphics often can be incorporated right into a fax document. Through the use of a print capture utility that redirects print output from applications to the fax server, fax documents can be transmitted directly from within a DOS or Windows application. In most cases, a cover sheet is automatically added to the fax file by the fax server software. The software also implements periodic cleanup procedures, which helps prevent the server's hard disk from running out of space.

Centralized management. Like any other network device, the fax server itself can be monitored and controlled via familiar SNMP commands entered at the management station. Via an SNMP-compliant management module, network line failures, internal errors, and fax port/board problems can be reported. The administrator can be warned through a series of notification methods that include a local beep, an SNMP trap sent to a remote management station, a hard copy report, an e-mail message, or a signal to a pager device.

In addition, the fax server usually comes with tools that allow administrators to obtain current and long-term status information regarding both inbound and outbound faxes. The types of data provided include what faxes were transmitted successfully, when they were sent, how many pages were sent, and if any retries were necessary. These logs help users track their individual faxes and allow administrators to keep track of overall usage. Some models offer an account billing module that tracks fax usage and generates charge-back and billing reports. The administration tools of some fax servers also permit manipulation of jobs in the queue, changes to priority settings, and cover page creation.

Summary. With the high cost of faxing, many users have turned instead to e-mail because it is virtually free over the Internet. Many documents are not in electronic form, however. In fact, they may contain annotations, signatures, drawings, or clippings from other sources that are important to retain in their original form, even if they can easily be put into electronic form for transmission as e-mail. Another consideration is that not everyone has an Internet connection to receive e-mail. All they might have is a stand-alone fax machine. Thus there continues to be heavy reliance on facsimile, not only among large companies, but telecommuters, home-based businesses, and individuals as well.

See also
ELECTRONIC MAIL
INTERNET FACSIMILE
UNIVERSAL MESSAGING

Federal Communications Commission

The Federal Communications Commission (FCC) is an independent federal agency in the United States, responsible directly to Congress. Established by the Communications Act of 1934, the Commission is charged with regulating interstate and international communications by radio, television, wire, satellite, and cable. Its jurisdiction covers the 50 states and territories, the District of Columbia, and the U.S. possessions overseas.

The FCC is directed by five commissioners appointed by the president and confirmed by the Senate for 5-year terms. The president designates one of the commissioners to serve as chairman, who presides over all FCC meetings. The commissioners hold regular open- and closed-agenda meetings and special meetings. By law, the commission must hold at least one open meeting per month. They also may act between meetings by "circulation," a procedure whereby a document is submitted to each commissioner individually for consideration and official action.

Certain other functions are delegated to staff units and bureaus, and to committees of commissioners. The chairman coordinates and organizes the work of the commission and represents the agency in legislative matters and in relations with other government departments and agencies.

Operating bureaus. At the staff level, the FCC is divided into operating bureaus and offices. Most issues considered by the Commission are developed by one of seven operating bureaus and offices (Fig. F1) organized by substantive area:

- The *Common Carrier Bureau* handles domestic wireline telephony.
- The *Mass Media Bureau* regulates television and radio broadcasts.
- The *Wireless Bureau* oversees wireless services such as private radio, cellular telephone, Personal Communications Service (PCS), and pagers.
- The *Cable Services Bureau* regulates cable television and related services.

- ➢ The *International Bureau* regulates international and satellite communications.
- ➢ The *Compliance and Information Bureau* investigates violations and answers questions.
- ➢ The *Office of Engineering and Technology* evaluates technologies and equipment.

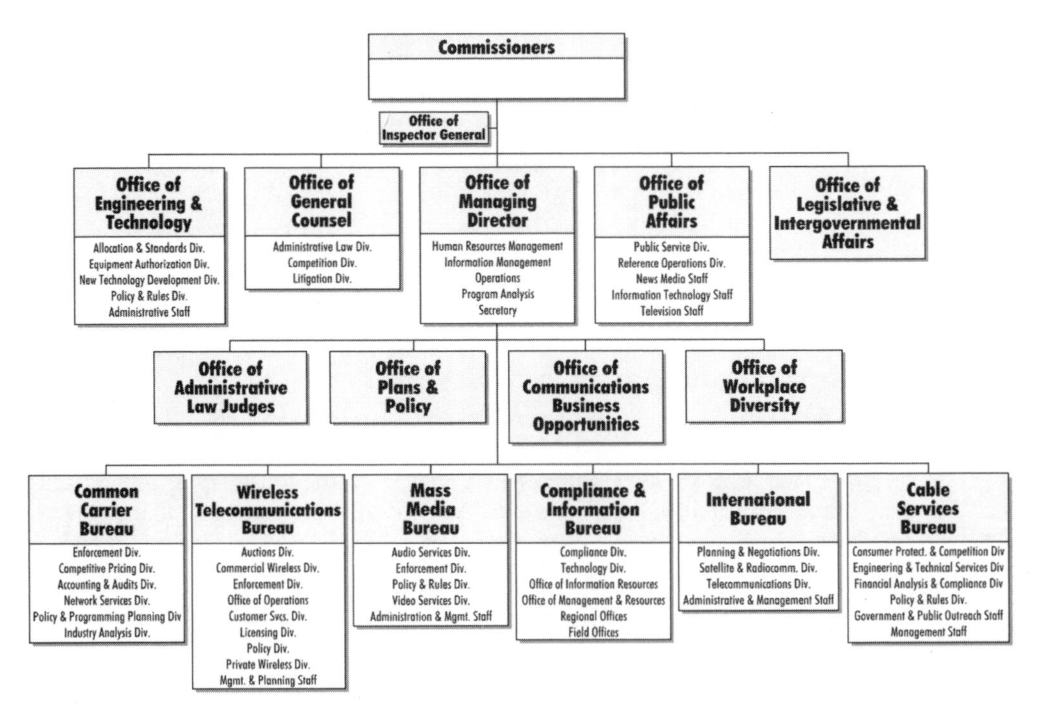

Fig. F1 *Organizational chart of the Federal Communications Commission.*

Other offices. In addition, the FCC includes the following other offices:

- ➢ The *Office of Plans and Policy* develops and analyzes policy proposals.
- ➢ The *Office of the General Counsel* reviews legal issues and defends FCC actions in court.
- ➢ The *Office of the Secretary* oversees the filing of documents in FCC proceedings.
- ➢ The *Office of Public Affairs* distributes information to the public and the media.
- ➢ The *Office of the Managing Director* manages the internal administration of the FCC.
- ➢ The *Office of Legislative and Intergovernmental Affairs* coordinates FCC activities with other branches of government.
- ➢ The *Office of the Inspector General* reviews FCC activities.

- The *Office of Communications Business Opportunities* provides assistance to small businesses in the communications industry.
- The *Office of Administrative Law Judges* adjudicates disputes.
- The *Office of Workplace Diversity* ensures equal employment opportunities within the FCC.

Among the FCC's major responsibilities for 1997 and beyond is implementing provisions of the Telecommunications Act of 1996, which is the first major overhaul of telecommunications law in almost 62 years. The goal of this new law is to let anyone enter any communications business and compete in any market.

See also

UNIVERSAL SERVICE

Federal Telecommunications System 2000

The Federal Telecommunications System 2000 (FTS2000) is an integrated communications network that is used by all agencies of the U.S. government. As such, it is the world's largest private network. AT&T, the primary service provider, completed its portion of the FTS2000 network in June, 1990. It serves 1.3 million users working for 92 government agencies in some 4200 locations scattered among the 50 states plus Alaska, Puerto Rico, Guam, and the Virgin Islands. MCI and Sprint also provide portions of the network.

In addition to flexible and customized management, control, administration, and billing, the FTS2000 Network provides the following services:

- Switched voice service
- Switched data service
- ISDN
- Packet-switched services
- Video transmission services
- Dedicated transmission service

All of these services support an extensive array of features, most of which are accessible by the individual users.

FTS2000 architecture. The FTS2000 network allows users to select any or all of a wide range of FTS2000 services through service delivery points (SDPs). An SDP is the combined physical and service interface between the network and the government's premises equipment, off-premises switching and transmission equipment, and other facilities. For Centrex service, for example, the SDP is the Local Exchange Carrier (LEC) central office. SDPs are used for billing purposes, and are also the monitoring points from which volume discounts are calculated based on monthly network usage.

The architecture of the FTS2000 Network (Fig. F2) consists of the following basic elements:

- Service nodes
- Network access
- Distributed network intelligence
- Transport

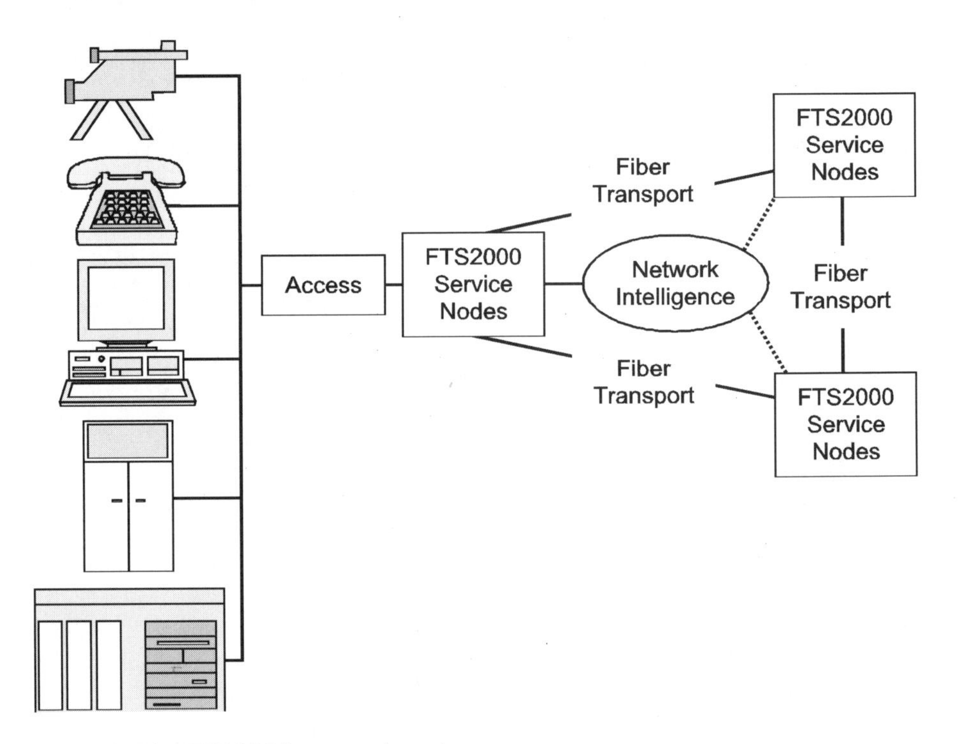

Fig. F2 *The FTS2000 network architecture.*

Service nodes. The service node contains components engineered to provide users easy and efficient access to each of the FTS2000 services. These components are as follows:

- Class 5 central office switch (5ESS)
- X.25 packet switch
- Digital Access and Crossconnect System (DACS)

To accommodate the circuit-switching requirements of both Switched Voice Service and Switched Data Service in a single system, AT&T uses its 5ESS switch in the service node. To meet X.25 packet switching needs, AT&T uses its 1PSS packet switch. Access to the 1PSS Switch is either direct or switched via the 5ESS Switch. Integrated access to the various switches and digital transmission channels is accomplished with AT&T's DACS in the service node. DACS is an intelligent facilities management system with channel drop-and-insert capabilities that increases the efficiency of digital circuit utilization and reduces facility administration and maintenance costs.

Network access. Network access provides connectivity between SDPs and the service nodes. The initial segment of access, from the government location to an FTS2000 access facility junction, is generally provided through the transmission facilities and serving wire centers of the regional holding companies (RHCs) or local exchange carriers (LECs). The access facility junction is the AT&T-provided in-

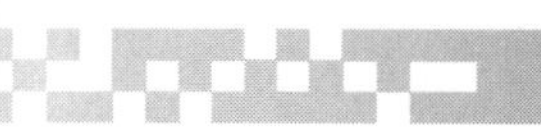

terface between AT&T and the Regional Bell Operating Company (RBOC) or LEC. This access facility junction provides standard functions required between a LEC and an interexchange carrier (IXC). As the prime contractor, AT&T is responsible for all access to the FTS2000 Network.

The connection to the FTS2000 Network from the SDP to the AT&T access facility junction is provided by the LECs using dedicated facilities to assure isolation of FTS2000 access lines from the public network, except in those cases where the small amount of voice traffic proves virtual on-net treatment to be more economical. The final segment of the access network is a dedicated, AT&T-provided circuit from the access facility junction to an FTS2000 node.

Off-net access to public network facilities is provided at all service nodes. Calls from an on-net location to an off-net location remain on the FTS2000 Network until those calls reach the service node nearest the call destination (referred to as "tail-end hop-off"). This maximizes the network utilization and high-performance transmission within the network.

Distributed intelligence. The service nodes are complemented by distributed intelligence consisting of network control points (NCPs), No. 2 signal transfer points (#2STPs), and signaling links. The NCPs share call-processing functions with the 5ESS switches, allowing the network to handle peak user demands. The NCP databases also store user information that provides customized services. The #2STPs are elements of the AT&T common channel signaling network and are also used to provide Signaling System 7 (SS7) signaling, which provides the network with ISDN capabilities.

Transport. Transport facilities are provisioned from existing AT&T fiber-optic facilities. FTS2000 is assigned exclusive use of selected 45 Mbps (DS3) segments of AT&T's nationwide fiber network.

Summary. Current FTS2000 contracts expire in 1998. The competition to provide the federal government's long-distance service began with the May, 1997, release of a request for proposals (RFP) from the U.S. General Services Administration's Federal Telecommunications Service 2001 (FTS2001) program. GSA plans to award multiple contracts for FTS2001, which are worth over US$5 billion over 4 years.

Fiber Distributed Data Interface

The Fiber Distributed Data Interface (FDDI) is a high-speed network that employs a counterrotating token-ring technology for fault tolerance. Originally conceived to operate over multimode fiber-optic cable, the standard has evolved to embrace single-mode fiber-optic cable, shielded twisted-pair copper, and even unshielded twisted-pair copper wiring. It is designed to provide high-bandwidth, general-purpose interconnection between computers and peripherals, including the interconnection of LANs (Fig. F3) and other networks, within a building or campus environment.

FDDI operation. A timed token-passing access protocol is used to pass frames of up to 4500 bytes in size, supporting up to 1000 connections over a maximum multimode fiber path of 200 km (124 miles) in length. Each station along the path serves as the means for attaching and identifying devices on the network, regenerating and repeating frames sent to it. Unlike other types of LANs, FDDI allows both asynchronous (time-insensitive) and synchronous (time-sensitive) devices to share

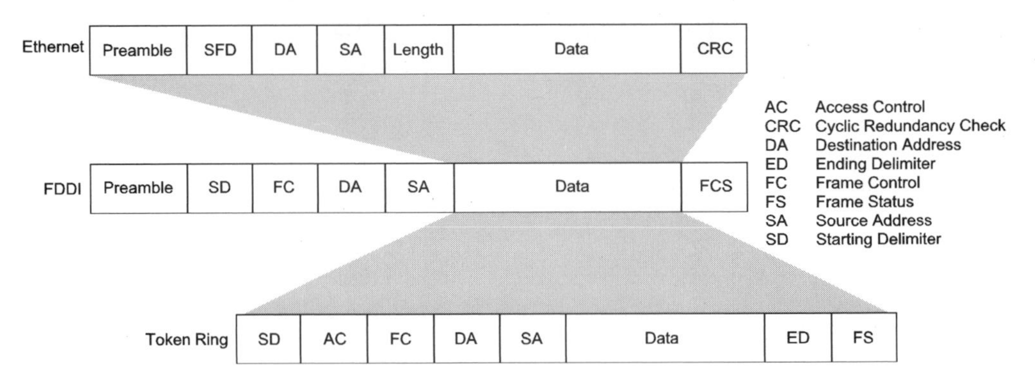

Fig. F3 *FDDI can carry Ethernet and Token-Ring frames as data, providing a multiprotocol backbone network.*

the network. Synchronous services (e.g., voice and video) are intolerant of delays and must be guaranteed a fixed bandwidth or time slot. Synchronous traffic is therefore given priority over asynchronous traffic, which can be delayed without degradation. FDDI stresses reliability and its architecture includes integral management capabilities, including automatic failure detection and network reconfiguration.

Any change in the network status, such as power-up or the addition of a new station, leads to a "claim" process during which all stations on the network bid for the right to initialize the network. Every station indicates how often it must see the token to support its synchronous service. The lowest bid represents the station that must see the token most frequently. That request is stored as the *Target Token Rotation Time* (TTRT). Every station is guaranteed to see the token within 2 × TTRT seconds of its last appearance.

This process is completed when a station receives its own claim token. The winning station issues the first unrestricted token, initializing the network on the first rotation. On the second rotation, synchronous devices may start transmitting. On the third and subsequent rotations, asynchronous devices may transmit if there is available bandwidth. Errors are corrected automatically via a beacon-and-recovery process during which the individual stations seek to correct the situation.

FDDI architecture. These processes are defined in a set of standards sanctioned by the American National Standards Institute (ANSI). The standards address four functional areas of the FDDI architecture (Fig. F4).

Physical Media Dependent. Data is transmitted between stations after converting the data bits into a series of optical pulses. The pulses are then transmitted over the cable linking the various stations. The PMD sublayer describes the optical transceivers, specifically the minimum optical power and sensitivity levels over the optical data link. This layer also defines the connectors and media characteristics for point-to-point communications between stations on the FDDI network. The PMD sublayer is a subset of the Physical layer of the OSI Reference Model, defining all of the services needed to transport a bit stream from station to station. It also specifies the cabling requirements for FDDI-compliant cable plant, including worst-case jitter and variations in cable plant attenuation.

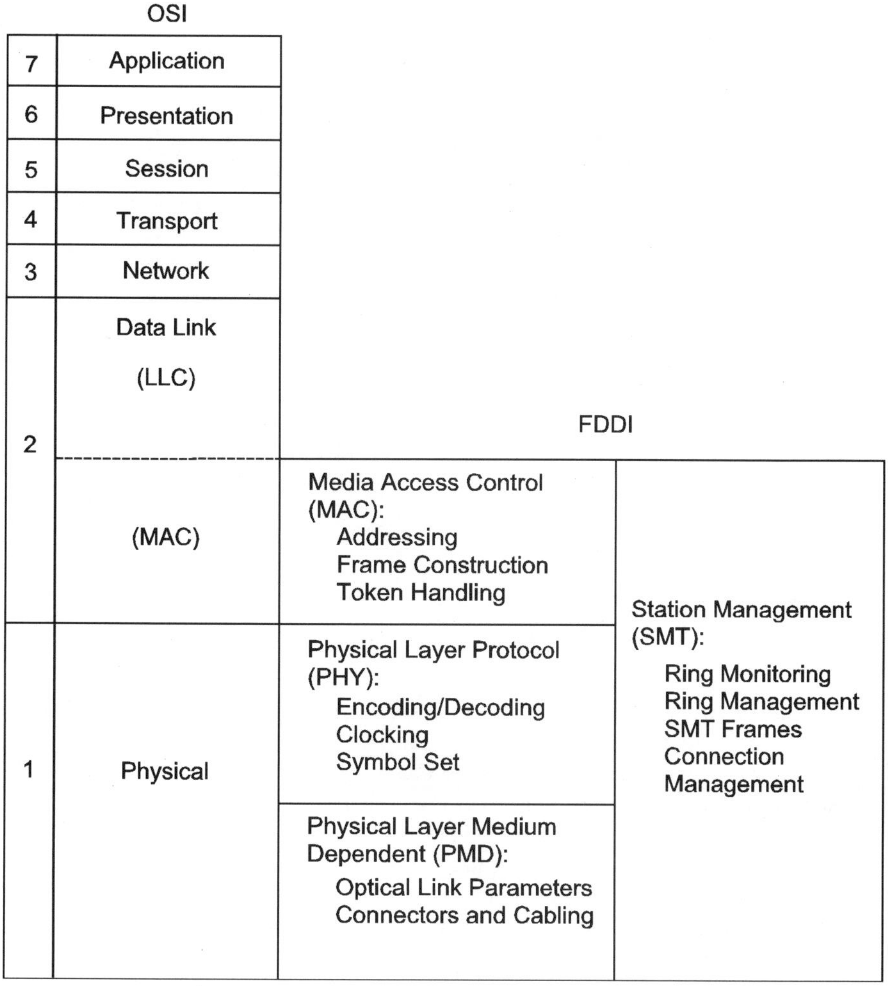

	OSI	FDDI	
7	Application		
6	Presentation		
5	Session		
4	Transport		
3	Network		
2	Data Link (LLC)		
	(MAC)	Media Access Control (MAC): Addressing, Frame Construction, Token Handling	Station Management (SMT): Ring Monitoring, Ring Management, SMT Frames, Connection Management
1	Physical	Physical Layer Protocol (PHY): Encoding/Decoding, Clocking, Symbol Set	
		Physical Layer Medium Dependent (PMD): Optical Link Parameters, Connectors and Cabling	

Fig. F4 *FDDI layers and their relationship to the 7-layer OSI Reference Model.*

Physical layer. The Physical Layer (PHY) protocol defines those portions of the physical layer that are media-independent, describing data encoding/decoding, establishing clock synchronization, and defining the handshaking sequence used between adjacent stations to test link integrity. It also provides the synchronization of incoming and outgoing code-bit clocks and delineates octet boundaries as required for the transmission of information to or from higher layers. These processes allow the receiving station to synchronize its clock to the transmitting station.

Media Access Control. FDDI's Data Link layer is divided into two sublayers. The Media Access Control (MAC) sublayer governs access to the medium. It describes the frame format, interprets frame content, generates and repeats frames, issues and captures tokens, controls timers, monitors the ring, and interfaces with station management.

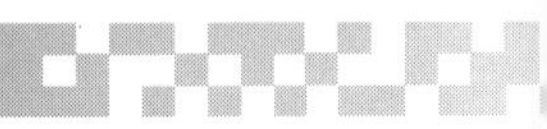

The Logical Link Control (LLC) sublayer, while not part of the FDDI standard, is required for proper ring operation and is part of the IEEE 802.2 standard. In keeping with the IEEE model, the FDDI MAC is fully compatible with the IEEE 802.2 Logical Link Control (LLC) standard. Applications that currently can interface to the LLC and operate over existing LANs, such as IEEE 802.3 CSMA/CD or 802.5 token ring, should be able to operate over FDDI networks.

The FDDI MAC, like the 802.5-defined token ring MAC, has two types of protocol data units, a frame and a token. *Frames* are used to carry data (such as LLC frames), while *tokens* are used to control a station's access to the network. At the MAC layer, data is transmitted in 4-bit blocks called *4B/5B symbols*. The symbol coding is such that 4 bits of data are converted to a 5-bit pattern; thus, the 100 Mbps FDDI rate is provided at 125 million signals per second on the medium. This signaling type is employed to maintain signal synchronization on the fiber. Two symbols carry a single octet of data.

Station Management. The Station Management (SMT) facility provides the system management services for the FDDI protocol suite, detailing control requirements for the proper operation and interoperability of stations on the FDDI ring. It acts in concert with the PMD, PHY, and MAC layers. The SMT facility is used to manage connections, configurations, and interfaces. It defines such services as ring and station initialization, fault isolation and recovery, and error control. SMT is also used for statistics gathering, address administration, and ring partitioning.

FDDI topology. FDDI is a token-passing ring network. Like all rings, it consists of a set of stations connected by point-to-point links to form a closed loop. Each station receives signals on its input side and regenerates them for transmission on the output side. Any number of stations (theoretically) can be attached to the network, although default values in the FDDI standard assume no more than 1000 physical attachments and a 200-km path.

FDDI uses two counterrotating rings, a primary ring and a secondary ring. Data traffic usually travels on the primary ring. The secondary ring operates in the opposite direction and is available for fault tolerance. If appropriately configured, stations may transmit simultaneously on both rings, thereby doubling the bandwidth of the network.

Three classes of equipment are used in the FDDI environment: single attached stations (SASs), dual attached stations (DASs), and concentrators (CONs).

A DAS physically connects to both rings, while a SAS connects only to the primary ring via a wiring concentrator. In the case of a link failure, the internal circuitry of a DAS can heal the network using a combination of the primary and secondary rings. If a link failure occurs between a concentrator and an SAS, the SAS becomes isolated from the network.

These equipment types may be arranged in any of three topologies: dual ring, tree, and dual ring of trees (Fig. F5). In the dual-ring topology, DASs form a physical loop, in which case all the stations are dually attached. In a tree topology, remote SASs are linked to a concentrator, which is connected to another concentrator on the main ring.

Any DAS connected to a concentrator performs as an SAS. Concentrators may be used to create a network hierarchy, which is known as a *dual ring of trees*. This topology offers a flexible, hierarchical system design that is efficient and eco-

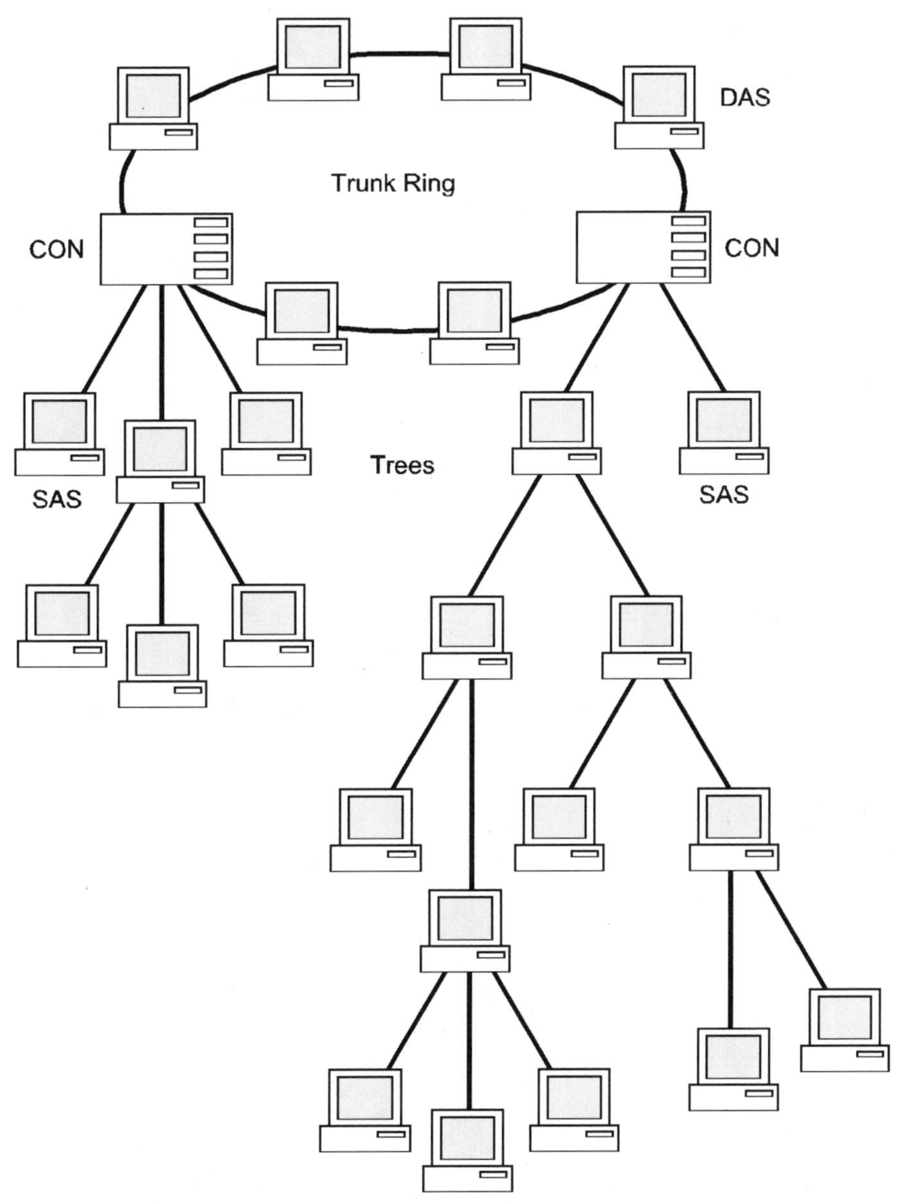

Fig. F5 *FDDI dual-ring topology with three types of interconnecting devices.*

nomical. Devices requiring highly reliable communications attach to the main ring, while those less critical attach to branches off the main ring. Thus, SAS devices can communicate with the main ring, but without the added cost of equipping them with a dual-ring interface or a loop-around capability that would otherwise be required to ensure the reliability of the ring in the event of a station failure.

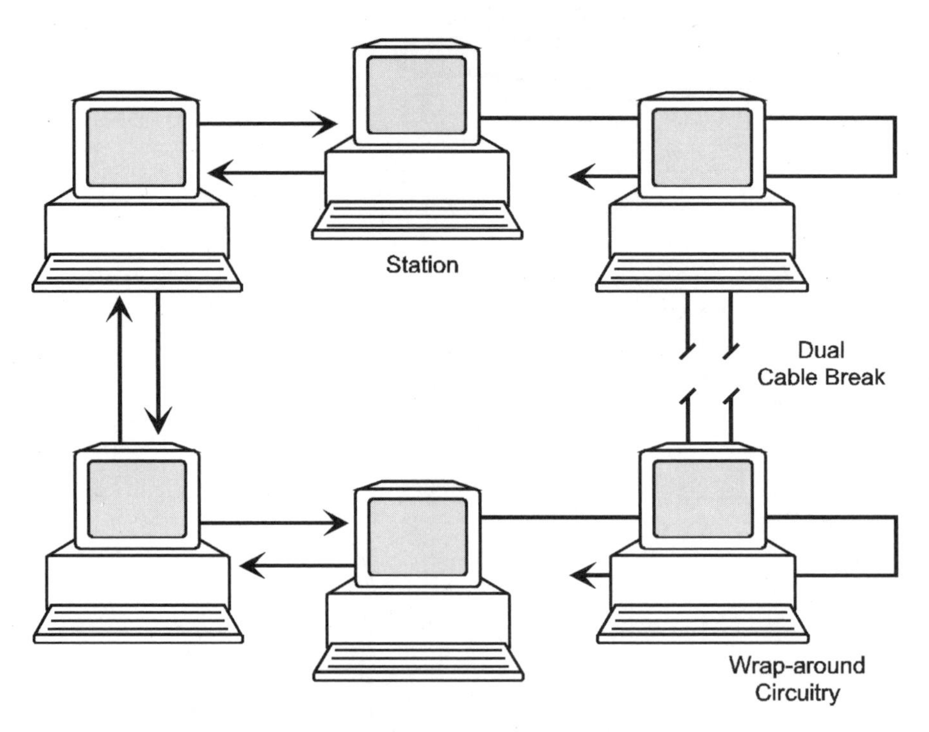

Fig. F6 *Self-healing capability of FDDI's dual-ring topology.*

Failure protection. FDDI provides an optional bypass switch at each node to overcome a failure anywhere on the node. In the event of a node failure, it is bypassed optically, removing it from the network. Up to three nodes in sequence may be bypassed; enough optical power will remain to support the operable portions of the network.

In the event of a cable break, the dual counterrotating ring topology of FDDI allows use of the redundant cable to handle normal 100 Mbps traffic. If both the primary and secondary cables fail, the stations adjacent to the failures automatically loop the data around and between rings (see Fig. F6), thus forming a new C-shaped ring from the operational portions of the original two rings. When the fault is healed, the network will reconfigure itself again.

FDDI concentrators normally offer two buses, which correspond to the two FDDI backbone rings. Fault tolerance is also provided for stations that are connected to the ring via a concentrator because the concentrator provides the loop-around function for attached stations.

Summary. An extension of FDDI, called FDDI-2, uses portions of its 100-Mbps bandwidth to carry voice and video, just like Asynchronous Transfer Mode (ATM) cell switching technology. FDDI is limited by distance, however, while ATM is a highly scalable broadband networking technology that spans both LAN and WAN environments. This means that ATM will eventually dominate corporate networks, especially for running multimedia applications.

See also

ASYNCHRONOUS TRANSFER MODE
ETHERNET (100BASE-T)

Fiber in the loop

Fiber in the loop (FITL) is a system in which services to contiguous groupings of residential and business customers are delivered using fiber-optic media in either all or a portion of the loop distribution network. FITL is an umbrella term that encompasses various systems, including:

- *Fiber to the curb (FTTC),* a technology that involves bringing fiber into the neighborhood and up to the curb, where signals would be carried to businesses and residences via the existing copper wiring.
- *Fiber to the home (FTTH),* a technology that involves bringing fiber all the way to the home.
- *Fiber-to-the-building (FTTB),* a technology that involves bringing fiber all the way to the building.

The newest FTTC systems are able to offer a dedicated rate of 52 Mbps on the downstream path and a dedicated rate of 1.6 Mbps on the upstream path. With this much bandwidth available, the network operator would be able to offer over the FTTC system multiple services in which the bandwidth will be shared among different applications within a home or building, including voice and video.

Summary. Many service providers have concluded that the cost of offering FTTC on a mass-market basis currently is out of reach due to the high cost of installing extra fiber and electronics in the local loop. Instead, hybrid fiber/coax (HFC) systems offer a more economical approach. Nevertheless, FTTC is expected to emerge as the ultimate architecture, at least in high-density metropolitan areas.

See also

HYBRID FIBER/COAX
SWITCHED DIGITAL VIDEO

Fiber optics technology

Fiber-optic transmission systems have been in commercial use for over 20 years. Fiber-optic cable provides many performance advantages over conventional metallic copper-based wire. These advantages make optical fiber the most advanced transmission medium available today, offering a low-cost alternative to satellite for international communications. At the same time, these performance advantages make optical fiber well suited to supporting emerging new applications such as interactive multimedia services delivered to the home.

Bandwidth capacity. The laser components at each end of the optical-fiber link allow for high encoding and decoding frequencies. For this reason, optical fiber offers much more bandwidth capacity than copper-pair wires. Data can be transmitted over optical fiber at multigigabit speeds. The speed barrier continues to be broken with new innovations in technology, such as wave division multiplexing (WDM), which can boost the transport rate of fiber to 160 Gbps and beyond.

Signal attenuation. Signal attenuation, measured in decibels (dB), refers to signal loss during transmission (i.e., when the signal received is not as strong as the signal transmitted). Signal attenuation is attributed to the inherent resistance of the transmission medium. For transmissions over metallic cable, loss increases with frequency as the signal radiates outward. The characteristics of optical fiber, however, are such that little or no inherent resistance exists. This low resistance allows the use of higher frequencies to derive enough bandwidth to accommodate thousands of voice channels. Whereas an analog line on the local loop has a frequency range of up to 4 kHz for voice transmission, a single optical fiber has a range of up to 3 GHz. Optical fiber's lower resistance also means that a constant signal level can be maintained over longer distances without the need for repeaters to regenerate the signals at various locations along the cable.

The extremely low signal attenuation of optical fiber also translates to higher transmission rates. Single-mode fiber, the most common type of cable, has transmission rates as high as 27 GHz, which is the equivalent of 400,000 simultaneous conversations, or all of the data contained in 2500 books of 400 pages each—every second.

Data integrity. *Data integrity* refers to a performance rating based on the number of undetected errors in a transmission. Once again, fiber optics surpass metallic cabling. A typical fiber-optic transmission system produces a bit error rate of less than 10^{-9}, while metallic cabling typically produces a bit-error rate of 10^{-6}.

Because of their high data integrity, fiber-optic systems do not require extensive use of the error-checking protocols common in metallic-cable systems. Because the error-checking overhead is eliminated, the data transmission rates are enhanced. In addition, because the required number of retransmissions is reduced with fiber, overall system performance is greatly improved.

Immunity to interference. Optical fibers are immune to electromagnetic and radio frequency interference (EMI and RFI), the principal sources of data errors in transmissions over metallic-cable systems. This immunity facilitates fiber installation, because fiber-optic cables need not be rerouted around elevators, machinery, auxiliary power generators, fluorescent lighting, or other potential sources of interference.

Fiber's immunity to interference makes it more economical to install, not only because less time is required to route the physical cables, but because there is no need to build special conduits to shield fiber from the external environment. In addition, because optical fibers do not generate the electromagnetic radiation that often causes crosstalk on metallic cables, multiple fibers can be bundled into a single cable to further simplify installation.

Security. Fiber is a more secure transmission medium than unshielded metallic cable because optical fibers do not use signal-radiating electromagnetic or radio frequency energy. In order to tap a fiber-optic transmission, the wire core must be physically broken and a connection fused to it. This procedure is routinely used to add nodes to the fiber cable, but it prohibits the transmission of light beyond the point of the break, and therefore makes unauthorized access easily detectable.

Durability. Optical fiber is not a delicate material; in fact, fiber's pull strength (the maximum pressure that can be exerted on the cable before damage occurs) is 200 pounds, or eight times that of Category 5 unshielded twisted-pair (UTP) copper wire. Fiber-optic cables are reinforced with a strengthening member inside the cable and a protective jacket around the outside of the cable. These reinforcements produce the same tensile strength as steel wire of an equal diameter.

The inherent strength of fiber, combined with the added reinforcement of being bundled into cable form, gives fiber-optic cables the durability necessary to withstand being pulled through walls, floors, and underground conduits without being damaged. In addition, fiber cables are designed to withstand higher temperatures than copper; this makes fiber networks better able to survive potentially disastrous fires. In typical operating environments, fiber also is more resistant to corrosion than copper wire and, consequently, has a longer useful life.

Types of fiber. There are two types of optical fiber: single-mode and multimode. Single-mode fibers transmit only one light wave along the core, while multimode fibers transmit many light waves. Single-mode fibers have lower signal loss and support higher transmission rates than multimode fibers; they are the type of optical fiber most often selected by carriers for use on the public network. Over 90 percent of fiber-optic cable installed by carriers is single-mode.

Multimode fibers have relatively large cores. Light pulses that simultaneously enter a multimode fiber can take many paths and may exit at slightly different times. This phenomenon, called *intermodal pulse dispersion,* creates minor signal distortion and thereby limits both the data rate of the optical signal and the distance that the optical signal can be sent without repeaters. For this reason, multimode fiber is most often used for short distances and for applications in which slower data rates are acceptable.

Multimode fiber can be further categorized as *step-index* or *graded-index.* Step-index fiber has a silica core encased with plastic cladding. The silica is denser than the plastic cladding; the result is a sharp, step-like difference in the refractive index between the two substances. This difference prevents light pulses from escaping as they pass through the optical fiber. Graded-index fiber contains multiple layers of silica at its core, with lower refractive indices toward the outer layers. The graded core increases the speed of the light pulses in the outer layers to match the rate of the pulses that traverse the shorter path directly down the center of the fiber.

The fibers in most of today's fiber-optic cable have an outside (or cladding) diameter of 125 microns. (A micron, abbreviated μm, is one-millionth of a meter.) The core diameter depends on the type of cable. The cores of multimode fibers comprise many concentric cylinders of glass; each cylinder has a different index of refraction. The layers are arranged so that light introduced to the fiber at an angle will be bent back toward the center. The bending results in light that travels in a sine-wave pattern down the fiber core, and allows an inexpensive, noncoherent light source to be used.

Almost all multimode fibers have a core diameter of 62.5 microns. Bandwidth restrictions of 200 to 300 MHz/km limit the maximum length of multimode segments to a few kilometers. Wavelengths of 850 and 1300 nm are used with multimode fiber-optic cable. Single-mode fiber consists of a single 8- to 10-micron core. This means that a carefully focused coherent light source, such as a laser, must be used to ensure that light is sent directly down the small aperture. Single-mode fiber is normally operated with light at a wavelength of 1300 nm.

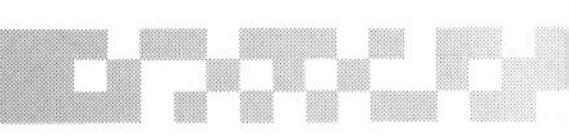

Summary. Fiber-optic cable is installed in carrier networks for its ability to carry voluminous amounts of data and large numbers of voice channels over great distances reliably and securely. Fiber is used for transoceanic links between international locations because it has more capacity than satellite links (which also makes fiber more economical), and because it offers less delay for voice conversations. Fiber is installed on much of the long-distance network in the United States and is being installed in the ring topology around major cities. Reliability is enhanced with dual rings that provide additional capacity for disaster recovery. Businesses use fiber to connect high-speed workstations, computers, and LANs in campus environments and office buildings. In some areas of the country, fiber is even being installed to the home to deliver entertainment services (i.e., video on demand) and support interactive multimedia applications. In other places, hybrid network architectures combining coaxial cable with fiber-optic backbones are being installed to provide advanced communications and entertainment services to the home more economically than fiber-only networks.

See also

FIBRE CHANNEL
SYNCHRONOUS OPTICAL NETWORK
WAVELENGTH DIVISION MULTIPLEXING

Fibre Channel

Fibre Channel is a high-performance interconnect standard designed for bidirectional point-to-point serial data channels between desktop workstations, mass storage subsystems, peripherals, and host systems. Serialization of the data permits much greater distances to be covered than parallel communications. Unlike networks in which each node must share the bandwidth capacity of the media, Fibre Channel devices are connected through a flexible circuit/packet switch capable of providing the full bandwidth to all connections simultaneously.

Advantages. Among the key advantages of Fibre Channel is speed: It is 10 to 250 times faster than typical LAN speeds. It can transmit at rates exceeding 100 Mbps, the equivalent of 60,000 pages of text per second, with the ability to transmit at up to 1 Gbps. Such speeds are achieved simply by transferring data between one buffer at the source device and another buffer at the destination device without regard for how it is formatted, whether cells, packets, or frames. What the individual protocols do with the data before or after it is in the buffer is not of consequence. Fibre Channel provides complete control over the transfer only and offers simple error checking.

Unlike many of today's interfaces, including the Small Computer Systems Interface (SCSI), Fibre Channel is bidirectional and can achieve 100 Mbps in both directions simultaneously. It is really a 200 Mbps channel if usage is balanced in both directions. Fibre Channel also overcomes the restrictions on the number of devices that can be connected, so that, for example, any number of SCSI devices can be accessed instead of the normal limit of 7 or 15.

Fibre Channel overcomes the distance limitations of today's interfaces. A fast SCSI parallel link from a disk drive to a workstation, for example, can transmit data at 20 Mbps, but it is restricted in length to about 20 m. In contrast, a quarter-speed

Fibre Channel link transmits information at 25 Mbps over a single, compact optical cable pair that can be up to as much as 10 km in length. This allows disk drives to be placed almost anywhere and makes for more flexible site planning.

Applications. The high-speed, low-latency connections that can be established using Fibre Channel make it ideal for a variety of data-intensive applications, including:

- *Backbones:* Fibre Channel provides the parallelism, high bandwidth, and fault tolerance needed for high-speed backbones. It is the ideal solution for mission-critical internetworking. The scalability of Fibre Channel makes it practical to create backbones that grow as one's needs grow, from a few servers to an entire enterprise network.
- *Workstation clusters:* Fibre Channel is a natural choice to enable supercomputer-power processing at workstation costs.
- *Imaging:* Fibre Channel provides the bandwidth on demand needed for high-resolution medical, scientific, and prepress imaging applications, among others.
- *Scientific and engineering:* Fibre Channel delivers the needed throughput for today's new breed of visualization, simulation, CAD/CAM, and other scientific, engineering, and manufacturing applications, which demand megabytes of bandwidth per node.
- *Mass storage:* Current mass storage access is limited in rate, distance, and addressability. Fibre Channel provides mass storage attachments at distances of up to several kilometers. Fibre Channel also interfaces with current SCSI, HIPPI, and IPI-3 connections, among others.
- *Multimedia:* Fibre Channel's bandwidth supports real-time videoconferencing and document collaboration among several workstation users, and is capable of delivering multimedia applications employing voice, music, animation, and video.

Topology. Fibre Channel uses a flexible circuit/packet switched topology to connect devices. Through the switch, Fibre Channel is able to establish multiple simultaneous point-to-point connections. Devices attached to the switch do not have to contend for the transmission medium as they do in a network. Through its intrinsic flow control and acknowledgment capabilities, Fibre Channel also supports connectionless traffic without suffering the congestion of the shared transmission media used in traditional networks.

The fabric relieves each Fibre Channel port of the responsibility for station management. All that a Fibre Channel port has to do is manage a simple point-to-point connection between itself and the fabric. If an invalid connection is attempted, the fabric rejects it. If there is a congestion problem en route, the fabric responds with a busy signal and the calling port tries again.

Fiber Channel layers. The Fibre Channel employs a five-layer stack that defines the physical media and transmission rates, encoding scheme, framing protocol and flow control, common services, and the upper-level applications interfaces.

- *FC-0* is the lowest layer, specifying the physical characteristics of the media, transmitters, receivers, and connectors that can be used with Fibre Channel, including electrical and optical characteristics, transmission rates, and other physical components of the standard.
- *FC-1* defines the 8B/10B encoding/decoding scheme used to integrate the data with the clock information required by serial transmission techniques. Fibre Channel uses 10 bits to represent each 8 bits of "real" data, requiring it to operate at a speed sufficient to accommodate this 25 percent overhead. The two extra bits are used for error detection and correction, known as *disparity control.* The 8B/10B encoding used in Fibre Channel is patented by IBM, and is the same one used in the company's 200 Mbps ESCON (Enterprise System Connection) interconnect system.
- *FC-2* defines the rules for framing the data to be transferred between ports, the different mechanisms for using Fibre Channel's circuit- and packet-switched service classes (discussed below), and the means of managing the sequence of data transfer. All frames belonging to a single transfer are uniquely identified by sequential numbering from 0 through *n,* allowing the receiver to tell not only if a frame is missing, but also which one it is.
- *FC-3* provides the common services required for advanced features such as striping (to multiply bandwidth) and hunt groups (the ability for more than one port to respond to the same alias address). A hunt group can be likened to a business that has 10 phone lines, but requires only a single number to be dialed. Whichever line is free will ring.
- *FC-4* provides seamless integration of legacy standards, as the layer that accommodates a number of other data communications protocols such as FDDI, HIPPI, IPI, SCSI, IP—as well as IBM's Single Byte Command Code Set (SBCCS) of the Block Multiplexer Channel (BMC), Ethernet, Token-Ring, and ATM.

Classes of service. To accommodate a wide range of communications needs, Fibre Channel provides different classes of service at the FC-2 layer.

- *Class 1* is a hard or circuit-switched connection that functions like today's dedicated physical channels. This service provides exclusive use of the connection for the duration of a session. It is used for time-critical, nonbursty traffic such as a link between two supercomputers.
- *Class 2* is a connectionless, frame-switched link that provides guaranteed delivery and confirms receipt of traffic. No dedicated connection is established between ports as in Class 1; instead, each frame is sent to its destination over any available route. This service is used for data transfers to and from a shared mass-storage system physically located at some distance from several individual workstations.
- *Class 3* is a one-to-many connectionless service that allows data to be sent rapidly to multiple devices attached to the fabric. Because no confirmation of receipt is given, this service is faster than Class 2 service. This service is used for real-time broadcasts and any other application that can tolerate lost packets.

- *Class 4* is a connection-based service that offers guaranteed fractional bandwidth and guaranteed latency levels. This is achieved by allowing users to lock down a physical path through the Fibre Channel switch fabric.

The same switching matrix can support multiple classes of service simultaneously, per the requirements of each application (Fig. F7).

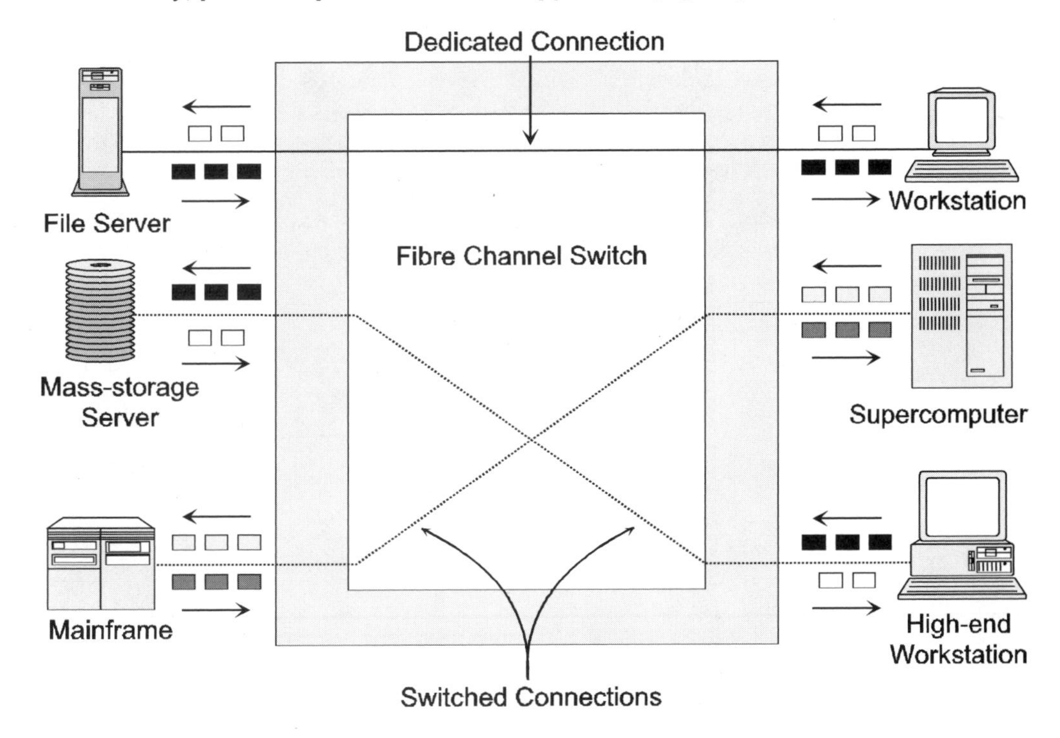

Fig. F7 *A Fibre Channel switching matrix supporting dedicated (Class 1) and switched (Class 2) connections.*

Summary. Fibre Channel is not a telecommunications-like solution for the wide area. It is a cost-effective, high-speed technology for transporting large volumes of data in the local area, where link distances do not exceed 10 km. Fibre Channel's ability to transfer data at speeds of up to 1 Gbps, securely and bidirectionally, make it an effective high-performance communications option for distributed computing environments, particularly those involving mass storage and clustering. While there is some overlap of capability between ATM and Fibre Channel, each can do something the other cannot. With ATM, it is the ability to span the wide area. With Fibre Channel, it is the ability to attach CPU and peripheral devices directly to a high-speed network infrastructure. This makes ATM and Fibre Channel complementary technologies.

See also

Asynchronous Transfer Mode
fiber optics technology
Synchronous Optical Network

Firewalls

Firewalls protect private networks against intruders. As more computers become networked, often providing access via the public Internet, the possibility of break-in attempts increases. Networks today contain valuable resources that provide access to services, software, and databases. For whatever reason, certain individuals exploit lax security at some network sites to download personal information, steal sensitive data, or even destroy files. Increasingly, therefore, companies are turning to firewalls to protect themselves against theft and malicious mischief. By occupying a strategic position on the network (Fig. F8), firewalls work to block certain types of traffic and allow the rest of the traffic to pass. They also provide tools that can be used to track down the source of the attempted security breach.

Types of firewalls. There are several types of firewalls, all of which may be bundled in a single firewall server and configured individually by the network administrator, often through a graphical user interface.

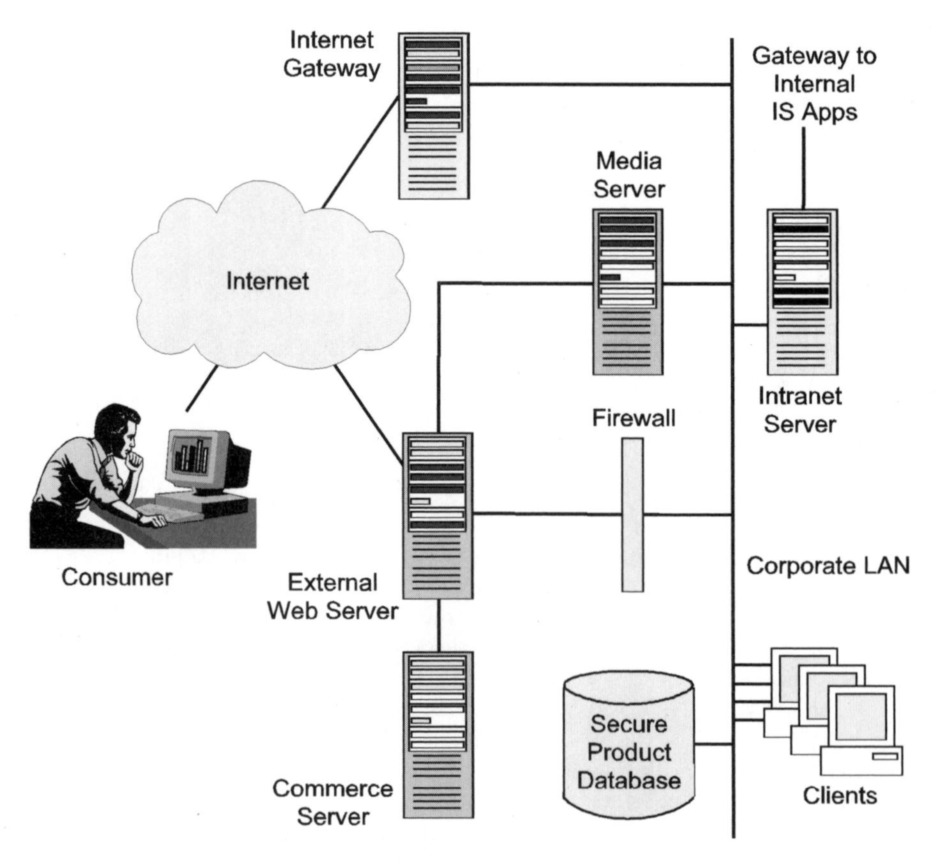

Fig. F8 *Firewalls guard resources on the corporate network from access by unauthorized persons on the greater Internet.*

Application-level gateways. Firewalls typically include support for several standard application servers, including mail, news, WWW, FTP, and DNS. Each application may even be compartmentalized from other firewall software, so that if an individual server is under attack, other servers/functions are not affected.

As packets move up the protocol stack, the various layers operate on data towards the inside of the packet. When the packet has reached the Application layer, operations are performed on the data payload of the packet. This is the information that the application program processes and displays in intelligible form to the user. At this layer, some very interesting things can be done from a security protection standpoint, such as:

- Virus scanning of incoming FTP files and e-mail to prevent the spread of infection.
- Control over what FTP commands the user is permitted to execute.
- Control over which commands are allowed to be executed for any particular service.
- Keyword checking of incoming e-mail to detect junk-mail ("spam") attributes.

Proxy servers. With proxies, the firewall acts as an intermediary for user requests, setting up a second connection to the desired resource either at the Application layer (an application proxy) or at the Session or Transport layer (a circuit relay). Most firewalls rely on proxies to perform their functions. This provides the means to keep someone from directly connecting from the outside to a service on the inside. For example, most of the firewalls provide a Telnet proxy that might operate as follows:

- When an external user tries to connect via Telnet to IP address 139.170.2.23 inside the 139.170 net, the IP packet gets routed to the proxy server.
- The proxy server goes into a Telnet session with the outside requester and establishes a Telnet session with the real 139.170.2.23 node inside the network.
- It then takes the contents of the Telnet packet sent from the outside, wraps it in new headers, and sends it to the node inside.

In this way, the proxy server maintains the Telnet session that it had established on behalf of the outside server.

Packet filters. Historically, routers have performed packet filtering at the Network layer. They control what addresses are allowed to engage in a session together (i.e., IP address A can connect to IP address B). Enhanced packet filtering techniques provide even more control, allowing the port to be specified as well (i.e., IP address A, port 23 can connect to IP address B, port 23). The basic intent is to keep unauthorized traffic from getting onto the corporate network.

A packet-filtering firewall works in a similar manner. It examines all the packets passed to it, then forwards them or drops them based on predefined rules (Fig. F9). The network administrator can control how packet filtering is performed, permitting or denying connections using criteria based upon the source and destination host or network and the type of network service (Fig. F10).

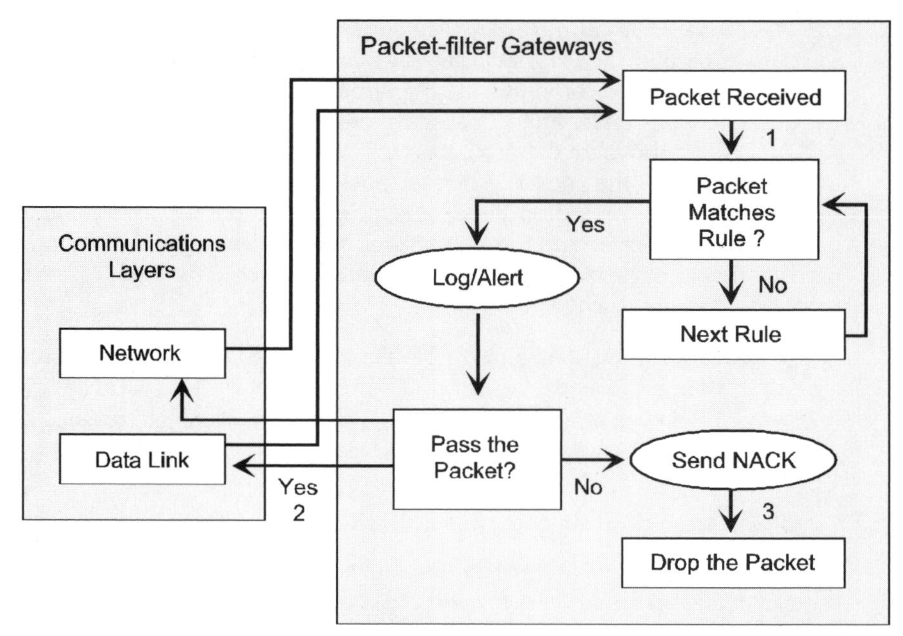

Fig. F9 *Operation of a packet-filtering firewall: (1) inbound/outbound packets are examined for compliance with company-defined security rules; (2) packets found to be in compliance are allowed to pass into the network; (3) packets that are not in compliance are dropped.*

Stateful packet inspection. A stateful inspection device examines the packets it sees, just like packet filters, but goes a step further: It remembers which port numbers are used by which connections and shuts down access to those ports after the connection closes.

Other security features. Firewalls often provide other security features, such as encryption, that enhance the primary capabilities discussed above. With the increasing use of Java applets and ActiveX controls on Web sites, more firewalls also are able to deny access to Web pages that contain these elements. Java applets and ActiveX controls are self-executing programs that not only can hide harmful viruses or agents, but also can potentially give remote users access to resources on private networks once they get inside.

IP address translation keeps IP addresses hidden to prevent crackers from gaining access to the corporate network through a technique called *IP spoofing,* where they impersonate authorized users. The firewall dynamically maps all internal IP addresses to one safe IP address on the firewall. Since the firewall hides internal IP addresses, the organization is no longer limited to using registered IP addresses.

Another security feature is the log file, which records connection requests and server activity. The information compiled in the log file can be used to identify possible security breaches. The files can be viewed from the console displaying the most recent entries and scrolls in real time as new entries come in (Fig. F11).

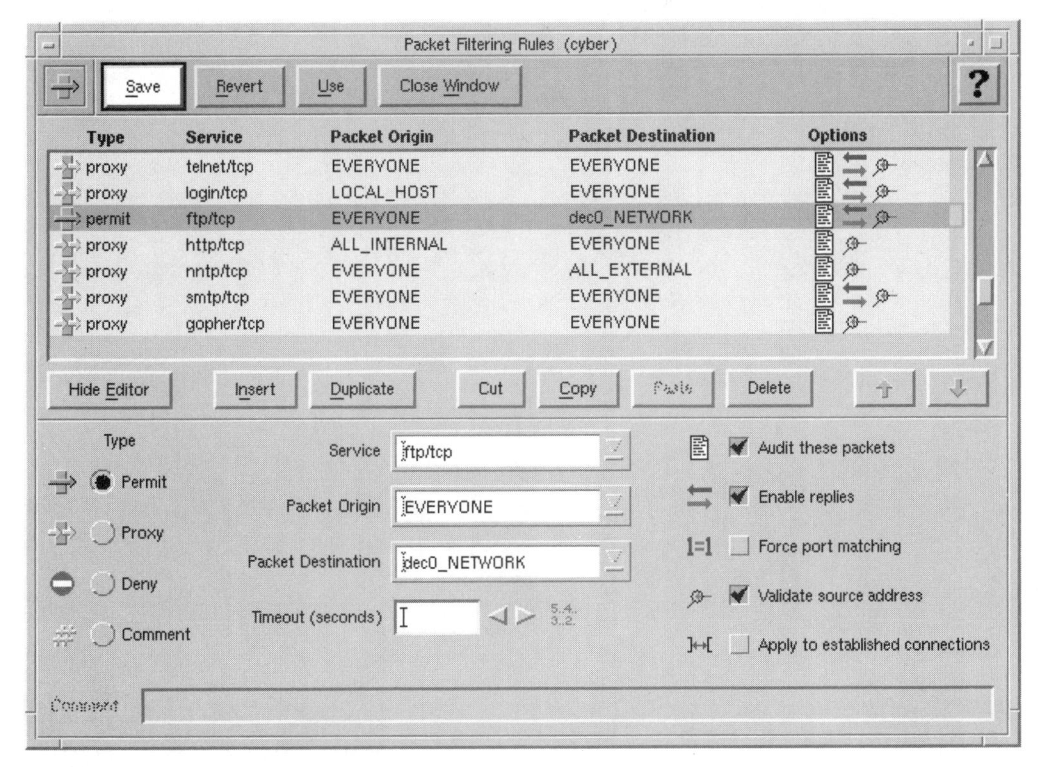

Fig. F10 *Like other products, CyberGuard's Firewall 3 allows the network administrator to control packet filtering, which permits or denies connections using criteria based upon the source and destination host or network and the type of network service.*

Many firewalls allow the system administrator to install, configure, and monitor network security from remote and local sites. Some even issue pager alerts (alarms) to notify the administrator of any remote or local security policy violation so prompt action can be taken to track down the source.

Summary. In years past, a strong, host-based security system would cut off most intruder attacks by not allowing launching points for illegal logins. As more businesses become networked, however, a host-oriented security system is not practical. Firewalls have now become the first line of defense against unlawful attacks. They reduce security risks to a single point by confining attackers to a narrow hole where there is more of a chance of catching or detecting them. In this way, network managers and security administrators have a better opportunity to maintain the integrity of the private network.

See also

NETWORK SECURITY

Fixed wireless access

Fixed wireless access technology provides a wireless link to the public switched telephone network (PSTN) as an alternative to traditional wire-based telephone ser-

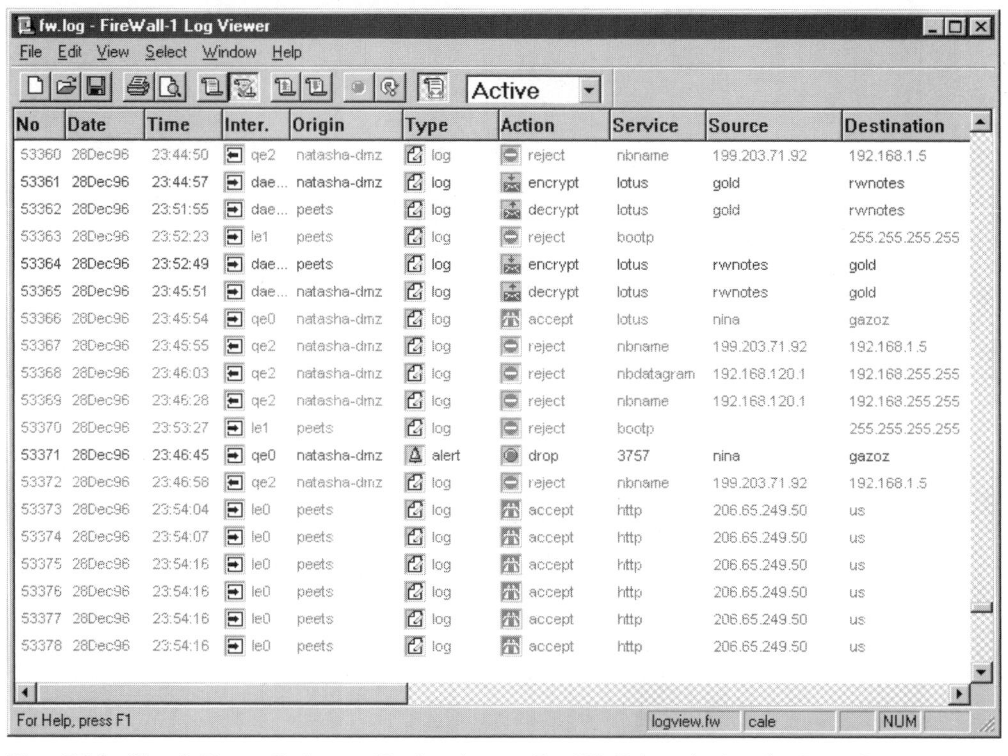

No	Date	Time	Inter.	Origin	Type	Action	Service	Source	Destination
53360	28Dec96	23:44:50	qe2	natasha-dmz	log	reject	nbname	199.203.71.92	192.168.1.5
53361	28Dec96	23:44:57	dae...	natasha-dmz	log	encrypt	lotus	gold	rwnotes
53362	28Dec96	23:51:55	dae...	peets	log	decrypt	lotus	gold	rwnotes
53363	28Dec96	23:52:23	le1	peets	log	reject	bootp		255.255.255.255
53364	28Dec96	23:52:49	dae...	peets	log	encrypt	lotus	rwnotes	gold
53365	28Dec96	23:45:51	dae...	natasha-dmz	log	decrypt	lotus	rwnotes	gold
53366	28Dec96	23:45:54	qe0	natasha-dmz	log	accept	lotus	nina	gazoz
53367	28Dec96	23:45:55	qe2	natasha-dmz	log	reject	nbname	199.203.71.92	192.168.1.5
53368	28Dec96	23:46:03	qe2	natasha-dmz	log	reject	nbdatagram	192.168.120.1	192.168.255.255
53369	28Dec96	23:46:28	qe2	natasha-dmz	log	reject	nbname	192.168.120.1	192.168.255.255
53370	28Dec96	23:53:27	le1	peets	log	reject	bootp		255.255.255.255
53371	28Dec96	23:46:45	qe0	natasha-dmz	alert	drop	3757	nina	gazoz
53372	28Dec96	23:46:58	qe2	natasha-dmz	log	reject	nbname	199.203.71.92	192.168.1.5
53373	28Dec96	23:54:04	le0	peets	log	accept	http	206.65.249.50	us
53374	28Dec96	23:54:07	le0	peets	log	accept	http	206.65.249.50	us
53375	28Dec96	23:54:16	le0	peets	log	accept	http	206.65.249.50	us
53376	28Dec96	23:54:16	le0	peets	log	accept	http	206.65.249.50	us
53377	28Dec96	23:54:16	le0	peets	log	accept	http	206.65.249.50	us
53378	28Dec96	23:54:16	le0	peets	log	accept	http	206.65.249.50	us

Fig. F11 *Check Point Software Technologies FireWall-1 includes the Live Connections Monitor, which gives network administrators the ability to view all currently active connections. The live connections are stored and handled in the same way as ordinary log records, but in a special file that is continuously updated as connections start and end.*

vice. Since calls and other information (e.g., data, images) are transmitted through the air, rather than through conventional cables and wires, the labor-intensive cost of providing and maintaining telephone poles and cables is avoided. The technology is especially useful in rural areas, and it provides more bandwidth (up to 128 kbps) than standard analog phone lines.

In order to provide fixed wireless service, a wireless access unit is installed on the exterior of a home or business (Fig. F12), which allows customers to originate and receive calls with existing analog telephones as they would with a wireline connection. This transceiver is positioned to provide an unobstructed view to the nearest base station receiver. Voice and data calls are transmitted from the transceiver at the customer's location to the base station equipment, which relays the call through carrier's existing network facilities to the appropriate destination. No investment in special phones or facsimile machines is required; customers use all their existing equipment.

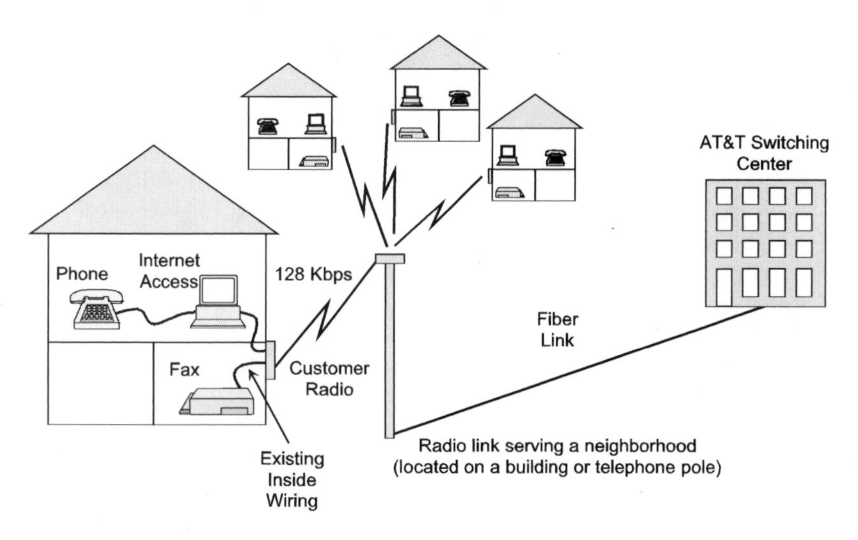

Fig. F12 *Fixed wireless access configuration.*

Various technologies are available for the wireless access units and network infrastructure:

➢ *GSM:* Available at 900, 1800, and 1900 MHz frequencies, the wireless access units use GSM's Enhanced Full Rate Codec (EFRC) for wireline-equivalent voice quality. These units also provide a variety of popular, revenue-generating GSM subscriber services, including calling line ID, call forwarding, call waiting, conferencing, and voice mail.

➢ *TDMA:* Wireless access units on 800 MHz operate in analog or TDMA digital mode, and can be used by the carrier to provide local phone service as an adjunct to existing mobile cellular services.

➢ *CDMA:* Wireless access units on 800 MHz and 1900 MHz operate in digital mode, and also can be used by the carrier to provide local phone service as an adjunct to existing mobile cellular services.

Summary. Fixed wireless access technology originated out of the need to contain carrier operating costs in rural areas, where pole and cable installation and maintenance are more expensive than in urban and suburban areas. However, wireless access technology also can be used in these areas to bypass the local exchange carrier for long-distance calls. Since the long distance carrier avoids having to pay the telephone company's local loop interconnection charges, the savings can be passed back to the customer.

See also

CELLULAR COMMUNICATIONS

Frame relay

In the past, WANs were built using low-speed analog facilities that were used primarily for voice traffic. For reliable data transmission, private and public packet-switched networks were used. The X.25 protocol was conceived in a networking environment of copper lines and electromechanical switches. The analog lines and equipment in use at the time were noisy and subject to a variety of other impairments that made the transmission of data difficult. To deal with this environment, packet switches were deployed using the X.25 protocol. X.25 was endowed with substantial error-correction capabilities so that any node on the network could request a retransmission of errored data from the node that sent it. Errors had to be detected and corrected within the network, since the user's equipment typically did not have the intelligence and spare processing power to devote to this task.

However, X.25's facilities for packetized data transmission, coupled with its inherent error-correction capabilities, impose an overhead burden that limits network throughput. This, in turn, limits X.25 to niche applications, such as terminal- to-host interactive services like point-of-sale transaction processing, where the reliable transmission of credit card numbers and other financial information—not speed—is the overriding concern.

With the trend toward the increasing use of digital facilities, there is less need for error protection. Fiber-optic networks have very low bit error rates, so there is even less of a need to burden the network with error-protection overhead. At the same time, users' end devices have a high level of intelligence, processing power, and storage, making them more adept at handling error control and diverse protocols. Consequently, the communications protocol used over the network may be scaled down to its bare essentials to greatly increase throughput. This is the idea behind frame relay, which can support voice traffic as well as data.

Comparison with X.25. Although X.25 has been around for a long time, its sensitivity to the protocols being transported consumes processing resources that can diminish performance. Frame relay operates virtually transparently to the carried load.

Whereas X.25 operates at the bottom three layers of the OSI Reference Model, frame relay operates at the first layer and the lower half of the second layer. This cuts the amount of processing by as much as 50 percent, improving network throughput by a factor estimated at between 3 and 10 times.

Although the frame relay network can detect errors, it does not correct them. Errored frames are simply discarded. When the receiving device detects missing frames, it can request a retransmission from the originating device, whereupon the appropriate frames are sent again.

Types of circuits. Packet networks support large numbers of nodes through the use of virtual circuits referred to as *logical channels.* The number of terminals that need to be accommodated on the network determines the number of logical channels. Unlike a physical channel, which is a port on a computer or multiplexer, a logical channel is merely a temporary connection that is made between portions of the network. The connection of a terminal to a host, for example, would require a logical channel at the host associated with one at the terminal.

The two primary types of virtual circuits supported by frame relay are *switched virtual circuits* (SVCs) and *permanent virtual circuits* (PVCs). SVCs are analogous to dial-up connections, which require path setup and teardown. A key advantage of SVCs is that they permit any-to-any connectivity. PVCs are more like dedicated private lines; once set, the predefined logical connections between sites stay in place. This allows logical channels to be dedicated to specific terminals. The SVC requires fewer logical channels at the host because the terminals contend for a smaller number of logical channels. Of course, it is assumed that not everyone will require access to the host at the same time.

Another type of virtual circuit, which is still being planned for frame relay, is the *multicast virtual circuit* (MVC), which is used to broadcast the same data to a group of up to 64 users over a reserved data link connection. This type of virtual circuit might be useful for expediting communications among members of a single workgroup dispersed over multiple locations, or to facilitate interdepartmental collaboration on a major project.

The same frame relay interface can be used to set up SVCs, PVCs, and MVCs. All three may share the same digital facility. It is even possible to bundle SVCs within PVCs to help avoid delays that can time out SNA sessions and disrupt voice traffic. In supporting multiple types of virtual circuits, frame relay networks provide a high degree of configuration flexibility, as well as more efficient utilization of available bandwidth.

Congestion control. Real-time congestion control must accomplish the following critical objectives in a frame relay network:

- Maintain high throughput by minimizing time-outs and out-of-sequence frame deliveries.
- Prevent session disconnects, unless required for congestion control.
- Protect against unfair users who attempt to hog the available network resources by exceeding their committed information rate (CIR), or established burst size.
- Prevent the spread of congestion to other parts of the network.
- Provide delays consistent with application requirements and service objectives.

In the frame relay network, congestion can be avoided through a control mechanism that provides *backward explicit congestion notification* (BECN) and *forward explicit congestion notification* (FECN), which are depicted in Fig. F13.

BECN is indicated by a bit in the data frame set by the network to notify the user that congestion avoidance procedures should be initiated for traffic in the opposite direction of the received frame. FECN is indicated by a bit in the data frame set by the network to notify the user that congestion avoidance procedures should be initiated for traffic in the direction of the received frame. Upon receiving either indication, the end point (i.e., bridge, router or other internetworking device) takes appropriate action to ease congestion.

The response to congestion notification depends upon the protocols and flow control mechanism employed by the end point. The BECN bit typically would be used by protocols capable of controlling traffic flow at the source. The FECN bit typically would be used by protocols implementing flow control at the destination.

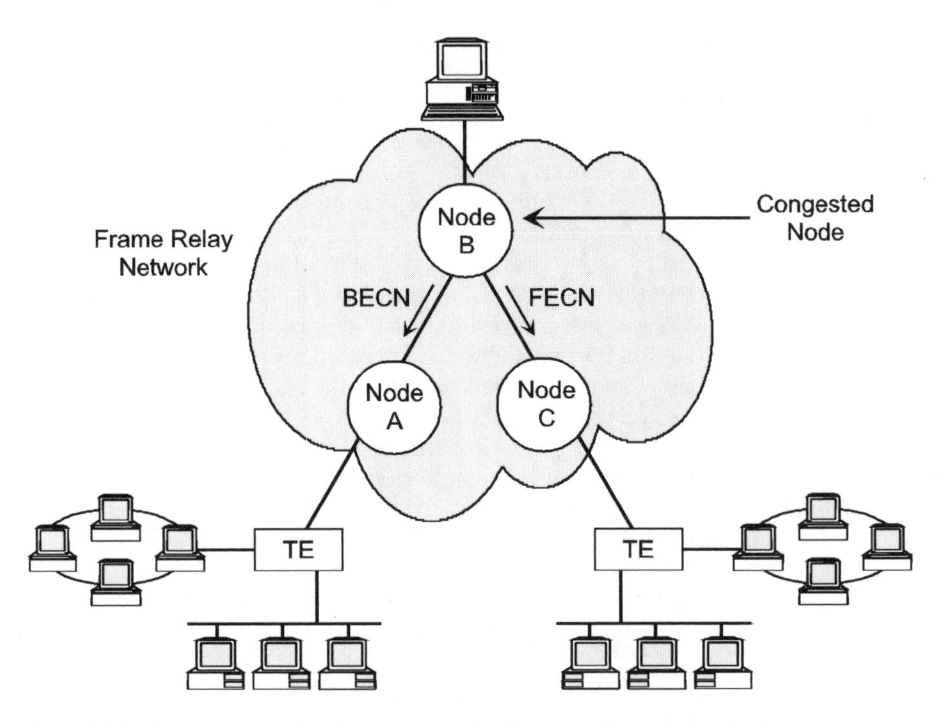

Fig. F13 *Congestion notification on the frame relay network.*

Upon receipt of a frame with the BECN bit set, the end point must reduce its offered rate to the CIR for that frame relay connection. If consecutive data frames are received with the BECN bit set, the end point must reduce its rate to the next "step" rate below the current offered rate. The step rates are 0.675, 0.50, and 0.25 of the current rate. After the end point has reduced its offered rate in response to receipt of BECN, it may increase its rate by a factor of 0.125 times current rate after receiving two consecutive frames with the BECN bit clear.

If the end point does not respond to the congestion notification, or the user's data flow into the network is not significantly reduced as a result of the response to the congestion notification, or an end-point is experiencing a problem that exacerbates the congestion problem, then congestion recovery procedures are invoked by the network. These procedures include discarding frames, in which case the end-to-end protocols employed by the end points are responsible for detecting and requesting the retransmission of missing frames.

Frame discard could be done on a priority basis; that is, a decision based on predetermined criteria is made on whether certain frames should be discarded in preference to other frames in a congestion situation. Frames are discarded based on their "discard eligibility" setting of 1 or 0, as specified in the data frame. A setting of 1 indicates that the frame should be discarded during congestion, while a setting of 0 indicates that the frame should not be discarded unless there are no alternatives.

The discard eligibility bit may be set in two ways. The user can declare whether the frames are eligible for discard by setting the discard eligibility bit in the data frame to 1. Or the network access interface may be configured to set the discard eligibility bit to 1 when the user's data has exceeded the CIR, in which case the data is considered excess and subject to discard.

Although BECN and FECN provide effective mechanisms for controlling network congestion, it will be 2–3 years before all vendors support these capabilities. Until then, the carriers will deal with the congestion problem by making more bandwidth available to customers (at no extra cost) to prevent congestion from occurring in the first place. Private network users, however, will have to order (and pay for) enough bandwidth to guard against congestion on their frame relay links.

Summary. The need for frame relay arose partly out of the emergence of digital networks, which are faster and less prone to transmission errors than older analog lines. Although the X.25 protocol overcomes the limitations of analog lines, it does so with a significant performance penalty due mainly to its error-checking capability, which relies on the store-and-forward method of transmission. In being able to do without this and other functions, plus support different types of virtual circuits, frame relay offers more efficient utilization of available bandwidth and, as a result, more configuration flexibility.

See also

Asynchronous Transfer Mode
Packet Switched Networks
Switched Multimegabit Data Services

Frequency Division Multiple Access

There are three basic multiple access schemes in use today for mobile communications systems:

- Frequency Division Multiple Access (FDMA), which serves the calls with different frequency channels.
- Time Division Multiple Access (TDMA), which serves the calls with different time slots.
- Code Division Multiple Access (CDMA), which serves the calls with different code sequences.

Of the three, FDMA is the simplest and still the most widespread technology in use today for mobile communications. For example, FDMA is used in the CT2 system for cordless telecommunications. The familiar cordless phone used in the home is representative of this type of system. It creates capacity by splitting bandwidth into radio channels in the frequency domain. In the initial call setup, the handset scans the available channels and locks onto an unoccupied channel for the duration of the call. Based on time division duplexing (TDD), the call will be split into time blocks that alternate between transmitting and receiving.

The traditional analog cellular systems, such as those based on the Advanced Mobile Phone Service (AMPS) standards, also use FDMA to derive the channel. In the case of AMPS, the channel is a 30-kHz "slice" of spectrum. Only one subscriber at a time is assigned to the channel. No other conversations can access the channel until the subscriber's call is finished, or until the call is handed off to a different channel in an adjacent cell.

The analog operating environment poses several problems. One is that the wireless devices are often in motion. Current analog technology does not deal with call hand-offs very well, as evidenced by the high incidence of dropped calls. This environment is particularly harsh for data, which is less tolerant of transmission problems than voice. Whereas momentary signal fade, for instance, is a nuisance in voice communications, it may cause a data connection to drop.

Another problem with analog systems is their limited capacity. To increase the capacity of analog cellular systems, the 30-kHz channel can be divided into three narrower channels of 10 kHz each. This is the basis of narrowband AMPS (N-AMPS) standard. However, this band-splitting technique incurs significant base station costs and its limited growth potential makes it suitable only as a short-term solution.

While cell subdivision often is used to increase capacity, this solution has its limits. Since adjacent cells cannot use the same frequencies without risking interference, a limited number of frequencies are being reused at closer distances, which makes it increasingly difficult to maintain the quality of communications. Subdividing cells also increases the amount of overhead signaling that must be used to set up and manage the calls, which can overburden switch resources. In addition, property for cell sites is difficult to purchase in metropolitan areas, where traffic is highest, and future substantial growth is anticipated.

These and other limitations of analog FM radio technology have led to the development of second-generation cellular systems based on digital radio technology and advanced networking principles. Providing reliable service in this dynamic environment requires that digital radio systems employ advanced signal processing technologies for modulation, error correction, and diversity. These are provided by TDMA and CDMA.

Summary. FM systems have supported cellular service for more than a decade, during which demand has finally caught up with the available capacity. Now first-generation cellular systems based on analog FM radio technology are rapidly being phased out in favor of digital systems that offer higher capacity, better voice quality, and advanced call-handling features. TDMA and CDMA are the primary technologies contending for acceptance among analog cellular carriers worldwide.

See also

CODE DIVISION MULTIPLE ACCESS
CORDLESS TELECOMMUNICATIONS
TIME DIVISION MULTIPLE ACCESS

Gateways

Gateways are used to interconnect dissimilar networks or applications. Gateways operate at the highest layer of the Open Systems Interconnection (OSI) reference model: the Application layer (Fig. G1). A gateway consists of protocol conversion software that usually resides in a server, minicomputer or mainframe or front-end device. Gateways interconnect disparate networks or media by processing the various protocols used by each so that information from the sender is intelligible to the receiver, despite differences in their networks or computing platforms.

For example, when an SNA gateway is used to connect an asynchronous PC to a synchronous IBM SNA mainframe, the gateway acts both as a conduit over which the computers communicate and as a translator between the various protocol layers. The translation process consumes considerable processing power, resulting in relatively slow transmission rates when compared with other interconnection

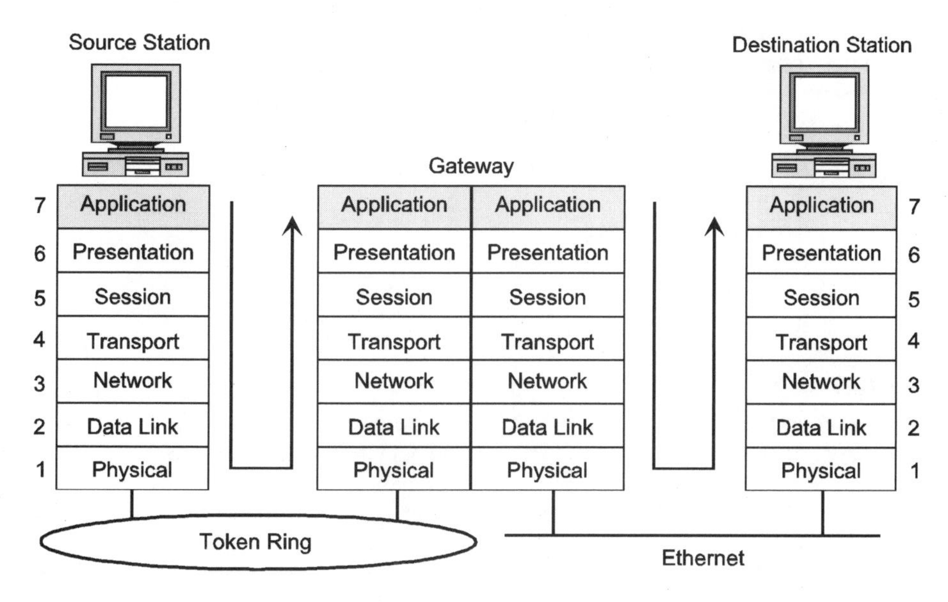

Figure G1 *Gateway functionality in reference to the OSI model.*

methods: hundreds of packets per second for a gateway versus tens of thousands of packets per second for a bridge.

In addition to its translation capabilities, a gateway can check on the various protocols being used, ensuring that there is enough protocol processing power available for any given application. It also can ensure that the network links maintain a level of reliability for handling applications in conformance to predefined error rate thresholds.

Gateways have a variety of applications. In addition to facilitating LAN workstation connections to various host environments, such as IBM SNA 3270 systems and IBM midrange systems, they facilitate connections to X.25 packet-switched networks. Other applications of gateways include the interconnection of various electronic mail systems, allowing mail to be exchanged between normally incompatible formats. This gateway function may be provided by servers equipped with the X.400 international messaging protocol.

In some cases, gateways can be used to consolidate hardware and software. An SNA 3270 gateway shared among multiple networked PCs can be used in place of IBM's 3270 Information Display System or many individual 3270 emulation products. Although the IBM system is a standard means of achieving the micro-to-host connection, it is expensive when used to attach a large number of stand-alone microcomputers. The relatively high connection cost per microcomputer discourages host access for occasional users and limits the central control of information.

If the microcomputers are on a LAN, however, one gateway can emulate a cluster controller and thereby provide all workstations with host access at a very low cost. Cluster controller emulators use an RS-232C or compatible serial interface to a host adapter or communications controller, such as an IBM 3720 or 3745. They can support up to 254 simultaneous sessions.

A relatively new type of gateway provides connections between the Internet and local telephone lines, enabling users to originate phone calls from their Internet-connected PCs to conventional telephones on the public network and vice versa. This arrangement allows users to leverage their existing Internet connections to save on long-distance and international call charges.

See also

BRIDGES
INTERNET TELEPHONY
OPEN SYSTEMS INTERCONNECTION (OSI)
REPEATERS
ROUTERS

General Mobile Radio Service

General Mobile Radio Service (GMRS) is a personal two-way radio communication service that can be used to facilitate the activities of the individual's immediate family members. The service has a communications range of 5 to 25 miles. Before any station transmits on any authorized GMRS channel, the user must obtain a license from the FCC.

A GMRS system consists of station operators, a mobile station (often comprising several mobile units), and sometimes one or more land stations. The classes of land stations are: base station, mobile relay station (also known as a repeater), and control stations. A small base station is one that has an antenna no more than 20 feet above the ground or above the tree on which it is mounted and transmits with no more than 5 W.

Users normally communicate over the general area of their residence, such as an urban or rural area. This area must be within the territorial limits of the 50 United States, the District of Columbia, and the Caribbean and Pacific Insular areas. In transient use, mobile station units from one GMRS system may communicate through a mobile relay station in another GMRS system with the permission of its licensee.

There are 23 GMRS channels. None of the GMRS channels is assigned for the exclusive use of any system. Licensees and license applicants must cooperate in the selection and use of the channels in order to make the most effective use of them and to reduce the possibility of interference.

Any mobile station or small base station in a GMRS system operating in the simplex mode may transmit voice-type emissions with no more than 5 W on the following 462.*x* MHz channels: 462.5625, 462.5875, 462.6125, 462.6375, 462.6625, 462.6875, and 462.7125 MHz. These channels are shared with the Family Radio Service (FRS).

Any mobile station in a GMRS system may transmit on the 467.675 MHz channel to communicate through a mobile relay station transmitting on the 462.675 MHz channel. The communications must be for the purpose of soliciting or rendering assistance to a traveler, or for communicating in an emergency pertaining to the immediate safety of life or the immediate protection of property.

Each GMRS system license assigns one or two of eight possible channels or channel pairs (one 462.*x* MHz channel and one 467.*x* MHz channel spaced 5 MHz apart), as requested by the applicant. Applicants for GMRS system licenses are advised to investigate or monitor in order to determine the best available channel(s) before making a selection. Each applicant must select the channel(s) or channel pair(s) for the stations in the proposed system from the following list:

- For a base station, mobile relay station, fixed station, or mobile station: 462.550, 462.575, 462.600, 462.625, 462.650, 462.675, 462.700 and 462.725 MHz.
- For a mobile station, control station, or fixed station in a duplex system: 467.550, 467.575, 467.600, 467.625, 467.650, 467.675, 467.700 and 467.725 MHz.

GMRS system station operators must cooperate in sharing the assigned channel with station operators in other GMRS systems by monitoring the channel before initiating transmissions, waiting until communications in progress are completed before initiating transmissions, engaging only in permissible communications, and limiting transmissions to the minimum practical transmission time.

Other services. There are other private Personal Radio Services for short-distance two-way voice communications. There is the Citizens Band (CB) Radio Service, over which users are authorized to operate an FCC type-accepted CB unit for communications covering a distance of 1 to 5 miles. No license document is issued.

There is also the Family Radio Service (FRS), over which users are authorized to operate an FCC-certified FRS unit for communications covering a distance of less than 1 mile. No license document is issued.

Summary. Any individual 18 years of age or older who is not a representative of a foreign government is eligible to apply for a GMRS system license. Application for a GMRS system license is made on FCC Form 574. A booklet entitled "Instructions for Completion of FCC Form 574" is helpful in preparing the application. This booklet can be obtained from the FCC's Consumer Assistance Branch by calling (800) 322-1117. There is a filing fee.

Generic digital services

Generic digital services (GDS) are dedicated digital private line services offered at transmission speeds of 2.4, 4.8, 9.6, 19.2, 56, and 64 kbps. They are aimed at analog service users seeking higher-quality transmission, and at AT&T's Dataphone Digital Service (DDS) customers seeking a more affordable digital service. With the exception of 64 kbps, all transmission speeds include a secondary channel for low-speed data transport applications, including diagnostic signaling.

With customer network reconfiguration (CNR), a user has the flexibility to alter his or her network from an on-premises network management terminal according to time-of-day volume needs, as well as to perform disaster recovery quickly. Alternatively, the user can opt to have the carrier reconfigure the network. These features are implemented at the carrier's digital crossconnect system (DCS). Through the DCS, multipoint and point-to-point topologies are supported.

Table G1 provides the brand names under which generic digital services are provided by the respective carriers.

Table G1. Generic Digital Services Brand Names.

Carrier	Service
Ameritech	Optinet
Bell Atlantic	Digital Connect service
Bell South	Synchronet
Nynex	DigiPath
Pacific Bell	Advanced Digital Network
SBC	MegaLink
US West	DigiCom

Applications. GDS can be integral to performing several vital day-to-day tasks for virtually any business, from specialized applications like medical imaging for the health care industry, to the routine information management tasks that are essential to the daily operations on all businesses. Table G2 summarizes the common applications of GDS in the education, banking/finance, and retail sectors.

Table G-2. Common GDS Applications.

Application	Education	Banking/Finance	Retail
File Transfer	•	•	•
Network Management	•	•	•
Disaster Recovery	•	•	•
Time of Day Routing	•	•	•
Electronic Funds Transfer		•	•
Automatic Teller Machines		•	
Order Entry			•
Inventory Management			•
Information Retrieval	•	•	•
CAD/CAM	•		
LAN-to-LAN Interconnection	•	•	•

Digital crossconnect (DCS) nodes located in the carrier's serving wire centers (SWCs) are used to support GDS. The crossconnect nodes facilitate network administration, disaster recovery, and remote testing. Through the DCS, users can make or order changes to their networks to meet daily, occasional, or seasonal data traffic needs. The carrier automatically takes steps to correct any errors that occur on the transmission links, with the goal of 99.96 percent error-free seconds, which equates to 3.5 hours of downtime per year.

GDS can connect to interLATA services provided by the user's long-distance carrier of choice, making possible connectivity to every central office served by a digital T-carrier facility. GDS also provides a migration path to more sophisticated technologies as data needs change, including T1 and ISDN. The carrier can provide all the equipment needed to use GDS, such as the CSU/DSUs that are required for the front end of the link(s), as well as bridges/routers for LAN-to-LAN interconnection. Alternatively, users can purchase or use their own equipment.

The multiplexing capability of the DSU makes it particularly compatible with multiple stand-alone terminals or applications supplied by different vendors. The banking industry is representative of this environment. All banks essentially run the same applications at each location, with each application being served by separate multipoint lines:

- A teller application running at 4.8 or 9.6 kbps
- Automated teller machines (ATMs) running at 1.2 or 24 kbps
- A platform application running at 4.8 or 9.6 kbps
- A security application running at 75 bps to 1.2 kbps

Costs for GDS at 56 kbps in many areas are similar to the costs of a voice-grade private line. In such cases, a multiport, multipoint DSU affords even great cost savings. A network of three multipoint lines with six drops each can be replaced with one 56 kbps multipoint line, using a DSU at each location to multiplex the applications. Many DSUs have software-selectable port rates of 75, 150, 300, and 600 bps, as well as 1.2, 2.4, 4.8, 9.6, 14.4, 19.2, 38.4, 56, and 64 kbps. These speeds can be used in any combination as long as they do not exceed the total bandwidth of 64 kbps.

Multidrop networks require a multipoint junction unit (MJU) to bridge or connect the individual circuit segments. The unit is composed of hardware and software integral to the generic service. Use of the MJU requires the customer to designate a control station, with the other stations on the multipoint circuit designated as remote stations. One unit can support one control station and four remote stations, but these can be cascaded to support additional remote stations.

Much of the cost saving associated with GDS is attributable to the elimination of traffic back-hauling. With hub-oriented services (e.g., DDS), the user's traffic must be routed through a special hub office. Because there are fewer DDS hubs nationwide (AT&T has about 100), longer circuit mileage is typical, which inflates the service cost. With GDS, the service is available from the serving wire centers of the local exchange carriers (about 20,000 among all carriers), so there is less of a need to back-haul traffic, typically 6 miles with GDS versus 60 miles with DDS.

Service components. Generic digital services are supported by the following network elements (Fig. G2): the local loop, a serving wire center, interoffice facilities, a digital crossconnect system, a multipoint junction unit, and a D4 channel bank equipped with an office channel unit data port.

Local loop. The local loop uses a four-wire, nonloaded copper pair that connects the customer premises to the local exchange carrier's serving wire center. The local loop is terminated at the customer's premises at the network interface.

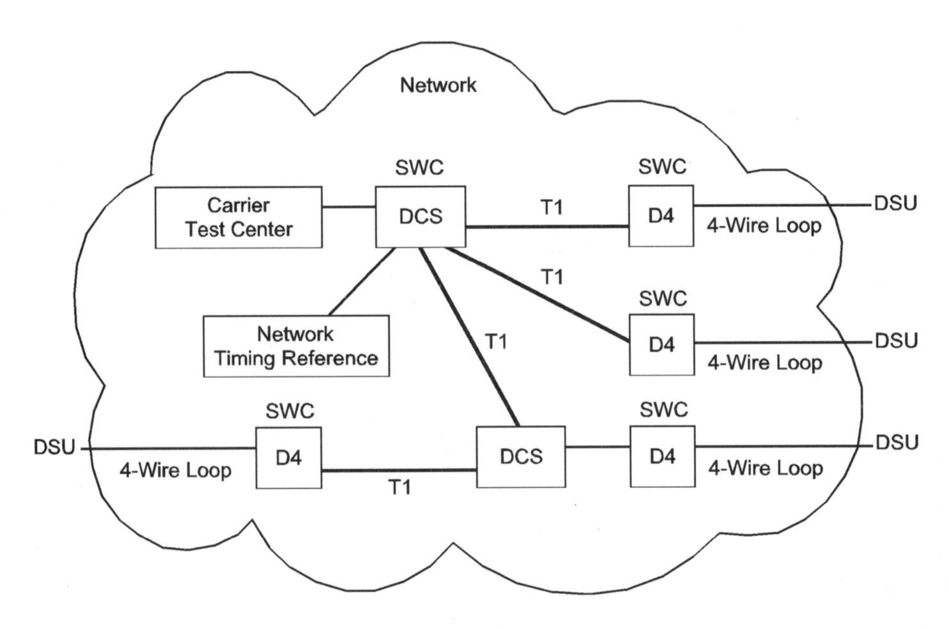

Figure G2 *Network elements associated with the provision and support of generic digital services (GDS).*

On the other side of the network interface is customer-owned network channel terminating equipment, which for GDS is a digital service unit (DSU). The functions of the DSU include generating and reconstructing the digital signal, signal encoding and formatting, timing extraction from the incoming signal (including sampling and loop timing), and generating and recognizing control signals.

Serving wire center. At the serving wire center, the local loop is terminated on an office channel unit data port on a D4 digital channel bank. The office channel unit data port's functions include transmitting outgoing loop signals to the customer station; reshaping, retiming, and regenerating the incoming loop signal; assembling the data into a format suitable for network and loop transmission; providing electrical current to the loop; and passing network-originated commands (e.g., loopbacks) to the DSU to implement diagnostics.

Interoffice facilities. Interoffice facilities carry 24 individual 64 kbps (DS0) channels on the 1.544 Mbps (DS1) transport facilities between the digital nodes. In addition to the cost of the local loop, the customer is charged for interoffice facilities between serving wire centers and the customer premises. The charges include a fixed element, generally called a *channel termination,* and a mileage-sensitive element. Mileage is calculated for the shortest airline distance between serving wire centers using vertical and horizontal coordinates. These rate elements consist of monthly recurring charges only.

Digital crossconnect system. The digital crossconnect system can electronically connect individual 64 kbps DS0 channels. These channels can be monitored, tested, and reconfigured from the local exchange carrier's network test systems. Users can tap into the DCS with on-premises network management terminals to reconfigure the channels.

D4 channel bank. The D4 channel bank is a time-division multiplexer (TDM) that combines 24 input signals for transport to a digital node over a T1 link. The channel bank's Office Channel Unit Data Port (OCUDP) performs several functions: transmitting outgoing loop signals to the customer station; reshaping, retiming, and regenerating the incoming loop signal; assembling the data into a format suitable for network and loop transmission; providing current to the loop; and performing the diagnostic and functional commands received from the network test center.

Secondary channel. The secondary channel gives customers an independent low-speed, derived data channel that operates in parallel with the primary data channel. It can be used to perform network surveillance on a continuous and nondisruptive basis, meaning that production data can continue to flow unaffected over the primary channel.

Although the secondary channel is used primarily for the exchange of test, diagnostic, measurement, and control messages between a central site and its remote terminals, it may also be used as a low-speed, general-purpose data channel. To take advantage of the secondary channel, the user must have DSUs that specifically support this capability. Table G3 summarizes the primary and secondary channels available with GDS.

Summary. A significant advantage of generic digital services is that bandwidth can be used more efficiently than with other types of digital services. The flexible multiplexing scheme improves the integration of different applications on one net-

Table G-3. Primary and Secondary GDS Channels.

Primary Channel	Secondary Channel
2.4 kbps	133 bps
4.8 kbps	266 bps
9.6 kbps	533 bps
19.2 kbps	1066 bps
56 kbps	2132 bps
64 kbps	Not Available

work, and the use of multiport, multipoint DSUs improves bandwidth efficiency even more. If the somewhat lower service quality and availability of GDS (compared to DDS) is acceptable, they are an excellent way to improve a network's efficiency and reduce costs.

See also

DIGITAL DATA SERVICES

Global Positioning System

The Global Positioning System (GPS) is a network of 24 Navstar satellites orbiting Earth at 11,000 miles. Originally established by the U.S. Defense Department at a cost of about US$13 billion, access to GPS is free to all users, including those in other countries. The system's positioning and timing data are used for a variety of applications, including air, land, and sea navigation, vehicle and vessel tracking, surveying and mapping, and asset and natural resource management. With military accuracy restrictions lifted in March, 1996, the GPS can now pinpoint the location of objects as small as a dime anywhere on the earth's surface.

The first GPS satellite was launched in 1978. The first 10 satellites were developmental satellites, called Block I. From 1989 to 1993, 23 production satellites, called Block II, were launched. The launch of the 24th satellite in 1994 completed the system. The satellites are positioned so that signals from six of them can be received nearly 100 percent of the time at any point on earth.

GPS components. The GPS consists of satellites, receivers, and ground control systems. The satellites transmit signals (1575.42 MHz) that can be detected by GPS receivers on the ground. These receivers can be mounted in ships, planes, and cars to provide exact position information, regardless of weather conditions. They detect, decode, and process GPS satellite signals to give the precise position of the user.

The GPS control or ground segment consists of five unmanned monitor stations located in Hawaii, Kwajalein in the Pacific Ocean, Diego Garcia in the Indian Ocean, Ascension Island in the Atlantic Ocean, and Colorado Springs, Colo. There is also a master ground station at Falcon AFB in Colorado Springs, and four large ground antenna stations that broadcast signals to the satellites. The stations also track and monitor the GPS satellites.

System operation. With GPS, signals from the satellites arrive at the exact position of the user and are triangulated. To triangulate, GPS measures distance using the travel time of a radio message from the satellite to a ground receiver. To measure travel time, GPS uses very accurate clocks in the satellites. Once the distance to a satellite is known, knowledge of the satellite's location in space is used to complete the calculation. GPS receivers on the ground have an "almanac" stored in their computer memory, which indicates where each satellite will be in the sky at any given time. GPS receivers calculate for ionospheric and atmospheric delays to further fine-tune the position measurement.

To make sure both satellite and receiver are synchronized, each satellite has four atomic clocks that keep time to within 3 ns, or three billionths of a second. For cost savings, the clocks in the ground receivers are made somewhat less accurate. To compensate, an extra satellite range measurement is taken. Trigonometry says that if three perfect measurements locate a point in three-dimensional space, then a fourth measurement can eliminate any timing offset. This fourth measurement compensates for the receiver's imperfect synchronization.

The ground unit receives the satellite signal, which travels at the speed of light. Even at this speed, the signal takes a measurable amount of time to reach the receiver. The difference between the time the signal is sent and the time it is received, multiplied by the speed of light, allows the receiver to calculate the distance to the satellite. To measure precise latitude, longitude, and altitude, the receiver measures the time it took for the signals from several satellites to get to the receiver (Fig. G3).

GPS uses a system of coordinates called the Worldwide Geodetic System 1984 (WGS-84). This is similar to the familiar latitude and longitude lines commonly seen on large wall maps. The WGS-84 system provides a built-in, standardized frame of reference, enabling receivers from any vendor to provide exactly the same positioning information.

GPS applications. Although the GPS system was completed only in 1994, it has already proven itself in military applications, most notably in Operation Desert Storm, where U.S. and allied troops faced a vast, featureless desert. Without a reliable navigation system, sophisticated troop maneuvers could not have been performed. This could have prolonged the operation well beyond the 100 hours it actually took.

With GPS, troops were able to go places and maneuver in sandstorms or at night when even the soldiers native to the area could not. Initially, more than 1000 portable commercial receivers were purchased for use in Desert Storm. The demand was so great that, before the end of the conflict, more than 9000 commercial receivers were in use in the Gulf region. They were carried by foot soldiers and attached to vehicles, helicopters, and aircraft instrument panels. GPS receivers were used in several aircraft, including F-16 fighters, KC-135 tankers, and B-52s. Navy ships used GPS receivers for rendezvous, mine sweeping, and aircraft operations.

Today GPS has become an important element of nearly all military operations and weapons systems. In addition, it is used on satellites to obtain highly accurate orbit data and to control spacecraft orientation.

While the GPS system originally was developed to meet the needs of the military community, new ways to use its capabilities are continually being found, from the exotic to the mundane. Among the former is the use of GPS for wildlife man-

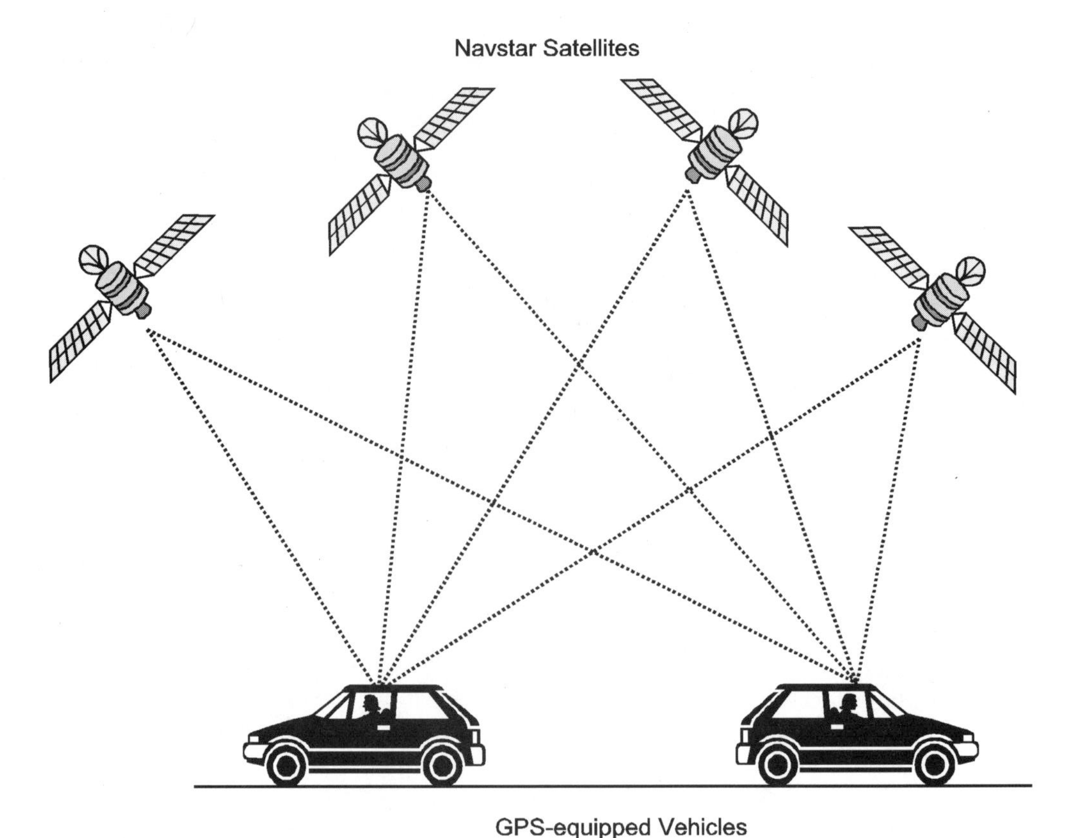

Figure G3 *Signals from four satellites, captured by a vehicle's onboard GPS receiver, are used to determine precise location information.*

agement. Endangered species such as Montana elk and Mojave Desert tortoises have been fitted with GPS receivers and tiny transmitters to help determine population distribution patterns and possible sources of disease. In Africa, GPS receivers are used to monitor the migration patterns of large herds for a variety of research purposes.

Handheld GPS receivers are now routinely used in field applications that require precise information gathering, including field surveying by utility companies, mapping by oil and gas explorers, and resource planning by timber companies.

GPS-equipped balloons are monitoring holes in the ozone layer over the polar regions, and air quality is being monitored using GPS receivers. Buoys tracking major oil spills transmit data using GPS. Archaeologists and explorers are using the system to mark remote land and ocean sites until they can return with proper equipment and funding.

Vehicle tracking is one of the fastest-growing GPS applications. GPS-equipped fleet vehicles, public transportation systems, delivery trucks, and courier services use receivers to monitor their locations at all times.

GPS is also helping save lives. Many police, fire, and emergency medical service units are using GPS receivers to determine the police car, fire truck, or ambulance nearest to an emergency, enabling the quickest possible response in life-or-death situations. GPS data will become more useful to consumers when it is linked with digital mapping. Accordingly, some automobile manufacturers are offering, as an option on new vehicles, moving-map displays guided by GPS receivers. The displays can even be removed and taken into a home to plan a trip. Some GPS-equipped vehicles give directions to drivers on display screens and through synthesized voice instructions. These features allow drivers to get where they want to go more rapidly and safely than has ever been possible before.

When GPS data is used in conjunction with geographic data collection systems, it is possible to instantaneously arrive at sub-meter positions together with feature descriptions to compile highly accurate geographic information systems (GIS). When used by cities and towns, for example, GPS can help in managing the kinds of assets listed in Table G4.

Table G-4. GPS Asset Management Applications.

Point Features	Line Features	Area Features
Signs	Streets	Parks
Manhole covers	Sidewalks	Landfills
Fire hydrants	Fitness trails	Wetlands
Light poles	Sewer lines	Planning zones
Storm drains	Water lines	Subdivisions
Driveways	Bus routes	Recycling centers

Some government agencies, academic institutions, and private companies are using GPS to determine the location of a multitude of features, including point features such as pollutant discharges and water supply wells, line features such as roads and streams, and area features such as waste lagoons and property boundaries. Before GPS, such features had to be located with surveying equipment, aerial photographs, or satellite imagery. With GPS, these features can now be located by a single operator using handheld equipment.

GPS and cellular. GPS technology is even being used in conjunction with cellular technology to provide value-added services. With the push of a button on a cellular telephone, automobile drivers and operators of commercial vehicles in some areas can talk to a service provider and simultaneously signal their position, emergency status, or equipment failure information to auto clubs, security services, or central dispatch services.

This is possible with Motorola's Cellular Positioning and Emergency Messaging Unit, which brings a new era of mobile security and tracking to those who drive automobiles or operate fleets. The system is designed for sale to systems integrators who configure consumer and commercial systems that operate via cellular telephony. The Cellular Positioning and Emergency Messaging Unit communicates

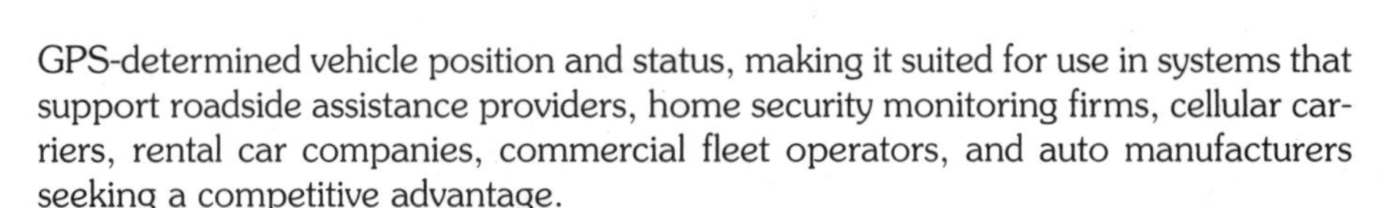

GPS-determined vehicle position and status, making it suited for use in systems that support roadside assistance providers, home security monitoring firms, cellular carriers, rental car companies, commercial fleet operators, and auto manufacturers seeking a competitive advantage.

Skytel, the provider of wireless messaging services, is marketing its GPS-based AutoLink system for automobiles. The AutoLink system provides automatic emergency response, theft deterrence, vehicle tracking and immobilization, two-way personal paging, remote vehicle unlocking, driver personalization, navigational guidance, and location-based information service.

Summary. Due to its accuracy, GPS is rapidly becoming the locational data collection method of choice for a variety of commercial, government, and military applications. GPS certainly has become an important and cost-effective method for locating terrestrial features too numerous or too dynamic to be mapped by traditional methods. Although originally funded by the U.S. Department of Defense, access to the GPS network is free to all users. This has encouraged applications development and created an entirely new consumer market, particularly in the area of vehicular location and highway navigation.

Global System for Mobile (GSM) telecommunications

Global System for Mobile (GSM) telecommunications (formerly known as Groupe Spéciale Mobile, for the group that started developing the standard in 1982) was designed from the beginning as an international digital cellular service. It was intended that GSM subscribers should be able to cross national borders and find that their mobile services crossed with them. Today GSM is well established in most countries, with the highest concentration of service providers and users in Europe.

Originally the 900 MHz band was reserved for GSM services. Since GSM first entered commercial service in 1992, it has been adapted to work at 1800 MHz for the Personal Communications Networks (PCN) in Europe, and at 1900 MHz for Personal Communications Services (PCS) in the U.S.

GSM services. GSM telecommunication services are divided into teleservices, bearer services, and supplementary services.

Teleservices. The most basic teleservice supported by GSM is telephony. There is an emergency service, where the nearest emergency-service provider is notified by dialing three digits (similar to 911). Group 3 fax, an analog method described in ITU-T recommendation T.30, also is supported by use of an appropriate fax adapter.

Bearer services. A unique feature of GSM compared to older analog systems is the Short Message Service (SMS). SMS is a bidirectional service for sending short alphanumeric messages (up to 160 bytes) in a store-and-forward manner. For point-to-point SMS, a message can be sent to another subscriber to the service, and an acknowledgment of receipt is provided to the sender. SMS also can be used in a cell broadcast mode for sending messages such as traffic updates or news updates. Messages can be stored in a smart card called the Subscriber Identity Module (SIM) for later retrieval.

Since GSM is based in digital technology, it allows synchronous and asynchronous data to be transported as a bearer service to or from an ISDN terminal. The data rates supported by GSM are 300 and 600 bps, as well as 1.2, 2.4, and 9.6 kbps. Data can use either the transparent service, which has a fixed delay but no guarantee of data integrity, or a nontransparent service, which guarantees data integrity through an Automatic Repeat Request (ARQ) mechanism, but with variable delay.

Supplementary services. Supplementary services are provided on top of teleservices or bearer services, and include such features as caller identification, call forwarding, call waiting, and multiparty conversations. There also is a lockout feature that prevents the dialing of certain types of calls, such as international calls.

Network architecture. A GSM network consists of the following elements: mobile station, base station subsystem, and mobile services switching center (MSC). Each GSM network also has an operations and maintenance center, which oversees the proper operation and setup of the network. There are two air interfaces: The Um interface is a radio link over which the mobile station and the base station subsystem communicate, and the A interface is a radio link over which the base station subsystem communicates with the MSC.

The mobile station. The mobile station (MS) consists of the radio transceiver, display and digital signal processors, and the Subscriber Identity Module (SIM).

The SIM provides personal mobility, so that the subscriber can have access to all services regardless of the terminal's location or the specific terminal used. By removing the SIM from one GSM cellular phone and inserting it into another GSM cellular phone, the user is able to receive calls at that phone, make calls from that phone, or receive other subscribed services. The SIM card may be protected against unauthorized use by a password or personal identity number.

An International Mobile Equipment Identity (IMEI) number uniquely identifies each mobile station. The SIM card contains an International Mobile Subscriber Identity (IMSI) number, identifying the subscriber and containing a secret key for authentication and other user information. Since the IMEI and IMSI are independent, this arrangement provides users with a high degree of personal mobility.

Base station subsystem. The base station subsystem consists of two parts: the base transceiver station (BTS) and the base station controller (BSC). These communicate across the A-bis interface, enabling operation between components made by different suppliers.

The base transceiver station contains the radio transceivers that define a cell and handles the radio-link protocols with the mobile stations. In a large urban area, there typically will be a number of BTSs to support a large subscriber base of mobile service users.The base station controller provides the connection between the mobile stations and the mobile service switching center (MSC). It manages the radio resources for the BTSs, handling such functions as radio channel setup, frequency hopping, and hand-offs. The BSC also translates the 13 kbps voice channel used over the radio link to the standard 64 kbps channel used by the land-based Public Switched Telephone Network (PSTN) or ISDN.

Mobile services switching center. The mobile services switching center (MSC) acts like an ordinary switching node on the PSTN or ISDN, and provides all the functionality needed to handle a mobile subscriber, such as registration, au-

thentication, location updating, hand-offs, and call routing to a roaming subscriber. These services are provided in conjunction with several other components, which together form the network subsystem. The MSC provides the connection to the public network (PSTN or ISDN) and signaling between various network elements that use Signaling System Number 7 (SS7).

The MSC contains no information about particular mobile stations. This information is stored in two location registers, which are essentially databases. The Home Location Register (HLR) and Visitor Location Register (VLR), together with the MSC, provide the call routing and roaming (national and international) capabilities of GSM.

The HLR contains administrative information for each subscriber registered in the corresponding GSM network, along with the current location of the mobile device. The current location of the mobile device is in the form of a Mobile Station Roaming Number (MSRN), which is a regular ISDN number used to route a call to the MSC where the mobile device is currently located. Only one HLR is needed per GSM network, although it may be implemented as a distributed database.

The Visitor Location Register (VLR) contains selected administrative information from the HLR, necessary for call control and provision of the subscribed services, for each mobile device currently located in the geographical area controlled by the VLR.

There are two other registers that are used for authentication and security purposes. The Equipment Identity Register (EIR) is a database that contains a list of all valid mobile equipment on the network, where each mobile station is identified by its IMEI. An IMEI is marked as invalid if it has been reported stolen or is not type-approved. The Authentication Center is a protected database that stores a copy of the secret key stored in each subscriber's SIM card, which is used for authentication.

Channel derivation and types. Because radio spectrum is a limited resource shared by all users, a method must be devised to divide up the bandwidth among as many users as possible. The method used by GSM is a combination of Time Division and Frequency Division Multiple Access (TDMA/FDMA).

The FDMA part involves the division by frequency of the total 25 MHz bandwidth into 124 carrier frequencies of 200 kHz bandwidth. One or more carrier frequencies are then assigned to each base station. Each of these carrier frequencies is then divided in time, using a TDMA scheme, into eight time slots. One time slot is used for transmission and one for reception by the mobile device. They are separated in time so that the mobile unit does not receive and transmit at the same time.

Within the framework of TDMA, two types of channels are provided: traffic channels and control channels. Traffic channels carry voice and data between users, while the control channels carry information that is used by the network for supervision and management. Among the control channels are the following:

- *Fast Associated Control Channel (FACCH):* This channel is created by robbing slots from a traffic channel to transmit power control and hand-off signaling messages.
- *Broadcast Control Channel (BCCH):* This channel continually broadcasts, on the downlink, information including base station identity, frequency allocations, and frequency-hopping sequences.

- *Stand-alone Dedicated Control Channel (SDCCH):* This is used for registration, authentication, call setup, and location updating.
- *Common Control Channel (CCCH):* This comprises three control channels used during call origination and call paging.
- *Random Access Channel (RACH):* This channel used to request access to the network.
- *Paging Channel (PCH):* This is used to alert the mobile station of an incoming call.

Authentication and security. Because radio signals can be accessed by virtually anyone, authentication of users, to prove that they are who they claim to be, is a very important feature of a mobile network.

Authentication involves two functional entities, the mobile unit SIM card and the Authentication Center (AC). Each subscriber is given a secret key, one copy of which is stored in the SIM card and the other in the Authentication Center. During authentication, the AC generates a random number that it sends to the mobile unit. Both the mobile unit and the AC then use the random number, in conjunction with the subscriber's secret key and an encryption algorithm called A3, to generate a number that is sent back to the AC. If the number sent by the mobile unit is the same as the one calculated by the AC, the subscriber is authenticated.

The calculated number also is used, together with a TDMA frame number and another encryption algorithm called A5, to encrypt the data sent over the radio link, preventing others from listening in. Encryption provides an added measure of security, since the signal is already coded, interleaved, and transmitted in a TDMA manner, thus providing protection from all but the most technically astute eavesdroppers.

Another level of security is performed on the mobile equipment, as opposed to the mobile subscriber. As mentioned earlier, each GSM terminal is identified by a unique International Mobile Equipment Identity (IMEI) number. A list of IMEIs in the network is stored in the Equipment Identity Register (EIR). The status returned in response to an IMEI query to the EIR is one of the following:

- *White-listed* indicates that the terminal is allowed to connect to the network.
- *Gray-listed* indicates that the terminal is under observation from the network for possible problems.
- *Black-listed* indicates that the terminal has either been reported as stolen, or it is not type-approved (i.e., the correct type of terminal for a GSM network). The terminal is not allowed to connect to the network.

Summary. Soon it will be possible to increase transmission speed up to 28.8 kbps, a significant leap from the 9.6 kbps circuit-switched data as it stands today. For even higher speeds, a new part of the GSM standard is being developed, known as High Speed Circuit Switched Data (HSCSD), which will boost user capacity up to 64 kbps. Under HSCSD, all eight voice channels provided by TDMA will be used to provide one 64 kbps circuit. This will allow GSM phones and other devices to handle advanced applications such as videoconferences, multimedia presentations, and medical scans.

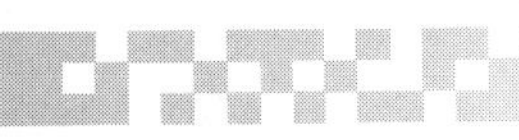

See also

CODE DIVISION MULTIPLE ACCESS
DIGITAL ENHANCED CORDLESS TELECOMMUNICATIONS
TIME DIVISION MULTIPLE ACCESS

Help desks

With personal computer hardware and software now permeating corporations, IS and telecom managers are faced with the task of providing troubleshooting assistance to users scattered throughout the organization. The consequences of not providing adequate levels of assistance are too compelling to ignore: lost corporate productivity, slowed responses to competitive pressures, and eventual loss of market share. One way to service the needs of a growing population of computer and communications users efficiently and economically is to set up a help desk.

Briefly, the help desk acts a central clearinghouse for support issues, and is manned by a technical staff that addresses support problems and attempts to solve them in house before calling in vendors or carriers. The help desk operator logs every call and, if possible, attempts to isolate the cause of the caller's problem. If the problem cannot be solved over the phone, the operator dispatches a technician and monitors progress to a satisfactory conclusion before closing out the transaction.

Help desk operators usually are able to answer from 50 to 70 percent of all calls without having to pass them to another authority. Aside from handling calls from users, help desks can provide such services as order and delivery tracking, asset and inventory tracking, preventive maintenance, and monitoring vendor performance.

Although the concept of the help desk originated in the mainframe environment, increased corporate reliance on LANs has expanded the role of the help desk into realms not ventured into by many mainframe professionals. Fortunately, there are now software packages available that assist with help desk administration. There also are problem determination tools, based on expert systems, that can quickly bring untrained help desk personnel up to competency.

A recent trend is to extend the availability of help desk support to the World Wide Web (WWW). With the ability to update transactions via the Web, field technicians now can update information in existing requests from any location where they have Web browser access. End users can perform a range of actions such as authorize a change request and add new information to an existing trouble ticket. Help desk applications and data generate Web forms (schemas) and hyperlinks dynamically (Fig. H1). This lets help desk managers focus on serving customers and improving business processes in the help desk system, instead of constantly maintaining static Web pages.

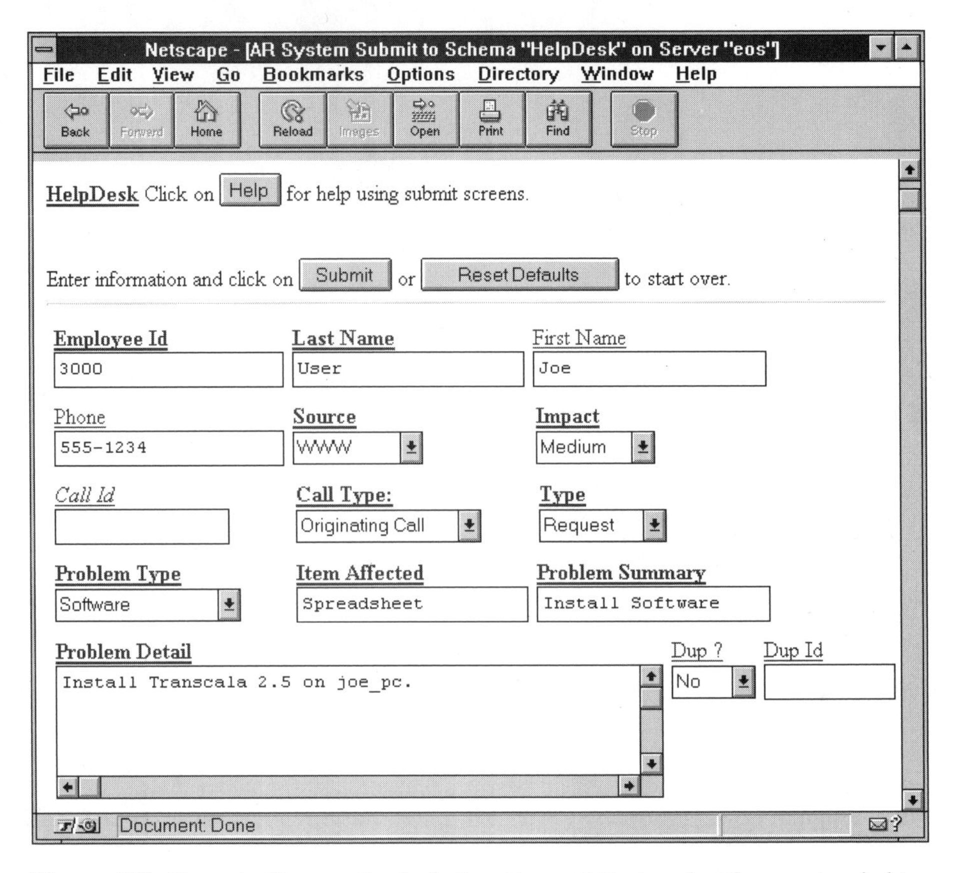

Figure H1 *Remedy Corporation's Action Request System has been extended to the Web with the ARWeb client, which lets organizations create a Web-based help desk that can be accessed by anyone with a Web browser. In this case, a user can access a Web form for reporting a problem to the help desk.*

Benefits. Establishing a centralized help desk to coordinate the resolution of system problems offers a number of benefits. Users have a single number to remember; support personnel are assured of an orderly, controlled flow of tasks and assignments; and management is provided with an effective means of tracking problems and solutions. The help desk provides users with a "warm and fuzzy" level of support. Knowing that someone is available to solve any problem, or even to help them find their way through unfriendly documentation, adds to an individual's confidence and willingness to learn new applications and office technologies.

Although a help desk costs money, it can pay for itself in ways that unfortunately can be hard to quantify. The fact is, most companies have millions of dollars invested in computer and communication systems. They also have millions of dollars invested in people. To ensure that both are utilized to best advantage, there should be an entity in place that is capable of solving the many and varied problems that inevitably arise.

Determining what level of support the help desk should provide can be difficult. One way to determine the proper level of support is to have a system that automatically tracks calls to the help desk and records what problems users encounter most. That way, people with appropriate expertise can be identified to lend assistance, or be recruited from outside the organization to fill in any gaps. Depending on the problems, in-house training might provide the dual benefits of helping users become more productive and reducing the support burden on IS staff.

Internal vs. external support. Assuming these benefits are attractive, the next step in setting up a help desk is to decide whether to provide the service internally or rely on an outside vendor. Another option is to offer the help desk as a billable corporate service. This can prevent the help desk from becoming deluged with calls from users who can just as easily look up the information in manuals or use the online help facilities that come with many application packages.

While it is important that the help desk staff have technical skills, "people" skills are much more useful. When a user calls with a problem, the support person must extract information from someone who does not know the technical jargon needed to reach a solution or how to use seemingly arcane DOS commands to isolate and solve problems.

The help desk operator can carry out all the maintenance activities that once required dispatching an on-site technician. Instead of the help staff or technician having to go to users' desks and physically look over their shoulders, a remote control product allows the help desk operator to stay put, hit a few keys, and instantly see a user's screen and control the keyboard. By taking control of the user's computer, the help desk operator often can assess the problem immediately.

An alternative to internally staffed help desks is subscription to a commercial service. Subscribers typically call an 800 number to get help in the use of DOS-based microcomputers, software programs, and peripherals. A variety of pricing schemes are available. An annual subscription, for instance, may entitle the company to unlimited advice and consultation. Per-call pricing also is available. The cost of basic services is between US$10 to US$15 per call, depending on call volume. Depending on the service provider, calls usually are limited to 15 or 20 minutes. If the problem cannot be resolved within that time, the charge for the call might even be waived.

Resolution can be anything, from talking the caller through a system reboot, to helping the user to ascertain that a micro-to-mainframe link is out of order and what must be done to get it back into service. Callers also can obtain solutions to common software problems, from how to recover an erased file, to importing files from one package to another, to modifying system configuration files when adding new applications packages or hardware.

Some hot line services even offer advice on the selection of software and hardware products, provide software installation support, and guide users through maintenance and troubleshooting procedures. Other service providers specialize in LANs of one type or another.

Because many callers are not familiar with the configuration details of their hardware and software, this information is compiled into a subscriber profile for easy access by online technicians. With this information readily available, the time spent with any single caller is greatly reduced. This helps to keep down the cost per call, which

is what attracts new customers. The cost of a customer profile varies greatly, depending on the size of the database that must be compiled. It might be based on the number of potential callers, or entail a flat yearly charge for the entire organization.

The operations of some service providers are becoming quite sophisticated. To service their clients, staff may access a shared knowledge base that could contain tens of thousands of questions and answers. Sometimes called an *expert system,* the initial knowledge base is compiled by technical experts in their respective fields. As online technicians encounter new problems and devise solutions, this information is added to the knowledge base for immediate access when the same problem is encountered at a later time.

Even users with access to vendors' free help line services can benefit from a third-party help line. This is because the third-party service may cover situations where more than one application or hardware platform is being used, whereas vendor help line services only provide assistance covering their own products. Moreover, most users rarely bother to phone the vendors because of constantly busy lines. So rather than supplanting vendor services and in-house support desks (which may be overburdened), third-party services can be used to complement them.

In addition to delivering a variety of help services, for an extra charge third-party service providers can issue call-tracking and accounting reports to help clients keep a lid on expenses for this kind of service and allocate expenses appropriately among departments and other internal cost centers.

Summary. The help desk can improve the quality of service while decreasing service costs. In addition, it frees up experts to work in other areas and provides consistent answers to questions. Some expert systems use the database to generate graphics and text reports on what types of hardware and software cause the most problems. This information can be used to guide product purchases and ascertain the response times of vendors. The continued growth of distributed computing over LANs and WANs, and the increasing complexity of hardware and software, will expand the role of the corporate help desk.

See also
ASSET MANAGEMENT
NETWORK MANAGEMENT SYSTEMS

Hertz

The frequency of electromagnetic waves (indeed, any waves) is expressed in cycles per second, now called *hertz* (Hz). An electromagnetic wave is composed of complete cycles. The number of cycles that occur each second is the measure of frequency, while the peak-to-peak distance of the waveform gives the amplitude of the signal (Fig. H2).

The top frequency of standard speech is about 3000 cycles per second, or 3 kilohertz (kHz). Some radio waves may have frequencies of many millions of hertz (megahertz, or MHz), and even billions of hertz (gigahertz, or GHz). Table H1 shows the range of radio frequencies and their band classifications.

The term *hertz* was adopted in 1960 by an international group of scientists and engineers at the General Conference of Weights and Measures, in honor of

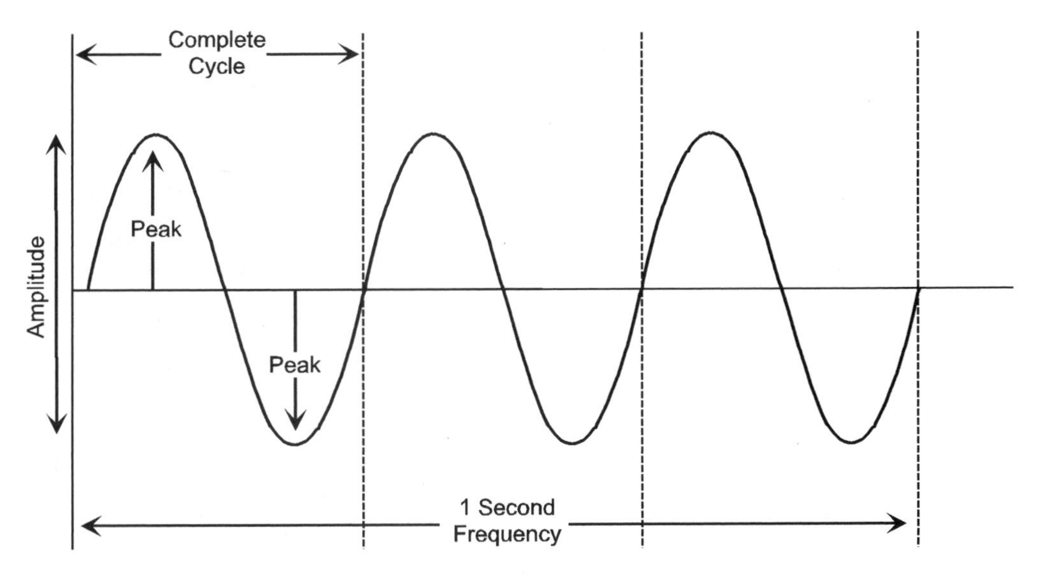

Figure H2 *Each cycle per second equates to 1 hertz (Hz). In this case, 3 cycles occur in 1 second, which equates to 3 Hz.*

Table H-1. Radio Frequency Bands.

Frequency	Band Classification
less than 30 kHz	Very low frequency (VLF)
30 kHz to 300 kHz	Low frequency (LF)
300 kHz to 3 MHz	Medium frequency (MF)
3 MHz to 30 MHz	High frequency (HF)
30 MHz to 300 MHz	Very high frequency (VHF)
300 MHz to 3 GHz	Ultrahigh frequency (UHF)
3 GHz to 30 GHz	Super high frequency (SHF)
more than 30 GHz	Extremely high frequency (EHF)

Heinrich R. Hertz (1857–1894), a German physicist. Hertz is best known for his discovery of electromagnetic waves, which had been predicted by British scientist James Clerk Maxwell in 1864. Hertz used a rapidly oscillating electric spark to produce ultrahigh frequency waves. These waves caused similar electrical oscillations in a distant wire loop. The discovery of electromagnetic waves and how they could be manipulated paved the way for the development of radio, television, radar, and other forms of telecommunications.

See also
DECIBEL

Hierarchical storage management

Hierarchical storage management provides a cost-effective and efficient solution to meet the demand for increased storage space. HSM uses two or more levels of storage; a three-level scheme is typical: on-line, off-line, and near- line. HSM came about because of the need to move low-volume and infrequently accessed files from disk, thus freeing up space. Although disks can be added as storage requirements increase, budget constraints often limit the long-term viability of this solution. In an HSM scheme, data can be categorized according to its frequency of usage and stored appropriately: online, near-line, or offline. Different storage media come into play for each of these categories; migration operations are under control of an HSM management system (Fig. H3).

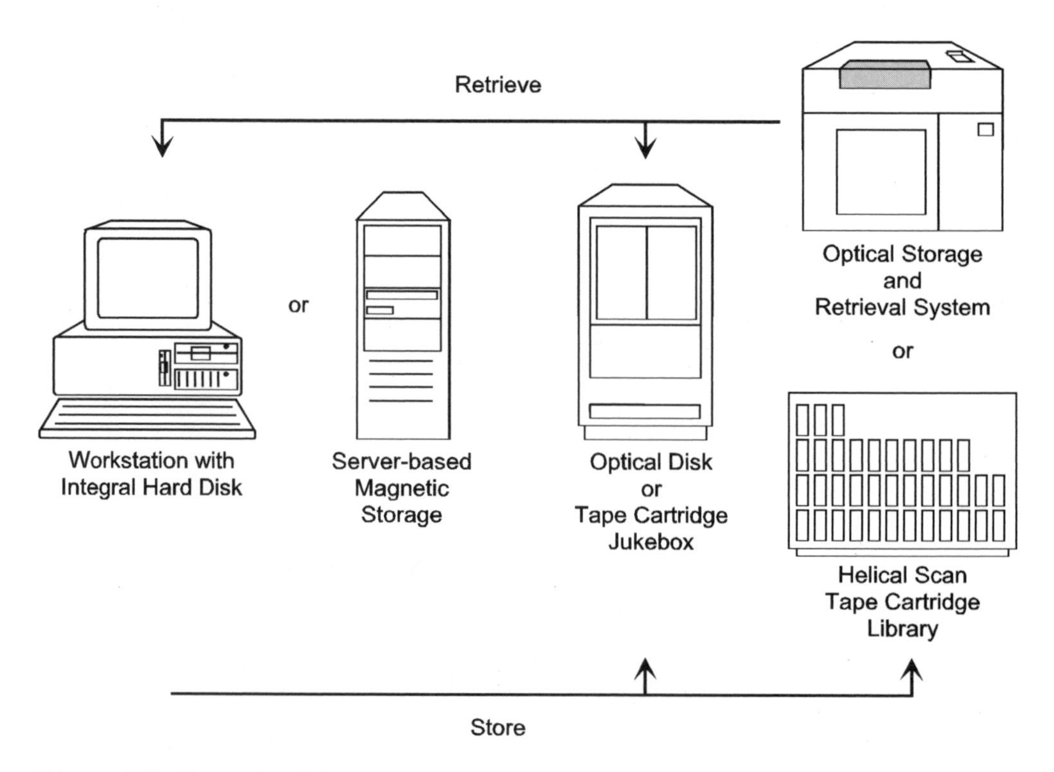

Figure H3 *Hierarchical data storage spanning magnetic disk, optical disk, and tape options.*

Frequently used files are stored online on local disk drives installed in a server or workstation. Occasionally used files are stored near-line on secondary storage devices such as rewritable optical disks installed in a server-like device called an *autochanger* or *jukebox*. Infrequently used files usually are migrated offline to tape cartridges that are stored in a tape jukebox or a library facility capable of holding hundred or thousands of tapes. The library facility uses sophisticated robotics to retrieve individual bar-coded cartridges and inserts them into a tape drive so that the data can be migrated to local storage media. The exchange time can be several minutes.

For organizations with mixed needs for online and offline storage, a near-line automated tape library offers the best compromise between price and performance. These systems bridge the gap between fast, expensive online disk storage and slow, high- capacity offline tape libraries. The exchange time is about 30 seconds.

A management system determines when a file should be transferred or retrieved, initiates the transfer, and keeps track of its new location. As files are moved from one type of medium to another, they are put into the proper directory for user access. The management system automatically optimizes storage utilization across different media types by removing files from one to the other until they are permanently archived in the most economical way, usually a tape library. At the same time, individual files or whole directories can be excluded from migration.

Data migration can be controlled according to such criteria as file size and last date of access. Files can also be migrated when the hard disk reaches a specified capacity threshold. For example, when magnetic disk storage reaches the established threshold of 80 percent, files are migrated to optical storage, freeing up magnetic storage until it reaches another specified threshold, say 60 percent.

Summary. Huge amounts of data—hundreds of gigabytes or even terabytes—cannot be managed efficiently without the help of HSM solutions. HSM balances the cost, capacity, and efficiency of different storage media according to the frequency of data usage. This provides all clients and servers with expanded online storage through the migration of files between magnetic disks, optical disks, and tape, according to predefined rules.

See also

REDUNDANT ARRAY OF INEXPENSIVE DISKS (RAID)
STORAGE MEDIA

High-definition television

High-definition television (HDTV) is a new kind of system that uses digital technology to improve both the production and reception of television programs. HDTV eventually will replace all existing analog television systems in use today. During the transition, the needs being met by the current system will continue to be met by HDTV.

HDTV offers several improvements on the current television standard developed by the National Television Standards Committee (NTSC). The NTSC standard consists of a picture with 525 lines of resolution, a 4:3 horizontal-to-vertical aspect ratio, and analog sound. HDTV offers 1124 lines of resolution, a panoramic 16:9 aspect ratio, and digital sound. HDTV also permits multiple digital audio tracks, which not only allows for stereo, CD-quality sound, but audio in several languages.

HDTV offers a full 60 frames per second temporal resolution, which is twice that offered under the NTSC standard. This allows very smooth motion and high picture clarity. The picture will be displayed in a way that is more like film, adding the feeling of realism to TV.

Development efforts. Research into HDTV was initiated in 1968 by Nippon Hoso and Kyokai (NHK), the Japan Broadcasting Corporation. This original system, called MUSE, was analog and required 36 MHz channels, six times wider than the U.S. 6 MHz standard. Japan and Europe further developed the system and each launched experimental services. In 1985 the quest for a single worldwide production standard was thwarted in a political struggle between Europe and Japanese competitors. But the development of HDTV continued, driven mostly by the fear that Japan would gain commercial domination of everything in the next century, particularly in the field of communications technology.

Many U.S. companies responded to this perceived threat and began developing their own versions of HDTV. The development of an HDTV standard had been underway in the U.S. since 1987. Initially all of the proposals were for analog transmissions, but in 1990 the FCC was confronted with four digital proposals among which it could not choose. Instead of selecting one of these, the Commission suggested the idea of a Grand Alliance of companies and research organizations that would all work together to come up with one standard for HDTV. Members of the so-called Grand Alliance included AT&T, General Instrument Corporation, the Massachusetts Institute of Technology, Philips Consumer Electronics, the David Sarnoff Research Center, Thomson Consumer Electronics, and Zenith Electronics Corporation. The Grand Alliance issued a proposal based on recommendations from the Advanced Television Systems Committee (ATSC), which included support for 18 different digital formats of information and MPEG 2 for compression. The result was a standard called Digital Television, or DTV, which was proposed to the FCC in 1996.

FCC approval. In the Telecommunications Act of 1996, Congress directed the FCC to issue licenses for digital television to incumbent television broadcasters. In April, 1997, the FCC issued its plan for the commercial implementation of DTV. The overarching goal of the plan is to provide for the success of free, local digital broadcast television.

To bolster DTV's chance for success, the FCC decided to let broadcasters use their channels according to their best business judgment, as long as they continue to offer the free programming on which the public has come to rely. Specifically, broadcasters must provide a free digital video programming service that is at least comparable in resolution to today's service and aired during the same time periods as today's analog service. The Commission will not require broadcasters to air high-definition programming or, initially, to simulcast their analog programming on the digital channel.

Broadcasters will be able to put together whatever package of digital product they believe will best attract customers, and to develop partnerships with others to help make the most productive and efficient use of their channels. These services could include data transfer, subscription video, interactive materials, audio signals, and whatever other innovations broadcasters can promote, and from which, profit. Giving broadcasters the flexibility in their use of their digital channels will allow them to put together the best mix of services and programming to stimulate consumer acceptance of digital technology and the purchase of new digital receivers.

The FCC requires the affiliates of the top four networks in the top 10 markets to be on the air with a digital signal by May, 1999. The top ten markets include 30 percent of television households. Affiliates of the top four networks in markets 11 to 30 must be on the air by November, 1999. The top 30 markets include 53 percent of television households. The FCC has set a target of 2006 as a reasonable date for the end of NTSC service. After that, there will be no analog TV.

Summary. The FCC will review the progress of DTV implementation every 2 years and make adjustments in the timetables, if necessary. Stations that cannot meet the target dates for DTV service due to unforeseen circumstances will be given extensions to the deadline. About 30 stations have committed to having DTV service in operation by October, 1998. It is anticipated that by Christmas of that year, consumers will be able to buy a digital TV for under US$2000.

Hubs

With the increasing complexity of today's networks, the conventional bus and ring LAN topologies have exhibited shortcomings. In the past, a fault anywhere in the cabling often brought down the entire network or a significant portion of it. This weakness was compounded by the fact that technicians could not readily identify the point of failure from a central administration point, which tended to prolong network downtime. This situation led to the development of the wiring hub in the mid-1980s.

Hubs are at the center of the star configuration, with the wires (i.e., segments) radiating outward to connect the various network devices, which could include bridge/routers that connect to remote LANs via the wide area network (Fig. H4). Wiring hubs physically convert the networks from a bus or ring topology to a star topology, while logically maintaining their Ethernet or Token-Ring characteristics. The advantage of this configuration is that the failure of one segment—shared among several devices or dedicated to just one device—does not necessarily impact the performance of other segments.

Not only do hubs limit the impact of cabling faults to a particular segment, they provide a centralized administration point for the entire network. If the wiring hub also employs some central processing units and management software to automate fault isolation, network reconfiguration, and statistics gathering operations, it is no longer just a hub, but an "intelligent hub" capable of solving a wide range of connectivity problems efficiently and economically.

Types of hubs. High-end hubs are modular in design, allowing the addition of ports, network interfaces, and special features as they are needed by the organization. These enterprise-level hubs can support networks that combine different LAN topologies and media types in a single chassis. Ethernet, Token-Ring, and FDDI networks can coexist in a single hub. LAN segments using twisted-pair wiring, coaxial cable, and optical fiber also can be interconnected through the hub.

For departments or workgroups, other hubs are available in fixed configurations that do not anticipate future growth. A variation of the fixed-configuration hub is the *stackable hub.* A unique feature of stackable hubs is that they can be in-

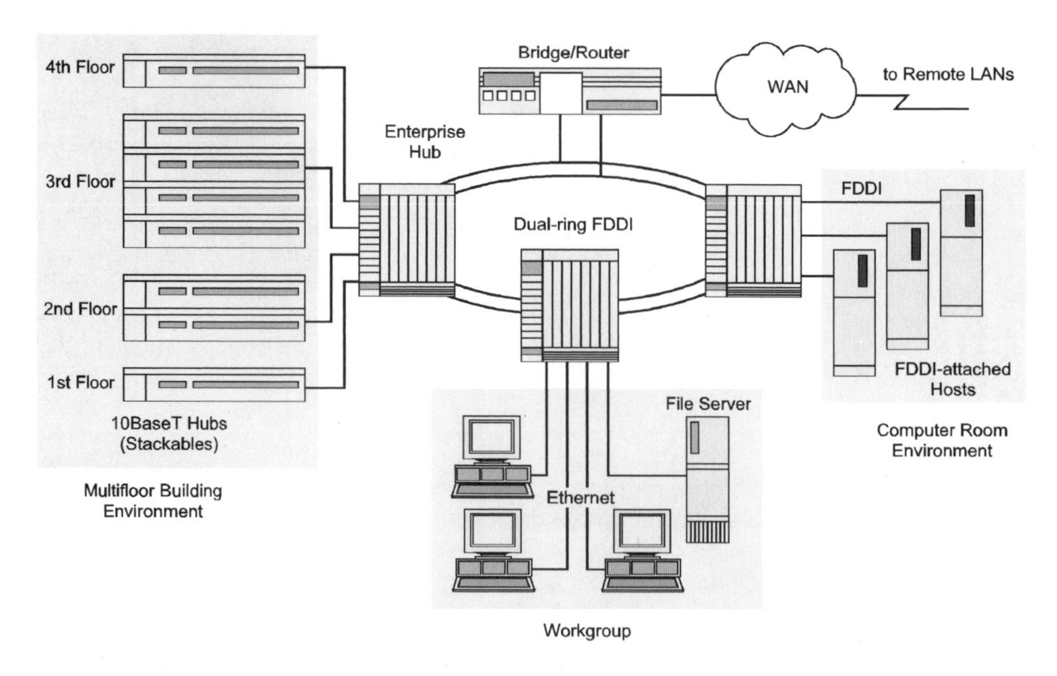

Figure H4 *Some connectivity options available through various types of hubs serving multifloor building, workgroup, and computer room environments.*

terconnected through a modular backplane interface. This offers managers the ability to expand economically their workgroup and departmental networks as needed. Whereas the high-end modular systems are used to build large-scale enterprise networks, stackables are designed for small to medium-sized networks.

A relatively new category of hub is the so-called "super hub." These are modular units that provide—in addition to 100 Mbps LAN support and integrated switching and routing—at least an uplink to a stand-alone ATM (Asynchronous Transfer Mode) switch, if not some level of integral ATM switching. Fully populated super hubs support in excess of 500 ports of mixed-media, shared, and switched connectivity over a gigabit-per-second backplane in a software-manageable, fault-tolerant, hot- swappable modular chassis that can cost well over US$100,000.

Hub components. Enterprise-level intelligent hubs contain four basic components: chassis, backplane, plug-in modules, and a network management system.

Chassis. The chassis is the hub's most visible component. It contains an integral power supply and/or primary controller unit and varies in the number of available module slots. The modules plug into the chassis and are connected by a series of buses, each of which may constitute a separate network or be integrated into one or more backbone networks. The chassis holds the individual modules. Each module fits into a chassis backplane socket, providing connectivity to other modules across the common backplane.

Backplane. The main artery of the hub is its *backplane,* a board that contains one or more buses that carry all communications between LAN segments. The hub's backplane is analogous to a PC bus through which various interface cards may be interconnected; the data path that carries traffic from card to card is often called a *channel.* Unlike the PC, though, the hub's backplane typically consists of multiple physical or logical channels. Minimally, the hub accommodates one LAN segment for each channel on the backplane.

Segmenting the backplane in this way allows multiple independent LANs or LAN segments to coexist within the same chassis. Usually there is a separate backplane channel to carry management information. The segmented backplane typically has dedicated channels for Ethernet, Token-Ring, and FDDI. Some hubs employ a multiplexing technique across the backplane to divide the available bandwidth into multiple logical channels. Some hubs support load sharing that allows network modules to select the backplane channel that will transport the traffic. Other hubs are designed to allow backplanes to be added or upgraded to accommodate network expansion and new technologies.

The potential bandwidth capacity of newer backplane designs supporting ATM switching is impressive, reaching several gigabits per second—more than enough to accommodate several Ethernet, Token-Ring, and FDDI networks simultaneously.

Modules. The functionality of hubs is determined by individual modules. The types of modules available depends on the hub vendor. Typically the vendor will provide multiuser Ethernet and Token-Ring cards, LAN management, and LAN bridge and router cards. The use of bridge and router modules in hubs overcomes the distance limitations imposed by the LAN cabling and facilitates communication between local and wide area networks.

There are even plug-in modules for terminal servers, communications servers, file/application servers, and SNA gateways. Hub vendors also offer a variety of WAN interfaces, including those for X.25, frame relay, ISDN, fractional T1, T1, T3, SMDS, and ATM. As many as 60 different types of modules might be available from a single hub vendor, many of them provided under third-party OEM, technology-swap, and other vendor-partnering arrangements.Modules plug into vacant chassis slots. Depending on the vendor, the modules can plug into any vacant slot, or into slots specifically devoted to their function. Hubs supporting any-slot insertion automatically detect the type of module that is inserted into the chassis and establish the connections to other compatible modules. In addition, many vendors offer hot-swappable capability that permits modules to be removed or inserted without powering down the hub.

Management system. Hubs occupy a strategic position on the network, providing the central point of connection for workstations, servers, hosts, bridges, and routers on the LAN and over the WAN. The hub's management system is used to view and control all devices connected to it, providing information that can greatly aid troubleshooting, fault isolation, and administration. The management tools typically fall into five categories: accounting management, configuration management, performance management, fault management, and security management.

Hub vendors typically provide proprietary management systems that offer value-added features that can make it easier to track down problem-causing work-

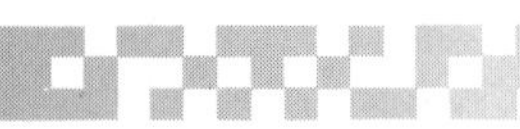

stations or servers. Most of these management systems support the Simple Network Management Protocol (SNMP), enabling them to be controlled and managed through an existing management platform such as IBM's SystemView for AIX, Hewlett-Packard's OpenView, and Sun's Solstice SunNet Manager. Some hubs have Remote Monitoring (RMON) embedded in the hub, making possible more advanced network monitoring and analysis up to the OSI Application layer.

Switching methods. The primary activity of hubs is switching. There are two types of switching: port switching and segment switching. *Port-switching* hubs let administrators assign ports to segments via network management software. In effect, these hubs act as software-controlled patch panels.

Segment-switching hubs treat ports as separate segments and forward packets from port to port. The manner in which the hubs accomplish this varies by vendor and has important implications that depend on the application. What all segment- switching hubs have in common is that they can substantially increase available network bandwidth. Acting like high-speed multiport bridges, segment-switching Ethernet hubs offer 10 Mbps to each port, for example.

The concept behind switching hubs is to improve upon bridging's ability to segment busy networks by providing multiple dedicated connections, while speeding up the store-and-forward method of packet passing that bridges have traditionally performed. A switch directly connects network segments; after decoding the address, the switch sends the packet directly to its destination. This technique is faster than the store-and-forward architecture used in 802.1-compliant bridges, but since the packet is transmitted immediately, most vendors offer no error checking. In contrast, 802.1-compliant bridging includes cyclical redundancy check (CRC) regeneration and filtering of bad packets.

In addition to traditional port switching, today's hubs support virtual LANs. Port switching enables existing networks to be segmented easily through software, eliminating the need to swap cables in the wiring closet and deploy personnel at multiple sites to implement changes. Vendors differentiate their port switching hubs from this basic capability, however, in the degree of freedom provided to move a port to any network operating through a hub, and in the number of networks supported within a hub.

With virtual LANs (VLANs), users can be assigned to specific LANs or segments based on the their MAC addresses, IP addresses, or some other identifier. VLANs allow the same quality and level of service to be delivered to users no matter where they move in the network.

Summary. Intelligent hubs are now the central point of control and management for the elements that make up departmental and enterprise networks. Hubs, which were developed in response to needs for structured wiring as networks became bigger and more complex, allow the wiring infrastructure to expand in a cost-effective manner as the organization's computer systems grow and move, and as interconnectivity requirements become more sophisticated.

See also

BRIDGES
ROUTERS
VIRTUAL LANs

Hybrid fiber/coax

As its name implies, *hybrid fiber/coax* (HFC) is the combination of optical fiber and coaxial cable on the same network. Such networks provide the foundation for future high-speed digital services to the home. A full HFC system could deliver:

- Plain old telephone service (POTS)
- 25 to 40 broadcast analog TV channels
- 200 broadcast digital TV channels
- 275 to 475 digital point-cast channels (that deliver programming at a time selected by the customer)
- High-speed, two-way digital link for Internet and corporate LAN access

HFC divides the total bandwidth into a downstream band (to the home) and an upstream band (to the hub). The downstream band typically occupies 50–750 MHz, while the upstream band typically occupies from 5–40 MHz.

Cable operators originally envisioned a coaxial tree-and-branch architecture to bring advanced services to the home. However, the capacity of fiber optic transmission technology led many cable operators to shift to an approach that combined fiber and coax networks for optimal advantage. Transmission over fiber has two key advantages over coaxial cable:

- A wider range of frequencies can be sent over the fiber, increasing the bandwidth available for transmission.
- Signals can be transmitted over greater distances without amplification.

Fiber to the neighborhood. A key disadvantage of fiber is that the optical components required to send and receive data are still too expensive to deploy to each subscriber. Cable operators therefore have adopted an intermediate approach known as *fiber to the neighborhood* (FTTN). In this approach, fiber reaches into the neighborhood and coaxial cable branches out to each subscriber. This arrangement increases the bandwidth that the plant is capable of carrying, while reducing both the total number of amplifiers needed and the number of amplifiers in cascade between the head end and each subscriber.

The total number of amplifiers is an important economic consideration because each amplifier must be upgraded or, more typically, replaced to pass the larger bandwidth that the fiber and shorter coaxial cable runs allow. The number of amplifiers in cascade is important for ensuring signal quality. Since each amplifier is an active component that can fail, the fewer amplifiers in cascade, the lower the chance of failure. Fewer amplifiers and shorter trees also introduce less noise into the cable signal. These improvements translate into higher bandwidth, better quality service, and reduced maintenance and operating expense for the cable operator.

HFC topology advantages. In HFC networks, fiber is run from a services distribution hub (i.e., head end) to an optical feeder node in the neighborhood, with tree-and-branch coax distribution in the local loop (Fig. H5). Two overriding goals in an HFC architecture are to minimize the fiber investment by distributing it over the maximum number of subscribers, and to use the upstream bandwidth efficiently for the highest subscriber fan-out.

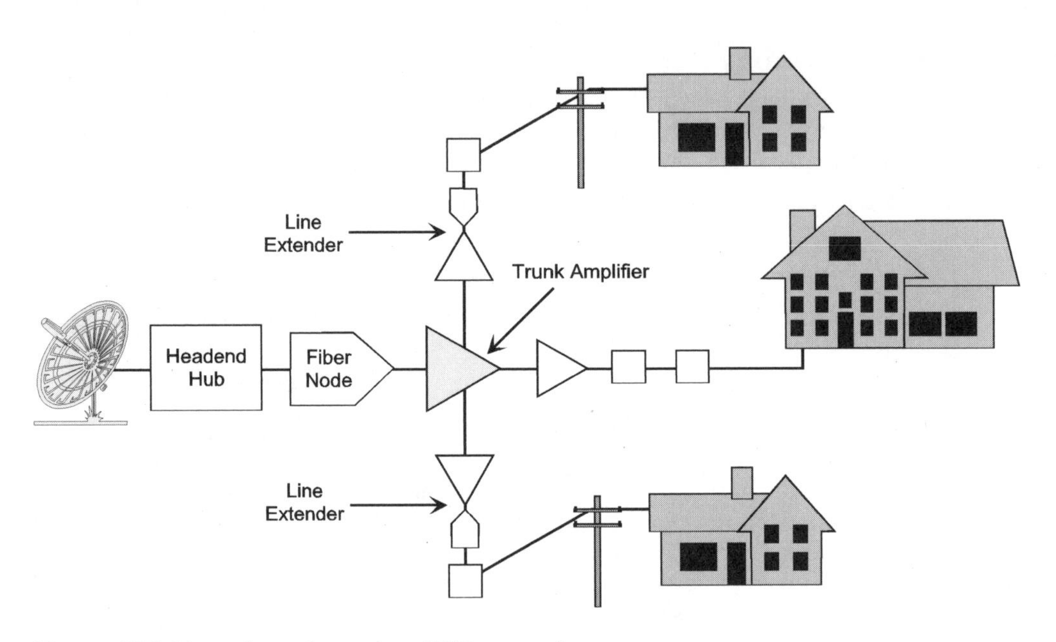

Figure H5 *Typical topology of an HFC network.*

There are approximately 200 million TV sets and 100 million VCRs deployed in the United States—and none of them are digital. This casts doubt about the immediate viability of all-digital *fiber to the curb* (FTTC) networks. Any serious attempt to provide video services must deal with this embedded base. Services that provide digital television signals require a separate digital decoder for each TV tuner, including the VCR, making FTTC an expensive proposition for subscribers, especially since the digital set-top converter would be required even to receive basic television channels. The advantage of HFC networks, however, is that they carry RF signals, delivering video signals directly to the home in the exact format that television sets and VCRs are designed to receive.

HFC networks also have the ability to evolve over time from a basic broadcast plant to a two-way network with interactive bandwidth equal to or even exceeding the stated goals of digital FTTC networks. This evolution is achieved as it is needed, by activating unused dark fiber and subdividing existing nodes to serve fewer homes per node.

During the initial stages of interactive services development, it is anticipated that consumer demand will be low, with early subscribers scattered throughout the service area. To provide services, FTTC network operators must adhere to standard network design concepts, which entails bringing fiber to every curb whether or not the household wants to subscribe to the service. With HFC networks, scattered demand can be accommodated by providing modems as needed. Using modems, digital bandwidth can be allocated flexibly on demand, with virtually arbitrary bit rates. This allows the service provider to add bandwidth capacity inexpensively as demand builds, spreading out capital expenditures to meet subscriber growth in any given service area.

For example, suppose the HFC network operator wants to supply digital HDTV service. Today's HFC network designs already can accommodate individual users of digital HDTV at 20 Mbps, even if the demand is scattered. Using QAM modulation, a single HDTV channel could be supplied in a standard 6 MHz channel slot within the digital band between 550 MHz and 750 MHz. HFC network equipment operates equally well with NTSC signals, which occupy 6 MHz channels, or PAL broadcast analog signals (the European standard), all on the same network. Channel assignments are completely arbitrary within the 750 MHz spectrum.

Emerging technologies. There are several new digital subscriber lines technologies available that increase the usable channel capacity of twisted-pair copper (POTS) telephone lines, enabling telephone companies to offer advanced communications services to the home over the existing local loop infrastructure. One of these technologies is Asymmetric Digital Subscriber Line (ADSL), which supports voice, data, and video. Although more investment- intensive, HFC systems are capable of delivering more channels and are more interactive than today's ADSL systems.

One of the major difficulties with HFC is that the cable system was never intended for reliable high-speed data transmission. Cable was installed with the aim of providing a low-cost conventional (analog) broadcast TV service. As a result of the original limited performance requirements, CATV networks are very noisy and communications channels are subject to degradation. In particular they can suffer badly from very strong narrowband interference called *ingress*. Ingress noise plays havoc with conventional, single-carrier modulation schemes such as Quadrature Amplitude Modulation (QAM) and Quadrature Phase Shift Keying (QPSK), which cannot avoid the noisy region.

There is a new modulation technique, Discrete Wavelet Multitone (DWMT). This is a wavelet transform–based multicarrier modulation scheme that provides better channel isolation, thereby increasing bandwidth efficiency and noise immunity. DWMT divides the channel bandwidth into a large number of narrowband subchannels, and adaptively optimizes the number of bits per second that can be transmitted over each subchannel. DWMT provides throughput equivalent to 32 DS0s (64 kbps each) over each megahertz of bandwidth on the coaxial cable. When a subchannel is too noisy, DWMT does not use it, thereby avoiding channel impairments and maintaining a reliable high-bit-rate throughput.

Using DWMT, higher data rates can be achieved (which translates to more channels) over longer distances. It uses the cable bandwidth and communications infrastructure more efficiently, and that allows cable operators to offer more services.

Summary. The growth of video and interactive communications services, coupled with developments in digital compression, have driven both CATV and telephone operators to seek effective ways to integrate interactive video and data services with traditional communications networks. With two-way communications, HFC allows cable operators to offer two-way communications, which will provide subscribers with telephone service over the cable, as well as full interactive access to broadband signals, including hundreds of channels of interactive TV, digital services, and more. By using fiber links from the central site to a neighborhood hub,

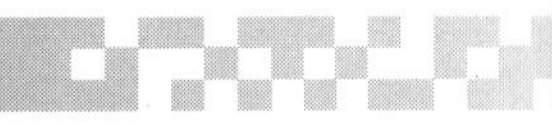

and coax cable from there to a few hundred homes, HFC provides an efficient and economical way to deliver the next generation of communications services while supporting current services.

See also

FIBER IN THE LOOP
DIGITAL SUBSCRIBER LINE TECHNOLOGIES

Infrared networking

Infrared technology is used to implement both wireless LANs and the wireless interface for connecting laptops and other portable machines to desktop computers equipped with infrared transceivers. Infrared LANs are proprietary in nature, so users must rely on a single vendor for all the equipment. The infrared interface for connecting portable devices with the desktop computer is standardized by the Infrared Data Association (IrDA).

Infrared LANs. Infrared LANs typically use the wavelength band between 780 and 950 nanometers (nm). This is due primarily to the ready availability of inexpensive, reliable system components.

There are two categories of infrared systems commonly used for wireless LANs. One is *directed infrared,* which uses laser beams to transmit data over 1–3 miles. This approach may be used for connecting LANs in different buildings. Although transmissions over laser beam are virtually immune to electromechanical interference and would be extremely difficult to intercept, such systems are not widely used because their performance can be impaired by atmospheric conditions, which can vary daily. Such effects as absorption, scattering, and shimmer can reduce the amount of light energy that is picked up by the receiver, causing the data to be lost or corrupted.

The other category of infrared systems is *nondirected,* which uses a less focused approach. Instead of a narrow beam to convey the signal, the light energy is spread out and bounced off narrowly defined target areas or larger surfaces such as office walls and ceilings.

Nondirected infrared links may be further categorized as either line-of-sight or diffuse (Fig. I1). Line-of-sight links require a clear path between transmitter and receiver, and generally offer higher performance.

The line-of-sight limitation may be overcome by incorporating a recovery mechanism in the infrared LAN, which is managed and implemented by a separate device, called a *multiple access unit* (MAU), to which the workstations are connected. When a line-of-sight signal between two stations is temporarily blocked, the MAU's internal optical link control circuitry automatically changes the link's path to get around the obstruction. When the original path is cleared, the MAU restores the link over that path. There is no data loss during this recovery process.

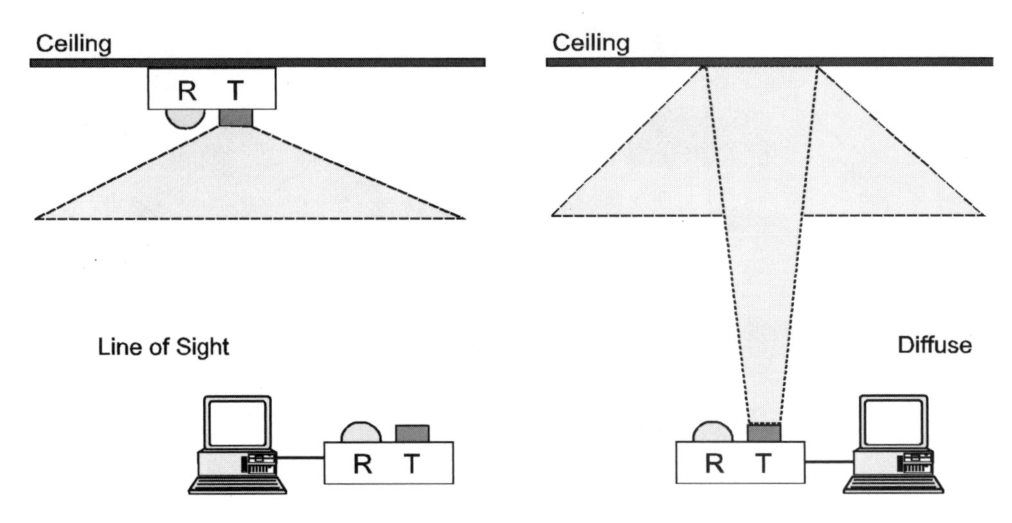

Fig. I1 *Line-of-sight vs. diffuse configurations for infrared links.*

Diffuse links rely on light bounced off reflective surfaces. Because it is difficult to block all of the light reflected from large surface areas, diffuse links are generally more robust than line-of-sight links. The disadvantage of diffused infrared is that a great deal of energy is lost and, consequently, the data rates and operating distances are much lower.

System components. Light-emitting diodes (LEDs) or laser diodes (LDs) are used for transmitters. LEDs are less efficient than LDs, typically exhibiting only 10 to 20 percent electro-optical power conversion efficiency; LDs offer an electro-optical conversion efficiency of 30 to 70 percent. LEDs are much less expensive than LDs, however, which is why most commercial systems use them.

Two types of low-capacitance silicon photodiodes are used for receivers, *positive-intrinsic-negative* (PIN) and *avalanche*. The simpler and less expensive PIN photodiode typically is used in receivers that operate in environments with bright illumination, whereas the more complex and more expensive avalanche photodiode is used in receivers that must operate in environments where background illumination is weak. The difference in the two types of photodiodes is their sensitivity.

The PIN photodiode produces an electrical current in proportion to the amount of light energy projected onto it. Although the avalanche photodiode requires more complex receiver circuitry, it operates in much the same way as the PIN diode, except that when light is projected onto it, there is a slight amplification of the light energy. This makes it more appropriate for weakly illuminated environments. The avalanche photodiode also offers a faster response time than the PIN photodiode.

Operating performance. Current applications of infrared technology yield performance that matches or exceeds the data rate of wire-based LANs: 10 Mbps for Ethernet and 16 Mbps for Token-Ring. Infrared technology has a much higher performance potential, however: Transmission systems operating at 50 and 100 Mbps already have been demonstrated.

Because of its limited range and inability to penetrate walls, nondirected infrared can be secured easily against eavesdropping. Even signals that go out windows are useless to eavesdroppers because they do not travel far, and may be distorted both by impurities in the glass and by its placement angle.

Infrared offers more immunity from electromagnetic interference than spread spectrum, which makes IR suitable for operation in harsh environments like factory floors. Because of its limited range and inability to penetrate walls, several infrared LANs can operate in different areas of the same building without interfering with each other. Since there is less chance of multipath fading (large fluctuations in received signal amplitude and phase), infrared links are highly robust.

Many indoor environments have incandescent or fluorescent lighting, which induce noise in infrared receivers. This is overcome by using directional infrared transceivers with special filters to reject background light.

Media access control. Infrared supports both contention-based and deterministic media access control techniques, making it suitable for Ethernet as well as Token-Ring and, eventually, FDDI LANs.

To implement Ethernet's contention protocol, Carrier Sense Multiple Access (CSMA), each computer's infrared transceiver typically is aimed at the ceiling. Light bounces off the reflector in all directions to let each user receive data from other users (Fig. I2). CSMA ensures that only one station can transmit data at a time. Only the station(s) to which packets are addressed can actually receive them.

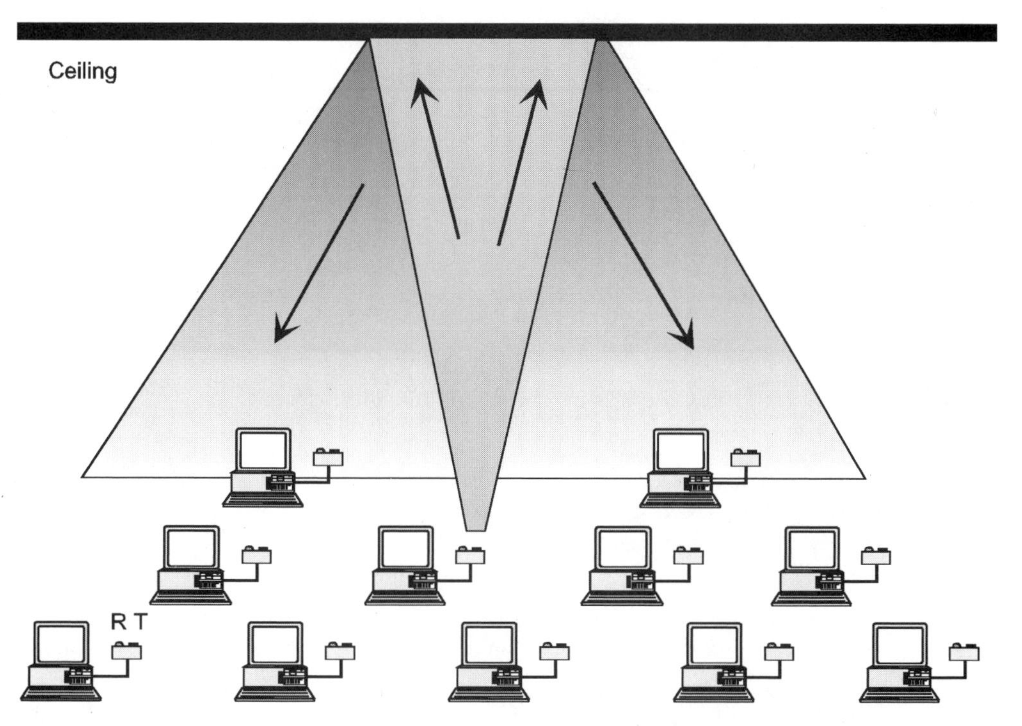

Fig. I2 *Ethernet implementation using diffuse infrared.*

Deterministic media access control relies on token-passing to ensure that all stations get a chance to transmit data in their turn. This technique is used in FDDI as well as Token-Ring LANs. In both types of LAN, each station uses a pair of highly directive (line-of-sight) infrared transceivers. The outgoing transducer is pointed at the incoming transducer of a station down the line, the wireless-infrared links among the computers eventually forming a closed ring (Fig. I3). With this configuration, much higher data rates can be achieved because of the gain associated with the directive infrared signals. This approach improves overall throughput because fewer bit errors will occur, minimizing the need for retransmissions.

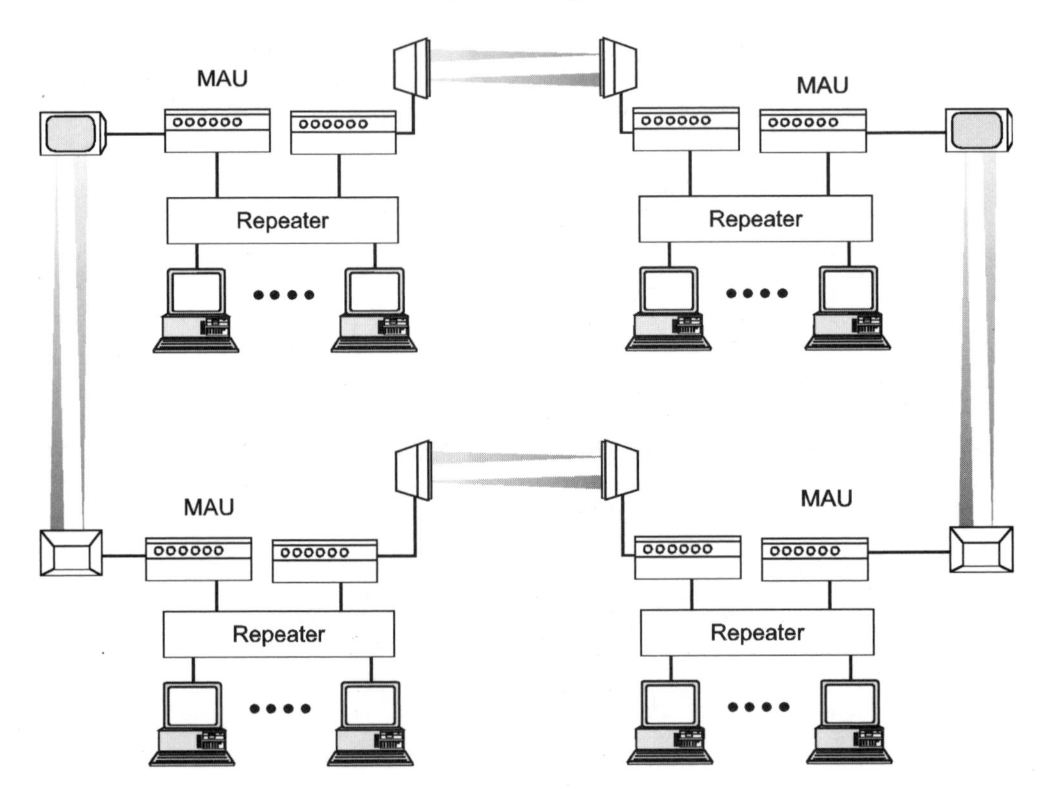

Fig. I3 *Token-Ring implementation using line-of-sight infrared.*

Infrared computer connectivity. Infrared products for computer connectivity conform to the standards developed by the Infrared Data Association (IrDA). The complete IrDA protocol suite contain contains five interdependent layers, as follows:

- *Infrared Physical Layer (IrPL):* This specifies infrared transmitter and receiver optical links, modulation and demodulation schemes, and frame formats.
- *Infrared Link Access Protocol (IrLAP):* This is responsible for link initiation, device address discovery, address conflict resolution, and connection

startup. It also ensures reliable data delivery and provides disconnection services.

- *Infrared Link Management Protocol (IrLMP):* This protocol layer allows multiple software applications to operate independently and concurrently, sharing a single IrLAP session between a portable PC and network access device.
- *Infrared Transport Protocol (IrTTP):* This is responsible for data flow control, and packet segmentation and reassembly.
- *Infrared LAN (IrLAN):* This is a protocol defining how a network connection is established over an IrDA link.

IrDA-standard infrared links are half-duplex with the maximum data rate of 115.2 kbps. There are IrDA high-speed extensions for 1.15 and 4 Mbps transmission. The hardware consists of an infrared transmit encoder/receiver decoder and the IR transducer, which consists of the output driver and infrared emitter for transmitting and the receiver/detector. The encoder/decoder interfaces to the UART (Universal Asynchronous Receiver/Transmitter), which already is present in most computers.

In addition to directly connecting portable and desktop computers, infrared links can be used to connect portable computers to the corporate LAN via a network access device. Although this also can be accomplished by inserting the portable computer into a docking device connected to the LAN, docking stations are typically based on proprietary technology and, in most cases, do not work with portable PCs from multiple vendors.

Summary. Infrared's primary impact will take the form of benefits for mobile professional users. It enables simple, point-and-shoot connectivity to standard networks, which streamlines users' workflow and allows them to reap more of the productivity gains promised by portable computing. At this writing, the Infrared Data Association has proposed a set of standards that will allow cellular phones and pagers to exchange information with PCs via infrared data communications ports.

Infrared also confers substantial benefits for network administrators. Infrared is easy to install and configure, requires no maintenance, and imposes no remote-access tracking hassles. It does not disrupt other network operations and it provides data security. Because it makes connectivity so easy, it encourages the use of high-productivity network and groupware applications on portables, thus helping administrators amortize the costs of these packages across a larger user base.

See also

SPREAD SPECTRUM RADIO

Institute of Electrical and Electronics Engineers

The Institute of Electrical and Electronics Engineers (IEEE), founded in 1884, is the world's largest technical professional society, with over 320,000 members who conduct and participate in its activities in 147 countries.

The technical objectives of the IEEE focus on advancing the theory and practice of electrical, electronics, and computer engineering and computer science, publishing nearly 25 percent of the world's technical papers in these fields. The

IEEE sponsors technical conferences, symposia, and local meetings worldwide. It also provides educational programs for members.

Technical societies. The IEEE consists of technical societies that provide publications, conferences, and other benefits to members within 37 specialized areas, from aerospace and electronic systems to vehicular technology. Each of these societies has technical committees that define and implement the technical directions of the society. For example, there are 19 technical committees within the Communications Society:

- Cable-Based Delivery and Access Systems
- Communications Software
- Communication Switching
- Communications Systems Integration and Modeling
- Communication Theory
- Computer Communications
- Enterprise Networking
- Gigabit Networking
- Information Infrastructure
- Interconnections in High-Speed Digital Systems
- Internet
- Multimedia Communications
- Network Operations and Management
- Optical Communications
- Personal Communications
- Quality Assurance Management
- Radio Communications
- Satellite and Space Communications
- Signal Processing and Communications Electronics
- Signal Processing for Storage
- Transmission and Access and Optical Systems

Standards Board. The IEEE Standards Board is responsible for all matters regarding standards in the fields of electrical engineering, electronics, radio, and the allied branches of engineering. Currently there are ten standing committees of the IEEE Standards Board:

Procedures Committee (ProCom). This committee is responsible for recommending to the IEEE Standards Board improvements and procedural changes to promote efficient discharge of responsibilities by the IEEE Standards Board, its committees, and other committees of the institute engaged in standards activities.

New Standards Committee (NesCom). This committee is responsible for ensuring that proposed standards projects are within the scope and purpose of the IEEE, that standards projects are assigned to the proper society or other organizational body, and that interested parties are appropriately represented in the development of IEEE standards. This committee examines project authorization requests and makes recommendations to the IEEE Standards Board regarding their approval.

Standards Review Committee (RevCom). This committee is responsible for reviewing submittals for the approval of new and revised standards, and for the reaffirmation or withdrawal of existing standards, to ensure that the submittals represent a consensus of the parties having a significant interest in the covered subjects. The committee makes recommendations to the IEEE Standards Board regarding the approval of these submittals.

Awards and Recognition Committee (ArCom). This committee is responsible for the administration of all awards presented by the IEEE Standards Board. It acts on behalf of the board to approve nominations for IEEE Standards Awards. It also submits nominations for standards awards sponsored by other organizations.

New Opportunities in Standards Committee (NosCom). This committee is responsible for identifying and exploring avenues for enhancing IEEE leadership in areas of new technological growth, and for recommending to the IEEE Standards Board actions to achieve this purpose.

Procedures Audit Committee (AudCom). This committee provides oversight of the standards development activities of the societies, their standards-developing entities, and the Standards Coordinating Committees (SCCs)[1] of the IEEE Standards Board.

Seminars Committee (SemCom). This committee provides oversight for the operation of the seminar program by providing technical expertise and support. The committee reviews and proposes new seminars to ensure that the topics covered are appropriate.

International Committee (IntCom). This committee is responsible for coordinating IEEE standards activities with non-IEEE standards organizations. The committee also assists in the adoption by IEEE of non-IEEE standards when appropriate.

Administrative Committee (AdCom). This committee acts for the Standards Board between meetings and makes recommendations to the Standards Board for its disposition at regular meetings.

Patent Committee (PatCom). This committee provides oversight on the use of any patents and patent information in IEEE standards. PatCom also reviews any patent information submitted to the IEEE Standards Board to determine conformity with the patent procedures and guidelines.

Summary. The IEEE is not a member of the International Electrotechnical Commission (IEC) or the International Organization for Standardization (ISO), because only countries (not standards bodies) can be members of IEC and ISO. When IEEE working groups need global participation in their projects, however, they can go through any IEC or ISO member country to make submissions to their IEC or ISO technical committees.

See also

AMERICAN NATIONAL STANDARDS INSTITUTE
INTERNATIONAL ELECTROTECHNICAL COMMISSION
INTERNATIONAL ORGANIZATION FOR STANDARDIZATION
INTERNATIONAL TELECOMMUNICATIONS UNION

1 When a proposed standard does not fall into the subject area covered by one of the technical societies, or a technical society cannot handle the workload, a Standards Coordinating Committee is established and coordinated by the IEEE Standards Board.

Integrated Services Digital Network

The Integrated Services Digital Network (ISDN) made its debut in 1980 with the promise of high-quality, ubiquitous switched digital service. Although ISDN was intended to become a worldwide standard to facilitate global communications, this was not to be the case. In the United States, nonstandard carrier implementations, incompatibilities between customer and carrier equipment, the initial high cost of special adapters and telephones, spotty coverage, and configuration complexity hampered user acceptance of ISDN through the mid-1990s.

Despite these implementation problems, ISDN offers tangible benefits over the existing public telephone network, including:

- Faster call setup and network response times,
- Increased network management and control facilities,
- Support for such advanced applications as videoconferencing,
- Improved configuration flexibility and additional restoral options,
- The elimination of delays in switching over new lines and services, and
- The ability to streamline networks through integration, thus reducing the complexity and cost of cabling and equipment.

By 1996, many of the problems with ISDN had been ironed out. The increasing popularity of the World Wide Web sparked consumer demand for ISDN as a means of accessing the Internet and improving the response time for navigating Web sites and viewing multimedia documents.

Applications. ISDN can be used for many applications, including videoconferencing, medical image transmission, the delivery of multimedia training sessions, and Internet access. Other applications of ISDN include temporarily rerouting traffic around failed leased lines and handling peak traffic loads (Fig. I4). ISDN also can play a key role in various applications of computer-telephony integration (CTI), telecommuting, and remote access (i.e., remote control and remote node).

For certain applications, ISDN can be even more economical than leased lines, which entail fixed monthly charges based on distance, whether or not they are fully used. For example, high-bandwidth applications that are used infrequently (e.g., videoconferencing and medical image transmission) might not justify the cost of purchasing leased lines that will go underutilized most of the time. Because users are typically billed for ISDN channels only when used, significant savings can accrue.

ISDN channels. ISDN is a circuit-switched digital service that comes in two varieties. The basic rate interface (BRI) provides two bearer (B) channels of 64 kbps each, plus a 16 kbps signaling (D) channel. The primary rate interface (PRI) provides 23 B channels of 64 kbps each, plus a 64 kbps D channel. Any combination of voice and data can be carried over the B channels. Additional channels can be created within the B channels through the use of compression.

ISDN was designed to be compatible with existing digital transmission infrastructures, specifically T1 in North America and Japan (1.544 Mbps), and E1 in Europe (2.048 Mbps). Because ISDN can evenly reduce both T1 and E1 into 64 kbps increments—a transmission rate that carriers typically employ for voice communications—64 kbps channels became the worldwide standard. The use of 64 kbps

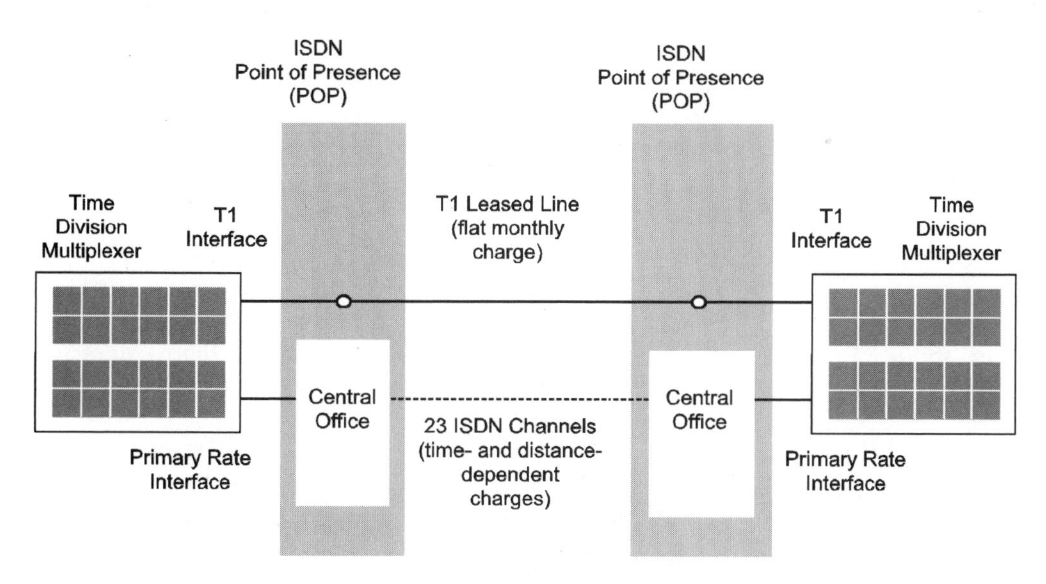

Fig. I4 *Among other applications, ISDN can be used to backup T1 leased lines in case of failure, provide an additional source of temporary bandwidth to handle peak traffic loads, or support special applications such as videoconferencing or medical imaging on an as-needed basis.*

channels also allows users to migrate more easily from private T1/E1 networks to ISDN and build hybrid networks consisting of both public and private facilities.

In both BRI and PRI, ISDN's separate D channel is used for signaling. As such, it has access to the control functions of the various digital switches on the network. It provides message exchange between the user's equipment and the network to set up, modify, and clear the B channels. The D channel also gathers information about other devices on the network, such as whether they are idle, busy, or off. The ability to check ahead to determine if calls can be completed can help conserve network bandwidth. If the called party is busy, for example, the network can be notified before telco resources are committed.

Whenever the D channel is not being used for signaling, it can be used as a bearer channel (if the carrier offers this capability as a service) over X.25 networks for point-of-sale applications such as automatic teller machines (ATMs), lottery terminals, and cash registers (Fig. I5).

With regard to ISDN PRI, there are two higher-speed transport channels called H channels. The H0 channel operates at 384 kbps and the H11 operates at 1.536 Mbps. These channels are used to carry multiplexed data, data and voice, or video at higher rates than that provided by the 64 kbps B channel. The H channels also are ideally suited for backing up FT1 and T1 leased lines. Eventually other high-speed transport channels will be added in support of Broadband ISDN.

Multirate ISDN lets users select appropriate increments of switched digital bandwidth on a per-call basis. Speeds up to 1.536 Mbps are available in increments of 64 kbps. Multirate ISDN is used mostly for multimedia applications and videoconferencing.

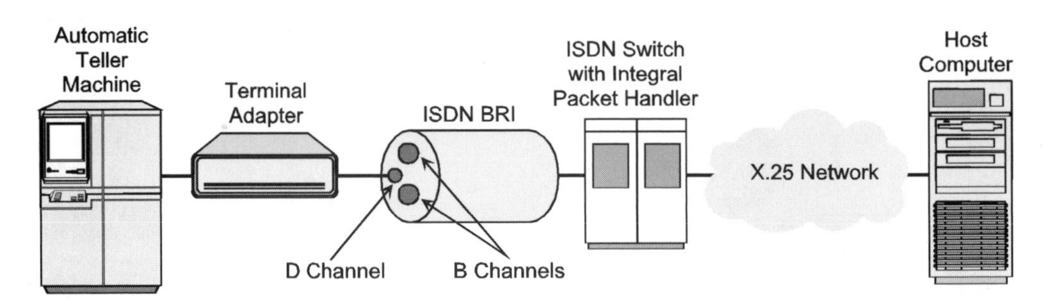

Fig. I5 *ISDN's D channel can be used for low-speed applications that require no more than 9.6 kbps of bandwidth, while the B channels carry other voice and data calls.*

ISDN architectural elements. The architectural elements of ISDN include several reference points that define network demarcations between the telephone company and the customer premises:

- *Reference point R* is the point separating non-ISDN (TE2) equipment and the terminal adapter (TA), which provides TE2 with ISDN compatibility.
- *Reference point S* is the point separating terminals (TE1 or TA) from the network terminal (NT2).
- *Reference point T* is the point separating NT2 from NT1. (It is not required if NT2 and NT1 functionality is provided by the same device.)
- *Reference point U* is the point separating the subscriber's portion of the network (NT1) from the carrier's portion of the network (LT).

There also are interfaces between various types of customer premises equipment. Network terminators provide network control and management functions, while terminal equipment devices implement user functions. Figure I6 illustrates the reference points and architectural elements of ISDN.

Although ISDN BRI and PRI services consist of different configurations of communications channels, both services require the use of distinct functional elements to provide network connectivity. One of these elements is the *Network Terminator 1* (NT1), which resides at the user's premises and performs the four-wire to two-wire conversion required by the local loop. Aside from terminating the transmission line from the central office, the NT1 device is used by the telephone company for line maintenance and performance monitoring.

Network Terminator 2 (NT2) devices include all NT1 functions in addition to protocol handling, multiplexing, and switching. These devices are usually integrated with PBX and key systems.

ISDN *Terminal Equipment* (TE) provides user-to-network digital connectivity. TE1 provides protocol handling, maintenance, and interfacing functions and supports such devices as digital telephones, data terminal equipment, and integrated workstations, all of which comply with the ISDN user-network interface. The large installed base of non-ISDN TE2 devices (e.g., telephones and microcomputers) can communicate with ISDN-compatible devices when users attach or install a *Termi-*

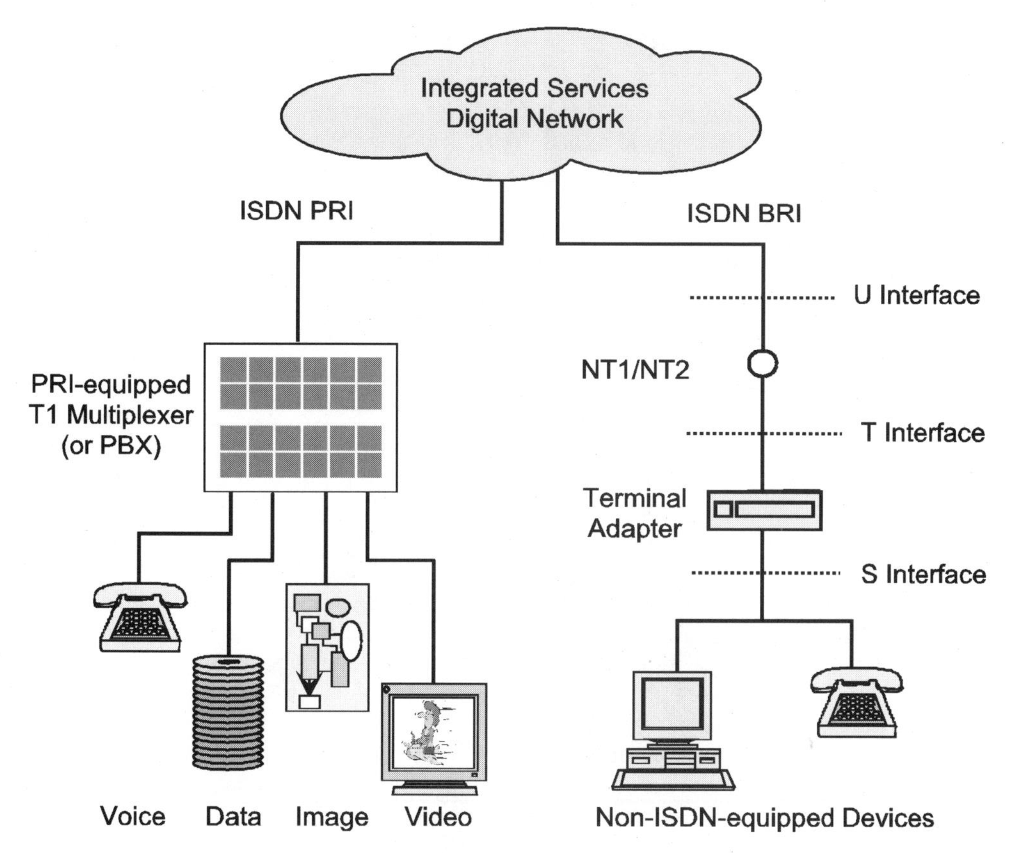

Fig. I6 *ISDN architectural elements and reference points.*

nal Adapter (TA) to or in the non-ISDN device. A TA takes the place of a modem. Users can connect a maximum of eight TE/TA devices to a single NT2 in a multidrop configuration.

Summary. Over the years, ISDN has been touted as a breakthrough in the evolution of worldwide telecommunications networks, the single most important technological achievement since the advent of the telephone network itself in the 19th century. Others disagree, pointing out that ISDN implementation has stalled and that other, more powerful telecommunications technologies such as frame relay and ATM are progressing so rapidly that ISDN is now obsolete.

The debate over ISDN's relevance to today's telecommunications needs must be put into the context of specific applications. Just like any other service, ISDN will be adequate to serve the needs of some users but not others. Meanwhile, ISDN will continue to evolve to include broadband services, eventually ending the debate about its obsolescence.

See also

MULTIPLEXERS

VIDEOCONFERENCING

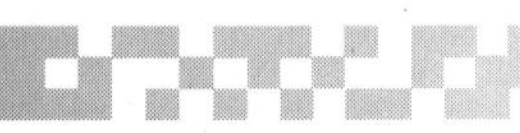

Intelligent call routing

With Internet usage growing at 10 percent per month, dial-up Internet access is beginning to seriously impact the public telephone network, which was never designed to carry a heavy load of data traffic. Switches engineered to handle 3-minute voice calls are now deluged with Internet calls lasting ten times longer, tying up local switches and, according to the telcos, threatening 911 emergency calls.

Although this effect is most apparent on switches directly serving Internet service providers (ISPs), the more serious problem occurs throughout the network on switches serving residential and business customers who are ISP subscribers. Even a modest percentage of modem calls from as few as one home in ten can lead to congestion in small residential communities.

While some vendors have come up with solutions to intercept ISDN calls before they tie up central office switch resources, ISDN represents less than 5 percent of all dial-up access to ISPs. And even though cable modems are already available, they will have little overall impact on total network traffic over the next few years. Modem usage, on the other hand, will continue to grow and be the mainstream technology to access the Internet for the foreseeable future.

Intelligent Call Routing (ICR) technology is designed to free up central office bottlenecks by intercepting modem calls before they hit the voice network, so they can be routed to a fast packet data network. Traditional solutions that reinforce switch capacity pose significant cost problems for carriers. Lower-cost ICR technology, combining telephone service and packet-mode Internet access, provides an attractive and economical alternative.

The ICR technology connects directly to subscribers, migrating a voice-only network to a multipurpose network, supporting both voice and data services. The technology supports features that include full connectivity for all users, all the time, as well as high-bandwidth modem capabilities.

Summary. ICR detects and redirects calls with long holding times to solve the congestion problem on the telephone network, while giving modem users prompt access to the Internet. In turn, this allows the telephone company to support data services without having to upgrade the local loop.

See also

INTEGRATED SERVICES DIGITAL NETWORK
INTERNET

Interactive voice response

Interactive voice response (IVR) is a technology that allows callers to obtain requested information stored in a corporate database. IVR technology uses the familiar telephone keypad as an information retrieval and data-gathering conduit. Recorded voice messages prompt and respond to the callers' inquiries or commands. Examples of IVR range from simply selecting announcements from a list of options stored in the computer, also known as *audiotext,* to more complex interactive exchanges such as querying a database for particular information.

The IVR system takes the call, asks questions with multiple-choice answers, and responds to the DTMF tone digit entered by the caller with the proper prerecorded announcement. This self-service solution provides requested information to callers services 24 hours per day, 7 days per week. This allows companies to meet the information needs of various constituents efficiently and economically—without having to devote staff to handle routine requests for information.

IVR can be an integral function of a PBX or ACD, or it can be provided as a service by a telecommunications carrier. Sprint's InterVoice multiple-application platform, for example, provides the means to offer Sprint business customers a variety of call-processing functions, including interactive voice response.

Applications. IVR systems can be used for a variety of applications. A brokerage firm can use an IVR system to take routine orders from investors who want to order corporate prospectuses. An investment fund can take routine requests for new account applications. A company can take routine requests from employees about their benefit plans. A help desk can use an IVR system to take routine questions from computer users, and step them through a preliminary troubleshooting process that reveals the most common hardware or software problems that they can correct themselves. Colleges can use IVR systems to answer routine questions about degree programs, the registration process, and fees.

Just about any organization can use an IVR system to meet the informational needs of callers. Businesses can improve customer service by offering 24-hour access to information, with the added benefit of providing consistent information and transaction capabilities to their customers through the simple use of a telephone. Large inventories of literature can be eliminated, not to mention the costs of storage and the staff to maintain it. Mailing expenses also can drop dramatically.

IVR systems even allow companies to bill callers for services. They can capture credit card information from callers with tone-dial phones, or voice-capture callers' names and addresses. For more complicated transactions, the IVR system also can provide the caller with the option of accessing a live operator.

Although IVR systems can deliver requested information via recorded voice announcements, sometimes the information is best delivered in printed form. Some IVR systems include a *fax on demand* capability that allows callers to select documents from a menu of available items that are described to them (Fig. I7). Callers can receive information at their fax machines instantly, or they have the option of scheduling delivery at times more convenient for them.

Process flow and navigation. Many IVR systems are designed for a particular application. The process flow and navigation of an IVR system can be illustrated by examining its role in supporting a typical employee benefit plan application.

Employees are provided with an enrollment worksheet that shows them available options for benefits for the upcoming plan year. This worksheet also has a PIN number printed on it for access into the IVR. Employees call an 800 number to access the system. The system will prompt them to enter their Social Security number and PIN. Any incorrect entries will be voiced back as incorrect, along with a message inviting the user to try again.

Built-in editing features are used to eliminate errors typically made by employees using a paper-based enrollment. During the call, employees are only presented with the options for which they are eligible. The IVR system validates each entry to ensure that the entry conforms to the plan's provisions.

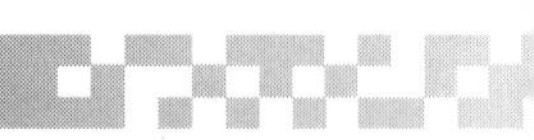

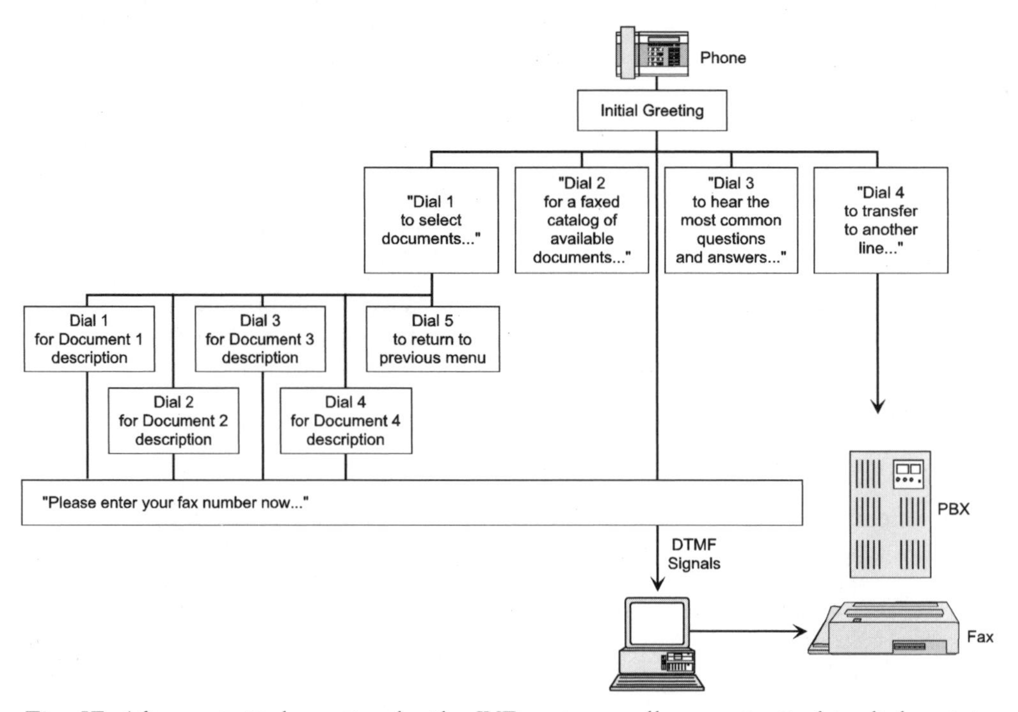

Fig. I7 *After an initial greeting by the IVR system, callers are invited to dial certain numbers for information that can be conveyed to them via recorded voice announcements or sent in printed form to their fax machines.*

If the employee is a first-time caller (no selections have been recorded yet), the system will guide him or her through the enrollment process. Included in the script are benefit option eligibility prompts using dynamic menuing for the following typical benefit plans:

- ➢ Medical options (only presents eligible options and HMOs)
- ➢ Medical coverage category (single, family, etc.)
- ➢ Dental options (only presents available options and DMOs)
- ➢ Dental coverage categories
- ➢ Vision options
- ➢ Vision coverage categories
- ➢ Life insurance options (only presents available coverages)
- ➢ LTD options (only presents available coverages)
- ➢ Dependent life
- ➢ Spending accounts
- ➢ Other plans

If the employee's total benefit coverage price is in excess of company provided flex dollars or credits, the IVR system can communicate any salary reduction or deduction impact. If the employee has an excess of credits, a choice of the credit allocation method, such as cash or transfer to a spending account, can be offered.

Many functions are offered with voice response systems, such as total pay period costs and modeling, PIN changes, current enrollment status inquiry, ordering of materials, primary care physician (PCP) data collection, dependent data collection, and new-hire enrollments. IVR voice scripts should be designed to communicate this information with ease and simplicity. Additional features—such as custom messages about proof of insurability, waiver of coverage forms, special instructions, and fax-back services—can be added to the voice response script.

Summary. IVR is becoming a vital communication tool that alleviates time and resource burdens currently facing many businesses. Easily accessible and accurate information is important in this increasingly competitive age, one marked by corporate staff reductions, streamlined business processes, and the need for cost containment. IVR technology uses the familiar telephone keypad for information retrieval and data gathering, essentially offering corporate constituents the means to serve themselves. Through a single toll-free number, the IVR system can bring together information retrieval, directory services, and transaction capabilities.

See also
AUTOMATIC CALL DISTRIBUTORS
CALL CENTERS
FACSIMILE

International callback service

Callback service is provided by U.S. international long-distance resellers as a means for their customers, located outside of the United States, to access U.S.-based international lines. Typically, a signaling call is placed by the originating caller overseas to the callback provider's switch located in the United States. If uncompleted-call signaling is used, the caller dials the provider's switch in the United States, waits a predetermined number of rings, and hangs up without the switch answering. The switch then automatically returns the call and, upon completion, provides the caller with a U.S. dial tone. All traffic is thus originated at the U.S. switch. The calls are billed at U.S. tariffed rates, which often are much lower than those of the originating country.

Legal challenge. In 1995, AT&T complained to the FCC that international callback services were illegal, arguing that uncompleted-call signaling is an unreasonable practice because it violates the federal wire fraud statute and constitutes theft of service. The service providers offering international callback contended that wire fraud cannot occur without a completed call and, further, that uncompleted-call signaling does not constitute wire fraud because there is no wrongful appropriation of money or property, based on the practice of international facilities-based carriers not to charge for uncompleted calls.

The FCC rejected AT&T's charge that uncompleted call signaling constitutes federal wire fraud in violation of Section 201(b) of the Communications Act. On reconsideration of the matter, the FCC even sought the opinion of the U.S. Department of Justice as to whether uncompleted-call signaling violates the federal wire

fraud statute. The department agreed with the FCC in its determination that uncompleted-call signaling does not constitute wire fraud, adding: "It appears...that the U.S. carriers have found and are legitimately exploiting a loophole in AT&T's tariff structure."

With regard to AT&T's contention that uncompleted-call signaling is an unreasonable practice under Section 201(b) because it constitutes theft of service, the FCC recognized that uncompleted-call signaling constitutes an uncompensated use of the network. In the system as currently structured by facilities-based carriers, however, customers do not expect to pay for an uncompleted call. Nor do carriers expect to be compensated. Because there is no expectation of payment for uncompleted calls, the failure to pay for those calls does not deprive carriers of anything they are otherwise due. Thus, the FCC did not find that any "property" had been "taken" from AT&T.

Summary. International callback is a popular service and has been expanded to include fax transmissions as well. It provides the means for users to save on international calls to the United States because, typically, calls are billed at a higher rate in other countries. While AT&T initially opposed international callback—before it finally was ruled as permissible by the FCC in June, 1995—AT&T now offers an international callback service of its own. Customers of its Software Defined Network (SDN) service can obtain alternate network access. Under this offering, employees of SDN user companies traveling overseas call a number in the United States to obtain a U.S. dial tone for a call to another country. In this way, employees can avoid expensive international direct-dial rates on calls that would otherwise originate overseas.

International Electrotechnical Commission

The International Electrotechnical Commission (IEC) was founded in 1906 as a result of a resolution passed at the International Electrical Congress held in St. Louis in 1904. The membership currently consists of 52 national committees, representing the electrotechnical interests in each country, from manufacturing and service industries to government, scientific, research and development, academic, and consumer bodies. Membership includes all the world's major trading nations and a growing number of industrializing countries.

The IEC promotes international cooperation on all questions of standardization and related matters, such as the assessment of conformity to standards, in the fields of electricity, electronics, and related technologies. It provides a forum for the preparation and implementation of consensus-based voluntary international standards, which serve as a basis for national standardization and as references when drafting international tenders and contracts. To fulfill its mission, the IEC publishes international standards and technical reports.

Organization. The governing authority of the IEC is the Council, which is the general assembly of the national committees, who are the commission's members. The IEC also comprises executive and advisory bodies.

Council. The IEC Council deals mainly with administrative matters and is assisted by the General Policy Committee (GAC). The decisions and policy of the IEC

Council are implemented under the supervision of the Management Board. The Council receives reports from the GPC, the Management Board, the Committee of Action, and the Conformity Assessment Board.

In addition, the President's Advisory Committee on Future Technologies (PACT) provides a link with private and public research and development activities, keeping the IEC informed of accelerating technological changes and the accompanying demand for new standards.

Committee of Action. The Council delegates the management of standards work to the Committee of Action (CA), the membership of which consists of representatives of 12 national committees.

Among the main tasks of the CA is to set up technical committees (TCs), follow up and coordinate the work of the TCs, and examine the need to undertake work in new fields.

Conformity Assessment Board. The Council delegates the overall management of the IEC's conformity assessment activities to the Conformity Assessment Board (CAB). The CAB provides a single coordinated contact point for high-level negotiations with other conformity assessment bodies at international and regional levels. Among the main tasks of the CAB are setting the IEC's conformity assessment policy so as to serve the present and future needs of international trade; monitoring the operation of IEC conformity assessment schemes by examining their continued relevance; and coordinating and interfacing with international and regional bodies on CA matters.

Central Office. The Central Office supports the TCs and their subcommittees (SCs), as well as the national committees, by ensuring the reproduction and circulation of working documents and the final texts of standards.

Technical committees and subcommittees. The technical work of the IEC is carried out by technical committees and subcommittees. The TCs prepare technical documents on specific subjects within their respective scopes, which are then submitted to the national committees for voting with a view to their approval as international standards.

If a technical committee finds that its scope is too wide to deal with all the items on its work program, it may set up SCs, defining in each case a scope covering part of the subjects dealt with by the main committee. The SCs report on their work to the parent TC.

In order to draft documents, a TC or SC may set up working groups (WGs) composed of a limited number of experts. An ad hoc group may be set up to examine a particular point and report on it to the TC or SC.

Summary. The IEC cooperates with numerous international organizations, particularly with the International Organization for Standardization (ISO) and the International Telecommunication Union (ITU). At the regional level, there is a joint working agreement with the European Committee for Electrotechnical Standardization (CENELEC), comprising 18 national committees, most of which are also IEC members, and a cooperation agreement with COPANT, the Pan American Standards Commission.

Close links also are in place with other bodies in non-electrotechnical areas, examples being the liaisons with the World Health Organization, the International La-

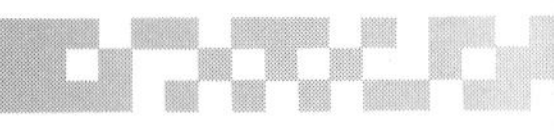

bor Office, the International Organization of Legal Metrology, and the International Atomic Energy Agency.

See also

INSTITUTE OF ELECTRICAL AND ELECTRONICS ENGINEERS
INTERNATIONAL ELECTROTECHNICAL COMMISSION
INTERNATIONAL ORGANIZATION FOR STANDARDIZATION
INTERNATIONAL TELECOMMUNICATIONS UNION

International Organization for Standardization

Established in 1947, the International Organization for Standardization (ISO)[2] is a nongovernmental, worldwide federation of national standards bodies from 100 countries. A member body of ISO is the national body "most representative of standardization in its country." Only one such body for each country is accepted for membership. The member body for the United States is the American National Standards Institute (ANSI).

ISO promotes the development of standardization and related activities to facilitate the international exchange of goods and services, and develops cooperation in the spheres of intellectual, scientific, technological, and economic activity. ISO's work results in international agreements that are published as international standards.

ISO is best known for its standards work in worldwide communications systems, particularly the development of the seven-layer Open Systems Interconnection (OSI) Reference Model. ISO is active in many other fields, however, including advanced materials, the environment, life sciences, urbanization and construction, and quality assurance. ISO covers all standardization fields except electrical and electronic engineering, which is the responsibility of the International Electrotechnical Committee (IEC). The work in the field of information technology is carried out by a joint ISO/IEC technical committee (JTC 1).

The technical work of ISO is highly decentralized, carried out in a hierarchy of about 2700 technical committees, subcommittees, and working groups. In these committees, qualified representatives of industry, research institutes, government authorities, consumer bodies, and international organizations come together as equal partners in the resolution of global standardization problems.

Standards development process. There are three main phases in the ISO standards development process. The need for a standard is usually expressed by an industry sector, which communicates this need to a national member body. The latter proposes the new work item to ISO as a whole. Once the need for an international standard has been recognized and formally agreed, the first phase involves definition of the technical scope of the future standard. This phase is usually carried out in working groups that comprise technical experts from countries interested in the subject matter.

2 *ISO* is a word, not an acronym for the organization. It is derived from the Greek *isos*, meaning "equal," which is the root of the prefix *iso-* that occurs in many other terms, such as *isometric* (of equal measure or dimensions) and *isonomy* (equality of laws, or of people before the law). The line of thinking from "equal" to "standard" led to the choice of ISO as the name of the organization.

Once agreement has been reached on which technical aspects are to be covered in the standard, a second phase is entered, during which countries negotiate the detailed specifications within the standard. This is the consensus-building phase.

The final phase comprises the formal approval of the resulting draft international standard. The acceptance criteria stipulate approval by two-thirds of the ISO members that have participated actively in the standards development process, and approval by 75 percent of all members that vote. Upon approval, the text is published as an ISO international standard.

Most standards require periodic revision. Several factors combine to render a standard out of date: technological evolution, new methods and materials, and new quality and safety requirements. To take account of these factors, ISO has established the general rule that all ISO standards should be reviewed at intervals of not more than 5 years. On occasion it is necessary to revise a standard more frequently.

To accelerate the standards process (handling of proposals, drafts, comment reviews, voting, publishing, etc.), ISO makes use of information technology and program management methods. To date, ISO's work has resulted in 9300 international standards, representing some 170,700 pages.

ISO structure. Like all standards bodies, ISO has an organizational structure that allows it to carry out its mission in the most effective way possible (Fig. I8).

General Assembly. The General Assembly meets once a year. Its agenda includes a multiyear strategic plan and financial status report.

Policy development committees. The General Assembly establishes advisory committees, called *policy development committees,* which are open to all member bodies and correspondent members.

- *Committee on Conformity Assessment (CASCO)* studies the means of assessing the conformity of products, processes, services, and quality systems to appropriate standards or other technical specifications. It prepares international guides relating to the testing, inspection, and certification of products, processes, and services, and the assessment of quality systems, testing laboratories, inspection bodies, certification bodies, and their operation and acceptance.
- *Committee on Consumer Policy (COPOLCO)* studies the means of assisting consumers to benefit from standardization, and the means of improving their participation in national and international standardization efforts. It promotes from the standardization point of view the information, training, and protection of consumers.
- *Committee on Developing Country Matters (DEVCO)* identifies the needs and requirements of the developing countries in the fields of standardization and assists those countries, as necessary, in defining these needs and requirements. Having done so, it recommends measures to assist the developing countries in meeting them.
- *Committee on Information Systems and Services (INFCO)* coordinates the activities of ISO and its members in relation to information services, databases, marketing and sales of standards, technical regulations, and related matters, including these services and products in electronic form.

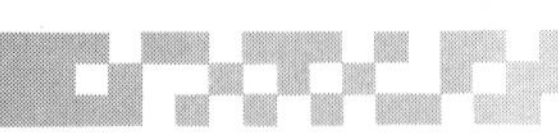

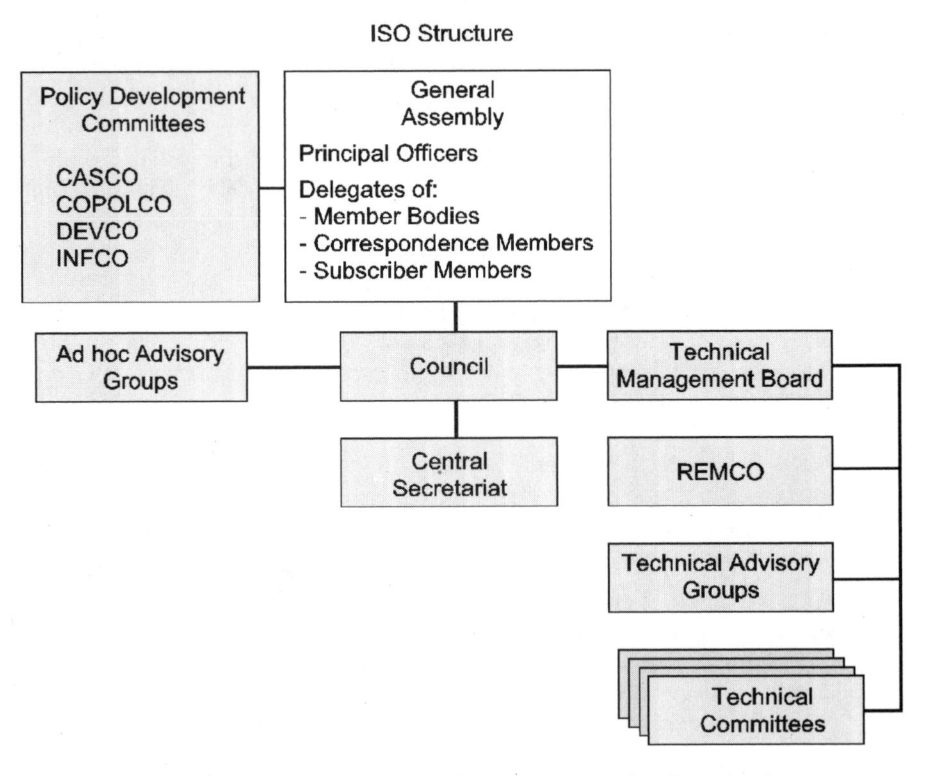

Fig. 18 *Structure of the International Organization for Standardization.*

Council. The operations of ISO are governed by its Council, consisting of the principal officers and 18 elected member bodies. The council appoints the treasurer, the 12 members of the Technical Management Board, and the chairmen of the policy development committees. It also decides on the annual budget of the Central Secretariat.

Central Secretariat. The Central Secretariat in Geneva acts to ensure the flow of documentation in all directions, clarifies technical points with secretariats and chairmen, and ensures that the agreements approved by the technical committees are edited, printed, submitted as draft international standards to ISO member bodies for voting, and then published. Although the greater part of the ISO technical work is done by correspondence, there are, on average, a dozen ISO meetings taking place somewhere in the world every work day of the year.

Ad hoc advisory groups. To advance the goals and strategic objectives of the organization, it may establish ad hoc advisory groups consisting of external executive leaders from organizations having a substantial interest in international standardization. Members of such groups may be invited to participate as individuals rather than as representatives of member bodies. Recommendations of such groups are made to the council for any subsequent action.

Technical Management Board. This group reports to and, when relevant, advises the Council on all matters concerning the organization, coordination, strategic planning, and programming of the technical work of ISO. It examines proposals for new fields of ISO technical activity, and decides on all matters concerning the establishment and dissolution of technical committees.

Committee on Reference Materials (REMCO). This committee establishes definitions, categories, levels, and classification of reference materials for use by ISO.

Technical advisory groups. Technical advisory groups (TAGs) are established, when necessary, by the Technical Management Board (and the IEC Committee of Action in cases of Joint ISO/IEC TAGs) to advise the board (and the IEC Committee of Action when relevant) on matters of basic, sectoral, and cross-sectoral coordination, coherent planning, and the needs for new work.

Technical committees. The ISO has over 2700 technical committees that propose standards in many diverse areas, from screw threads and fasteners to in vitro diagnostic test systems and elevating work platforms. The technical committee on Information Technology alone has issued 1216 ISO standards.

Summary. ISO does not work alone in international standardization. It collaborates very closely with its partner, the IEC. An agreement reached in 1976 defines their respective responsibilities: The IEC covers the field of electrical and electronic engineering, and all other subject areas fall under the purview of ISO. When necessary, attribution of responsibility for work programs to ISO or IEC is made by mutual agreement. In specific cases of mutual interest, joint technical bodies or working groups are set up. Common working procedures ensure efficient coordination and the widest possible global application.

Although ISO and the IEC are not part of the United Nations, they have many technical liaisons with various specialized UN agencies. Several UN agencies are actively involved in international standardization, among them the International Telecommunications Union, the World Health Organization, the Food and Agriculture Organization, and the International Atomic Energy Agency.

See also
American National Standards Institute
International Electrotechnical Commission
International Telecommunications Union

International Telecommunications Union

The International Telecommunications Union (ITU) is an international organization within which governments and the private sector coordinate global telecommunications networks and services. ITU activities include the coordination, development, regulation, and standardization of telecommunications and organization of regional and world telecom events. Founded in Paris in 1865 as the International Telegraph Union, the International Telecommunication Union took its present name in 1934 and became a specialized agency of the United Nations in 1947. Currently the ITU is headquartered in Geneva.

The ITU adopts international regulations and treaties governing all terrestrial and space uses of the frequency spectrum, as well as the use of the geostationary satellite orbit, within which countries adopt their national legislation. It also develops standards to facilitate the interconnection of telecommunication systems on a worldwide scale, regardless of the type of technology used. Spearheading telecom-

munications development on a world scale, the ITU fosters the development of telecommunications in developing countries by establishing medium-term development policies and strategies in consultation with other partners in the sector, and by providing specialized technical assistance in the areas of telecommunication policies, the choice and transfer of technologies, management, financing of investment projects and mobilization of resources, the installation and maintenance of networks, the management of human resources, and research and development.

In essence, the Union's mission covers the following domains:

- *Technical domain:* To promote the development and efficient operation of telecommunication facilities to improve the efficiency of telecommunication services, their usefulness, and their general availability to the public.
- *Development domain:* To promote and offer technical assistance to developing countries in the field of telecommunications, to promote the mobilization of the human and financial resources needed to develop telecommunications, and to promote the extension of the benefits of new telecommunications technologies to people everywhere.
- *Policy domain:* To promote at the international level the adoption of a broader approach to the issues of telecommunications in the global information economy and society.

The ITU comprises 187 member countries and 363 members, which include scientific and industrial companies, public and private operators, broadcasters, and regional/international organizations that participate in the three sectors.

Structure and functioning. The ITU comprises a Plenipotentiary Conference, a Council that acts on behalf of the Plenipotentiary Conference, world conferences on international telecommunications, a General Secretariat, the Radiocommunication Sector, a Telecommunication Standardization Sector, and a Telecommunications Development Sector.

Plenipotentiary Conference. The Plenipotentiary Conference, the supreme authority of the ITU, adopts the fundamental policies of the organization and decides on the organization and activities of the ITU in a treaty known as the International Telecommunication Constitution and Convention. The Plenipotentiary Conference is composed of delegations representing all members and is convened every 4 years. Conferences normally are limited to 4 weeks and focus on long-term policy issues. In this respect, Plenipotentiary Conferences take decisions on draft strategic plans submitted by the ITU Council outlining the objectives, work programs, and expected outcome for each constituent of the union until the following conference.

The ITU Council. The ITU Council is composed of 46 members of the Union elected by the Plenipotentiary Conference with due regard to the need for equitable distribution of the seats on the council among all five regions of the world: the Americas, Western Europe, Eastern Europe and Northern Asia, Africa, Asia, and Australasia.

The role of the council is to consider, in the interval between two Plenipotentiary Conferences, broad telecommunication policy issues in order to ensure that the ITU's policies and strategy fully respond to the constantly changing telecommunication environment. In addition, the council is responsible for ensuring the efficient coordination of the work of the ITU and for exercising an effective financial control over the General Secretariat and the three sectors.

World Conferences on International Telecommunications. World Conferences on International Telecommunications are empowered to revise telecommunications regulations. They establish the general principles that relate to the provision and operation of international telecommunications services offered to the public, as well as the underlying international telecommunication transport means used to provide such services. They also set the rules applicable to administrations and operators in respect of international telecommunications.

World Conferences on International Telecommunications are open to all ITU Member Administrations, and to the United Nations and its specialized agencies, regional telecommunication organizations, intergovernmental organizations operating satellite systems, and the International Atomic Energy Agency.

General Secretariat. In addition to handling all the administrative and financial aspects of the ITU's activities, including provision of computer services, the work of the General Secretariat essentially covers:

- Publication and distribution of information on telecommunication matters,
- Organization and provision of logistic support to the ITU's conferences,
- Coordination of the work of the ITU with the United Nations and other international organizations,
- Public relations, including relations with members, industry, users, press, and academia,
- Organization of the World and Regional TELECOM Exhibitions and Forums, and
- Electronic information exchange and access to ITU documents, publications, and databases.

Radiocommunication Sector. The role of the Radiocommunication Sector is to ensure the rational, equitable, efficient, and economical use of the radio frequency spectrum by all radiocommunication services (including those using the geostationary-satellite orbit), and carry out studies without limit of frequency range on the basis of which recommendations are adopted. Subjects covered include:

- Spectrum utilization and monitoring
- Inter-service sharing and compatibility
- Science services
- Radio wave propagation
- Fixed satellite service
- Fixed service
- Mobile services
- Sound broadcasting
- Television broadcasting

The Radiocommunication Sector operates through Radio Conferences held every 2 years, along with a Radiocommunication Assembly supported by study groups (legislative functions), an Advisory Group (strategic advice), and a Bureau headed by a Director (administrative functions)

Telecommunications Standardization Sector. The Telecommunication Standardization Sector studies technical, operating, and tariff questions and issues recommendations on them with a view to standardizing telecommunications on a worldwide basis, including recommendations on interconnection of radio systems in public telecommunication networks, and on the performance required for these interconnections. Technical or operating questions specifically related to radiocommunication come within the purview of the Radiocommunication Sector.

The Telecommunication Standardization Sector operates through World Telecommunication Standardization Conferences supported by study groups (legislative), an Advisory Group on Standardization (strategic advice), and a Standardization Bureau headed by a Director (administrative).

Telecommunications Standardization Study Groups are groups of experts in which administrations and public/private sector entities participate. Their focus of work is on standardization of telecommunication services, operation, performance and maintenance of equipment, systems, networks and services, tariffs principles, and accounting methods.

Although they are not binding, ITU Recommendations generally are complied with because they guarantee the interconnectivity of networks and technically enable services to be provided on a worldwide scale.

Activities of the telecommunication standardization sector cover:

- Telecommunication services and network operation
- Telecommunication tariffs and accounting principles
- Maintenance
- Protection of outside plant
- Data communication
- Terminal for telematic services
- Switching, signaling, and human-machine language
- Transmission performance, systems and equipment
- ISDN

Telecommunications standardization conferences are held every four years. An additional conference may be held at the request of one quarter of the membership, provided a majority of the members agree. Telecommunications Standardization Conferences approve, modify, or reject draft standards called Recommendations (because of their voluntary character), and approve the program of work. On that basis, they also decide which study groups to maintain, set up, or abolish.

Telecommunication Development Sector. The role of the Telecommunication Development Sector is to discharge the ITU's dual responsibility as a specialized agency of the United Nations and executing agency for implementing projects under the United Nations development system or other funding arrangements. The aim is to facilitate and enhance telecommunications development by offering, organizing, and coordinating technical cooperation and assistance activities.

The objectives of the Telecommunication Development Sector are to:

- Raise the level of awareness of decision-makers concerning the important role of telecommunications in the national economic and social development program, and provide information and advice on possible policy and structural options.
- Promote the development, expansion, and operation of telecommunication networks and services, particularly in developing countries.
- Enhance the growth of telecommunications through cooperation with regional telecommunications organizations, and with global and regional development financing institutions.
- Activate the mobilization of resources to provide assistance in the field of telecommunications to developing countries by promoting the establishment of preferential and favorable lines of credit, and cooperating with international and regional financial and development institutions.
- Promote and coordinate programs to accelerate the transfer of appropriate technologies to the developing countries in the light of changes and developments in the networks of the developed countries.
- Encourage participation by industry in telecommunication development in developing countries, and offer advice on the choice and transfer of appropriate technology.
- Offer advice, and carry out or sponsor studies, as necessary, on technical, economic, financial, managerial, regulatory, and policy issues, including studies of specific projects in the field of telecommunications.

Summary. The ITU endeavors to respond quickly to the requirements of emerging services and market expectations. By serving as the focal point for coordination with other organizations, forums, and consortia worldwide, consumers are eventually provided with access to an increasing range of interoperable products and services. At the same time, the risk of market chaos is greatly reduced, which benefits the economies of all countries.

See also

AMERICAN NATIONAL STANDARDS INSTITUTE
BELLCORE
FEDERAL COMMUNICATIONS COMMISSION
INSTITUTE OF ELECTRICAL AND ELECTRONICS ENGINEERS

Internet

The Internet consists of tens of thousands of interconnected packet-switched networks worldwide, all of which use the Internet protocol (IP). The Internet has developed largely without any central plan, and no single entity can control or speak for the entire system.

The technology of the Internet allows new types of services to be layered on top of existing protocols. Numerous users can share physical facilities, and the mix of traffic through any point changes constantly through the actions of a distributed network of thousands of routers.

For purposes of understanding how the Internet works, three basic types of entities can be identified: end users, Internet service providers, and backbone providers. Figure I9 shows the general relationships of these entities.

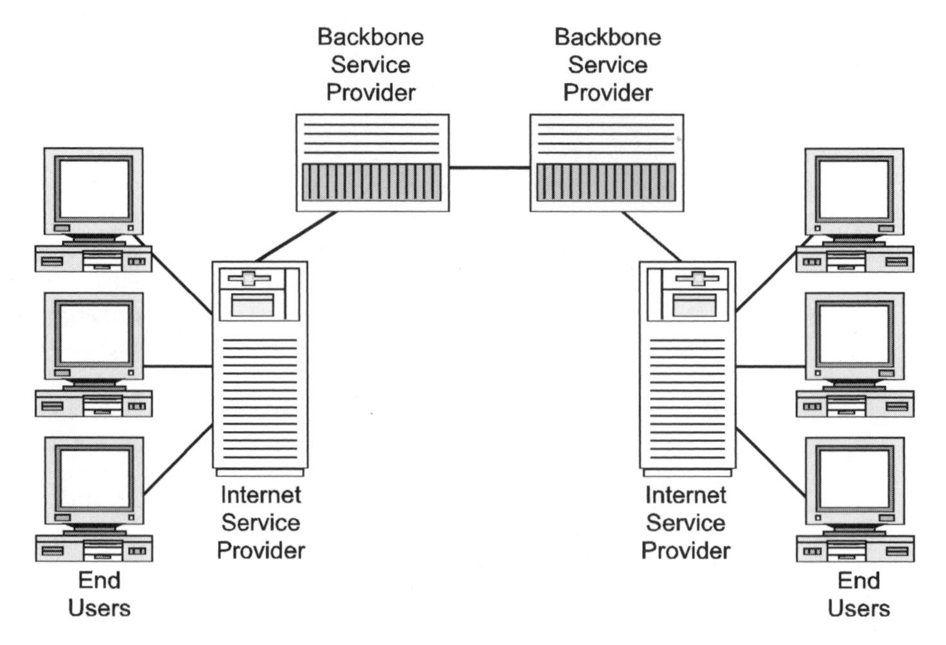

Fig. I9 *Conceptual overview of the Internet.*

End users access and send information either through individual connections or through organizations such as universities and businesses. End users include both those who use the Internet primarily to receive information, and content creators who use the Internet to distribute information to other end users. Internet service providers (ISPs)—such as Netcom, PSI, and America Online—connect those end users to Internet backbone networks. Backbone providers such as MCI, UUNet, and Sprint route traffic between ISPs, and interconnect with other backbone providers.

The actual architecture of the Internet is far more complex. Backbone providers typically also serve as ISPs; for example, MCI offers dial-up and dedicated Internet access to end users, but also connects other ISPs to its nationwide backbone. End users such as large businesses may connect directly to backbone networks, or to access points where backbone networks exchange traffic. ISPs and backbone providers typically have multiple points of interconnection, and the interrelationships between these providers are changing over time. Since the Internet has no central point, individual transmissions may be routed through multiple different providers.

End users may access the Internet though several different types of connections. Most residential and small business users have dial-up connections, which use analog modems to send data over the plain old telephone service (POTS) lines of local exchange carriers (LECs) to ISPs. Larger users often have dedicated connections using high-speed ISDN, frame relay, or T1 lines between a local area network at the customer's premises and the Internet. Although the vast majority of Internet access today originates over telephone lines, other types of communications companies—including cable companies, terrestrial wireless, and satellite providers—also provide Internet access.

Internet history. The roots of the current Internet can be traced to ARPANET, a network developed in the late 1960s with funding from the Advanced Research Projects Administration (ARPA) of the United States Department of Defense. ARPANET linked together computers at major universities and defense contractors, allowing researchers at those institutions to exchange data. As ARPANET grew during the 1970s and early 1980s, several similar networks were established, primarily between universities. The TCP/IP protocol was adopted as a standard to allow these networks, comprising many different types of computers, to interconnect.

In the mid-1980s, the National Science Foundation (NSF) funded the establishment of NSFNET, a TCP/IP network that initially connected six NSF-funded national supercomputing centers at a data rate of 56 kbps. NSF subsequently awarded a contract to a partnership of Merit (one of the existing research networks), IBM, MCI, and the state of Michigan to upgrade NSFNET to T1 speed (1.544 Mbps) and to interconnect additional research networks. The NSFNET backbone, completed in 1988, initially connected 13 regional networks. Individual sites such as universities could connect to one of these regional networks, which then connected to NSFNET, so that the entire network was linked together in a hierarchical structure. Connections to the federally subsidized NSFNET generally were free for the regional networks, but the regional networks typically charged smaller networks a flat monthly fee for their connections.

The military portion of ARPANET was integrated into the Defense Data Network in the early 1980s, and the civilian ARPANET was taken out of service in 1990. But by that time NSFNET had supplanted ARPANET as a national backbone for an Internet of worldwide interconnected networks. In the late 1980s and early 1990s, NSFNET usage grew dramatically, jumping from 85 million packets in January, 1988, to 37 billion packets in September, 1993. The capacity of the NSFNET backbone was upgraded to handle this additional demand with the addition of T3 (45 Mbps) lines.

In 1992, the NSF announced its intention to phase out federal support for the Internet backbone, and encouraged commercial entities to set up private backbones. Alternative backbones already had begun to develop because NSFNET's acceptable use policy, rooted in its academic and military background, did not permit the transport of commercial data. In the 1990s, the Internet expanded beyond universities and scientific sites to include businesses and individual users connecting through commercial ISPs and consumer online services.

Federal support for the NSFNET backbone ended on April 30, 1995. The NSF continues to provide funding to facilitate the transition of the Internet to a pri-

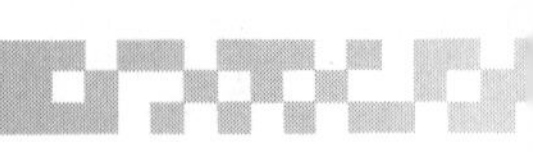

vately operated network. The NSF supports the three priority Network Access Points (NAPs), in Northern California, Chicago, and New York, at which backbone providers exchange traffic with one another, as well as a routing arbiter to facilitate traffic routing at these NAPs. The NSF funds the VBNS (Very High Speed Backbone Network Service), a noncommercial, research-oriented backbone operating at 155 Mbps over optical fiber.

The NSF also provides transitional funding to the regional research and educational networks, as these networks are now required to pay commercial backbone providers rather than receiving free interconnection to NSFNET. Finally, the NSF remains involved in certain Internet management functions, through activities such as its cooperative agreement with SAIC Network Solutions, Inc., to manage aspects of Internet domain name registration.

Since termination of federal funding for the NSFNET backbone, the Internet has continued to evolve. The largest private backbone providers have negotiated bilateral peering arrangements to exchange traffic with one another, in addition to multilateral exchange points such as the NAPs. Several new companies have built nationwide backbones. Despite this increase in capacity, Internet usage has increased even faster.

Operating characteristics. The fundamental operational characteristics of the Internet are that it is a distributed, interoperable, packet-switched network. A *distributed* network has no one central repository of information or control, but consists of an interconnected web of host computers, each of which can be accessed from virtually any point on the network. Routers throughout the network regulate the flow of data at each connection point and reroute data around points of congestion or failure.

An *interoperable* network uses open protocols so that many different types of networks and facilities can be transparently linked together, and allows multiple services to be provided to different users over the same network. The Internet can run over virtually any type of facility that can transmit data, including copper and fiber-optic circuits of telephone companies, coaxial cable of cable companies, and various types of wireless connections. The Internet also interconnects users of thousands of different local and regional networks, using many different types of computers. The interoperability of the Internet is made possible by the TCP/IP protocol, which defines a common structure for Internet data and for the routing of that data through the network.

A *packet-switched* network means that data transmitted over the network is split up into small chunks, or packets. Unlike circuit-switched networks such as the public switched telephone network (PSTN), a packet-switched network is connectionless. In other words, a dedicated end-to-end transmission path does not need to be opened for each transmission. Rather, each router calculates the best routing for a packet at a particular moment in time, given current traffic patterns, and sends the packet to the next router. Thus, even two packets from the same message may not travel the same physical path through the network. This mechanism is referred to as *dynamic routing*. When packets arrive at the destination point, they must be reassembled; packets that do not arrive for whatever reason generally must be re-sent.

This system allows network resources to be used more efficiently, as many different communications can be routed simultaneously over the same transmission facilities.

Addressing. When an end user sends information over the Internet, the data is first broken up into packets. Each of these packets includes a header that indicates the point from which the data originates and the point to which it is being sent, as well as other information. TCP/IP defines locations on the Internet through the use of IP numbers. These numbers include four address blocks consisting of numbers between 0 and 256, separated by periods (for example, 160.130.0.252). Internet users generally do not need to specify the IP number of the destination site because IP numbers can be represented by alphanumeric domain names such as *fcc.gov* or *ibm.com.* Domain name servers throughout the network contain tables that cross-reference these domain names with their underlying IP numbers.

Some top-level domains (such as *.uk* for Britain) are country-specific; others, such as *.com,* are generic and have no geographical designation. The domain name system was originally run by the United States Department of Defense, through private contractors. In 1993, responsibility for nongovernmental registration of generic domains was handed over to the NSF. The NSF established a cooperative agreement with Network Solutions, Inc. (NSI), under which NSI handles registration under these domains. NSI currently charges $50 per year to register a domain name; a portion of this money goes to NSI to recover their administrative costs, and a portion goes into an "Internet intellectual infrastructure fund." The cooperative agreement is scheduled to end in mid-1998. Country-specific domains outside the United States are generally handled by registration entities within those countries.

Services on the Internet. The actual services provided to end users through the Internet are defined not through the routing mechanisms of TCP/IP, but depend instead on higher-level application protocols such as HyperText Transport Protocol (HTTP), File Transfer Protocol (FTP), Network News Transport Protocol (NNTP), and Simple Mail Transfer Protocol (SMTP). Because these protocols are not embedded in the Internet itself, a new Application layer protocol can be operated over the Internet through as little as one server computer that transmits the data in the proper format, and one client computer that can receive and interpret the data. The utility of a service to users, however, increases as the number of servers that provide that service increases.

By the late 1980s, the primary Internet services included e-mail, Telnet, FTP, and Usenet news. E-mail, which is still the most widely used Internet service, allows users to send text-based messages to one another using a common addressing system. Telnet allows Internet users to log into other proprietary networks, such as library card catalogs, through the Internet, and to retrieve data as though they were directly accessing those networks. FTP allows users to download files from a remote host computer onto their own systems. Usenet newsgroups allow users to post and review messages on specific topics.

Despite the continued popularity of some of these services, in particular news and e-mail, the service that has catalyzed the recent explosion in Internet usage is the World Wide Web. The Web has two primary features that make it a powerful,

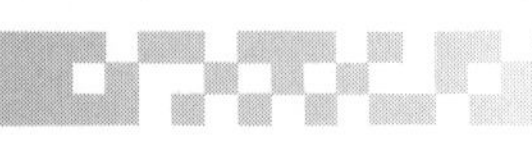

full-service method of accessing information through the Internet. First, Web clients, or *browsers,* can combine text and graphical material, and can incorporate all of the other major Internet services such as FTP, e-mail, and news into one standard interface. Second, the Web incorporates a hypertext system that allows individual Web pages to provide direct links to other Web pages, files, and other types of information. Thus, complex services such as online shopping, continuously updated news information, and interactive games can be provided through the Internet over a nonproprietary system. The Web is the foundation for virtually all of the new Internet-based services currently being developed.

Management. As noted, no one entity or organization governs the Internet. Each facilities-based network provider that is interconnected with the global Internet controls the operational aspects of its own network. No one can even be sure about the exact amount of traffic that passes across the Internet, because each backbone provider can only account for its own traffic and there is no central mechanism for these providers to aggregate their data.

Despite all this, the Internet does not operate in an environment of pure chaos. If certain functions (such as domain name routing and the definition of the TCP/IP protocol) were not coordinated, traffic would never be able to pass seamlessly between different networks. With tens of thousands of different networks involved, it would be impossible to ensure technical compatibility if each network had to coordinate such issues with all others.

These coordinating functions traditionally have been performed by an array of quasi-governmental, intergovernmental, and nongovernmental bodies. The United States government has in many cases handed over responsibilities to these bodies through contractual or other arrangements. In other cases, entities have simply emerged to address areas of need.

The broadest of these organizations is the Internet Society (ISOC), a nonprofit professional organization founded in 1992. ISOC organizes working groups and conferences, and coordinates some of the efforts of other Internet administrative bodies. Internet standards and protocols are developed primarily by the Internet Engineering Task Force (IETF), an open international body composed mostly of volunteers. The work of the IETF is coordinated by the Internet Engineering Steering Group (IESG) and the Internet Architecture Board (IAB), which are affiliated with ISOC. The Internet Assigned Numbers Authority (IANA) handles Internet addressing matters under a contract between the Department of Defense and the Information Sciences Institute at the University of Southern California.

The Internet today. As of January, 1997, there were over 16 million host computers on the Internet, more than ten times the number of hosts 5 years earlier. Several studies have produced different estimates of the number of people with Internet access, but the numbers are clearly substantial and growing. Some industry studies estimate the number of subscribers in the United States at between 40 and 50 million. Although the United States is still home to the largest proportion of Internet users and traffic, more than 175 countries are now connected to the Internet.

The Internet market comprises several segments, including network services (such as ISPs), hardware (such as routers, modems, and computers), software (such as server software and other applications), enabling services (such as directory and tracking services), expertise (such as system integrators and business consultants), and content providers (including online entertainment, information, and shopping). There are now some 3000 Internet access providers in the United States, ranging from small start-ups, to established players such as Netcom and AT&T, to consumer online services such as America Online. According to various industry studies, the Internet market exceeded US$1 billion in 1995, and is expected to grow to between US$22 and US$25 billion dollars in the year 2000.

Internet trends. Estimates from many sources suggest as many as half a billion people will use the Internet by the year 2000. As the Internet grows, methods of accessing the Internet will also expand and fuel further growth. Today, most users access the Internet through either universities, corporate sites, dedicated ISPs, or consumer online services. Telephone companies, whose financial resources and network facilities dwarf those of most existing ISPs, have also entered the Internet access market and are serving businesses and residential customers.

At the same time as these new access technologies are being developed, new Internet clients are also entering the marketplace. Low-cost Internet devices such as WebTV and its competitors allow users to access Internet services through an ordinary television for a unit cost of less than US$200, far less than a personal computer. Various other devices, including network computers (NCs) for business users, and Internet-capable video game stations for consumers, promise to reduce the up-front costs of Internet access even further.

An important trend in recent years has been the growth of intranets and other corporate applications. Intranets are internal corporate networks that use the TCP/IP protocol of the Internet. These networks are either completely separate from the public Internet, or are connected through firewalls that allow corporate users to access the Internet but prevent outside users from accessing information on the corporate network. Corporate users are often ignored in discussions about the number of households with Internet access. However, these users represent a substantial portion of Internet traffic. In addition, intranets generate a tremendous amount of revenue, because companies tend to be willing to pay more than individual users in order to receive a higher level of service.

Summary. Limited government intervention is a major reason the Internet has grown so rapidly in the United States. The Telecommunications Act of 1996 adopts such a position. The 1996 act states that it is the policy of the United States "to preserve the vibrant and competitive free market that presently exists for the Internet and other interactive computer services, unfettered by federal or state regulation," and the FCC has a responsibility to implement that statute.

See also

INTRANETS
TCP/IP
WORLD WIDE WEB

Internet facsimile

To help contain telecommunications costs, companies and individuals are leveraging their existing Internet connections in order to fax documents. There are several ways to send faxes over the Internet: subscribe to a mail-to-fax gateway service, use a technique called remote printing, buy Windows 95 application software, or use the do-it-yourself method.

Mail-to-fax gateways. With a mail-to-fax gateway service, subscribers send faxes as they would e-mail. This type of service is of particular value to users in other countries because it allows them to bypass their often costly PTT networks to send faxes to the United States using their existing Internet connections. The service can save users in other countries as much as 80 percent on faxes to the United States.

The service provider's gateway accepts e-mail from the sender with an attachment that contains an image of the document. Attachment support varies among service providers. Most support PostScript, HTML, TIFF, JPEG, and GIF files as attachments. The gateway then routes the document to an Internet server closest to the recipient. The server dials the local number of the recipient's fax machine to deliver the document.

Internet delays are minimized by the architecture of the service provider's network (Fig. I10). After the initial transmission from the sender's desktop to the near-

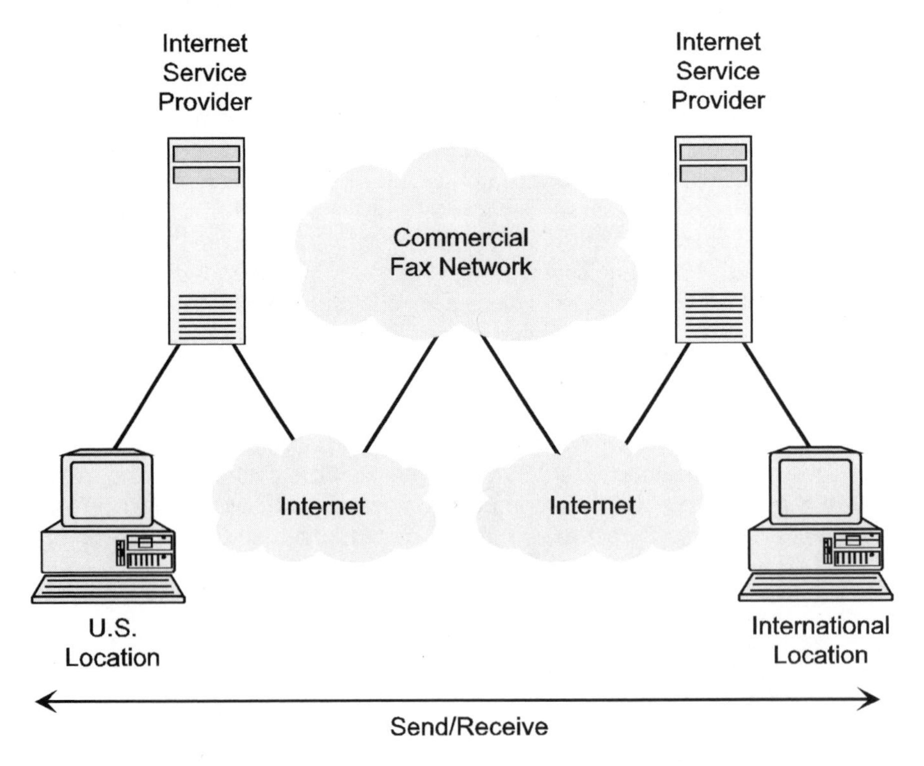

Fig. I10 *Architecture of a typical fax network. The network links to the Internet for broader reach and cost savings for users, avoiding expensive PTT service charges normally associated with fax transmissions.*

est gateway on the Internet, the service provider uses multiple types of networks, as well as the Internet, to dynamically select the best overall route for delivering each fax. The use of multiple networks gives the service provider broad geographical coverage.

To send e-mail to the service provider, the sender enters the fax number of the remote device in the "To:" field of the e-mail program using the following format:

faxnumber@faxnet.com

where *faxnumber* is the telephone number of the remote fax machine (including country code and area code, if any) and *faxnet* is the name of the service provider.

As documents traverse the Internet, confidentiality is assured by encryption supplied by the service provider. At the receiving fax machine, the document arrives just as it would from any other fax machine. When the fax is successfully delivered, the service provider sends back a delivery notice stating the time the fax was received, the e-mail address of the sender and receiver, the subject, number of pages delivered, delivery attempts, called fax machine identifier, and a fax document reference number. If delivery cannot be completed, an error message explaining the reason is returned to the sender via e-mail.

Some service providers offer a send-again capability in case the remote fax machine is busy; offline fax queuing, allowing the sender to assemble multiple faxes for transmission at a convenient time; a log of sent, delivered, and undelivered faxes; and the ability to import address book information from other applications. A broadcast capability allows subscribers to send the same document to multiple recipients listed in a group name on file with the service provider. The document merges with information from the broadcast fax group before the service provider routes it to all of the listed fax numbers.

Remote printing. Mail-to-fax gateway services are enhanced versions of the original experiment in remote printing, which has been available to Internet users on a limited basis since mid-1993.[3] The purpose of the experiment is to integrate the e-mail and facsimile communities, providing a way for e-mail users to send documents to the fax machines of people who do not have e-mail. The arrangement is called *remote printing* because the remote fax machine prints out the document.

Working together, many servers cooperatively provide remote printing access to the international telephone network, allowing people to send faxes via e-mail to the calling areas of participating servers equipped with special fax spooling software. The general-purpose Internet e-mail infrastructure takes care of all the routing, delivering the document to the appropriate server (i.e., gateway) for distribution within a predefined calling area.

The remote printing facility is not available in as many locations as the commercial services and it is not intended for heavy usage. In fact, a site administrator may choose to impose a usage limit on a daily or monthly basis. Such limits are intended to balance the desire to encourage legitimate users with the need to prevent abuse.

3 The two people who started the remote printing experiment are Carl Malamud of the Internet Multicasting Service, a nonprofit organization, and Marshall Rose of Dover Beach Consulting, Inc.

Commercial services have no usage restrictions; they simply bill subscribers accordingly. But the remote printing facility is free and requires no software other than an e-mail program. For a list of locations, by country code and area code, served by the remote printing facility, address e-mail to:

tpc-coverage@town.hall.org

To take advantage of the remote printing facility, users send electronic mail to an address that includes the phone number associated with the target facsimile device. Using the Domain Name System (DNS), the Internet message-handling infrastructure routes the message to a remote printer server, which provides access to facsimile devices within a specified range. The message is imaged on the target remote printer (i.e., fax machine) and an acknowledgment is sent back to the sender of the message via e-mail.

To send a document via the remote printer facility, the following format is used on the "To:" line of the message:

remote-printer.Joe_Smith@14157772525.iddd.tpc.int

where *remote-printer* identifies the kind of access and *Joe_Smith* is the name of the recipient. Next comes the familiar @ sign, followed by the recipient's fax number. A dot separates the fax number from the acronym *iddd,* which stands for International Direct Dialing Designator. After the next dot comes *tpc.int,* which is the Internet subdomain.

With this information, the message will be routed to a remote printer server, which will transmit it as a fax to the recipient. After delivery, an acknowledgment message is sent back to the originator via e-mail.

Windows fax software. There are some inexpensive stand-alone alternatives to faxing over the Internet. Typically these are installed as printer drivers on desktop computers, allowing documents to be faxed from any Windows application. The fax software itself usually supports directories, offline queuing, status messages, and the creation of cover sheets. Some allow received faxes to be redirected to another location. At the destination end, the fax is received via e-mail as an attachment.

Attachment support usually is limited with stand-alone products, and some of these products are not capable of reaching conventional fax machines unless the document goes through a mail-to-fax gateway service. The recipient opens his or her e-mail as usual, and uses an appropriate image viewer to read or print out the attachment.

Mainstream fax software packages, such as versions 7.5 and later of the Symantec (formerly Delrina) *WinFax Pro,* are beginning to support fax transmissions over the Internet. After preparing the document for faxing, the user is given a choice of delivery methods: through a normal phone line or through the Internet. To the user, transmitting a fax appears to be the same for either method. Behind the scenes the process differs, however. Once a user has chosen to send a fax over the Internet, WinFax Pro 7.5 compresses and encrypts the fax and sends it through a service provider called NetCentric, whose server uses intelligent routing algorithms to route the fax over the Internet at the lowest cost possible to the recipient.

Throughout the entire process, WinFax Pro 7.5 provides real-time status messages to the user.

Do-it-yourself. Anyone with an Internet connection, e-mail software, and a scanner can send faxes over the Internet without the aid of extra-cost mail-to-fax gateway services or special fax software. Documents are simply scanned, saved into a graphic format that can be opened by the recipient, and sent as e-mail attachments. This homegrown approach does not provide the bells and whistles of professional services and products, but the quality of the received faxes is the same.

Summary. Cost saving is the principle reason for turning to Internet-based fax solutions. Various methods are available to accomplish this. The choice will depend on several factors, including geographical reach, feature requirements, the number of faxes sent per month, cost per fax (if any), and ease of use.[4]

See also

ELECTRONIC MAIL
FACSIMILE
INTERNET

Internet telephony

One of the most important applications to hit the Internet in recent years is software that allows users to communicate with one another as they would over the telephone. Internet telephony software makes it possible for users to engage in long-distance conversation between virtually any locations in the world without regard for per-minute usage charges. In most cases, all that is needed is an Internet access connection and a computer equipped with telephony software, sound card, microphone, and speakers or headset.[5]

Telephony software compresses the user's voice and packetizes it for transmission over the Internet. On the Internet, these IP packets are treated as ordinary data, indistinguishable from any other kind of data. This means voice is subject to the inherent variable delay of the Internet and, in extreme cases, dropped packets that result in clipped speech. Consequently, the quality of phone calls over the Internet is significantly lower than calls placed over commercial, telco-operated circuit-switched networks. At its best, the quality of a voice conversation over the Internet is only as good as that offered by today's cell phones.

As the growth of the cellular industry demonstrates, however, people are willing to give up a significant level of quality in exchange for other benefits. In the case of cellular, the benefit is the ability of a mobile user to make a call from virtually anywhere. In the case of Internet telephony, the benefit is a vastly lower price, especially on international calls. While Internet telephony users may have to sacrifice quality, they do not have to go without advanced call handling features. Users can

4 For more detailed information on Internet facsimile, see my book, *The Totally Wired Web Toolkit,* published by McGraw-Hill in 1997 (ISBN 0-07-044434-X).

5 For more detailed information on Internet telephony, see my book, *The Totally Wired Web Toolkit,* published by McGraw-Hill in 1997 (ISBN 0-07-044434-X).

implement such features as call waiting, call screening, call transfer, call conferencing, and voice mail—again without having to pay extra monthly charges to a local telephone company or long-distance carrier. Most software also permits the user to adjust microphone and speaker volume during conversations. Some products allow the user to adjust the voice sampling rate and compression ratio to suit the modem speed and line conditions.

Internet telephony software is being incorporated into other types of products, including some text chat products, Web browsers, e-mail programs, and multiuser games. Voice-only conferencing is also an option found in some videoconferencing applications that run over the Internet. In addition, there is server-based Internet telephony software that obviates the need for its installation at the desktop.

Technology. Most voice conversations over the Internet rely on PCs equipped with special software and hardware. As the user speaks into the sound card's microphone, the Internet phone software samples the incoming audio signal, compresses it, and transmits the packets via TCP/IP over the Internet to the remote party. At the other end, the packets of compressed audio are received and pieced together in the right order. The audio is then decompressed and sent to the sound card's speaker for the other party to hear.

Much of the Internet's inherent delay is compensated for by the compression algorithm. As the packets are decompressed and the signals are being played, more compressed packets are arriving. This gives the illusion of real-time conversation. If packets do not arrive within the allotted time, they are simply dropped. If only a small percentage of the packets are dropped, say 2 to 5 percent, the users at each end may not notice the gaps in their conversation. If more packets are dropped, speech will be clipped. Some vendors use predictive analysis techniques in their voice compression/decompression algorithms to reconstruct lost packets, thereby minimizing the problem of clipped speech. As much as 20 percent of total packets can be lost and the conversation will still be intelligible.

Software vendors use a variety of compression algorithms to minimize bandwidth consumption over the Internet. Among the most effective compression methods is GSM (Global System for Mobile communications), which originated as the European standard for digital cellular. GSM provides nearly a 5:1 compression of raw audio with an acceptable loss of audio quality on decompression. Another popular compression standard for Internet telephony is TrueSpeech, developed by the DSP Group, Inc., in Santa Clara, California. It can provide upwards of 18:1 compression of raw audio with an imperceptible loss of audio quality on decompression.

Some products support multiple audio compression algorithms. GSM compression is used when it is installed on a 486-based computer and TrueSpeech when it is installed on a Pentium-based computer. Offering a high compression ratio makes TrueSpeech more CPU-intensive than GSM, so it requires a faster processor to compress the same audio signal in real time. There are also proprietary compression algorithms that are optimized for low-bandwidth conditions, as slow as 6.72 kbps. Some vendors offer their own boards to offload compression/decompression chores from the computer's main processor.

Basic operation. Placing calls on the Internet is as easy as clicking on a name in the public "white pages" or a private "phone book" entry. White pages are simply public directories that are located at one or more servers on the Internet, while a phone book is a private list of names and addresses that resides on the user's own PC.

The public directories are maintained by the various software vendors and are organized by name and topic of interest, making it easy for users to strike up a conversation with like-minded and willing participants. Depending on vendor, some directories are posted on special servers, while others are posted on Web pages. In most cases, when the user opens the telephony software, it automatically connects with a directory server (Fig. I11) and registers the user. This allows other registered users to find people to talk to. The directory is updated periodically and sent to every registered computer, reflecting changes as people enter and leave the network.

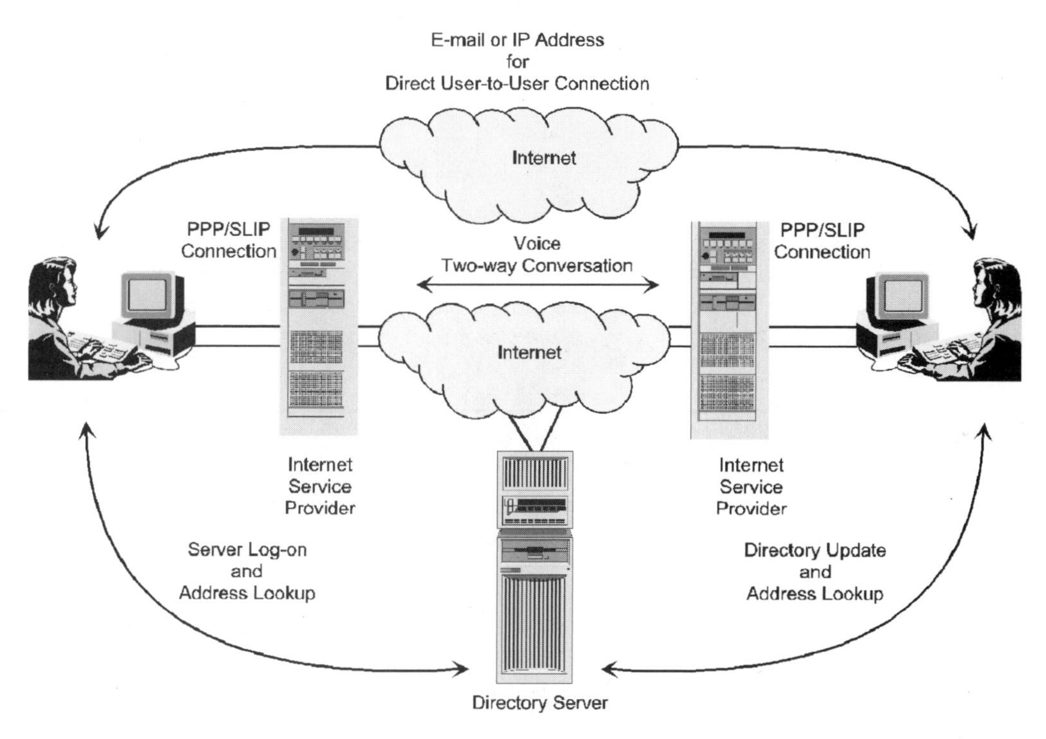

Fig. I11 *With most phoneware products, users login to a directory server and receive a list of other registered phoneware users. The connection is then user-to-user, bypassing the vendor's directory server. Most phoneware also allows the user to enter an IP or e-mail address to establish a user-to-user connection from the start without having to first go through a directory server. This has the added benefit of privacy; since the person's name and address do not show up on a public directory, they will not be bothered by unwanted calls.*

There are commercial gateway services that allow Internet calls to be placed to conventional telephones on the public switched network. The call goes as far as it can on the Internet and then hops off to the public switched network at the gateway closest to the call's destination. The gateway then dials the local phone number and con-

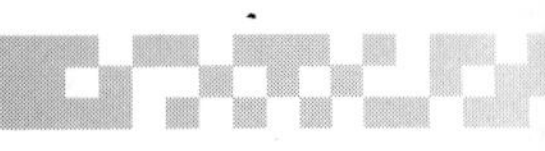

nects the two parties. Calls from PCs to cellular phones also are possible. These gateways also can be used on corporate intranets, providing employees with economical calling and conferencing services. Some gateways offer on-demand calling records, billing information, and directory assistance.

Summary. With continual improvements in compression and wider deployment of bandwidth reservation protocols by router vendors, Internet telephony may well become, in time, a competitive alternative to traditional circuit-switched voice telephony. It is already being used in electronic commerce, customer service, and help desk applications, and has prompted the creation of several virtual telephone companies that use the Internet as a backbone network for commercial service. Major carriers such as AT&T, MCI, and Sprint offer Internet-based calling services and, in the process, have legitimized this form of communication.

See also

TELECONFERENCING
VIDEOCONFERENCING

Intranets

An intranet is a private TCP/IP network that usually supports the same protocols and services as the public Internet, including e-mail, news, and Web-like services. Companies build intranets for improving internal communication, distributing information, and allowing more employees to access legacy data. With links to the public Internet, intranets have allowed companies to reach new potential customers, enter untapped markets, and experiment with electronic commerce. Coupled with such innovations as the Java development language and new "zero-administration" net-centric computers, intranets can become the base around which businesses can reinvent themselves.

Intranet benefits. The foremost benefit that a company can derive from an intranet is more cost-effective communications. Attaining cost-effective communications entails making information directly accessible to people who need it without overwhelming the people who do not need it. Intranets provide direct access to information, so people can easily find what they need without involving anyone else, either for permission or direction on how to navigate through the information. At the same time, companies can protect their information from people not entitled to access it.

From the perspective of managing information, an intranet can extend the reach of distribution and simplify logistics. For example, it can become quite cumbersome to maintain the distribution list for a typical quarterly status report sent to a large mailing list. If anything changes before the next report is due, an update can be developed easily and posted on an intranet, giving everyone access to the new information.

Publishing information on an intranet is quite simple, especially since intranets use the same protocols as the greater Internet, including the HyperText Transport Protocol (HTTP) used by the World Wide Web (WWW). There are a number of Web publishing tools, including some that quickly turn individual documents into the HyperText Markup Language (HTML) format.

Not only has Web publishing become much easier, users also are finding it easier to search for the right document just by using keywords. It is no longer necessary for employees to ask someone for copies of a document or request that their names be put on a distribution list. There is also more control over what is seen. If the big picture is all a person wants, then only that level of information is delivered. If detailed information is desired, however, such search mechanisms as boolean parameters, context sensitivity, and fuzzy logic may be employed. Of course, the user also has the option of accessing greater levels of detail by following the hypertext links embedded in documents.

Of note is that intranets may introduce new chores in managing information. For example, ensuring that all departments have the same updated versions of information requires synchronization across separate departmental servers, including directories and security mechanisms. Providing varying levels of information access to different audiences—engineering, manufacturing, marketing, human resources, suppliers, customers—is also an issue. And, although hypertext links facilitate information search and retrieval, the links must be maintained to ensure integrity as information changes or is added to the database. Fortunately, there are tools that database administrators can use to identify broken links so appropriate corrective steps can be taken.

Enhanced, timely information exchange within the organization is another benefit of intranet adoption. As people from various organizations, functions, and geographic locations increasingly work together, the need for real-time collaboration becomes paramount. Teams need to share information, review and edit documents, and incorporate feedback, as well as reuse and consolidate prior work efforts. Intranets that allow collaboration to occur without paper or copies of files can save hours and even days in a project schedule.

Electronic collaboration eliminates many hurdles such problems as distance between coworkers, multiple versions, and paper copies of information, and the need to integrate different work efforts. The intranet becomes a unifying communications infrastructure that greatly simplifies system management tasks and makes it easy to switch between internal and external communications.

Companies are also finding that, by extending their intranets beyond their immediate boundaries, they are able to communicate more directly and efficiently with the communities with which they do business. Establishing electronic connections to suppliers and partners can result in key savings in time and money in communicating inventory levels, tracking orders, announcing new products, and providing ongoing support. Intranets are being used in a range of application areas to leverage access to existing information and extend a company's reach to employees, partners, suppliers, and customers.

As companies progress in such methods of interaction, more sophisticated intranetworking can result in increased responsiveness and shortened order fulfillment time to customers. Suppliers track inventory levels directly, reduce delays in order fulfillment, and save costs of maintaining inventory. More companies are developing this form of information exchange, where the electronic capability actually drives the process.

Security. Increasing the number of people who have access to important data or systems can make a company's information technology infrastructure vulnerable to attack if precautions are not taken to protect it. Integrating security mechanisms into an intranet minimizes exposure to misuse of corporate data and to overall system integrity. A secure intranet solution implies seamless and consistent security function integrated between desktop clients, application servers, and distributed networks. It should include policies and procedures, the ability to monitor and enforce them, and robust software security tools that work well together and do not leave any gaps in protection.

The following basic functions are necessary for broad security coverage:

- *Access control software* allows varying degrees of access to applications and data.
- *Secure transmission mechanisms* like encryption impede outside parties from eavesdropping or changing data sent over a network.
- *Authentication software* validates that the information that appears to have been originated and sent by a particular individual was actually sent by that person.
- *Repudiation software* prevents people who have bought merchandise or services over the network from claiming they never ordered what they received.
- *Disaster recovery software and procedures* assist in recovering data from a server that experiences a major fault.
- *Antivirus software* detects and removes viruses before they cause damage.

Intranets that extend beyond organizational or company boundaries might require integration among various security systems. In addition, special firewall software might be required to prevent attacks from malicious hackers on the Internet.

Development. Costs are an important consideration when developing an intranet. Beyond the list prices for hardware and software components lie the less obvious costs of administration, maintenance, and additional applications development.

The skill sets that are required for developing an intranet are varied and quite specialized. They include technical people with a knowledge of system and network architectures, an understanding of IP, and experience in developing applications with such tools as Java, ActiveX, and Perl. There is also a need for creative people, particularly graphic artists and HTML coders who excel at making the content visually compelling through the integration of images and text.

The cumulative efforts of many people can go into the initial development and implementation of a corporate intranet. However, many of these people may be peripherally involved. For example, the same network managers and technical staff who keep the division's network up and running by default keep the intranets up and running, since the intranets all might run off the same server.

The daily maintenance of, say, two intranets might require only the part-time efforts of a few people from the marketing and technical groups. The caliber of skills these individuals bring to the task is what makes all the difference, rather than the number of people. While it takes people with specialized skills to develop an intranet, it takes a different set of skills to sustain it. Companies usually deal with this situation by recruiting multifunctional people—those who can apply what they normally do on the job to the medium of the intranet.

Extranets. A variation of the corporate intranet is the "extranet." An *extranet* is a collaborative TCP/IP network that brings together suppliers, distributors, application developers, and customers to achieve common goals via the Web. The concept is totally different from a public Web site or intranet, which are focused around an individual organization's objectives. Among the activities conducted over an extranet are the delivery of product availability and pricing information, custom product configuration, price checking, real-time order entry, and order status inquiry.

Summary. Corporate intranets are becoming as significant to the telecommunications industry as the PC has become to the computer industry. An intranet fundamentally changes the way people in large organizations communicate with one another. In the process, intranets can improve employee productivity and customer response. Intranets also are being used to connect companies with their business partners, allowing them to collaborate in such vital areas as research and development, manufacturing, distribution, sales, and service. Various tools are used for these purposes, including interactive text, conferencing (audio and video), file sharing, and whiteboarding. In fact, anything that can be done on the public Internet also can be done on a private intranet easily, economically, and securely.

See also

INTERNET
TCP/IP
WORLD WIDE WEB

Inverse multiplexing

Inverse multiplexing allows users to dial up appropriate increments of bandwidth to support a given application and pay for the number of local access channels only when they are set up and ready to send voice, data, or video traffic. On completion of the transmission, the channels are taken down and carrier billing stops. This method of access obviates the need for overprovisioning the corporate network to support temporary applications.

Inverse multiplexing can be implemented in customer premises equipment (CPE) or as a carrier-provided service. Either way, the advantages of inverse multiplexing include the immediate availability of extra bandwidth when needed, which eliminates of the need for standby links that are billed to the user whether or not they are fully used. This adds up to significant cost savings.

The inverse multiplexer gathers data from a bandwidth-intensive application. The information is then divided among multiple 56/64 kbps or 384 kbps channels that are dialed up as needed and aggregated to achieve what is, in effect, a higher-speed link. The inverse multiplexer synchronizes the information across the channels and transmits it via switched public network services to a similar device at the remote location. There, the data is received as a single data stream (Fig. I12).

Some inverse multiplexers can be configured to support multiple applications simultaneously. For example, the inverse multiplexer that can be used to link multiple applications at a single site to the public network via a T1 or ISDN access facility, also can link a PBX to a virtual private network (VPN), a router to a fractional-T1 network, and a video codec to a switched digital service. This capability appeals to users who want to spread the cost of a T1 access line across mul-

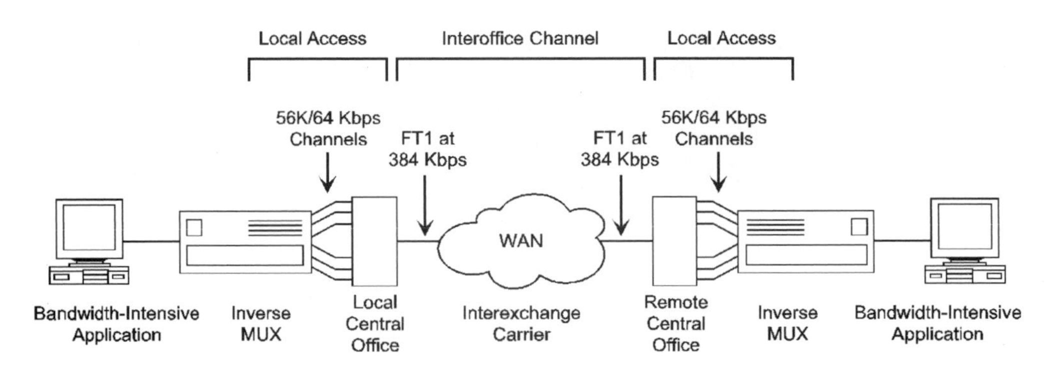

Fig. I12 *A simple inverse multiplexer configuration for a point-to-point videoconference or image transfer.*

tiple applications. Some products allow users to switch multiple applications on a call-by-call basis over different carriers' services simultaneously. While some inverse multiplexers interface only to switched services, others can access both switched and dedicated communications facilities.

Another capability of some inverse multiplexers is the transport of bandwidth-intensive data across multiple T1 circuits to achieve a fractional-T3 circuit. T3-level inverse multiplexers are intended for applications that require transport between the T1 and T3 rates of 1.544 Mbps and 44.736 Mbps. As many as 28 T1 circuits can be aggregated to achieve the desired bandwidth. Dynamic bandwidth capabilities allow users to add or delete T1 circuits based on application needs and traffic priorities. A user, for example, could send Token-Ring or Ethernet transmissions across the wide area network at native speeds.

System management. The system management interface usually consists of a microcomputer equipped with software that allows the network manager to define and monitor traffic flow, bandwidth requirements, access line quality, and various configuration parameters.

Through this interface administrative functions such as the creation of call profiles also are performed. A *call profile* is a file that contains the parameters of a particular data call so that a similar call can be quickly reestablished at another time simply by loading the call profile. Usually the call profile function includes a factory-loaded profile that acts as the template for creating and storing user-defined call profiles. Because each data call could involve as many as 25 separately configurable parameters, the use of call profiles can save a lot of time. Users typically load or edit a call profile using keyboard commands to the management software on the microcomputer.

Inverse multiplexer management interfaces often support remote management facilities. This capability allows a network administrator at a central location to configure, test, and otherwise manage other inverse multiplexers at remote locations in much the same way as is currently offered by the in-band management systems of some T1 multiplexers. This is accomplished by the management interface reserving a certain amount of the network bandwidth, usually not more than 2 percent, as a subchannel to implement remote management.

Most inverse multiplexers can be remotely monitored and controlled via SNMP. This usually is accomplished with SNMP agent software included with the product. The agent collects detailed error statistics, utilization ratios, and performance histories that can be retrieved for analysis.

Standards. The Bandwidth On Demand Interoperability Group (BONDING), formed in late 1991, has defined interoperability standards for inverse multiplexers. There also is a set of international standards for bandwidth-on-demand services called Global Bandwidth on Demand (GloBanD).

The BONDING specification describes four modes of inverse multiplexer operation:

- *Mode 0* allows inverse multiplexers to receive two 56 kbps calls from a video codec and initiate dual 56 kbps calls to support a video conference.
- *Mode 1* allows inverse multiplexers to spread a high-speed data stream over multiple switched 56/64 kbps circuits. Because this mode does not provide error checking, the inverse multiplexers operating in this mode have no way of knowing if one of the circuits in a multicircuit call has failed. In this case, it is up to the receiving node to detect that it has not received the full amount of data and must request more bandwidth.
- *Mode 2* adds error checking to each 56/64 kbps circuit by stealing 1.6 percent of the bandwidth from each circuit for the passage of information that detects circuit failures and reestablishes links.
- *Mode 3* uses out-of-band signaling for error checking, which may be derived from a separate dial-up circuit or the unused bandwidth of an existing circuit.

When establishing calls, inverse multiplexers at both ends first determine whether they can interoperate using the vendor's proprietary protocol. If not, this means that the inverse multiplexers of different vendors are being used and that they should use the BONDING protocol to interoperate.

Summary. The inverse multiplexer allows network managers to match bandwidth to the application. These devices (or a carrier-provided service) provide a degree of configuration flexibility that cannot be matched in efficiency or economy using any other technology. With inverse multiplexers, organizations no longer have to overprovision their networks to handle peak traffic or run occasional high-bandwidth applications. Instead, they can order bandwidth only when it is needed and, in the process, save on line costs.

See also

MULTIPLEXERS

ISO 9000 quality standard

ISO 9000 is a globally recognized standard for quality management and quality assurance. ISO 9000 certification provides assurance to an organization's global customer base that the processes involved in the design, development, manufacture, installation, service, and support of its products adhere to the most stringent and comprehensive quality standards.

The ISO 9000 system, established in 1987, is made up of a series of standards and supplementary guidelines created by the International Organization for Standardization. The quality standards are generic in nature and can be applied across industry lines. The ISO 9000 quality standard and certification process has been adopted by more than 90 countries. Services also can be certified as compliant with ISO 9000 standards. In this case, the certified company must conduct an annual satisfaction survey of all contract customers, recording and tracking the quality of after-sales service and support. The service provider's policies and procedures must be well documented and distributed, and adhere to the same quality standards.

Even software can qualify for ISO certification. A special program called TickIT is an expansion of ISO 9000 guidelines focusing on the design, development, and verification of software. ISO 9001 and TickIT quality standards are assessed by SGS International Certification Services, Inc., a company established in more than 140 countries.

See also

INTERNATIONAL ORGANIZATIONS FOR STANDARDIZATION

Java

Java is a relatively new programming language that originated with Sun Microsystems. Currently it is the fastest-growing programming language for cross-platform networks, particularly those that call for a thin client model. Java actually is a scaled-down version of the C++ programming language that omits many seldom-used features, while adding an object orientation. Java provides a cleaner, simpler language that can be processed faster and more efficiently than C or C++ on nearly any microprocessor.

Whereas C or C++ source code is optimized for a particular model of processor, Java source code is compiled into a universal format. It writes for a virtual machine in the form of simple binary instructions. Compiled byte code is executed by a Java runtime interpreter, performing all the usual activities of a real processor, but within a safe, virtual environment instead of a particular computer platform. This allows the same Java applications to run on all platforms and networks, eliminating the need to port an application to different client platforms. In fact, Java applications can run anywhere the virtual machine software is installed, including any Java-enabled browser, such as Microsoft Internet Explorer and Netscape Navigator.

The use of Java allows remote users, mobile professionals, and network managers to access corporate networks, systems, and legacy data through Java-based applications that are downloaded to the remote computer only when needed. This not only gives users access to the information they need, but simplifies the software maintenance tasks of network administrators, since emulators and other applications reside only in one place: a Web server.

In most cases the applications are stored in cache on a hard disk at the client location; in others, they are stored in cache memory. Either way, the application does not take up permanent residence on the client machine. Since applications are delivered to the client only as needed and all maintenance tasks are performed at the server, users are assured of access to the latest application release level. This not only saves on the cost of software, it permits companies to get away with cheaper computers, since every computer need not be equipped with the resources necessary to handle every conceivable application. At the same time, there is no sacrifice in the ability of users to do their work while away from the office.

Network management applications. Many interconnect vendors are using Java for building network management applications that can be accessed through

Web browsers. Through hypertext-linked home pages set up by the vendor, network managers can use their Java-enabled Web browsers to launch various network management applications. Routers, switches, hubs, multiplexers, CSU/DSUs —virtually any network device—can be configured, monitored, and troubleshot in real time from any location. Applications that provide trend analysis and network reports, access to the vendor's technical support, and on-line documentation are also integrated through the Web browser so that configuration changes and network planning can be accomplished using real data instead of guesswork.

One such network management framework, NetDirector@Web from Newbridge Networks, integrates core services such as discovery, topology, and event management offered by open platforms such as HP OpenView and IBM NetView, and provides distributed network directory services that can be exploited by applications for policy-based management.

The NetDirector home page provides a directory for the network, hyperlinking all of the company's VIVID family devices to simplify network navigation. The home page reflects the status of all discovered VIVID devices to show, at a glance, the health of the devices and other useful information such as firmware version and events. The network manager can manage the network from home or on the road by hot-linking to the devices. The home page also provides a method for the administrator to specify management policies, such as upgrading firmware and software across multiple devices throughout the network, or defining network behavior in the event of a broadcast storm.

The Web-based applications are bundled with NetDirector, the Newbridge Networks enterprise management solution that integrates with HP OpenView on Solaris, HP-UX, and Windows NT platforms.

Because of Java's real-time capabilities, changes in the network status are reflected immediately, without requiring the network manager to reload Web pages. Other Web management offerings only provide static, HTML-based configuration reports for various network devices. In addition, Java applets are loaded dynamically from NetDirector@Web servers so that the user does not have to preinstall or continually update the network management software on the system being used to manage the network.

Among the Java-based applications that run under the NetDirector@Web framework is VitalStat, a network diagnostic and analysis tool. VitalStat analyzes baseline response times and other performance characteristics. When deviations are detected, VitalStat diagnoses the problem; attempts to isolate whether the cause is application-, server-, or network-related; and recommends or initiates appropriate corrective actions. VitalStat has a Java-powered Web interface and provides any-time, anywhere management access via a standard Web browser.

VitalStat uses intelligent agents that run in the network elements. As a result, it can follow the same path an end-user station uses to access a server in order to detect and diagnose problems. This allows more accurate problem determination for intelligent reporting back to the network administrator. While policies can be configured centrally and reports viewed from the VitalStat graphical user interface, the actual event detection, analysis, and response can be addressed seamlessly by the agents themselves without user intervention. Figures J1, J2, J3, and J4 illustrate some of the reports available.

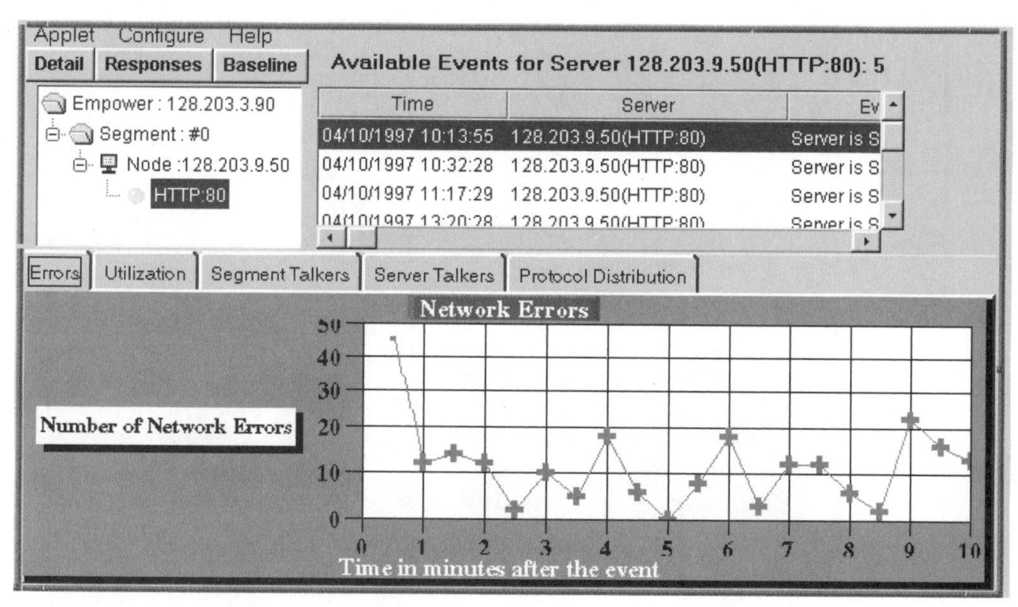

Figure J1 *VitalStat shows the number of errors for selected servers at each node by time in minutes after the event.*

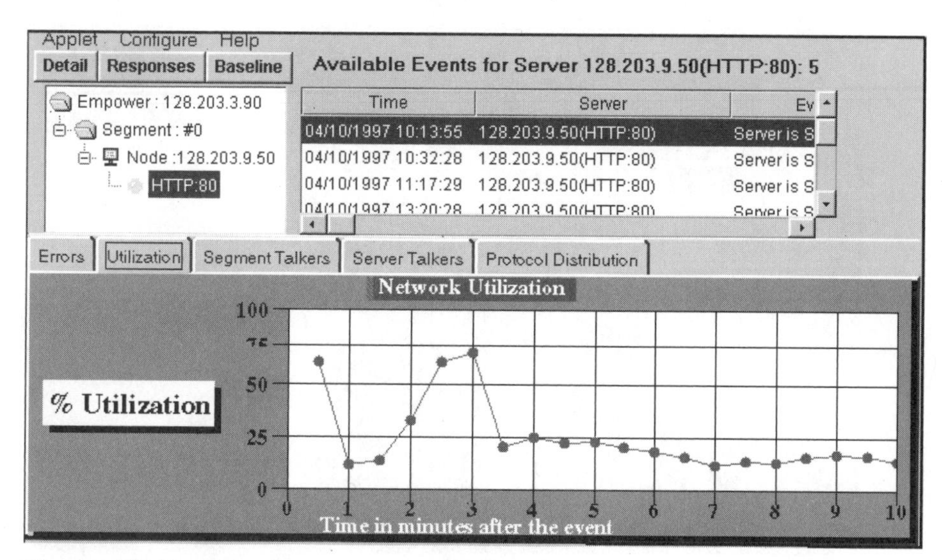

Figure J2 *VitalStat shows percent utilization for selected servers at each node. This information can be used for performance baselining.*

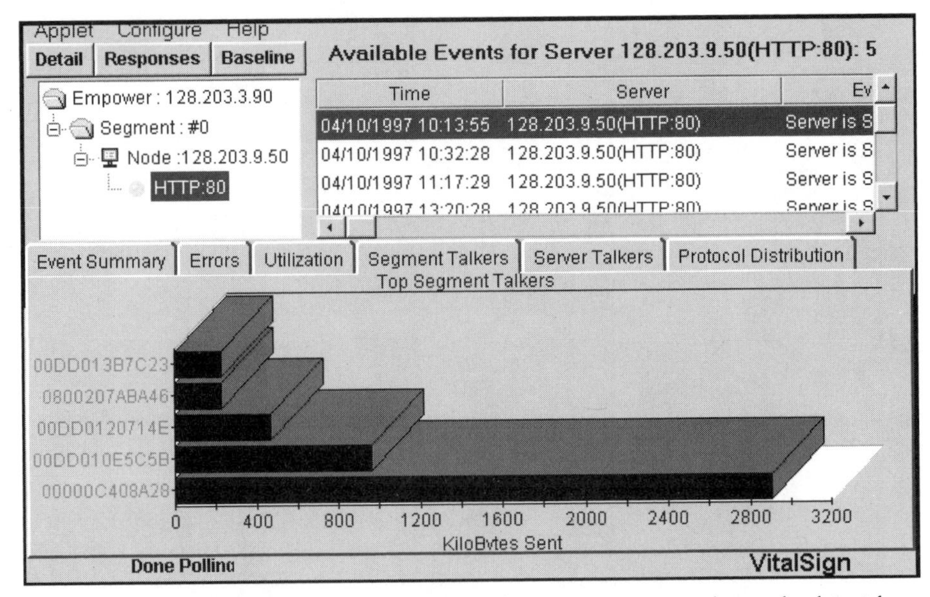

Figure J3 *VitalStat uses SNMP's RMON (remote monitoring) standard to identify a server's top talkers by MAC address in terms of kilobytes sent.*

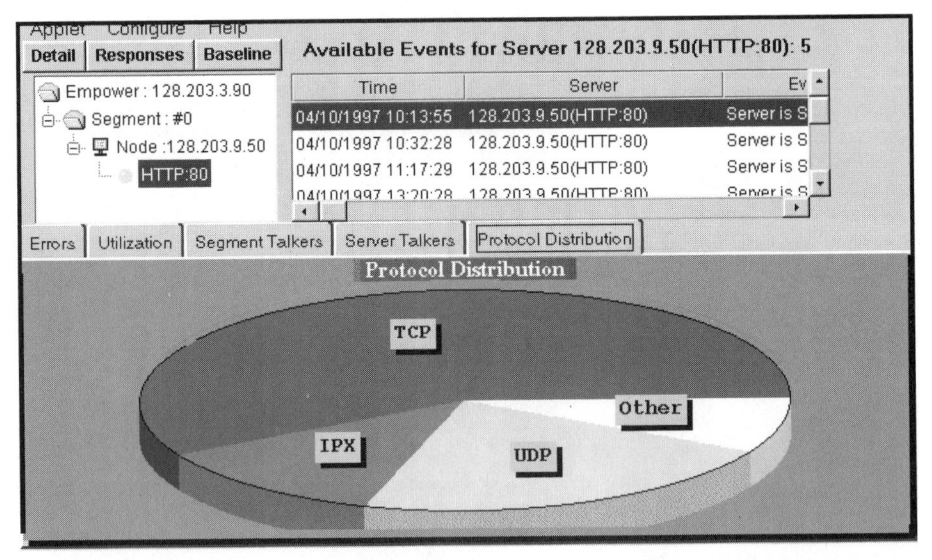

Figure J4 *VitalStat identifies how the traffic of a selected server is distributed by protocol. Among the protocols identified are IP, IPX, TCP, and UDP.*

NetDirector@Web uses several levels of security to access the network from any location at any time. The first level of security is a firewall. Because NetDirector@Web is targeted for management of an intranet within a firewall, the firewall prevents unauthorized users outside the intranet from accessing the network.

In addition, through an access list, administrators can define a list of allowed users/hosts that have access to various management functions. This type of security provides greater control than offered by SNMP community strings. Telnet, FTP, and Web connections are secured through host access security and by user name and password authentication. Only the "root" user with a valid password is given access to certain administrative functions. Java provides still another level of security, limiting the operating system and resources that the application can access. Java also has virus protection to prevent viruses from attaching themselves to the applications.

Summary. The early success of the C++ programming language owes a great deal to its ability to access legacy code written in C. Similarly, Java preserves much of C++ and offers a number of compelling benefits: It is object-oriented, portable, and relatively easy to master and maintain. Once written, Java applications can run unchanged on any operating system that has a Java interpreter. These and other benefits of Java can greatly speed the development cycle for Web-based applications, including those for remote network management.

The applications themselves are accessed only when needed, with the most up-to-date version downloaded to the client's cache. When the client disconnects from the network, the application is flushed from the cache, conserving limited system resources. This is the basis for network computing, a new paradigm that, in essence, treats the Internet as the computer. Major network management vendors are expected to issue Java versions of their products in the near future, making management data readily accessible from any Java-enabled workstation in the enterprise.

See also

NETWORK MANAGEMENT SYSTEMS
NETWORK COMPUTING
WORLD WIDE WEB

Jitter

While *delay* is the time it takes to get a unit of information from source to destination through a network, *jitter* is the variance of the delay. Both can have potentially disrupting effects on the application running over the network.

Delay and jitter were not important aspects of computer networks in the past. It did not matter, for example, if file transfers or e-mail took half a second longer, independent of the total transfer time (delay). Similarly, it did not matter if, on a particular file transfer, 70 percent of the data was sent during the first half of the transfer and 30 percent in the last half (jitter).But jitter and delay matter when it comes to two-way or multiway conversations and conferences: They must have low delay and jitter to support the natural interaction among participants, since even fraction-of-a-second pauses can be disruptive to a conversation.

Multimedia applications that combine audio and video content are even more sensitive to delay and jitter. To prevent dropouts in an audio stream or jerkiness in

video, jitter must be low. For one-way broadcasts, buffering can be used in the end stations to decrease the effect of jitter, but only at the cost of increased delay. While this delay is acceptable for one-way broadcasts, it is not acceptable for two-way conversation.

The electrical pulses sent through a network are normally sent at very specific intervals of time. The repeaters, bridges, and switches on the network contain buffers to accept the signal from one side and send it back out the other side at a tightly controlled speed. This results in a clean output signal that can be received by the next piece of equipment on the network.

Every piece of equipment has a narrow tolerance within which it operates, however. If a signal passes through several pieces of equipment or cable, these tolerances can add up and cause the resulting signal to shift in phase compared to what was originally sent. This makes it difficult or impossible for the next device on the network to lock onto the signal and causes errors (Fig. J5).

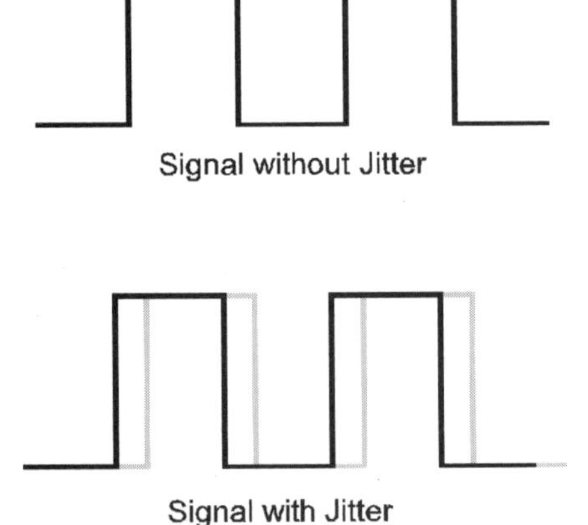

Figure J5 *A digital signal with no jitter (top) and the same signal with jitter (bottom).*

Summary. As clock speeds in computers and data rates in communications systems increase, timing budgets become tighter and the need to measure and characterize jitter becomes more critical. Oscilloscopes offer various tools to measure jitter. On new networks with no performance history, a device called a *jitter generator* can be used to add controlled jitter to digital signals for the purpose of stressing the connections to PSTN, cellular and PCS base stations, satellite modems, and microwave links. Not only does this ensure that the connections are error-free before being turned over to user traffic, but once the jitter parameters are known, lower-cost oscilloscopes can be used for periodic quality checks.

See also

LATENCY

Key telephone systems

Traditionally, key telephone systems functioned like private branch exchanges (PBXs), but offered only a subset of the available features. Because such systems were more economical than PBXs, small and midsize businesses found them very appealing, especially for the key system's configuration flexibility, which allowed multiple lines to be terminated at each telephone. Within certain limits, each line could be answered from any multiline telephone in the system by pressing the appropriate line key.

Since the introduction of the first key telephone system by Bell Telephone in 1938, key systems have evolved from cumbersome units with toggle switches to today's sophisticated, electronic systems. Key systems have advanced to the point where they can be incrementally equipped with a wealth of call processing and system management features, rivaling those of higher-cost PBXs.

Evolution of key systems. The first standard key systems were electromechanical devices, designated *1A key telephones,* that used adjunct key units and toggle switches for line control. These 1A systems had clearly defined system components and wiring schemes, and offered an attractive alternative to earlier custom-built telephone systems that lacked standardization.

In 1953, Bell replaced its 1A systems with 1A1 systems, which packaged line control equipment into the telephone set. Line status indicators on the 1A1 telephones lit steadily to show a line in use, and flashed to indicate held calls. Current key system standards evolved from early 1A and 1A1 system designs.

In 1963, further technological improvements led to the introduction of 1A2 systems, which used plug-in modules for line control, signaling, and intercom functions. Inter-Tel offered the first solid-state enhanced electromechanical 1A2 key system in 1975.

Electromechanical 1A2 key systems were a major improvement over the earlier 1A and 1A1 systems. The older systems were hardwired, but 1A2 systems were built around packaged components that made installation and maintenance much easier. These components included:

- *Key Service Unit:* A wall-mounted or freestanding cabinet that contains the key telephone units and circuit packs that control access to central office (CO) lines, Centrex lines, or PBX lines. The KSU also manages signaling units for private-line, intercom, paging, and music-on-hold connections.

- *Key Telephone Unit:* Supports CO line connections and provides connections for individual stations and intercom paths, as well as paging and music-on-hold equipment within the system. KTUs also provide tone and ringing generation, lamp indicator control, and dialed-digit detectors and decoders.
- *Main Distribution Frame:* Provides the primary connection points for the KSU and is the termination point for all cables and wires within the key system, including external CO lines and internal station lines.
- *Power Supply:* Converts commercial voltages to meet the system's unique operating requirements. The power supply can be housed within the KSU or in a separate cabinet near the commercial ac power source.
- *Station Equipment:* Consists of desktop phones and attendant stations. Each line key on an electromechanical key telephone requires a minimum of three wires for operation: one for the talk path, one for hold, and one for the line lamp.

Stored program control. While technical enhancements were continually added to 1A2 key systems, they were rapidly replaced by stored program control (SPC) electronic key systems, which were introduced in the mid-1970s. Electronic key systems are similar to 1A2 systems but offer several improvements, including reduced cabling requirements, support for more features, and greater configuration flexibility.

The most significant difference is that electronic systems replace the relay-based KSU and KTUs used in electromechanical systems with an electronic KSU and printed circuit boards (PCBs). The electronic KSU architecture typically consists of a card cage with a prewired backplane; specialized PCBs plug into the backplane to control various system functions. The modular design of the electronic KSU facilitates the installation and removal of PCBs, allowing new features to be added to the system without wiring changes.

Electronic station sets also use specialized circuits for control. Whereas 1A2 station equipment requires several wire pairs for each key, electronic key systems require only two or three wire pairs per station regardless of the number of keys.

Station sets have increased in intelligence to the point that individual telephones can be programmed by users for personal features such as speed dialing. Many station sets also include liquid crystal displays (LCDs) to identify calling parties, display messages, and guide users through feature implementation procedures.

Unlike electromechanical systems, electronic systems do not require time-consuming and difficult wiring changes to manage system reconfiguration and expansion. Features therefore can be modified from a station set keypad, from a dedicated console, or display monitor in a centrally administered system.

Basic key system call-handling features (including call forward, conference, speed dial, and last number redial) are accessible from electronic station sets. Many sophisticated system features and functions also are accessible from the station keys, including call accounting and station message detail recording (SMDR), automatic number identification (ANI), and least-cost routing (LCR). Auxiliary jacks provide connection to answering machines, facsimile machines, and other peripheral devices.

More key systems are now being programmed by PCs via a computer telephony interface (CTI). With open application programming interfaces (APIs), users can connect the key system with voice mail, call processing software, and other products that come from third-party manufacturers, including fax machines.

More key systems are being equipped to address the growing problem of toll fraud, such as giving the administrator the ability to program the system to automatically turn off whenever invalid attempts are made to access the system. The system also can be programmed to notify designated individuals via phone or pager when a toll fraud attempt is made.

Some key systems offer T1 and ISDN interfaces and support wireless technology for in-building mobile communication.

Summary. Key systems have advanced to the point where they are now "KSU-less." Instead of a KSU with a central processor, KSU-less systems incorporate system and feature control directly within the telephone sets. The use of very large scale integration (VSLI) circuit technology eliminates the need for a separate equipment cabinet. Not only do non-KSU systems retain the multiline capability, they also can be used with Centrex, allowing users to program one or two keys for Centrex features.

See also

CENTREX
PRIVATE BRANCH EXCHANGE (PBX)

Kiosks

Kiosks were introduced for a variety of applications in the mid-1980s, but failed to catch on in the United States until the recession of the late 1980s and early 1990s. At that time, kiosks became a convenient and economical way for businesses to disseminate information about their products and services without burdening overworked staff.

The earliest kiosks merely deluged users with facts that were not necessarily relevant to their information needs, however. There was very little interaction that let the user specify the kind of information he or she wanted. There was no provision for graphics, let alone video, to hold the attention of users. And if the kiosk broke down, the operator would not know about it until a technician made a scheduled visit.

Today multimedia techniques are being used not only to attract users to the kiosk, but, once there, to keep their attention until the entire message is delivered. There also is more opportunity for user interaction, which is important from an entertainment perspective as well as for getting the message across. Kiosks are now a common fixture in hotels, airports, convention centers, shopping malls, stores, travel stops, and hospitals.

Applications. At airports, for example, kiosks are also used to issue airline tickets. At the kiosk, the traveler passes his or her credit card through the magnetic stripe reader to pay for the ticket. The ticket and boarding pass are issued at the kiosk via the integral printer. With ticket and boarding pass in hand, the passenger can immediately board the plane.

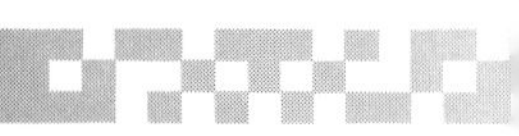

In a typical financial application, a bank would set up a kiosk in a convenient lobby location or other public access point to allow customers to see if they qualify for a car loan and how they might manage the payments over 3, 4, or 5 years. Rather than take up a bank officer's time, customers would approach the kiosk and score themselves on various factors that lead to loan qualification based on the inputs he or she provides. Also available are the current interest rates and payment schedules. The user then can approach the bank officer for the loan application or comparison shop elsewhere.

Many banks are using kiosks for issuing statements and for allowing customers to make account inquiries. A bank cannot provide these types of services with an automatic teller machine (ATM) because the dozen or so people standing in line will not tolerate waiting that long.

Government agencies also are experiencing similar concerns. For example, nearly every motor vehicle department in the country is being hamstrung by budget cuts and staff reductions, while the demand for services continues to rise. This is departments to look at ways to streamline work flow and automate such routine services as license renewal and vehicle registration.

One way to do this is by locating kiosks at shopping malls and other public access points, where people can renew their licenses. The user selects "renewal" from the terminal display, pays the renewal fee with a credit card, and walks away with a temporary license in a matter of minutes. Via a modem connection, the motor vehicle department's host computer can collect information from the kiosks at the end of the day for processing. Within 10 days a permanent license is mailed to each driver.

In automating and decentralizing service delivery, the kiosks greatly reduce the cost per transaction and reduce the tension level of overworked motor vehicle office personnel, who previously had to face long lines of often angry and frustrated customers.

Other government applications that lend themselves to the use of multimedia kiosks are the handling of court information and the payment of fines via credit card, as well as agency procedural and directory information.

System components. The multimedia terminals combine video capabilities and touch-screen technology in ergonomic, kiosk-type packaging. Bundled into the kiosks are such components as central processor, credit card reader, optical disk, printer, loudspeaker, and proximity detector that automatically starts the application when someone comes within a few feet of the system. A custom-designed runtime program presents information in text, graphic, or video formats for photo-quality images or full-motion video displays.

Retrieving accumulated user input information is a function of the application. Typically, a dial-up connection will transfer all accumulated information stored on the kiosk's hard disk for further processing by the host (Fig. K1). Alternatively, the application could be designed to sort information into separate files based on inputs from users. That way, the host can be more selective in its retrieval request. For example, a file may contain the number of users who made a specific touch selection during the course of the day. This touch selection may indicate a product preference or a request for specific information by mail.

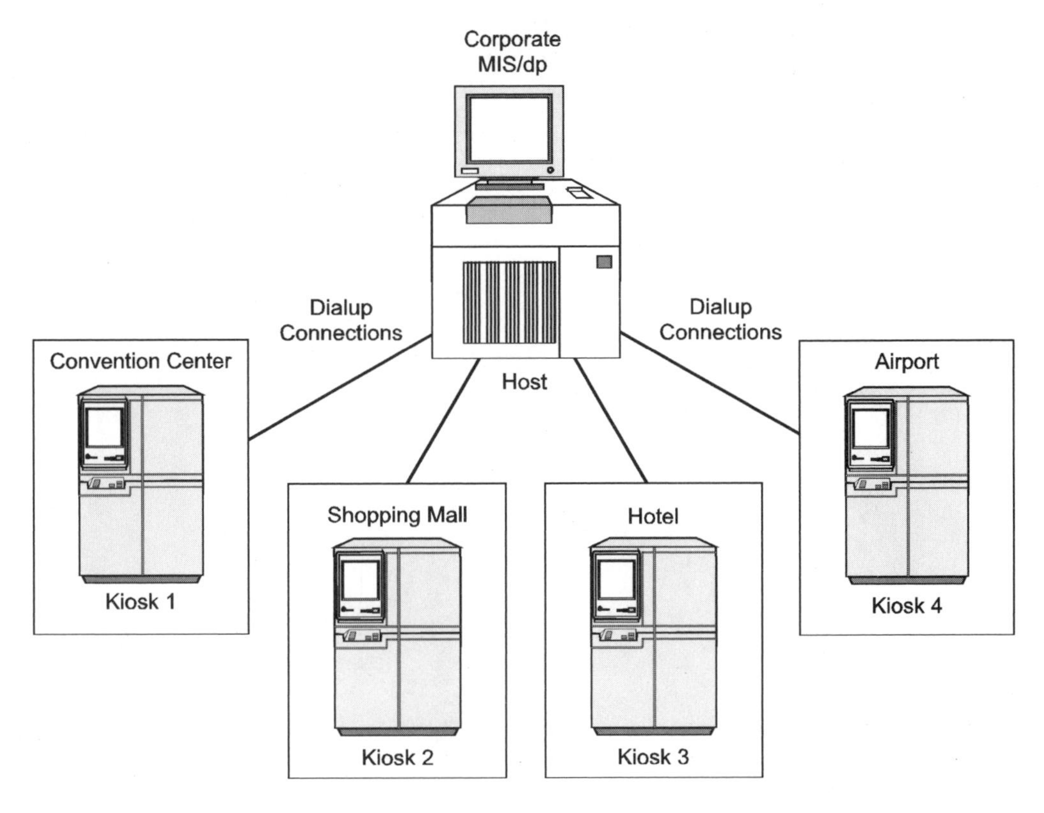

Figure K1 *Kiosks in a distributed environment.*

Network capabilities. The kiosk can be equipped with a leased-line modem, giving it the ability to be connected to the host for polling over a multidrop leased line. That way, accumulated information from each unit can be retrieved automatically upon request by the host.

At the same time, dial-up or leased-line connections enable the kiosks to be updated with new information from a central management facility. Instead of dispatching a technician to exchange CD-ROMs containing new video presentations, for example, the master station can download the information to the hard disk in each kiosk. The video and images are compressed for transmission and storage using such standard methods as JPEG or MPEG.

Kiosks can be integrated into nearly any type of host environment, including IBM 3270, SNA/SDLC, and Unix. The system can be networked in the X.25 as well as TCP/IP environments, or run over Ethernet and Token-Ring LANs.

Remote diagnostics can be implemented over dial-up lines via RS-232C interfaces. A dial-up connection can be used to gather alarms that report an out-of-paper condition, a full hard disk, no touch-screen activity, or that the central processor is down. The system's integral diagnostic functions tell an on-site technician or remote operator what components are not functioning properly.

Summary. Kiosks are used to assist businesses and government agencies with a variety of product and service support requirements, while offering consumers a convenient and private method of access. When networked together through a central management facility, kiosks are an efficient and economical vehicle for the delivery of information and services.

See also

MULTIMEDIA NETWORKING

LAN switching

When shared LANs become too slow, performance can be improved by creating segments linked together by bridges. Bridges keep local traffic on a particular LAN segment, while allowing packets destined for other segments to pass through in a process called *filtering*. But no matter how many segments are created, LAN performance tends to diminish, if only because more users are continually added to the network. The greater the number of workstations simultaneously accessing the LAN, the smaller each workstation's available bandwidth becomes.

With a switched LAN (Ethernet or Token-Ring), each user can have access to the network at full native speed instead of having to share it with multiple users. Dedicated LAN links improve network performance for all users, allowing them to be more productive and making the network easier to manage. In some cases, switched LANs provide enough performance improvements to hold off purchases of more expensive ATM or FDDI networks.

LAN switching is implemented in conjunction with an intelligent wiring hub or a dedicated LAN switch. Bandwidth can be controlled by restricting access to any logical segment to only authorized members of the workgroup (Fig. L1). Creation of these virtual workgroups also provides security, since packets (both broadcast and multicast) will be seen only by authorized users within that virtual workgroup. While users of one logical network cannot access another logical network, thus enforcing security, multiple virtual workgroups still can share centralized resources such as e-mail servers. Highly secure logical segments can be established and modified from an SNMP-based network management station and encompass any device on the network.

Port vs. segment switching. There are two basic types of switching hubs: port-switching and segment-switching. *Port-switching hubs* let administrators assign ports to segments via network management software. In effect, these hubs act as software-controlled patch panels. *Segment-switching hubs* treat ports as separate segments and forward packets from port to port. How they accomplish this varies by vendor and has important implications depending on the application. What all segment-switching hubs have in common is that they can substantially increase available network bandwidth. Acting like high-speed multiport bridges, segment-switching Ethernet hubs, for example, offer 10 Mbps to each port.

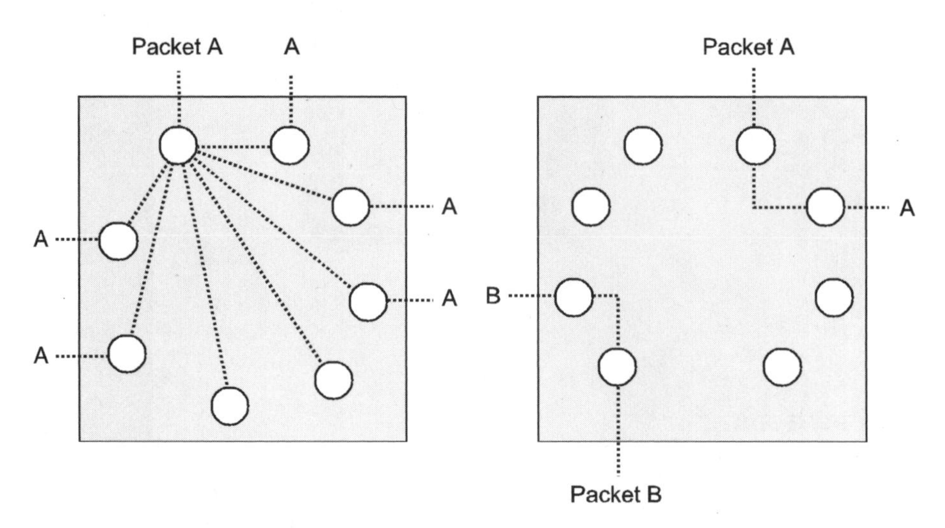

Figure L1 *Standard hub versus switching hub. A standard hub (left) broadcasts packets to all ports, while a switching hub (right) supports virtual connections to limit traffic to only specific addressees.*

Ethernet switching. In the traditional Ethernet environment, stations contend with one another for access to the network, a process that is controlled by a statistical arbitration scheme. Each station "listens" to the network to determine if it is idle. Upon sensing that no traffic currently is on the network, the station is free to transmit. If the channel is already in use, the station backs off and tries again later. If multiple stations sense that the network is idle and send out packets simultaneously, a *collision* occurs, which corrupts the data. When a collision is detected, the stations back off and try again at staggered, random intervals. This process is called *Carrier Sense Multiple Access with Collision Detection* (CSMA/CD).

The problem with CSMA/CD is that collisions force retransmissions, which causes the network to slow down. This in turn results in less bandwidth availability for all users. A related problem with Ethernet is that a station's packets are broadcast automatically to all other stations. This means that all other stations are aware of the packets, even though only one can actually read them. The broadcast nature of Ethernet increases the likelihood of collisions.

In Ethernet switching, the MAC-layer address determines to which hub or switch port the packet will go. Since no other ports are aware of the packet's existence, the stations do not have to be concerned about whether their packets will collide with data from other stations as they transmit toward the hub or switch. In Ethernet switching, a virtual connection is created between the sending and receiving ports. This dedicated connection remains in place only long enough to pass the packet between the sending and receiving stations.

If a station has packets for a busy port, that station's port momentarily holds them in its buffer. When the busy port becomes free, a virtual connection is established and the packet is released from the buffer and sent to the newly freed port. This mechanism works well unless the buffer gets filled, in which case packets are lost. To avoid this, some vendors offer a throttling capability. When a port's buffer begins to fill up, that port begins to send packets back to the workstation. This slows the workstation's transmission speed and evens out the "pressure" at the port. Because no packets are lost, many more packets make it through the hub than would otherwise be possible under heavy traffic conditions.

Some LAN switching products offer a choice of different packet switching modes: *cut-through* and *store-and-forward.* Cut-through speeds frame processing by beginning the transmission on the destination LAN before the entire frame arrives in the input buffer. Store-and-forward works like a traditional LAN bridge: Frame forwarding begins only after the entire frame arrives in the input buffer. Store-and-forward mode results in a greater performance hit but ensures error-free delivery. Some vendors support automatic switching between cut-through and store-and-forward modes. With this technique, called *error-free cut-through switching,* the switch changes from cut-through to store-and-forward mode if the percentage of bad packets flowing through the switch exceeds a predefined threshold.

Some LAN switching devices support only one address per port, while others support 1500 or more. Some devices are capable of dynamically learning port addresses and allowing or disallowing new port addresses. Disallowing new port addresses enhances hub security: In ignoring new port addresses, the corresponding port is disabled, preventing unauthorized access.

Some vendors offer full-duplex Ethernet connections, providing each user with 20 Mbps of dedicated bandwidth (send and receive) over unshielded twisted-pair wiring. With half-duplex signaling and collision detection disabled, one pair of wires can transmit at 10 Mbps while the other pair receives at 10 Mbps. This creates a collision-free connection that can ease bottlenecks between similarly equipped switching hubs or servers.

Token-Ring switching. LAN switching is also available for the Token-Ring environment. Token-Ring switching is a technology for dedicating 4 Mbps or 16 Mbps of bandwidth to each user on a LAN. Software-controlled switching of individual ports into any one of a number of token rings eliminates the need to patch network cables physically to change LAN configurations and provides centralized, per-port control over Token-Ring LAN configurations.

Like switched Ethernet LANs, switched Token-Ring LANs can be configured to operate in full duplex to send and receive data simultaneously. The total per-port capacity of each switch module would then be 32 Mbps.

Associated with Token-Ring switching is the capability to protect the ring automatically from potential disruptions. Although some vendors offer the ability to mix 4 Mbps and 16 Mbps on the same module, the ports can detect and deny entry to any device attempting to connect to a port with a transmission rate different from the port's predefined configuration. This prevents a device configured for 16 Mbps transmission from accessing a port configured for 4 Mbps transmission, for example.

Summary. Enterprise-level LAN switches are available that provide seamless switching among fast LAN technologies including FDDI, Fast Ethernet (100Base-T), and 100VG-AnyLAN, while providing a safe migration from these and existing Ethernet and Token-Ring LANs to ATM. There are even gigabit switches for sup-

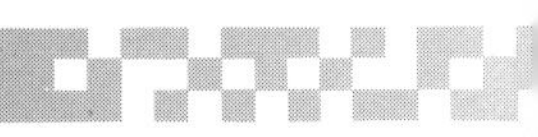

porting large departments in enterprise LANs. Combined with LAN bandwidth management intelligence, LAN switches provides users with the ability to build and manage reliable, high- performance switched intranetworks.

See also
ETHERNET
HUBS
TOKEN-RING

Latency

Latency is the amount of delay that affects all types of communications links. Delay on telecommunications networks is usually measured in milliseconds (ms), or thousandths of a second.

A rule of thumb used by the telephone industry is that the round-trip delay for a telephone call should be less than 100 ms. If the delay is much more than 100 ms, participants think they hear a slight pause in the conversation and use it as the opportunity to begin speaking. But by the time their words arrive at the other end, the other speaker has already begun the next sentence and feels that he or she is being interrupted. When telephone calls go over satellite links, the round-trip delay is typically about 250 ms and conversations become full of awkward pauses and accidental interruptions. Latency affects the performance of applications on data networks as well. On the Internet, for example, excessive delay can cause packets to arrive at their destination out of order, especially during busy hours. The reason packets may arrive out of sequence is that they can take different routes on the network. The packets are held in a buffer at the receiving device until all packets arrive and are put in the right order. While this does not affect e-mail and file transfers, which are not real-time applications, excess latency does affect multimedia applications in that it causes voice and video components to arrive out of synchronization.

If the packets containing voice or video do not arrive within a reasonable time, they are dropped. When packets containing voice are dropped, a condition known as *clipping* occurs, which is the cutting off of the first or final syllables in a conversation. Dropped packets of video cause the image to be jerky. Excessive latency also causes the voice and video components to arrive out of synchronization with one another, causing the video component to run slower than the voice component; in a video conference, for example, a person's lips will not match what he or she is really saying

The problem of latency can be addressed by assigning a class of service or quality of service (QoS) to a multimedia application, which identifies it as being time-sensitive and requiring priority over other, less time-sensitive data types. Quality of service be handled by the network (i.e., routers, hubs, switches), the operating system, or a combination of both hardware and operating system working together.

See also
JITTER

Leasing

Network managers today are not only responsible for selecting equipment that best satisfies corporate communications requirements, but increasingly are called upon to recommend the most cost-effective way to procure that equipment. Although the methods for technology procurement have become more creative in recent years, mostly in the structuring of contract terms and conditions, the fundamental decision still boils down to one of lease or purchase. Leasing often can provide organizations with many financial and nonfinancial benefits.

Financial incentives. There are a number of financial reasons for considering leasing over purchasing, such as when a company cannot secure credit for an installment loan on the system and has no available line of credit, or cannot make the down payment required for an installment loan. Leasing can improve a company's cash position, since costs are spread over a period of years. Leasing can free up capital for other uses, and even cost-justify technology acquisitions that would normally prove too expensive to purchase. Leasing also makes it possible to procure on short notice equipment that has not been planned or budgeted for.

A purchase, on the other hand, increases the debt relative to equity, and worsens the company's financial ratios in the eyes of investors, creditors, and potential customers. An operating lease on rental equipment can reduce balance sheet debt, since the lease or rental obligation is not reported as a liability.

So at the least, an operating lease represents an additional source of capital and preserves credit lines. Beyond that, leasing can help companies comply with the covenants in loan agreements that restrict the amount of new debt that can be incurred during the loan period. The purpose of such provisions is to make sure that the company does not jeopardize its ability to pay back the loan. But in providing additional capacity for acquiring equipment without violating loan agreements or hurting debt-to-equity ratios, leasing allows companies to "have their cake and eat it, too."

With major improvements in technology becoming available every 12 to 18 months, leasing can prevent a company from becoming saddled with obsolete equipment. This means that the potential for losses when replacing equipment that has not been fully depreciated can be minimized by leasing rather than purchasing. Furthermore, with rapid advancements in technology and consequent shortened product life cycles, it is becoming more difficult to sell used equipment. Leasing eliminates this problem too, since the leasing company owns the equipment.

For organizations concerned with controlling staff size, leasing also minimizes the amount of time and resources spent in cost-justifying capital expenditures, evaluating new equipment and disposing of old equipment, negotiating trade-ins, comparing the capabilities of vendors, performing reference checks on vendors, and reviewing contractual options. There also is no need for additional administrative staff to keep track of configuration details, spare parts, service records, and equipment warranties. And because the leasing firm usually is responsible for installing and servicing the equipment, there is no need to spend money on skilled technicians or for outside consulting services. Lease agreements may be structured to include ongoing technical support and even a help hot line for end users.

This brings up another advantage to leasing: It can minimize maintenance and repair costs. Because the lessor has a stake in keeping the equipment functioning properly, it usually offers on-site repair and the immediate replacement of defective components and subsystems. In extreme cases, the lessor may even swap out the entire system for a properly functioning unit.

Although contracts vary, maintenance and repair services that are bundled into the lease can eliminate the hidden costs often associated with an outright purchase. When purchasing equipment from multiple vendors, often the user will get bogged down in processing, tracking, and reconciling multiple vendor purchase orders and invoices to obtain a complete system or network. Under a lease agreement, the leasing firm provides a single-source purchase order and invoicing. This cuts down on the user's administration, personnel, and paperwork costs.

Finally, leasing usually allows more flexibility in customizing contract terms and conditions than normal purchasing arrangements, simply because there are no set rates and contracts when leasing. Unlike many purchase agreements, each lease is negotiated on an individual-case basis. The items that typically are negotiated in a lease are the equipment specifications, schedule for upgrades, maintenance and repair services, and training. Another negotiable item has to do with the end-of- lease options, which can include signing another lease for the same equipment, signing another lease for more advanced equipment, or buying the equipment. Many lessors will allow customers to end a lease ahead of schedule without penalty if the customer agrees to a new lease on upgraded equipment.

Nonfinancial incentives. There also are some very compelling nonfinancial reasons for considering leasing over purchasing. In some cases, leasing can make it easier to try new technologies, or the offerings of vendors that would not normally be considered. After all, leases always expire or can be canceled (a penalty usually applies), but few vendors are willing to take back purchased equipment.

Leasing permits users to take full advantage of the most up-to-date products at the least risk and often on very attractive financial terms. This kind of arrangement is particularly attractive for companies that use technology for competitive advantage because it means that they can continually upgrade by renegotiating the lease, often with little or no penalty for terminating the existing lease early. Similarly, if the company grows faster than anticipated, it can swap the leased equipment for an upgrade.

It must be noted, however, that many computer and communications systems are now modular in design, so that the fear of early obsolescence might not be as great as it once was. Nevertheless, leasing offers an inducement to try vendor implementations on a limited basis without committing to a particular platform or architecture, and with minimal disruption to mainstream business operations.

Companies that lease the equipment they need can avoid a problem that invariably affects companies that purchase equipment: how to get rid of outdated equipment. Generally, no used equipment is worth more on a price/performance basis than new equipment, even if it is functionally identical. Also, as new equipment is introduced, it erodes the value of older equipment. These byproducts of improved technology make it very difficult for users to unload older, purchased equipment.

With equipment coming off lease, the leasing company assumes the responsibility for finding a buyer. Typically the leasing company is staffed with marketers who know how and where to sell used equipment. They know how to prospect for customers for whom state-of-the-art technology is more than they need, but a secondhand system might be a step up from the 7-year-old hardware they are currently using.

There also is a convenience factor associated with leasing, since the lessee does not have to maintain detailed depreciation schedules for accounting and tax purposes. Budget planning also is made easier, since the lease involves fixed monthly payments. This locks in the pricing over the term of the lease, allowing the company to know in advance what its equipment costs will be over a particular planning period.With leasing, there also is less of an overhead burden to contend with. For example, there is no need to stockpile equipment spares, subassemblies, repair parts, and cabling. It is the responsibility of the leasing firm to keep inventories up to date. Their technicians (usually third-party service firms) make on-site visits to swap boards and arrange for overnight shipping of larger components when necessary.

Leasing also can shorten the delivery lead time on desired equipment. It might take as long as 8 weeks or longer to obtain the equipment purchased from a manufacturer. In contrast, it could take from 1 to 10 working days to obtain the same equipment from a leasing firm. Often the equipment is immediately available from the leasing firm's lease/rental pool.

For businesses that need equipment that is not readily available, some leasing firms will make a special procurement and have the equipment in a matter of 2 or 3 days, if the lease term is long enough to make the effort worthwhile.

Many leasing companies offer a master lease, giving the customer a preassigned credit limit. All of the equipment the customer wants goes on the master lease and is automatically covered by its terms and conditions. In essence, the master lease works like a credit card.

Summary. Leasing is another form of buying on credit: payments are made monthly and the total price includes interest on the principal amount. Equipment leasing is done routinely by computer vendors, telecommunications carriers, disaster recovery providers, and systems and network integrators, as well as third-party service firms and leasing companies that specialize in financing computer systems and communication networks. For network managers, there are commercial software packages available that automate the lease-versus-purchase decision process. Often the company leasing the equipment also will provide asset management software to help customers keep track of all the items under lease.

Alternatively, asset management may be provided to the customer as a service. Some asset management services go beyond traditional lease-versus-purchase analysis to include an analysis of the complete life cycle costs of technology assets, from initial acquisition through disposal, weighing financial implications with such issues as usage, software, support, and maintenance.

See also

ASSET MANAGEMENT

Local telecommunications charges

The telecommunications bill from the local exchange carrier (LEC) actually represents charges for a number of services, some of which may come from several service providers. Most of these charges are totaled and consolidated onto a single phone bill, making it easier for customers to reconcile and providing a convenient method of payment.

There are a number of basic charges that are itemized in the monthly statement from the LEC.

Local service. LECs operate central office facilities, which carry all local inbound and outbound traffic and, in some instances, long-distance, Internet, and telemetry traffic as well. Companies typically receive an itemized statement of outbound calls for each month. As automatic number identification (ANI) and caller identification services are enhanced, inbound traffic information also is being made available.

With passage of the Telecommunications Act of 1996, competitive local exchange carriers (CLECs) can apply to the state public service or public utilities commissions (PUCs) and the FCC to provide local exchange services. Upon approval of the application and subsequent interconnection arrangements, these services will be billed by the carrier providing the local service. Whatever the source of local service, the outbound call detail statement is a valuable analytical tool. It allows anomalies to be noted and investigated. Unverified charges then can be brought to the attention of the LEC, and credits sought.

Special central office services. With the increasing implementation of Advanced Information Network (AIN) services, the LECs have begun offering a variety of enhanced services to customers. These can include general business solutions, voice services, data services, and value-added services. All can be priced on a usage basis or a flat fee, depending on the type of service.

General business solutions. Examples of business solutions include contingency planning, disaster recovery through bandwidth-on-demand, videoconferencing, and LAN interconnection for client-server computing. Some LECs are offering ATM-based services to extend voice, video, and data to the desktop for a wide range of business applications.

Voice services. Voice service options range from premises-based to central office–based offerings. These can include services such as Centrex, cellular communications, and digital transport services. Some local exchange carriers also offer selected, integrated services targeted at the small business market. These services include:

- *Centrex:* In essence, this is a service that offers the features of a PBX/ACD, but without the customer having to invest in on-premises switching and call-routing equipment. Centrex charges are based on line usage and choice of feature package. Various types of Centrex service are available, including basic, key/hybrid Centrex, and ISDN. Enhanced Centrex offerings include voice mail, electronic mail, message center support, modem pooling, and packet-switched data transmission.
- *Cellular communications:* This option lets users add wireless business communications to existing Centrex, PBX, or key systems, converting a corporate facility into an intracompany cellular system. Wireless services can include services ranging from personal communications for individuals, to a complete network of wireless and data communications for an enterprise.
- *Digital transport:* This option provides switched exchange access services, such as direct inward dialing, one-way incoming or outgoing PBX trunks, two-way incoming/outgoing PBX trunks, and custom 800 services.

Voice enhancements. These services include features such as automatic call-back, call forwarding, remote call forwarding, voice mail, call waiting, call redirection, caller ID, caller ID with name, speed calling, three-way calling, automated voice messaging, and the many other services that have become an integral part of doing business. Pricing usually consists of a one-time setup charge and a fixed monthly charge thereafter.

Voice equipment. The LECs also provide equipment that can be purchased or leased. The equipment can include such items as Centrex telephone sets, PBX systems, key systems, single- and multiline telephone sets, and cellular telephones. If the equipment is leased, the monthly bill will include these charges.

Data services. The LECs have become increasingly involved in providing data services, both switched and dedicated-line. These usually cover the entire data transmission bandwidth and can range from less than 9600-bps switched services to SONET and ESCON-based dedicated services.

Value-added services. Value-added services can include options such as telemetry (for remote security and environmental monitoring), telemanagement, asset management products and services, and Centrex management services. Some LECs provide tools needed for on-site control over an organization's voice and data communications, as well as equipment and facilities management. These can include features such as automated call attendant services to handle incoming calls with the purpose of improving customer service, or interactive voice response service to automate routines such as confirming an account balance, checking the status of an order, or registering for classes. Centrex management features can include options such as automatic route selection, uniform call distribution, Centrex call management, and wireless Centrex services.

Internet services. Many LECs also now offer access to the Internet. Typically costs depend on the type of access chosen. Most carriers provide both switched or dedicated services for Internet access. Switched services include X.25, 56 kbps, and ISDN. Dedicated services include T1 and frame relay. Costs include a monthly charge that increases as transmission speed increases, and a one-time setup charge. Charges to access Usenet newsgroups are incurred based on the number of users, while electronic mail pricing can include a setup charge per mailbox, and a monthly fee per mailbox. Most LECs provide free Internet browsers.

Directory services. The LEC also provides separate line-item billing for directory services. These charges can be broken into two categories: directory assistance and directory advertising.

Directory assistance. Directory assistance charges typically are summarized and reflected as a single line item. Most businesses are given a fixed number of directory assistance calls per month, with each subsequent call billed at a fixed rate (usually between US$0.60 and US$0.75 per call). To reduce directory assistance charges, telecom managers can employ a variety of cost-control methods, including:

- Provide telephone directories for frequently called areas.
- Set up system speed-dial numbers or individual speed-dial lists for users through the telephone system.
- Program the telephone system to block access to directory assistance.

Directory advertising. The telephone company typically charges for two different types of directory listings: yellow pages listings and white pages listings. Directory charges usually are summarized and reflected as a line item on the LEC bill.

Yellow pages advertising tends to be more elaborate and expensive than white pages advertising. Many LECs now also provide interactive yellow pages, which offer enhanced search functions by a variety of criteria. Often the telecom manager has no authority in the advertising area because the company's marketing department makes deals directly with the LEC customer service representative. The telecom manager's job is simply to review the ad charges and note any variations in the billing.

White pages listings often are provided by the LEC at no charge if the listings are relatively simple. However, listings of entire departments, individual branches, and the like usually result in a charge on the billing statement.

Government-mandated services. One or two line items on the local exchange bill may represent legislatively mandated services, such as emergency services and relay center services. Most LEC bills include a fixed monthly charge for Emergency (or Enhanced Emergency) 911 services. This is generally a parochial issue, determined and authorized by the municipal, county, or state government. In most cases, the charge should be no more than pennies per month, per line.

The second such line item is for support of relay center services for the hearing-impaired, or for other services related to the Americans with Disabilities Act. This includes a small surcharge on every business line. The purpose of this surcharge is to offset the costs for providing communications services for the disabled. As with the 911 surcharge, the ADA surcharge should be no more than pennies per month, per line.

Long-distance services. Because almost every business in the United States is located in an "equal access area," virtually every company can designate the long-distance carrier(s) to appear on its 1+ routes. The LEC routes all of the company's long-distance calls to the designated carrier. The LEC also acts as the billing agent for the designated 1+ long-distance carrier. Typically the LEC does nothing more than post the long-distance carrier's billing information on the local exchange bill and act as a collection agent for the long-distance service provider.

With passage of the Telecommunications Act of 1996, LECs can apply to the state PUCs and FCC to provide conventional long-distance service. Upon approval of the application and subsequent interconnection arrangements with other carriers, these long-distance services will be billed by the carrier providing the local service.

Some LECs provide long-distance cellular service, as well as local cellular service. Businesses can have their cellular calls appear on the same telephone bill as other types of charges, including Internet access.

Special calling services. A wide variety of special calling services (i.e., 976-type or 900-type services) are currently available. Calling these information services, which offer details on everything from Dow-Jones Reports to psychic readings, is rarely useful in the business environment, so charges for these services should be monitored carefully.

The LEC typically acts as a billing agent for special calling service providers, and posts charges received from the service provider to the local exchange bill. These charges can be eliminated by programming the telephone system to block access to numbers with 976 and 900 prefixes.

Nearly all local exchange carriers provide a telephone calling card to which all calls can be charged. Calling card charges appear on the company's LEC bill, and vary according to usage.

Professional services. All LECs also typically offer a broad range of technical support services designed for companies that outsource many technical project components. These professional services can range from program management to temporary staffing consultants and training. Usually professional services are billed on an as-needed basis, and typically are used when a new service or a new technology is being implemented.

Other charges. The bill from the local exchange carrier may also include some or all of the following charges.

Inside wiring maintenance charges. Many state public utility commissions have passed rules that transfer ownership of inside wire from the local exchange service provider to the owner of the building. As an option, customers can elect to maintain inside wire themselves, or opt for the LEC's maintenance service for a nominal monthly charge, which is reflected as a fixed cost on a single line item on the billing statement.

Installation and engineering charges. LECs typically charge for the installation, setup, and testing of new lines and equipment, as well as for engineering services. Engineering services can include everything from site preparation (including heating, ventilation and air conditioning, if necessary) to traffic analysis, network modeling, and periodic fine-tuning of the communications system.

Network access charges. The LEC collects a network access charge mandated by the Federal Communications Commission. This can amount to US$6 per business line (US$3.50 for residential lines), which goes to the maintenance of the LEC's network. The purpose of this charge is to ensure that the LECs do not cut back on network maintenance as a way of improving their financial performance, especially as the local exchange opens up to competition. By having this money available, the LECs cannot use competitive pressures as an excuse for not being able to properly maintain the network.

Other charges and credits. The "Other Charges and Credits" section on the billing statement is a catchall category that includes charges and/or credits for service orders and setup fees for new features provided by the LEC. All of these services tend to have one-time, nonrecurring fees, which can be waived during special promotions.

Summary. The Telecommunications Act of 1996 clears the way for competition in the provision of local services. Eventually there will be multiple sources of local telecommunications services in a truly competitive market. When this happens, the cost of local telecommunications services will drop considerably. Competition also means that customers will have an expanded menu of choices available, and have

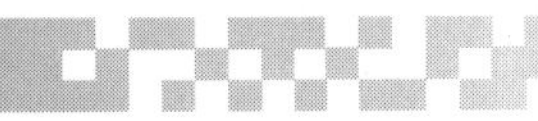

the charges for all equipment, services, and features appear on the same telephone bill, making it easier to reconcile and providing a convenient method of payment. As the market for local services becomes more competitive, the consolidated phone bill also will become an invaluable aid to comparison shopping.

See also

LONG-DISTANCE TELECOMMUNICATIONS CHARGES

Local Access and Transport Areas

Local Access and Transport Areas were an outcome of AT&T's divestiture agreement with the U.S. Justice Department in 1984. At that time, not only did AT&T divest itself of the Bell Operating Companies (BOC), but a total of 197 LATAs were created to define the service boundaries of the local exchange carriers.

Each LATA is identified regionally by a 3-digit number, which bears no relation to area codes. A state can have several LATAs or just one. In a few cases, some LATAs can cross state lines. Calls within a LATA (intraLATA) may be either local or long-distance calls, depending on distance. Calls between LATAs (interLATA) are considered long-distance calls.

Local calls within a LATA typically are made within a neighborhood or town, although this can differ from state to state. These calls are carried by the local telephone company. Calls within the same LATA that are of long enough distance to be a toll call also are carried by the local telephone company. In most states, a long-distance carrier can now carry these calls. Initially this was done by dialing the 5-digit code of the long-distance carrier, 10*xxx*, followed by the phone number. In most LATAs today, users have the option of presubcribing to a preferred long-distance carrier so that the 5-digit code does not have to be dialed.

When a call crosses LATAs, they are handled by long-distance carriers, even if the same local phone company services each LATA. With the Telecommunications Act of 1996, however, the local exchange carriers can qualify to handle interLATA calls by meeting the requirements of a 14-point competitive checklist and receiving approval of the state PUC and FCC. The Telecommunications Act of 1996 not only preserves the LATA concept, but gives the FCC exclusive authority over LATA boundaries.

Summary. Subscribers can choose a preferred long-distance carrier for intraLATA service and reap several advantages. By combining all toll calls (intraLATA and interLATA) in one bill, corporate and residential subscribers can qualify for greater volume discounts and realize such savings more quickly. In addition, the features used on long distance calls (such as call accounting, account codes, and geographic restrictions) can be applied to the intraLATA local toll calls.

See also

LOCAL TELECOMMUNICATIONS CHARGES
LONG-DISTANCE TELECOMMUNICATIONS CHARGES

Long-distance telecommunications charges

Long-distance carriers, also known as *interexchange carriers* (IXCs), offer a variety of network services that can be tailored to meet the specific needs of small to large businesses, telecommuters, mobile professionals, and self- employed people working out of offices in their homes. Many of these offerings can be aggregated together for call volume discounts and are consolidated on a single monthly billing statement.

Long-distance service charges. The continued pace of innovation has produced an enormous choice of long-distance services. They are delivered over dedicated private lines, the switched public network, cellular facilities, satellite, or a combination of these. Each service has various features and options, the choice of which influences costs and volume discounts. Billing statements consolidate all call activity, making it convenient to track and control long-distance telecommunications costs.

Outbound services. The major long-distance carriers offer outbound calling services, traditionally known as *Wide Area Telecommunications Service,* or more commonly, *WATS.* This type of service is basically a bulk-rate toll service priced according to call distance, or rate band. Introduced 36 years ago by AT&T, traditional WATS has been replaced by more flexible and manageable virtual WATS-like services, in which billing is based on time of day and call duration, as well as distance. Discount plans based on call volume and the number of corporate locations enrolled in the plan are available.

Most carriers can provide detailed billing reports available on customer request, and in a choice of media. Such reports provide specific call detail and, in some instances, exception reporting, which flags unusual usage patterns, anomalies in billing, or significant changes in calling patterns.

Inbound services. Most IXCs also offer 800 service and the newer 888 service, both of which are also known as *inbound WATS.* These services are bulk-rate inbound toll service available in various areas. Carriers offer either switched-access or dedicated-access 800 and 888 service. Switched service enables 800 and 888 calls to be received over regular telephone lines; dedicated service provides a private line that is digitally connected from the carrier network to the business network. Each service provides several options. For example, users can geographically screen their calls, or block calls from certain parts of the country; other services automatically route calls to specific locations based on customer-specified requirements.

The larger IXCs now also offer 800 service for international calls. Callers from international locations use country-specific numbers to route calls to a subscriber's access line in the United States. There also is personal 800 service for individuals who work out of the home.

Billing for 800 services is based on several factors: destination of the call, time of day when the call is placed, duration of the call, and selection of switched versus dedicated access. Most carriers have discount plans based on time-of-day and day-of-week, in addition to offering volume usage discounts; similar discount plans and feature options are available for international 800 services. Detailed billing reports for 800 services, available on customer request, provide specific call detail and exception reporting. The carriers also offer the option to receive call detail information in real time, or on a daily or monthly basis via PCs. Such services can be used to measure marketing responses, track lost calls, and gauge the effectiveness of call center operations.

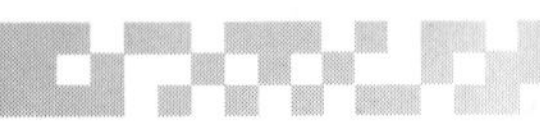

Virtual network services. The three major long-distance carriers (AT&T, MCI, and Sprint) all offer virtual private network (VPN) services. While each offering has its own name, the three services work in essentially the same way. Each carrier provides feature and option packages, aimed at companies of various sizes, which combine outbound and inbound services, and offer varying levels of service and discounts based on combined usage.

Basically, VPN services provide capabilities similar to private lines (including line conditioning, error testing, and high-speed full-duplex transmission with line quality adequate for data) over the public switched network. VPNs provide on-demand dial-up circuits, or bandwidths that can be dynamically allocated, reducing or eliminating the need for fixed point-to-point private lines. Most VPNs support analog data transmission speeds up to 28.8 kbps, and end-to-end digital data transmission starting at 56 kbps. Most carriers also provide international VPN capabilities with larger foreign countries. The international VPNs provide point-to-point, two-way calling capabilities for voice, fax, and data.

Billing is based on the call destination, time of day, and call duration. All of the VPN carriers also have discount plans for volume usage and plans based on the day-of-week usage. There is even the option for cellular access, with all cellular calls added to the total volume discount. All IXCs can provide detailed billing reports for VPN service. VPN call detail reports provide specific information about each call and, in some instances, report exceptions as well. The bill consolidates call activity and charges for all corporate locations on the VPN.

Special line charges. Long-distance carriers offer special access facilities, including (but not limited to) tie lines, foreign exchange (FX) lines, off-premises extensions (OPX), T-carrier facilities, frame relay, SONET, ATM, international facilities, and other such services. In addition to the provision of the interexchange portion of the facility, the long-distance carriers also provide coordination for the installation of these facilities with the LECs and the local carriers at international locations. When digital service to a foreign country is not available, the carriers also offer voice-grade private line service. This type of service supports lower-speed applications such as facsimile and electronic mail.

Charges for special access facilities usually appear as a single line item under the special access line heading on the long-distance billing statement. Itemization of these charges is available on a separate facilities record, which can be obtained from the long-distance carrier upon request. This record lists each individual facility by circuit number and describes all associated components and related equipment. Typically it also identifies the circuit end points by location address and serving central office.

Cellular services. Like the LECs, the IXCs offer cellular services for voice and data. The charges for cellular services can be included on the same billing statement as other types of services and, in many cases, can be included with other types of calls in the volume discount plan.

Some cellular services are data-oriented and are billed according to the number of packets sent and received, as well as call duration and distance of the call. Voice calls are also billed by call duration and distance of the call. The bill shows total air time charges and call details, and might also contain charges for such optional services as call waiting, roaming, and voice mail. In some cases there might be the option for unlimited weekend calling for a flat fee. And if the individual or company has elected to insure the phone against theft or loss, the monthly insurance charge also will appear on the billing statement.

Network management services. As networks have become more complex, the larger carriers have begun offering end-to-end network management services for multivendor environments. These services include management of private-line and switched services, WANs, LANs, mainframe connections, and networking equipment such as routers, switches, PBXs, and modems. There are even managed SNA services that entail the carrier managing and maintaining a frame relay network that has been specifically designed and performance-tuned to carry delay-sensitive SNA traffic. These management services are provided 24 hours per day, 7 days per week.

These are essentially custom services, since each company will have its own connectivity, performance, and management requirements. Accordingly, the network management service is priced on an individual company basis, with costs varying by the size of the network, the services included, and the equipment to be managed. Once the total cost is determined, the customer is billed a fixed monthly amount. As changes are requested by the customer, these are added to the billing statement.

Satellite services. A few long-distance carriers also provide satellite-based end-to-end network services. The satellite service usually is available on a dedicated or shared basis to receive and transmit data, voice, and video signals, with transmission speeds ranging from 56 kbps to 1.544 Mbps. Costs vary by the type of service chosen, but typically include a monthly service charge and transmission cost. If equipment is purchased, such as Very Small Aperture Terminals, or VSATs, there is a one-time equipment charge. Often this type of equipment can be leased as well. In that case, fixed monthly payments will be shown on the bill, along with the transmission charges.

Calling card services. All of the IXCs now also offer long-distance calling cards, and, like all of the other services, a wide variety of options are available that affect billing. Standard calling cards allow long-distance calls to be made by dialing an account number and PIN. Prepaid calling cards are available that let users control long-distance calling time, thereby containing service charges. Charges for this service are based on long-distance usage and options chosen with the calling card.

Conferencing services. The IXCs offer conferencing services that include both videoconferencing and telephone conferencing. Switched video services provide videoconferencing and related options such as video applications help desks, multipoint conferencing, digital video broadcast services, and connections to other carrier's networks. Some carriers also provide equipment options that are either sold or leased. Most carriers also provide an operator as needed to assist with setting up and connecting all parties to the conference. Telephone conferencing can support communication among hundreds of locations simultaneously. Monthly billing options for both types of conferencing include call detail by department, division, or user name; a telemanagement report that summarizes conference usage company-wide; and usage-analysis reports, including a sort by cost center, for charge-back purposes.

Messaging services. IXC-provided messaging services allow for either caller-paid or sponsor-paid live or recorded one-way messages, two-way conversa-

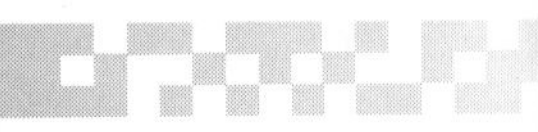

tions, and electronic counting for opinion polls, with a wide range of options available for each type of service. Other messaging services may include electronic delivery of news clippings, e-mail, fax delivery, and interactive voice services. All carriers are also now providing Electronic Data Interchange (EDI) services, which allow companies to electronically exchange business documents with trading partners. The billing charges for these messaging services are based on the options chosen, bundling with other service, and usage.

Other charges and credits. The catchall category, "Other," on the long-distance billing statement includes charges or credits for installation, service orders, engineering charges, and other such services provided by the long-distance carrier. These tend to be one-time, nonrecurring service fees and will show up on the billing statement only once, unless arrangements have been made to spread the costs over multiple billing periods.

Summary. In the current competitive environment, telecommunications companies are learning to differentiate themselves in a variety of ways. In addition to services, features, and options, they are seeking to differentiate themselves by the billing services they can offer. In response to customer demand for more sophisticated billing options, the LECs and long-distance carriers are now viewing their billing systems not only as basic collection tools, but also as full-fledged information systems that can be tailored to the needs of their customers. In the near future, customers will even be able to view their bills via the Internet, just as some credit card issuers now allow customers to view their monthly statements on the Internet with their Web browsers.

See also

CALLING CARD
LOCAL TELECOMMUNICATIONS CHARGES
VIRTUAL PRIVATE NETWORKS
WIDE AREA TELECOMMUNICATIONS SERVICES (WATS)

Managed SNA services

As companies move from IBM's host-centric (i.e., mainframe) to the distributed computing environment (i.e., local area networks), the major carriers are offering wide area network services that address the specific needs of SNA users. These managed SNA services operate over frame relay connections and are especially suited to organizations with many remote locations that must be tied into one or more hosts. Among the advantages of managed SNA services are:

- Frame relay's permanent virtual circuits (PVCs) replace expensive SDLC and BSC multidrop networks between the host and branch offices.
- Consolidating connections through frame relay eliminates costly serial line interface coupler (SLIC) ports on front-end processors (FEPs), while increasing performance.
- WAN access extends the useful lives of SDLC/BSC controllers and 3270 terminals.
- The availability and reliability of SNA connections is increased by allowing controllers to take advantage of WAN connections with multiple host paths.

A managed frame relay service includes the leased or purchased frame relay access devices (FRADs), which transport SNA traffic over the PVCs. FRADs are more adept than routers at congestion control and prioritization. Some FRADs multiplex multiple SNA/SDLC devices onto a single PVC, instead of requiring a separate PVC for each attached device, resulting in even greater cost savings. The FRADs encapsulate SNA/LLC2 frames with minimal overhead and allow traffic to be selectively prioritized, ensuring that mission-critical SNA data arrives in a timely fashion. Most FRADs perform local polling acknowledgment of keep-alive frames to minimize the risk of timing out SNA sessions. Some services permit NetView visibility and control to be extended to the attached SNA devices.

For a legacy SNA shop that does not have the expertise or resources, a managed SNA service can be an economical interconnectivity option. Although frame relay networks are much more difficult to configure, administer, and troubleshoot than private lines, the carrier assumes these responsibilities. In addition, the carrier provides design and reconfiguration assistance to ensure that mission-critical applications are providing the highest level of performance, reliability and availability.

There is legitimate concern that frame relay is not yet very effective in guarding against the loss of data during congestion conditions. Although warnings of impending congestion are relayed from the carriers' frame relay switches to the FRADs on the edges of the network, they cannot force them to adjust the frame rate. If the FRADs do not react to these congestion indicators, the carrier networks can discard frames to reduce congestion. This is usually not a serious issue, however. The carriers routinely overprovision their frame relay networks to guard against congestion and possible frame loss, and they will continue to do so to attract new SNA customers.

The frame relay services themselves are priced attractively. On average, the frame relay service costs about 25 percent less than the equivalent private network. In some cases, discounts of up to 40 percent are possible. Penalty fees are waived if upgrading the network entails having to break contracts for other services or leases on hardware. In some cases it might be possible to get the first month of the managed SNA service free, so the old services can continue to be used until the changeover to frame relay is complete.

See also

FRAME RELAY
ADVANCED PEER-TO-PEER NETWORKING (APPN)
ADVANCED PROGRAM-TO-PROGRAM COMMUNICATIONS (APPC)

Microwave communications

A microwave is a short radio wave, in the length range from 1 mm to 30 cm. Because microwaves can pass through the ionosphere, which blocks or reflects longer radio waves, microwaves are well suited for satellite communications. This reliability makes microwave well suited to terrestrial communications as well.

Much of the microwave technology in use today for point-to-point communications was derived from radar developed during World War II. Initially these microwave systems carried multiplexed speech signals over common carrier and military communications networks; today, however, they are used to handle all types of information—voice, data, facsimile, and video—in either an analog or digital format.

The first microwave transmission occurred in 1933, when European engineers succeeded in communicating reliably across the English Channel, a distance of about 12 miles (20 km). In 1947, the first commercial microwave network in the United States came online. Built by Bell Laboratories, this system connected New York to Boston with 10 relay stations carrying television signals and multiplexed voice conversations.

A year later, New York was linked to San Francisco via 109 microwave relay stations. By the 1950s, transcontinental microwave networks were routinely handling over 2000 voice channels on hops averaging 25 miles (41.5 km). By the 1970s, not a single telephone call, television show, telegram, or data message crossed the country without spending some time on a microwave link.

Over the years, microwave systems have matured to the point that they have become major components of the nation's public switched network and essential mechanisms that private organizations use to satisfy internal communications requirements. Microwave systems can even exceed the 99.85 percent reliability standard set by the telephone companies for their phone lines.

Microwave applications. Early technology limited the operations of microwave systems to radio spectrum in the 1 GHz range, but due to improvements in solid-state technology, today's commercial systems are transmitting in the 40 GHz region. In recognition of these changes, the FCC recently adopted rules allowing the use of spectrum above 40 GHz.

This spectrum offers a variety of possibilities, such as use in short-range, high-capacity wireless systems that support educational and medical applications, and wireless access to libraries or other information databases. Short-haul microwave communications equipment also is routinely used by hotel chains, CATV service providers, and government agencies. Corporations are making greater use of short-haul microwave, especially for extending the reach of LANs in places where the cost of local T1 lines is prohibitive. Common carriers use microwave systems for backup in the event of fiber-optic transmission equipment failures, and in terrain where laying fiber-optic cable is not economically feasible. Cellular service providers use microwave to interconnect cell sites, with one another and to the regular telephone network. Some interexchange carriers (and corporations) even use short-haul microwave to bypass local exchange carriers to avoid paying local access charges.

Network configurations. There are about 25,000 microwave networks in the United States alone. There are basically two microwave network configurations: point-to-point and point-to-multipoint.

Point-to-point microwave products and systems meet a variety of low- and medium-density communications requirements. These range from simple links to more complex extended networks, such as:

- Sub-T1/E1 data links
- Ethernet/Token-Ring LAN extensions
- Low-density digital backbone for wide area mobile radio and paging services
- PBX/OPX/FX voice, fax, and data extensions
- Facility-to-facility bulk data transfer

Point-to-multipoint microwave systems provide communications between a central command and control site and remote data units. A typical radio communications system provides connections between the master control point and remote data collection and control sites. In the United States, this type of radio system must have a minimum of four remote locations. Repeater configurations also are possible. The basic equipment requirements for a point-to-multipoint system include:

- *Antenna:* An omnidirectional antenna is used for the master, and a highly directional antenna for the remotes, aimed at the master station's location.
- *Tower:* A structure such as a tower or mast supports the antenna and transmission line.
- *Transmission line:* A low-loss coaxial cable connects the antenna and the radio.
- *Master station radio:* Interfaced with the central computer, it transmits and receives data from the remote radio sites and can request diagnostic information from remote transceivers. The master radio also can serve as a repeater.

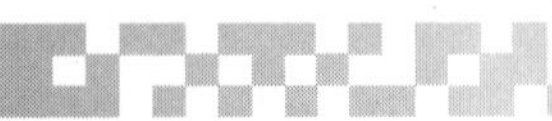

- *Remote radio transceiver:* Interfaced to the remote data unit, it receives from and transmits to the master radio.
- *Personal computer:* It can be connected to the master station's diagnostic system, either directly or remotely, for control and collection of diagnostic information from master and remote radios.

Wireless cable. Cable system operators traditionally have used microwave transmission systems to link cable networks and transport signals from point to point, but not to the home. These Cable Antenna Relay Services (CARS) have experienced declining usage as cable operators have deployed more fiber optics in their transmission systems. Improvements in microwave technology, however, and the opening of new frequencies for commercial use, have contributed to a resurgence in short-haul microwave. In the broadcast industry, short-haul microwave is often referred to as "wireless cable."

Wireless cable has two key advantages. One is availability: It can be made available in areas of scattered population and other areas where it is too expensive to build a traditional cable station. The other is affordability: Due to the lower costs of building a wireless cable station, savings can be passed on to subscribers.

MMDS. For television broadcast there is the multichannel, multipoint distribution service (MMDS), an analog technology that uses omnidirectional microwave facilities and 8–10 m satellite earth stations (dishes) to transmit 33 channels. Of these, 20 are reserved for education, so only 13 channels are available for commercial use. Spectrum for MMDS is being auctioned by the FCC. MMDS license holders generally negotiate with the holders of the educational channels for the rights to their surplus capacity.

MMDS is a fairly new service that developed from MDS (multipoint distribution service), which could only send one or two channels. The FCC originally thought MDS would be used primarily to send business data. Since MDS's creation in the early 1970's, however, the service has become increasingly popular in sending entertainment programming. Because the FCC does not regulate the content of the transmission, alternative uses were not prohibited.

If line-of-sight is available, wireless cable is cost-competitive with traditional wireline cable systems. With MMDS, instead of stringing miles of coaxial cable to reach the subscriber, channels are broadcast by microwave to small antennas (approximately the size of an open newspaper) on the roofs of subscribers' homes or businesses. Line-of-sight is aided by positioning transmitting antennas atop the tallest building. Inside the customer premises, the subscriber has a device known as a down-converter. This unit usually includes a built-in addressable decoder and a VHF/UHF tuner. This gives it the ability to tune in broadcast channels without having to use up valuable MMDS channels. It also allows pay-per-view services and simplifies channel blocking and premium channel activation/deactivation.

MMDS is used not only for delivering entertainment programs, but also for its ability to link remote school districts, increase interactive learning, and offer community education programs. Recognizing that MMDS requires lower microwave frequencies to maximize broadcast coverage, the FCC opened the 2 to 12 GHz range to wireless cable operators in 1992.

Analog MMDS has been around for many years, but has been relegated to niche markets; it was primarily used as an economical way to get analog channel capacity to rural areas. With an average of 25 analog channels available in each licensed area, MMDS does not provide much competition in the cities and suburbs, where 60-plus analog channels are available from many cable systems. But with the advent of compression technologies, these 25 channels became 100 channels. Digital MMDS today is capable of challenging wireline cable operators, especially in urban areas where there are more potential subscribers per transmitter.

Although digital MMDS does not support emerging interactive video applications (it is used for one-way broadcast only), some regional telephone companies and cable firms have made substantial investments in digital MMDS. They see the technology as an economical means of providing large numbers of video channels quickly to their subscribers.

Not only does digital MMDS provide 100 channels of programming, the antennas are small and cheap, which eliminates zoning problems. In some cases the antenna can be mounted indoors and aimed out the window. Although direct broadcast satellite (DBS) is becoming popular, the advantage of MMDS is that it offers local channels. Digital MMDS does share one key problem with DBS, however: As noted, it requires a clear line of sight.

LMDS. Local Multipoint Distribution System (LMDS) is an emerging technology that will compete effectively with MMDS. LMDS evolved out of the need to transmit a large number of broadband TV channels in a spectrum that is becoming increasingly congested. Like MMDS, it is a multicell, point-to-multipoint distribution system, but one which operates at millimeter wavelengths in the 28 GHz band.

While MMDS is now being used for so-called wireless cable TV transmissions into homes, it is only a one-way service with no interactive capability at present. For this and other reasons, LMDS has far greater potential than MMDS.

LMDS uses super high frequency (SHF) microwaves in the 28 GHz range to send and receive two-way broadband signals in an area, or *cell,* approximately 3 to 6 miles in diameter. It works essentially like the narrowband operations of cellular telephone systems. However, the video, voice, and data broadband LMDS signals are capable of two-way operation. LMDS has a range of 6 miles and has much more channel capacity than MMDS, up to 100 television channels as well as interactive video services.

Along with providing higher-quality video than MMDS due to its use of sophisticated modulation techniques, LMDS also is capable of operating without having a direct line of sight with the receiver. This feature, highly desirable in urban areas, is achieved by having transmitted signals arrive at the receiver via a number of different paths. To minimize interference between adjoining cells, a combination of techniques is used, including horizontal/vertical polarization and frequency interleaving in diagonally proximate cells.

MVDS. In Europe, LMDS operates under a different name and a different frequency. Currently under development in the United Kingdom, the Microwave Video Distribution System (MVDS) has been licensed to operate in the 40.5 to 42.5 GHz band. The allocation of higher frequency bands displays a trend that is now becoming familiar as a result of increasing levels of spectrum congestion. (As noted, the FCC has recently opened the 40 GHz band for public use in the United States.)

MVDS cells are even smaller than LMDS cells, typically 3 km. Again this is due to the lack of availability of higher-power millimeter wave sources, and also to losses that result from gaseous absorption (i.e., pollution) and rain attenuation. To overcome these losses and obtain acceptable signal quality, a receiver incorporating a low-conversion-loss mixer and an antenna providing 32 dB gain is used.

Summary. Microwave transmission is now almost exclusively a short-haul medium, while optical fiber and satellite have become the long-haul transmission media of choice. Short-haul microwave is now one of the most agile and adaptable transmission media available, with the capability of supporting data, voice, and video. It also is used to back up fiber-optic facilities and to provide communications services in locations where it is not economically feasible to install fiber.

See also

DIRECT BROADCAST VIA SATELLITE (DBS)
FIBER OPTICS TECHNOLOGY
SATELLITE COMMUNICATIONS

Modems

A modem, or *modulator-demodulator,* converts the digital signals generated by a computer into analog signals suitable for transmission over dial-up telephone lines or voice-grade leased lines. Another modem, located at the receiving end of the transmission, converts the analog signals back into digital form for manipulation by the data recipient. Although long-distance lines are digital, most local lines are not, which explains why modems are often required to transfer files, access bulletin boards, send electronic mail, and connect to host computers from remote sites. Of course, where digital local lines are available, ISDN can be used to achieve higher and more reliable transmission than modems currently can provide. For most PC users, however, ISDN is still prohibitively expensive, not available, or too complicated to configure.

Modems are packaged as internal, external, or rack-mount models. Internal modems are inserted into a vacant expansion bus slot of a computer, while external desktop modems connect to a computer's RS-232 serial port. Rack-mount modems are full or half cards housed in an equipment frame located in a wiring closet. From there, cables connect to each PC's serial port. Modem cards also may be installed in a communications server attached to the corporate LAN so they can be shared by many users.

Modem manufacturers are continually redesigning their products to incorporate the latest standards, enhance existing features, and add new ones. The advancement of modulation techniques, error correction, data compression, and diagnostics are among the continuing activities of modem manufacturers.

Modulation techniques. Modems use modulation techniques to encode the serial digital data generated by a computer onto the analog carrier signal. The simplest modulation techniques rely on two signal characteristics to transmit information: frequency shift keying and phase shift keying. Frequency shift keying (FSK) is similar to the frequency modulation technique used to transmit FM radio signals. By forcing the signal to shift back and forth between two frequencies, the modem is able to encode one frequency as a 1 and the other as a 0. This modulation technique was widely used in the early 300-bps modems. At higher rates, FSK is too vulnerable to line noise to be effective.

Phase shift keying (PSK), another early modulation technique, uses shifts in a signal's phase to indicate 1s and 0s. The problem with this method is that *phase* refers to the position of a waveform in time; therefore, the data terminal clocks at both ends of the transmission must be synchronized precisely.

Another method, known as differential phase shift keying (DPSK), uses the phase transition to indicate the logic level. With this scheme, it is not necessary to assign a specific binary state to each phase; it only matters that some phase shift has taken place. The telephone bandwidth is limited, however, so it is only possible to have 600 phase transitions per second on each channel, thus limiting transmission speeds to 600 bps. To increase speed, it is possible to expand DPSK from a two-state to a four-state pattern represented by four two-bit symbols (known as dibits) as follows:

- Maintain the same state (0,0)
- Shift counterclockwise (0,1)
- Shift clockwise (1,0)
- Shift to the opposite state (1,1)

Other modulation techniques, such as trellis encoding, are much more sophisticated. Trellis encoding uses a 32-bit constellation with "quintbits" to pack more information into the carrier signal and offer more immunity from noise. The use of quintbits offers 16 extra possible state symbols. These extra transition states allow dial-line modems to use transitions between points, rather than specific points, to represent state symbols. The receiving modem uses probability rules to eliminate illegal transitions and obtain the correct symbol. This gives the transmission greater immunity to line impairments. Additionally, the fifth bit can be used as a redundant bit, or *checksum,* to increase throughput by reducing the probability of errors.

By increasing the number of points in the signal constellation, it is possible to encode greater amounts of information to increase the modem's throughput. This is because ever-slighter variations in the phase-modulated signal may be used to represent coded information, which translates into higher throughput. Some modem manufacturers use constellations consisting of 256 or more points.

Transmission techniques. Modems support two types of transmission techniques: asynchronous or synchronous. The user's operating environment determines whether an asynchronous or synchronous modem is required. During asynchronous transmission, start and stop bits frame each segment of data during transfer to distinguish each octet (byte) from the one preceding it. Synchronous transmission transfers data in one continuous stream; the transmitting and receiving DTE (data terminal equipment) therefore must be synchronized precisely in order to distinguish each character in the data stream. Most mainframes and minicomputers use synchronous protocols, whereas PC-to-PC communications are typically asynchronous. Some types of modems support both types of transmission.

Modem speed. For many users, the most important modem characteristic is data rate. Modems that adhere to the V.34 standard offer speeds up to 28.8 kbps, while modems that offer speeds of up to 33.6 kbps adhere to the V.34+ standard. The quality of the connection has a lot to do with the actual speed of the modem. If the connection is noisy, for example, the modem may have to step down to a lower speed to continue transmitting data. Some modems are able to sense improvements in line quality and can automatically step up to higher data rates as line quality improves.

In 1997, a new class of modems became available that offer data transmission rates of up to 56 kbps. The modems are based on new technology that exploits the fact that for most of its length, an analog modem connection is really digital. When an analog signal leaves the user's modem, it is carried to a phone company central office where it is digitized. If it is destined for a remote analog line, the signal is converted back to analog at the central office nearest the receiving user. The conversion is made at only one place: where the analog line meets the central office. During the conversion, noise is introduced that cuts throughput. But the noise is less in the other direction, from digital to analog, allowing the greater downstream throughput (Fig. M1).

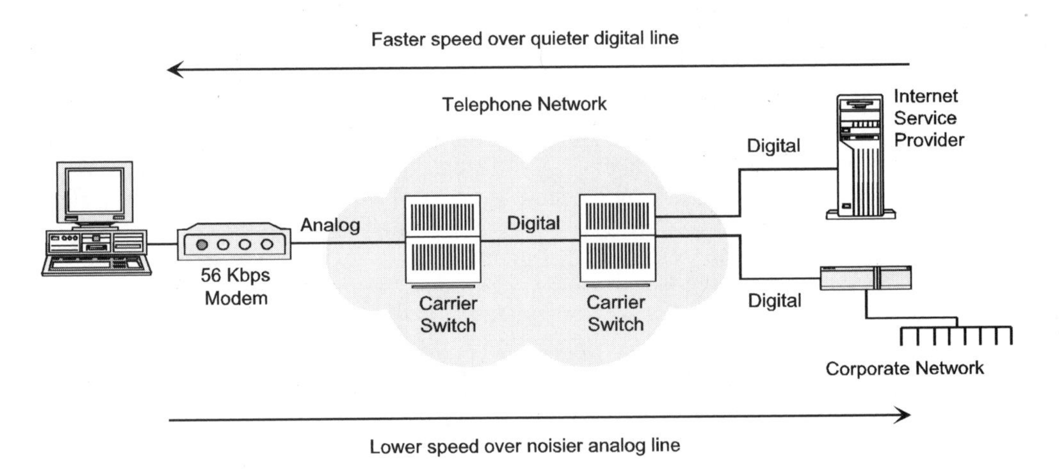

Fig. M1 *New 56 kbps modems can send data at top speed, but only from a digital source. Since it is already in digital form, the traffic is free of impairments from noise introduced when an analog modem signal is made digital within the carrier network. From an analog source, the top speed is quite a bit lower than 56 kbps, since the traffic is subject to impairment from noise.*

It also is possible to tie together two or three 56 kbps modems to achieve a combined data rate of 156 kbps over dial-up lines (Fig. M2). When two or three modems in a modem pool device are used for Internet access, for example, they call the Internet service provider simultaneously and share the downloading of Web pages, resulting in throughput of up to 156 kbps. Download time can be cut by as much as two-thirds. Although the Internet service provider does not have to do anything differ-

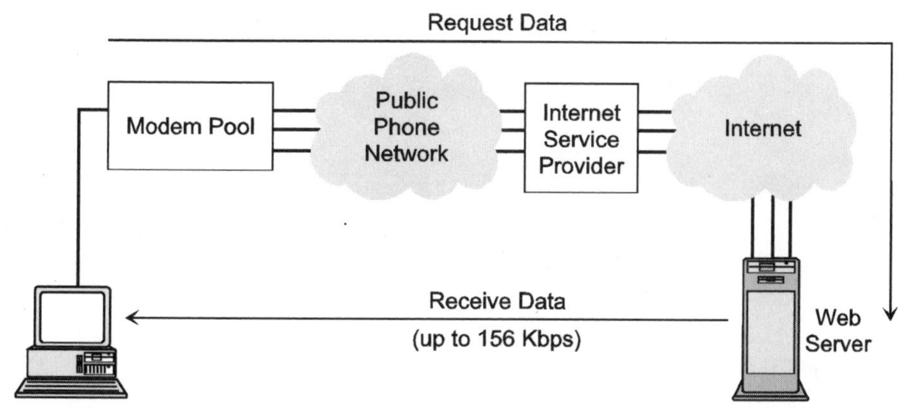

Fig. M2 *The modem pool device intitiates two or three simultaneous calls, dividing up the separate TCP/IP sessions, making Web page downloads go much faster. The Web server sends the requested page to the modem pool device, and the page is assembled on the user's screen.*

ent as far as hardware is concerned—except have enough 56 kbps modems—it must allow users to establish multiple sessions with a single user ID and password.

Modem features. Most modems come equipped with the same basic features, including error correction and data compression. In addition, they have features associated with the network interface, such as flow control and diagnostics. There also are various security features that are implemented by modems.

Error correction. Networks often contain disturbances with which modems must deal, or in some cases, overcome. These disturbances include attenuation distortion, envelope delay distortion, phase jitter, impulse noise, background noise, and harmonic distortion, all of which negatively affect data transmission. To alleviate the disturbances encountered when transferring data over dial-up lines and leased lines without line conditioning, most products include an error correction technique in which a processor puts a bit stream through a series of complex algorithms prior to data transmission.

The most prominent error correction technique has been the Microcom Networking Protocol (MNP), which uses the cyclic redundancy check (CRC) method for detecting packet errors, and requests retransmissions when necessary.

Link Access Procedure B (LAP-B), a similar technique, is a member of the High Level Data Link Control (HDLC) protocol family, the error-correcting protocol in X.25 for packet-switched networks. LAP-M is an extension to that standard for modem use and is the core of the ITU error-correcting standard, V.42. This standard also supports MNP Stages 1 through 4. Full conformance with the V.42 standard requires that both LAP-M and MNP Stages 1 through 4 are supported by the modem. Virtually all modems currently made by major manufacturers conform to the V.42 standard.

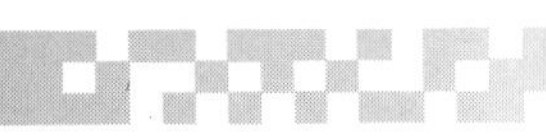

The MNP is divided into nine classes. Only the first four deal with error recovery, which is why only those four are referenced in V.42. The other five classes deal with data compression. The MNP error recovery classes are as follows:

- MNP Classes 1 to 3 packetize data and, Microcom claims, ensure 100 percent data integrity.
- MNP Class 4 achieves up to 120 percent link throughput efficiency via Microcom's Adaptive Packet Assembly and Data Phase Optimization, which automatically adjusts packet size relative to line conditions and reduces protocol overhead.

Data compression. With the adoption of the V.42*bis* recommendation by the ITU in 1988, there is a single data compression standard: Lempel-Ziv. This algorithm compresses most data types, including executable programs, graphics, numerics, ASCII text, or binary data streams. Compression ratios of 4:1 can be achieved, although actual throughput gains from data compression depend on the types of data being compressed. Text files are the most likely to yield performance gains, followed by spreadsheet and database files. Executable files are most resistant to compression algorithms because of the random nature of the data.

Diagnostics and other features. Most modems perform a series of diagnostic tests to identify internal and transmission line problems. Most modems also offer standard loopback tests, such as local analog, local digital, and remote digital loopback. Once a modem is set in test mode, characters entered on the keyboard are looped back to the screen for verification.

Most modems also include standard calling features such as automatic dial, answer, redial, fallback, and call-progress monitoring. Calling features simplify the chore of establishing and maintaining a communications connection by automating the dialing process. Telephone numbers can be stored in nonvolatile memory.

Other standard modem features commonly offered include fallback capability and remote operation. Fallback allows a modem to automatically drop, or fall back, to a lower speed in the event of line noise, and then revert to the original transmission speed after line conditions improve. Remote operation, as the name implies, allows users to activate and configure a modem from a remote terminal.

Security. Modems that offer security features typically provide two levels of protection: password and dial-back. The former requires the user to enter a password, which is verified against an internal security table. The dial-back feature offers an even higher level of protection. Incoming calls are prompted for a password, and the modem either calls back the originating modem using a number stored in the security table or prompts the user for a telephone number and then calls back.

Security procedures can be implemented before the modem handshaking sequence, rather than after it. This effectively eliminates the access opportunity for potential intruders. In addition to saving connection establishment time, this method uses a precision, high-speed analog security sequence that is not even detectable by advanced line monitoring equipment.

Transmission facilities. Modems support two types of lines: leased or dial-up. The primary difference between the two deals with the procedure for establishing a connection as opposed to the line itself.

Dial-up lines. Dial-up lines are used for typical telephone service. These lines usually connect to a small modular wall jack called an RJ-11, and a companion plug, which is inserted into the jack to establish a connection to the telephone or modem. When a voice-grade analog line is used with a modem, a short cord with an RJ-11 on both ends is inserted into the jack and into the modem.

Although the RJ-11 modular-jack connection is common for low-speed modem connection, problems may arise with regard to the consistency of the signals transmitted over the line. To ensure a consistently high signal level, a special data line jack, such as RJ-41 or RJ-45, can be installed. These data line jacks are designed specifically to operate with modem circuitry and are often used in leased-line environments to help maintain the quality of the transmitted signal.

Leased lines. Leased lines are available in two- and four-wire versions. Four-wire leased lines differ from their two-wire counterparts in terms of cost (four-wire lines are more expensive) and the mode of modem operation supported. Not all modems can support leased lines. An effective way of determining whether a modem can support leased-line connection is to examine the way the telephone line is connected. Modems designed for two- and four-wire leased-line operation have two sets of terminal screws with which to attach the two pairs of lines.

To sustain the optimal performance of leased-line modems, the lines may be specifically selected for their desirable characteristics. This is an extra-cost service called *line conditioning,* which is provided by the carriers on a best-effort basis. AT&T's D6 line conditioning, for example, addresses phase jitter, attenuation distortion, and envelope delay distortion, all of which can impair transmission at data rates approaching 19.2 kbps.

The monthly charges and installation cost of D6 conditioning is higher than for other levels of conditioning, if only because there are fewer wire pairs available that exhibit the higher immunity from generic noise and nonlinear distortion in high-density locations. Immunity from noise and nonlinear distortion occurs on copper pairs more by chance than design because they may have external causes and be beyond the control of the carrier. This means that numerous wire pairs must be tested before those having the desired characteristics can be identified and put into service as *conditioned lines.*

Wireless links. Wireless modems are required to transfer data over public wireless services and private wireless networks. These modems come in a variety of hardware configurations: stand-alone, built-in, and laptop-type removable PCMCIA (PC Card). The newer modems are programmable and therefore capable of being used with a variety of wireless services using different frequencies and protocols. There are even modems that mimic wireline protocols, allowing existing applications to be run over the wireless network without modification.

Single-frequency modems. Private wireless networks operate in a range of unique frequency bands to ensure privacy. Using radio modems operating over dedicated frequencies within these frequency bands also permits the transmission of business-critical information without interference problems. Furthermore, the strategic deployment of radio modems can provide metropolitan area coverage without the use of expensive antenna arrays.

Such modems are designed to provide a wireless, protocol-independent interface between host computers and remote terminals located as far away as 30 miles.

Most provide a transmission rate of at least 19.2 kbps point-to-point in either half- or full-duplex mode. Some radio modems even support point-to-multipoint radio network configurations, serving as a virtual multidrop radio link that replaces the need for expensive, dedicated lines (see Fig. M3). In this configuration, one modem is designated as the master, passing polling information and responses between the host and terminals over two different frequencies.

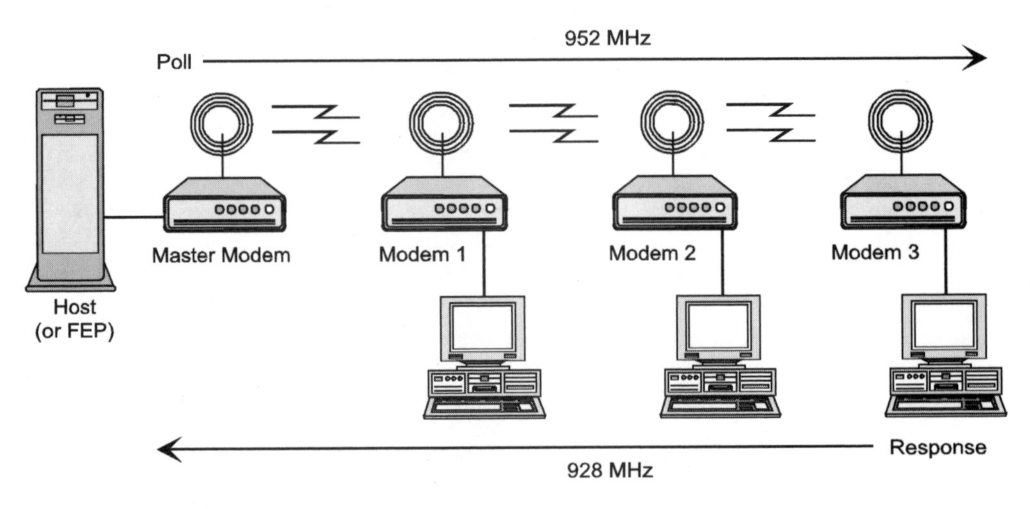

Fig. M3 *A typical multidrop radio modem configuration.*

In multidrop configurations, a given radio network is capable of supporting one type of asynchronous or synchronous polling protocol. Since such modems perform no processing or interpreting of the protocol, the host (or front-end processor) must generate all required protocol framing, line discipline, node addressing, and data encapsulation. Depending on vendor, these modems also might be equipped with an integral repeater to maintain signal integrity over longer distances.

Multifrequency modems. Regardless of the transmission technology or the hardware configuration used, the modem must be tuned to the frequency of the service provider's wireless network to operate properly. Until recently, modems were offered in different versions, depending on the destination wireless network. This delayed product development and inflated the cost of manufacturing, which was passed on to users in the form of higher prices for equipment. To overcome these problems, chip manufacturers have developed programmable chip sets that are not limited to a specific network's radio frequency. Newer wireless modems are computer-configurable: Within specified frequency ranges, the transmit and receive frequencies are independently selectable via software.

Multimedia modems. Not only can modems be programmed for multifrequency use, they can provide seamless integration of multiple media—wireline and wireless—through a common programmable interface. This is accomplished with a chip set that supports both wireline and wireless communications. Special software used with the chip set provides a method for connecting cellular phones to modems, which is important because cellular phones lack dial tones and other features used by modems on the wireline phone network. The software makes it appear that those features exist.

PCMCIA (PC Card) modems. The advantages of PCMCIA (now referred to as PC Card) technology now extend to modems. These modems are increasingly being used for connecting portable computer users with hosts and LANs at corporate locations. A variety of such modems is available for portable computers, allowing users to access the Internet, e-mail services, and facsimile. These credit card-sized devices use the Type II slot that now comes standard with most portable computers. There are also cards that connect to various wireless messaging services, including those based on Cellular Digital Packet Data (CDPD) technology. They have a built-in antenna and can even act as stand-alone receivers when the computer is shut off. Some cards are programmable, allowing users to access or receive messages from different wireless services.

There are CDPD modems that work with any DOS- or Windows-based computer, supporting the V.22*bis*, V.23, V.23*bis*, V.42*bis*, Group 3 fax, and V.17 wireline fax and data protocols, plus Microcom's MNP-10 cellular protocol or Paradyne's Enhanced Cellular Throughput (ETC) protocol. Some modems can automatically identify the type of modem protocols used at the receiving end and adjust their own operation accordingly.

The latest PC Card trend melds a 10Base-T Ethernet LAN adapter and a fax/modem on a single card. Some manufacturers have even added wireless capabilities to such cards, providing support for one- and two-way paging as well as direct cellular phone connection. Requiring only one slot, these multifunction cards are an ideal choice for subnotebook computers and personal digital assistants (PDAs). Of note, however, is that these multifunction cards do not yet support all the special features typically found in single-function cards.

Cable modems. While today's computers powered by Pentium and PowerPC chips are better equipped than ever to handle multimedia and video, they face a network bottleneck that, in most cases, offers top speeds of no more than 14.4 kbps or 28.8 kbps. In order to deliver entertainment and information services to PCs and television set-top boxes, a new type of modem—the cable modem—has emerged. Using traditional cable TV facilities, these modems can deliver speeds of up to 1000 times that of today's analog modems.

The cable modems developed by leading vendors use different modulation schemes for upstream and downstream data, error correction, interface, and protocols. Zenith, for example, uses 16 vestigial sideband modulation (VSB) to deliver up to 40 Mbps of data on the downstream channel, and employs Code Division Multiple Access (CDMA) frequency-agile technology for upstream data. Scientific Atlanta, on the other hand, provides 27 Mbps over the downstream data channel using 64/256 quadrature amplitude modulation (QAM) technology, while quadrature phase-shift keying (QPSK) is used for the upstream data channel.

Multifunction modems. Multifunction modems use programmable digital signal processing (DSP) technology to turn a computer into a complete desktop message center, allowing the user to control telephone, voice (recording and playback), fax, data transfers, and e-mail. Typical features include multiple mailboxes for voice mail, caller ID support, call forwarding, remote message retrieval, phone directory, and contact database.

In some cases the modem is actually on a full-duplex sound card. By plugging in speakers and a subwoofer, the user can even enjoy a stereo-sound speaker phone. A separate connection to a CD-ROM player allows the user to work at the computer while listening to music. However, these DSP-based products cannot be used as modems and sound cards simultaneously, since the processor can take on only one identity at a time.

With DSP, the modem can be easily upgraded to the latest communications standards, and new capabilities can be added simply by loading additional software. For example, a 14.4 kbps modem can be upgraded to 28.8 kbps by installing new software instead of having to buy new hardware. Likewise, 28.8 kbps modems can be upgraded to 33.6 kbps in the same way, often at no extra charge from the vendor.

Multimedia modems use digital simultaneous voice and data (DSVD), which enables the user to send voice and data at the same time over a single telephone line. The biggest advantage of DSVD is that users no longer need to interrupt telephone conversations or install a separate line to transmit data or receive faxes. Multimedia modems typically include full-duplex speakerphone, fax, modem, and 16-bit stereo audio capabilities.

Another new way vendors are packaging modems is by integrating them with ISDN terminal adapters. This allows users to communicate with conventional dial-up services and also take advantage of ISDN when possible, all without cluttering the desktop or having to use up scarce slots in the PC.

Soft modems. The next step in modem packaging is to eliminate the need for dedicated hardware altogether. Intel's Native Signal Processing (NSP) initiative, for example, embeds software emulation of modem and sound card hardware in the Pentium chip itself. The idea is to allow any Windows application to have access to features implemented in a special driver—to send and receive e-mail and faxes or to play sound files—with no modem or specialized hardware, aside from the main CPU.

Motorola takes a different approach, offering a line of host signal processor (HSP) modems, which rely heavily on software and the processing power of the host PC. But rather than relying on its own digital signal processor, this type of modem is based on a cheaper Application Specific Integrated Circuit (ASIC) and takes advantage of a PC's central Pentium chip for the processing power. HSP technology has been around for several years, but has only recently become feasible because of the growing availability of faster desktop computers based on the Pentium chip.

Summary. Once thought to be an outdated technology that would be supplanted by ISDN terminal adapters connected to digital lines, modems are not only growing in use, but are undergoing a surge in innovation as well. Higher-speed modems, the advent of cable modems that work over CATV networks, and new technologies that rely more on a computer's CPU for carrying out modem functions, have all combined to breathe new life into this market segment. Although today's new-generation modems will not replace the need for ISDN in all cases, they certainly

will give many potential ISDN users reason to consider the modem as a viable near-term alternative.

See also
INTEGRATED SERVICES DIGITAL NETWORK

Multimedia networking

Multimedia networking is concerned with the delivery of real-time applications over LANs and WANs. Because multimedia applications may comprise several data types (i.e., text, voice, images, video), all of which may be stored in different locations, the problem is how to ensure their synchronized delivery. Even when only one data type is involved, such as a video stream, the problem is how to ensure that its delivery is smooth and not interrupted whenever the network gets clogged with other kinds of traffic. For example, if the audio portion of a video conference does not arrive at the same time as the video components, the lip movements of participants will not be synchronized with their actual conversations.

The problem of synchronization can be addressed by assigning a class of service or quality of service to a multimedia application, which identifies it as being time-sensitive and requiring priority over other, less time-sensitive, data types. While standards-based ATM networks are designed to integrally support quality of service for multimedia applications, the older installed base of Ethernet networks does not.

Multimedia over LANs. To produce integrated, high-quality graphics and sound, multimedia applications need regular, predictable data delivery. But Ethernet delivery timing is not very predictable. Not only must workstations contend for network access, but Ethernet transmission under a full load is subject to excess latency and jitter. Latency is the delay between the time data is transmitted and the time it is received at the destination. Jitter is the uncertainty in the arrival time of a packet, or the variability of latency. Although video compression techniques such as MPEG are essential to high-quality multimedia, they are not enough to overcome these problems.

Even the limitations of Ethernet can be surmounted to support multimedia applications. A variety of techniques can be used to overcome excess packet delay, including the use of star-wired switching configurations and vendor-specific enhancements to Ethernet that are transparent to each end-system adapter and backward compatible with existing Ethernet adapters.

One of these enhancements is a traffic control algorithm implemented at the hub, which allows each Ethernet segment to operate at more than 98 percent efficiency and, even under full load, to service a mix of real-time and conventional data traffic. The use of such traffic control algorithms provides predictable LAN transmission by regulating the flow of traffic on the link to minimize jitter. The result is increased Ethernet predictability in support of real-time multimedia applications, even at 10 Mbps. Because IEEE 802.3 Ethernet is essentially speed-independent, the same enhancements work at the Fast Ethernet speed of 100 Mbps.

A serious limitation of Ethernet for multimedia transmission is that it offers no priority access scheme. All traffic must contend for access on a best-efforts basis, causing delay in getting data onto the network in the first place. Some vendors offer a method of prioritizing traffic over Ethernet to deliver QoS to applications. In one scheme, traffic can be prioritized as either high or low, with high-priority traffic allowed onto the network before low-priority traffic. When a workstation receives a video stream from the server, the data is temporarily held in the workstation's buffer. Through flow control techniques, the video stream is released for viewing at a consistent bit rate. This has the effect of smoothing out any jerkiness in the video and allows the audio to synchronize better with the video. The whole process provides the illusion of real-time multimedia transmission. As the buffer empties, it is continually being replenished by the server as it contends for priority access to the network.

Similar priority schemes can be applied to Token-Ring LANs with even better results. This is because Token-Ring LANs provide each workstation with guaranteed access to the transmission facility. With each workstation given its turn to access the transmission facility for a given length of time, latency is predictable and jitter becomes much less of a problem. Natively, therefore, Token-Ring is more adept at handling multimedia applications than Ethernet.

Multimedia over the Internet. The problem of delay is more serious when one tries to run multimedia applications over TCP/IP networks, including corporate intranets and the public Internet. A key disadvantage of the Internet Protocol (IP) is that it does not allocate a specific path or amount of bandwidth to a particular session. The resulting delay can vary wildly and unpredictably, disrupting real-time applications. A number of solutions have been proposed which attempt to perform resource setup functions similar to that of the Q.93x signaling protocol used in the public circuit-switched network. There, a dedicated path is set up between two parties, which stays in place for the duration of the conversation. This results in excellent speech quality with little or no delay.

A similar mode of operation can be applied to TCP/IP networks with the addition of certain protocols to the routers. One of the most promising of these is RSVP, the resource ReSerVation Protocol, developed by the Internet Engineering Task Force (IETF).

As an Internet control protocol, RSVP runs on top of IP to provide receiver-initiated setup of resource reservations on behalf of an application data stream. When an application requests a specific quality of service for its data stream, RSVP is used to deliver the request to each router along the path(s) of the data stream and to maintain router and host states to support the requested level of service. In this way, RSVP essentially allows a router-based network to mimic a circuit-switched network on a best-efforts basis.

At each node, the RSVP program applies a local decision procedure, called *admission control,* to determine if it can supply the requested QoS. If admission control succeeds, the RSVP program in each router passes incoming data packets to a packet classifier that determines the route and the QoS class for each packet. The packets are then queued as necessary in a packet scheduler that allocates resources for transmission on the particular link. If admission control fails at any node, the RSVP program returns an error indication to the application that originated the request.

The advantage of RSVP is that it will work with any physical network architecture. In addition to Ethernet, it will run over other popular networks such as Token-Ring and FDDI, as long as IP is the underlying network protocol. This makes RSVP suitable for company-wide networks as well as the Internet, providing end-to-end service between them.

RSVP also meshes well with the next generation of IP (version 6, or IPv6), allowing users to set up end-to-end connections with a specified amount of flow control for a given time period. This is made possible by the ability of IPv6 to label packets in traffic patterns, making it easier to identify packets that belong to particular traffic flows for which the sender requests special handling. This means that time-sensitive services such as real-time video and voice will get the special handling they require along the route path.

IPv6 could take a few years to be widely implemented throughout the Internet. Until IPv6 is widely implemented, many vendors do not see much sense in supporting RSVP.

Role of middleware. *Middleware* can play a key role in implementing multimedia applications over LANs, particularly in enabling Windows applications to run over ATM. The middleware runs invisibly under the Application layer and above the Network layer (Fig. M4), bridging the gap between ATM standards and the de facto Microsoft Windows standards. In this scenario, middleware brings ATM quality of service directly to the application, resulting in high-quality voice, video, text, and images to the desktop, while leveraging the inherent management, security, and authentication features of the existing network operating system (NOS).

Users on ATM-attached PCs still run the same applications, and they access the LAN server as before, except that they can run networked multimedia. QoS is carried out automatically by the middleware based on the type of traffic. Predetermined delay and bandwidth parameters are allocated to the applications according to their bandwidth and delay (measured in milliseconds) needs. For example, MPEG traffic may be assigned 1.5 Mbps of bandwidth, while an audio file is granted 5 Mbps. For a H.320-based videoconference, 384 kbps is assigned, with a delay of 200 to 300 ms. All of these services—including applications running over TCP/IP, IPX, NetBIOS protocols—can be delivered with guaranteed end-to-end quality using the existing desktop wiring infrastructure. Furthermore, no changes to existing hardware and software are required.

To implement videoconferencing, for example, each PC must have an ATM adapter with a direct connection to an H.320-based codec interface board. An H.320-compliant multimedia gateway server connects local ATM workgroups to remote sites through ISDN for videoconferencing across the WAN.

The multimedia gateway server is an integrated hardware and software package that upgrades a designated PC to an ATM/ISDN gateway with BRI or PRI connections. An ISDN connection is made via the server rather than directly at the PC level. The server initiates external calls over the WAN through ISDN's Q.931 call setup and management signaling protocol. Videoconferencing PCs communicate with one another or with the gateway server through an ATM connection to a media switch. The ATM adapter and middleware provide ISDN redirection, setting up calls from one PC to another for videoconferencing.

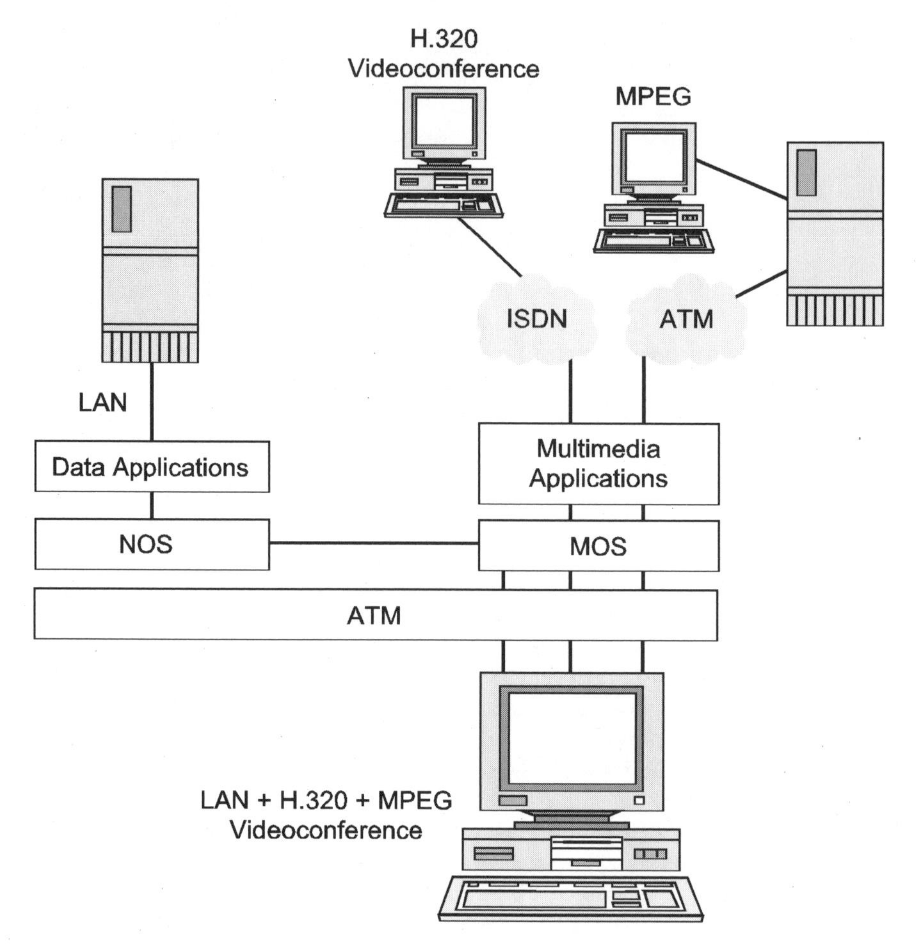

Fig. M4 *First Virtual's Multimedia Operation Software (MOS) allows Windows applications to run over ATM. A predetermined class of service is assigned according to the type of data stream detected by the MOS, in this case LAN data, MPEG, or an H.320 videoconference.*

The portion of the middleware that resides on the end user's PC sits between the applications and the LAN emulation software, where it sets up real-time calls for voice, video, and audio. The software redirects real-time data streams generated by Windows programs from the local disk drive to a destination across the ATM network. In essence, this redirection "spoofs" or fools applications into thinking they are executing locally.

When a DOS or Windows program requests a file, the middleware consults a map to determine if the file is stored locally, on the LAN server, or on the media server. If the file is stored on the LAN server, the request is processed normally through the appropriate protocol stack. The LAN emulation software on the adapter sets up an ATM call to the LAN server and segments the IP or IPX packets into 53-byte cells and sends them off. At the other end, the ATM-to-Ethernet adapter at the remote port of the media switch performs cell-to-packet reassembly and passes them on to the local LAN server.

A component of the middleware also resides on a media server, where its primary function is to handle client requests and retrieve data. When the client requests a file stored there, the middleware sets up an ATM call to the media server and passes the request along to it. At the core of this middleware component is a real-time kernel that performs call scheduling. This function enables the media server to support up to multiple users simultaneously.

Summary. These are only a few of the ways multimedia networking is being implemented. There is no question that many areas can benefit from multimedia, enhancing collaborative efforts in online engineering, manufacturing, and health care, to name only a few. Multimedia can even facilitate the real-time integration of multiple sites into a "virtual corporation." The technologies, standards, and products for delivering multimedia solutions are rapidly falling into place with an eye toward ATM. Isochronous Ethernet is also a viable solution for multimedia applications, with the added benefit of interconectivity over the WAN through ISDN.

See also

ASYNCHRONOUS TRANSFER MODE
ETHERNET (ISOCHRONOUS)
VIDEOCONFERENCING

Multiplexers

Multiplexers first appeared in the mid-1980s when private networks became popular. These devices combine voice, data, and video traffic from various input channels so they can be transmitted over a single, higher-speed digital link, usually a T1 line. At the other end, another device separates the individual lower-speed channels, sending the traffic to the appropriate terminals. Multiplexers allow businesses to reduce telecommunications costs by making the most efficient use of the leased line's available bandwidth. Since the line is billed at a flat monthly rate, this is incentive enough to load it with as much traffic as possible. Using different levels of voice and/or data compression, the channel capacity of a leased line can easily be doubled or quadrupled to save even more money.

There are several types of multiplexing in common use today. On private leased lines, the dominant technologies are time division multiplexing (TDM) and statistical time division multiplexing (STDM). Each lends itself to particular types of applications. TDM multiplexers are used when most of the applications must run in real time, including voice, videoconferencing, and multimedia. When an input device has nothing to send, however, its assigned channel is wasted. With STDM, if a device has nothing to send, the channel it would have used is taken by a device that does have something to send. If all channels are busy, input devices wait in queue until a channel becomes available. STDM multiplexers are used in situations where efficient bandwidth usage is valued and the applications are not bothered by delay.

Although used mostly on private networks, both types of multiplexers can interface with the Public Switched Telephone Network (PSTN) as well. For example, if a private T1 line degrades or fails, the multiplexer can be configured to automatically switch traffic to an ISDN link until the private line is restored to service.

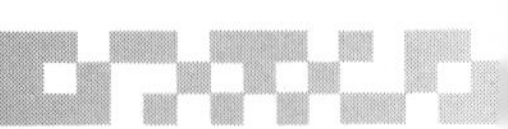

Time Division Multiplexing. With TDM, each input device is assigned its own time slot, or *channel,* into which data or digitized voice is placed for transport over a high-speed link. When a T1 line is used, there are 24 channels, each of which operates at 64 kbps. The link carries the channels from the transmitting multiplexer to the receiving multiplexer, where they are separated out and sent on to assigned output devices. If an input device has nothing to send, the assigned channel is left empty.

The TDM manages access to the high-speed line and cyclically scans (or polls) the terminal lines, extracts bits or characters, and interleaves them into the assigned time slots (e.g., frames) for output to the high-speed line. The multiplexer includes channel cards for each low-speed channel and its associated device, a scanner/distributor, and common equipment for the high-speed line. The low-speed channel cards handle the data and control signals for the terminal devices. They also provide storage capacity through registers that provide bit or character buffering for placing or receiving data from the time slots in the high-speed data stream.

The TDM's scanner/distributor scans and integrates information received from the low-speed devices into the message frame for transmission on the high-speed line, and also distributes data received from the high-speed line to the appropriate terminals at the other end.

The common equipment provides the logical functions used to multiplex and demultiplex incoming and outgoing signals. It contains the necessary logic to communicate with both the low-speed devices and the high-speed device. It also generates data, control, and clock signals, which ensure that the time slots are perfectly synchronized at both ends of the link.

When digital lines are used on the network side, the TDM is connected to a CSU/DSU (channel service unit/digital service unit), a required network interface for carrier-provided digital facilities, which typically is provided as a plug-in card by most multiplexer vendors. The CSU is positioned at the front end of a circuit to equalize the received signal, filter both the transmitted and received waveforms, and interact with the carrier's test facilities. The DSU element transforms the encoded waveform from alternate mark inversion (AMI) to a standard business equipment interface, such as RS-232 or V.35. It also performs data regeneration, control signaling, synchronous sampling, and timing.

Operation. TDM technology supports asynchronous, synchronous, and isochronous data transmission. Asynchronous data transfer requires the framing of each character by a start bit and stop bit. This allows the originating terminal to control the timing of each transmitted character. Synchronous data transfer timing is controlled by the multiplexer. Terminals send synchronous blocks of data framed by characters. Bits within a block are synchronized to clock signals generated by the TDM. Synchronous terminals operate at higher speeds than their asynchronous counterparts, but both are multiplexed in a similar manner.

Isochronous transmission supports multimedia applications where voice and other data must arrive together. In this type of transmission, the individual terminals generate their own clock signals, with all clocks running at the same nominal rate. Isochronous multiplexers provide some buffering and rate adjustment to compensate for slight variations among the clock rates.

A TDM samples data from each terminal input channel and integrates it into a message frame for transmission over the high-speed line. Message frames consist of time slots, and each time-slot position is allocated to a specific terminal. Interleaving is the technique that multiplexers use to format data from multiple devices for aggregate transmission over the link (Fig. M5).

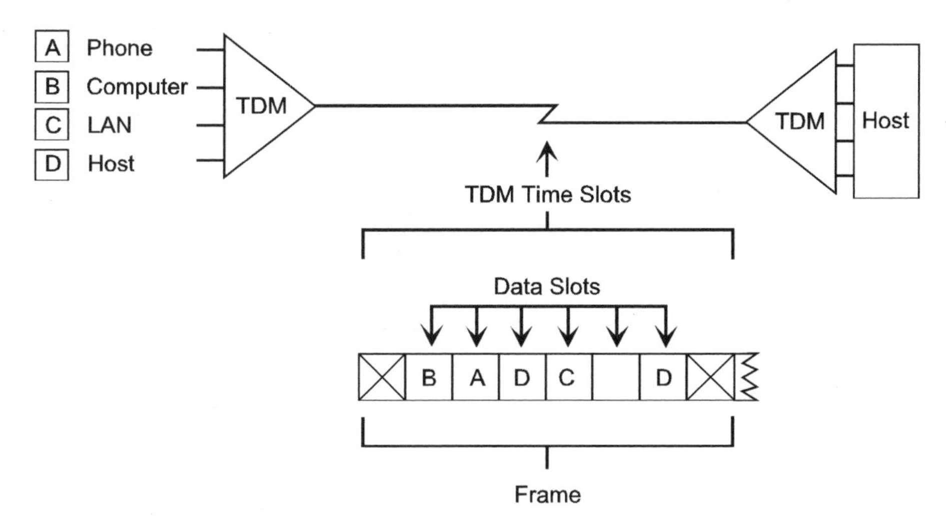

Fig. M5 *Data from multiple input sources is interleaved by the TDM multiplexer for transmission over the high-speed link. Note the empty time slot. If a device has nothing to send, this amount of bandwidth goes unused.*

Most of the market leaders offer multiplexers that will interface with public networks via the byte-interleaving technique. Some vendors support both bit and byte interleaving, enabling their products to readily interconnect with both private facilities for maximum efficiency (i.e., bit) and public switched services for increased connectivity (i.e., byte). This configuration flexibility enable companies to take advantage of both environments according to shifting economics or application needs.

Features. High-end T1 multiplexers provide numerous standard and optional features, including:

- *Drop-and-insert:* With this feature, a multiplexer is able to accept a high-speed composite data stream from another multiplexer, demultiplex (remove) a portion of the data stream, modify it (e.g., add additional data frames), and transmit the altered high-speed data stream to a third multiplexer.
- *Bypass:* A complementary feature, this allows a multiplexer to pass through a high-speed data stream without modification.
- *Subrate data multiplexing:* This provides programmable data rates at 56 kbps and below for both synchronous and asynchronous data. Subrate channels, together with normal channels (e.g., DS0), are carried over the

same physical T1 facility. To maintain transparency, the subrate multiplexing technique accommodates independent clocking of the transmit and receive data.

- *ISDN:* Via plug-in cards, many T1 vendors support primary rate ISDN (23B+D). In addition to the twenty-three 64-kbps B (bearer) channels and one 64-kbps D channel for out-of-band signaling, there are two high-capacity ISDN channels that can be supported by the multiplexer: the 384-kbps H0 channel and the 1.536-Mbps H11 channel. These channels are best suited for LAN interconnection, high-speed data applications, videoconferencing, and private network restoral.
- *LAN adapters:* Optional plug-in LAN adapters are available with most multiplexers. A 10Base-T module, for example, allows Ethernet traffic to be combined with voice, synchronous data, and video traffic over a fractional or full T1 link.
- *Bridges and routers:* To interconnect LANs, multiplexers can accommodate plug-in bridge/router modules. A variety of protocols are supported, including TCP/IP and Novell's IPX protocol.
- *Frame relay:* Frame relay interfaces give users added flexibility in tying branch office and workgroup LANs to the backbone network. With frame relay, users can create a virtual packet network that can be overlaid on a high-end T1 or statistical multiplexer network. The subnetwork can expand and contract to use available network resources. Some vendors even support voice traffic over the frame relay network. A separate voice compression module digitizes analog voice for priority transmission.
- *ATM:* Low-speed ATM interfaces for T1 multiplexers are now available. With constant-bite-rate (CBR) capability, the ATM interface can handle delay-sensitive applications such as videoconferencing. When the videoconference is over, the CBR capability can be disabled and the bandwidth made available to other data applications.
- *SMDS:* Some T1 multiplexers can be equipped with the DQDB (distributed queue dual bus) SNI (subscriber-network interface) and DXI (data exchange interface) for SMDS. In addition, the ports can be individually configured to convert SMDS to frame relay or ATM, or to convert frame relay to ATM.
- *Testing:* TDMs support local loopback tests on both the high- and low-speed ports. They also support remote loopback tests for the low-speed ports at the other end of the link. System diagnostic tests can be performed when the loopback tests are combined with character generation and error detection capabilities. Some multiplexers automatically "busy-out" individual remote low-speed ports when a failure is detected on a low-speed modem or computer port.
- *SNMP:* With SNMP support, the network manager can configure, monitor, and control T1 multiplexers from the same SNMP console that controls the LAN devices. A management information base (MIB) for the T1 multiplexer gives administrators at a remote site the same configuration flexibility they would have if they were at the device's control panel. MIBs give network

administrators access to, and control of, every configurable element of the device. GET messages let a network administrator receive status information from a device, TRAP messages report alarm conditions, and SET messages reconfigure network devices.

System management. A number of advanced system management features are available with T1 multiplexers. To help network managers effectively control system resources, some multiplexer management systems maintain a detailed physical inventory of every card in the network. When new cards are plugged into a node, the card ID information is automatically reported to the network management system, ensuring that equipment inventory is always current. This information includes card slot, card type, serial number, hardware revision level, and firmware revision level.

The network management systems of some multiplexers provide centralized order entry and order tracking. When a request is received for network service, for example, it is necessary to generate a network order for more bandwidth, which can be put into service immediately or at a specific time.

The alarm filter allows network operators to monitor alarms in real time and set filters to determine what information should be displayed at the network management system console. Most network management systems provide a visual and audible indication of most recent alarm summary information, and track and update alarm conditions automatically in the database. A level of severity can be assigned to each alarm type: critical, major, minor, alert, and ignore.

The network management database contains an event log. This information is date/time stamped and describes how the event was created. For example, if a technician pulls a card out of a multiplexer, a state change (event) is registered. Another state change (event) is registered when a new card is inserted. Other types of events include:

- Operator actions
- Alarms
- Status changes
- Order installation and removal
- System errors
- Capacity

The event log provides a chronological history of all activity within the network, giving the network manager a single source of information to track and analyze network performance.

Security. A multilevel password capability allows unique, individual access to the network. On an individual password basis, the system administrator can restrict access of any menu, submenu, or operating screen in the network management system. For example, a network supervisor may have access to all screens and functions, while an order entry clerk may have access only to the order management area to add or view orders, not to modify or delete them.

Redundancy. The redundant components of a multiplexer can include CPUs, buses, trunk cards, power supplies and fans, and network management systems. If the primary CPU/bus pair fails, for example, the secondary CPU/bus pair will take over nodal processing and clocking functions. When the switch takes place, the network management system is notified and an alarm is presented to the operator. Each CPU maintains a mirror image of information contained in the redundant CPU, so in the event of switchovers, information is not lost. Switchover from an active to a redundant CPU will not corrupt active circuits. The same applies to other redundant system components.

Management reports. Both standard and user-customized reports are available from most network management systems. Standard reports can be modified to facilitate network maintenance, capacity planning, inventory tracking, cost allocation, and vendor relations. In addition, ad hoc reports can be created from the management system's database using Structured Query Language (SQL).

Statistical Time Division Multiplexing. STDM is a more efficient multiplexing technique in that input/output devices are not assigned their own channels. If an input device has nothing to send at a particular time, another input device can use the channel. This uses the available bandwidth more efficiently. An STDM can be purchased as a stand-alone device or it can be an add-on feature to a TDM, providing service over one or more assigned channels.

Operation. STDM operation is similar to that of TDMs; the high-speed side appears very much like a TDM high-speed side, while the low-speed side is quite different. The STDM allocates high-speed channel capacity based on demand from the devices connected to the low-speed side. This allocation by demand (or contention) provides more efficient use of the available capacity on the high-speed line. In the variable-allocation scheme of an STDM frame, the time slots do not occur in a fixed sequence.

An STDM increases high-speed line usage by supporting input channels whose combined data rates would exceed the maximum rate supported by the high-speed port. When any given channel is idle (not sending or receiving data), input from another active channel is used in the time slot instead. The STDM has the option of turning off the flow of data from a sender if there is insufficient line capacity and then turning the flow back on when the capacity becomes available.

Features. TDMs and STDMs share many of the same operational and management features. Among the features found in STDMs are:

- *Data Compression:* Like TDMs, STDMs support techniques for compressing data so that they can actually transmit fewer bits per character. Data compression shrinks the time slot for the STDM and allows it to transmit more time slots per frame.
- *Error Detection and Correction:* While TDMs detect and flag errors, STDMs are able to correct them. The sending STDM stores each transmitted data frame and waits for the receiving STDM or computer to acknowledge receipt of the frame. A positive acknowledgment (ACK) or negative acknowledgment (NAK) is returned. If an ACK is received, the STDM discards the stored frame and continues sending the next frame. If a NAK is returned, the STDM retransmits the questionable frame and any subsequent frames. The process is repeated until the problematic frame is accepted, or a frame retransmission counter reaches a predetermined number of attempts and activates an alarm.

For applications such as asynchronous data transmission, where error detection is not performed as part of the protocol, STDM error detection and correction is a valuable feature. For protocols such as IBM's binary synchronous control (BSC), however, which contains its own error-detection algorithm, STDM-performed error control adds additional delays and redundancy that may not be appropriate for the application.

Several throughput enhancements are available for STDMs; they can be added later when needs change. These features include:

- *Per-channel compression,* in which each channel has its own compression table.
- *Fast packet technology,* which increases throughput by sending part of a frame before the entire frame is built.
- *Data prioritization* inserts shorter interactive frames between larger variable run-length frames.
- *Run-length compression* removes redundant characters from the transmission to improve performance.

Summary. Despite the continuous price reductions on leased lines since the 1980s, businesses are always looking to cut the cost of telecommunications. One of the most effective ways to do this is through the deployment of multiplexers that increase the bandwidth utilization of leased lines. Such lines are billed at a flat monthly rate by the carriers, no matter how little or how much traffic is actually carried over them. Not only can the business save money by using multiplexers, the cost of the devices themselves can be recovered in a matter of a few months out of the money saved. Using different levels of voice and/or data compression, the channel capacity of a leased line can be easily doubled or quadrupled to save even more money, enabling the business to recover the cost of the equipment even faster.

See also

INVERSE MULTIPLEXING

WAVELENGTH DIVISION MULTIPLEXING

Multiprotocol label switching

Lately there is much dissatisfaction with the performance of IP networks. This dissatisfaction has been brewing for the last two years and finally has reached the point where something must be done. The problem is that IP networks are being asked to do things for which they were not designed, and, as more users require access to the Internet and corporate intranets, IP networks are facing a serious performance barrier that threatens to bog down all users. Several factors have combined to inhibit network performance.

As computing power continues to increase, this drives the development of bandwidth-intensive, mission-critical, desktop applications. Demand for higher bandwidth, in turn, requires higher forwarding performance (packets per second) by routers, for both multicast and unicast traffic. In addition, other new applications, particularly multimedia, challenge the network infrastructure with new requirements, such as quality of service. And the emergence of intranets and Web-based computing has resulted in more widely distributed and unpredictable traffic flows, which makes for inefficient network operation.

The growth of the Internet is also stressing the scaling properties of the underlying routing system. The ability to contain the volume of routing information maintained by individual routers, and the ability to build a hierarchy of routing knowledge, are the keys to supporting a quality, scalable routing system.

About half a dozen vendor proposals are being offered that will provide users with the performance enhancements that come with Data Link layer (layer 2) switching, while retraining the services that routers perform, such as traffic prioritization, policy management, and security. These approaches entail configuring switches and routers to isolate particular groups of users or ports based on Data Link and Network layer infrastructure variables such as MAC address, switch port, Network layer subnet address, or even source/destination address flows. These techniques have a variety of names, such as *virtual LAN* (VLAN), *IP switching, Fast IP, IP Navigator,* and *tag switching.* All are based on adding functionality to the devices at the edge of the network and at the network core, to let them monitor more intelligently the traffic flows across the network and make decisions about what policies are needed to support them.

The Internet Engineering Task Force (IETF) is working to determine which of these concepts will be included in its much-anticipated *Multiprotocol Label Switching* (MPLS) standard. The stress is on "multiprotocol" because its techniques are applicable to any Network layer protocol. In this discussion, however, the focus is on the use of IP as the Network layer protocol. The reason is that IP networks will be among the first to gain from an MPLS standard issued by the IETF.

Approaches to MPLS. Ipsilon Corp. generally is credited with calling attention to the key issue of speeding IP performance. However, other leading internetworking vendors soon followed with their own solutions to the problem. At this writing, the IETF is still evaluating the merits of each approach for inclusion in its MPLS standard, but the consensus is that Cisco's tag switching will carry the day. Further information can be found at the following sites on the World Wide Web:

Ipsilon IP switching. Although routing protocols were regularly augmented throughout the 1980s and 1990s, and various packet-queuing advances were made to improve performance, vendors generally took the view that the best way to improve IP performance across the WAN was to build bigger and faster conventional routers. In many networks, this remains the best approach.

But a new approach to improving performance emerged in 1996 from Ipsilon, which proposed an architecture that could distinguish short- and long-lived flows in IP traffic, and map the long-lived flows to ATM virtual channels implemented in ATM switching hardware. This concept of *IP switching* addresses the need for faster throughput between separate subnets. IP switching retains the control functions of routing while leveraging the wire-speed forwarding capabilities of switching.

Ipsilon currently supports IP switching on top of ATM hardware, using two protocols it has defined: the *Ipsilon Flow Management Protocol* (IFMP) and the *General Switch Management Protocol* (GSMP), both of which have been submitted to the IETF as RFCs.

IFMP enables communication between IP devices, such as host stations and IP switches, by associating IP flows with ATM virtual channels. A *flow* is a sequence of IP packets between a particular source and destination that share the same type of protocol, type of service, and other characteristics, as determined by information in the packet header. Once a flow is identified and isolated to a particular I/O channel, the IP switch sets up a connection using a flow classification scheme. In this case, a cut-through connection is used. The IP switch automatically shifts between standard store-and-forward, packet-by-packet router operation, and cut-through switching operation based on the type of traffic it sees.

GSMP controls the underlying ATM switch hardware, allowing an IP switch controller to perform such functions as Establish and Release Virtual Channels, Manage Switch Ports, and Request Configuration Information. GSMP also supports a scheme whereby a priority can be assigned to a connection when it is established. Ipsilon has defined a broad priority system, which can be configured based on whatever IP information (address, subnet, network, protocol, application, or I/O port) is deemed relevant.

The most obvious benefit of IP switching is its performance. Ipsilon claims IP switches can forward packets at the rate supported by the underlying switch engine, which is as fast as 5.3 million packets per second for Ipsilon switches and 18 million packets per second for IP switches from Digital Equipment Corporation, which licenses Ipsilon technology. (These performance claims have not yet been independently verified, however.)

For the biggest performance boost, as many ATM-attached nodes as possible should support IFMP and as much network traffic as possible should traverse the IP switches.

Network traffic also needs to match the flows that the IP switch expects to see. If the organization's traffic patterns change over time, the IP switches will have to be tuned accordingly. Also, because few devices support IFMP, the network may need to use Ipsilon's PC-based IP switch gateway to get legacy LAN traffic onto the IP switch backbone.

As a routing solution, an IP switch is less expensive per port than a traditional router, although it does not support the same array of protocols. Critics of IP switching question its ability to scale, however, because it relies on setting up virtual circuits and in the wide area; IP switches would need to set up hundreds of connections per second. Ipsilon claims that IP switching can be deployed so that multiple flows are aggregated onto a single connection in the center of a network, thus conserving switch resources. Conversely, at the edge of the network, flows and their specific handling can be more granular.

Of note is that IP switching can support QoS without requiring that applications or operating systems be rewritten to explicitly request QoS, as is the case with the Resource Reservation Protocol (RSVP) and ATM-based QoS schemes. IP switching can support QoS based on implicit IP information (such as an IP address or TCP/User Data Protocol port), which can be configured into the switch.

IP switching supports native IP multicast, including standard IP multicast protocols such as the Distance Vector Multicast Routing Protocol (DVMRP) and Inte-

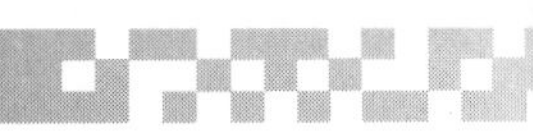

rior Gateway Routing Protocol (IGRP), via flow classification. Short-lived multicast traffic is forwarded by the IP routing software running on IP switches and is not switched.

Although IP switching is a proprietary technology, it is also an open technology. This means other vendors can license all of it or just pieces of it. 3Com, for example, will use its Fast IP to communicate across a LAN and use IFMP only to communicate from its routers to Cascade WAN switches. Cascade will then use its IP Navigator technology for communication across the WAN.

Currently, Ipsilon implements IP switching over ATM, but its technology is positioned for use in LANs and WANs. Accordingly, the company plans to implement IP switching over gigabit Ethernet and frame relay switches.

Cabletron VLAN. LAN switching increasingly is being used to improve network performance. LAN switches are based on layer 2 and act as fast bridges. A growing number of switches now support virtual LANs (VLANs), allowing network administrators to segment their networks into logical workgroups or broadcast domains to restrict traffic congestion. Layer 2 VLANs are mapped to switch ports or MAC addresses. However, routers are required to interconnect VLANs. With the trend toward incorporating a routing function in switches, allowing them to operate at both Data Link and Network layer levels, network administrators can now interconnect VLANs using the router function built into the switch.

This is the strategy of Cabletron. At the low end, Cabletron supports layer 2 switched networks that can be organized into VLANs. As the need arises, these VLANs can be scaled to a virtual switched network based on layer 2 switching and layer 3 routing. Cabletron does not use conventional routers in its switched network architecture because they would slow performance. Instead, Cabletron employs distributed routing functionality through its network to speed packet delivery. Cisco Systems also provides layer 2 VLANs, which it interconnects via its routers; however, its VLAN implementation differs from that of Cabletron in that it does not currently embrace switched virtual networking.

3Com Fast IP. 3Com's Fast IP combines the control-policy function of routing with the wire-speed forwarding performance of switching. 3Com's approach is to preserve the LAN router's role in filtering initial session requests between clients and servers in a network, while offloading the subsequent packet forwarding to faster LAN switches. According to 3Com, this improves network performance by up to 500 percent. Fast IP is the only IP switching solution that can be implemented across multiple network technologies, including Ethernet, Fast Ethernet, gigabit Ethernet, FDDI, Token-Ring, and ATM.

Unlike other IP switching solutions, Fast IP is initiated by the end system or "desktop" and not by a router or switch. Fast IP is initiated from the end system through a NHRP-based (Next Hop Resolution Protocol) request and response technique, which is part of 3Com's DynamicAccess technology. This technique is based on source and destination MAC addresses, as well as source and destination IEEE 802.1Q VLAN tags. The VLAN assignments, which often might correspond to subnet assignment or other grouping techniques, are learned by all attached switches via IEEE 802.1p advertisements. This allows LAN switches to provide the same level of broadcast and flooding containment as IP routers.

Once the request is initiated, a router is used to forward the request on to the destination while applying common filter/firewall policies. Upon receipt of the request, the destination system/server responds back to the end system via high-performance switches, not a router. The switches use 802.1Q-based tagging/flooding techniques to reach the destination, since it is not locally known. Once the response is received, the end system then uses the high-performance switched path for all subsequent data flows to and from the server, eliminating the dependence on the router for data flow. This technique also interoperates with all of today's existing layer 2 and layer 3 networking devices.

3Com's Fast IP solution is focused on increasing network performance in building and campus infrastructures using LAN and ATM. It gives desktops and servers an active role in requesting the services they need and streamlines the communication paths through the infrastructure.

Cisco's tag switching. Because Cisco Systems has some 80 percent of the installed base of routers, its tag switching solution is likely to be standardized by the IETF's Multiprotocol Label Switching (MPLS) working group. For this reason, it merits a closer look.

Cisco describes tag switching as a protocol-independent technology approach for large ISPs and carriers. Its primary benefits are route aggregation, improved performance, and simplification of the view a router has of a complex network. Today, all routers must know about and talk to all other routers on a network. Tag switching networks allow for the building of ATM or frame relay backbones to which routers can be attached, without the routers needing to know the end-route details.

To achieve this simplification, routes are aggregated to tags with varying granularity. For example, a single tag might represent one application-to-application flow or hundreds of routes, effectively reducing the complexity within the tag switching environment.

Under Cisco's scheme, devices called *tag switches* route packets at the Network layer. Each tag switch maintains a table of tag-to-route bindings. A tag has only local significance; a tag switch tracks incoming and outgoing tags for all routes it can reach, and it swaps an incoming tag with an outgoing tag as it forwards packet information.

Since tag switches do not need to read as far into a packet as a traditional router does, this allows them to forward packets more quickly. In addition, tag switching involves only one reference to memory, whereas a traditional router performs 4 to 16 memory references during a single IP routing lookup. Tag switches forward packets on a hop-by-hop basis, however, just like traditional routers.

Tag switching is a flexible scheme in that the tags can represent destination-based routing, QoS, multicasting, or other information. For example, tags could be used to manually define routes for load sharing, to establish a secure path, or for QoS support.

Cisco has even proposed a multilevel system of tags to indicate route information within a routing domain (interior routing) and across domains (exterior routing). This decoupling of interior and exterior routing means tag switches in the middle of a routing domain would need to track less routing information. That, in turn, helps the technology scale to handle large networks, such as the Internet.

Tag switching could provide a similar benefit to corporate users that have large ATM-based backbones with routers as edge devices. As the network grows and more routers are added, in this design each router might need additional memory to keep up with the increasing size of the routing tables. Tag switching alleviates this problem by having the ATM switches use the same routing protocols as routers. In this way, the routers on the edge of the backbone and the ATM-based tag switches in the core would maintain summarized routing information and only need to know how to get to their nearest neighbor—not to all peers on the network.

Because it is essentially a routing scheme, tag switching does not explicitly support VLANs. Tag switches will make a forwarding decision based on the tag and ignore 802.1q VLAN tags (unless specifically configured to recognize them).

Tag switching also offers benefits to Internet service providers and enterprise network administrators. It allows layer 2 switches to participate in layer 3 routing. This increases network scalability because it reduces the number of routing peers that each edge router must deal with. It also enables new traffic tuning mechanisms in router-based networks by integrating virtual circuit capabilities available previously only in layer 2 fabrics. With tag switching, packet flows can be directed across the router network along predetermined paths, similar to virtual circuits, rather than along the hop-by-hop routes of normal routed networks. This allows routers to perform advanced traffic management tasks, such as load balancing, in the same manner as ATM or frame relay switches.

Finally, tag switching can be applied not only to the IP networks, but to any other Network layer protocol. This is because tag switching is independent of the routing protocols employed. While the Internet runs on IP, a lot of campus backbone traffic is transported on protocols such as IPX, making a pure IP solution inadequate for most organizations.

Extensions to tag switching allow tag-switched networks to set up multiple paths across the network and offer differentiated QoS to packet flows. In this way, networks can support real-time packet traffic in support of multimedia applications.

Two mechanisms are needed to provide a quality of service to packets passing through a router or a tag switch. First, the packets must be categorized into different classes. Second, the handling of the packets must be such that the appropriate QoS characteristics (bandwidth, loss, etc.) are provided to each class.

Tag switching provides an easy way to mark packets as belonging to a particular class after they have been classified the first time. Initial classification is done using information carried in the Network- or higher-layer headers. A tag corresponding to the resultant class would then be applied to the packet. Tagged packets can then be efficiently handled by the tag switching routers in their path without needing to be reclassified.

Ascend IP Navigator. Cascade Communications developed IP Navigator, a carrier-class technology for delivering IP scalability to the wide area network. The company was acquired by Ascend Communications in July, 1997. Like tag switching, Ascend's IP Navigator architecture also focuses on the WAN, but it addresses the virtual circuit (VC) scaling issues differently. IP Navigator addresses the VC scaling issue at the Physical instead of Network layer. IP Navigator controls the number of VCs by defining a new type of VC, a multipoint-to-point (MPT), tree-structured VC. A single MPT VC's trunk is associated with each edge WAN switch, dramatically reducing the total potential number of VCs in the core to *N*, where *N* is the number of edge switches (each with many sites) instead of *N2*, where this *N2* is the square of the number of attached sites.

With IP Navigator, the WAN core now needs only layer 1 switches, which minimizes the number of layer 3 routing hops across the WAN, as well as significantly reducing end-to-end latency and jitter. Also, because the number of MPT VCs is kept relatively small, it becomes practical to later add additional MPT VCs dedicated to guaranteed levels of service.

Migration and integration strategies. Tag switching is not constrained to a particular Network layer protocol; it is a multiprotocol solution. The forwarding component of tag switching is simple enough to facilitate high-performance forwarding, and may be implemented on high-performance forwarding hardware such as ATM switches. The control component is flexible enough to support a wide variety of routing functions, such as destination-based routing, multicast routing, hierarchy of routing knowledge, and explicitly defined routes. By allowing a wide range of forwarding granularities that could be associated with a tag, both scalable and functionally rich routing are provided.

Because tag switching is performed between a pair of adjacent tag switches, and because the tag binding information could be distributed on a pair-wise basis, tag switching could be introduced in a fairly simple, incremental fashion.

For example, once a pair of adjacent routers are converted into tag switches, each of the switches would tag packets destined to the other, thus enabling tag switching at the other switch. Since tag switches use the same routing protocols as routers, the introduction of tag switches has no impact on routers. In fact, a tag switch connected to a router acts just like a router from the router's perspective.

As more and more routers are upgraded to enable tag switching, the scope of functionality provided by tag switching widens. For example, once all the routers within a domain are upgraded to support tag switching, it becomes possible to start using the hierarchy of routing knowledge function.

At the wide area, 3Com and Cascade have agreed to understand each other's Network to Data Link layer translations through use of Ipsilon's IFMP. This implicit endorsement of IFMP is great news for Ipsilon. Whether it is a good idea to extend the 3Com approach through the WAN will be determined by the number of machines and the possibilities for misbehaving protocols. Chances for protocol misbehavior increase as the overall layer 2 connected network becomes more complicated.

With regard to integration, 3Com, Cascade, and IBM have announced their commitment to work together on the industry's first and only fully interoperable LAN-to-WAN IP switching solution. This will connect Fast IP to more than 70 percent of all corporate data networks with the rapidly emerging public switched Internet, cost-effectively accelerating the performance of corporate networks end-to-end across LANs and the Internet.

Summary. The Internet has become a standard means of conducting day-to-day business for many corporations who find that the ubiquitous connectivity, wealth of information, and user-friendly application access that it affords can significantly enhance productivity. At the same time, corporate intranets are being built for conducting internal company business, using the same technology and providing similar enhanced productivity benefits. Besides business use, leisure-time access to the Internet by individuals looking for information or entertainment accounts for a growing percentage of IP usage. And new users join the Internet community by the hour.

More users, more complex applications, and increased traffic volumes have combined to put incredible strains on IP backbone networks. From a performance standpoint, the Internet and, increasingly, corporate intranets, are clearly in trouble. The routers running on these networks are not keeping up with the increased processing and traffic loads. Although Cisco's tag switching architecture is a promising solution, it faces potential hurdles, such as receiving widespread vendor support. Tag switching requires all devices in the network to support proprietary Cisco protocols such as the tag distribution protocol (TDP) to realize any particular benefit. In the Internet, there are several vendors that must agree to support these protocols. This is unlikely, as these vendors are competitors. Eventually, however, there must be a convergence of approaches that allows the full benefits of each to be realized.

See also

MULTIMEDIA NETWORKING
TAG SWITCHING

Network agents

Intelligent agents are special programs that accomplish specified tasks by executing commands remotely. Network managers can create and use intelligent agents to execute critical processes that include performance monitoring, fault detection and restoral, and asset management. The agent-manager concept is not a new one. The manager-agent relationship is intrinsic to most standard network management protocols, including the Simple Network Management Protocol (SNMP) used to manage TCP/IP networks. In fact, SNMP agents are widely available for all kinds of network devices, including bridges, routers, hubs, multiplexers, and switches.

In the SNMP world, agents respond to polls from a management station that requests information on the operational status of the various devices on the network. Based on that information, agents are then directed by the management station to get more data, set management variables, or generate traps when specified events occur. To retrieve the collected data, however, the agents must be polled by central management software, a process that increases network traffic. On wide area networks, which increasingly are being burdened with multimedia and other delay-sensitive applications, traffic from continuous polling and the resultant data transfers can degrade network performance to deteriorate. So-called "intelligent agents" address this problem.

What makes these agents so smart is the addition of programming code containing rules that tell them exactly what to do, how to do it, and when to do it. In essence, the intelligent agent plays the dual role of manager and agent. Under this rules- based scheme, polling is localized, events and alarms are collected and correlated, various tasks are automated, and only the most relevant information is forwarded to the central management station (Fig. N1). In the process, network traffic is greatly reduced and problems are resolved more quickly.

Applications. With intelligent agents, management hierarchies can be established with flexible management responsibility among distributed network management personnel. This allows problems to be passed upward as necessary from one level in the hierarchy to another for rapid resolution of severe problems. This capability, based on a set of predetermined policies and rules, helps organizations take full advantage of their management personnel's expertise, regardless of where they might be located.

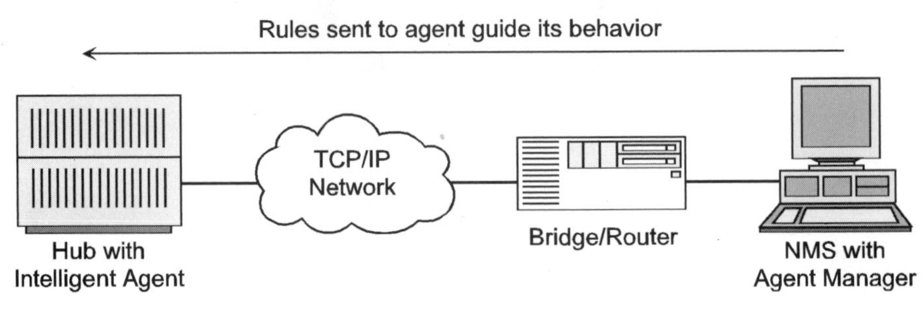

Fig. N1 *Rules sent to the agent tell it what to do and when to do it, so that only relevant information is sent to the network management system.*

For example, an international bank might have the need to control computing services for hundreds of branch offices throughout Europe. To eliminate the high expense of maintaining local operators at each branch office, the bank can deploy intelligent agents at its branches, each of which is monitored by operators at one of four regional management centers. To complete the management hierarchy, the bank can establish one European escalation center to handle any problems the regional centers cannot resolve.

Any problem that the intelligent agents cannot resolve at the branch level is sent automatically to the appropriate operator at a regional management center. If a redundant link fails to come online as a result of service degradation on the primary link, for example, an alert would be sent to a network specialist, along with all the relevant performance data gathered by the agent. If the regional operator cannot solve the problem, it is forwarded to the European escalation center for resolution, along with the supporting data.

In addition to playing a key role in coordinating problem resolution among levels of a management hierarchy, intelligent agents allow organizations to establish management centers-of-expertise in different geographical locations. The intelligent agents would differentiate among the various types of problems and send them to the most appropriate experts, regardless of location. For example, if an organization's database specialists reside in a New York management center, all database-related problems occurring nationwide in the United States could be directed to the management consoles at the New York site. Alternatively, all network-related events could be directed to the Chicago site, where there are network specialists.

Management solutions based on the use of intelligent agents also allow organizations to juggle the shifting of management control and the flow of information among multiple management centers. A large company, for example, might have major management centers in New York and Los Angeles that share responsibility for keeping a nationwide network running. This type of organization can preconfigure its management environment so that, based on the time of day, control is shifted automatically from one management center to another.

As the management shift ends in New York, for example, control can pass to the management center in Los Angeles. With this action, each distributed intelligent agent would stop reporting status information to New York and begin reporting status information to the Los Angeles management center, which then would be responsible for problem resolution. Applied to a global corporation, this capability provides around-the-clock worldwide support with only one operations shift working at any given time.

Data collection. Many data collection tasks can be automated using intelligent agents. Specifically, intelligent agents can provide valuable automated assistance in the areas of performance management, fault management, capacity planning and reporting, and security management.

Performance management. Network performance monitoring can help determine network service level objectives by providing measurements to help managers understand typical network behavior and normal periods. The following capabilities of intelligent agents are particularly useful for building a network performance profile:

- *Baselining and network trending* identifies the true parameters of the network by defining typical and normal behavior. Baselining also provides long-term measurements to check service level objectives and show out-of-norm conditions, which, if left unchecked, may adversely impact the productivity of network users.
- *Application usage and analysis* identifies the overall load of network traffic; what times of the day certain applications load the network; which applications are running between critical servers and clients; and what their load is throughout the day, week, and month.
- *Client-server performance analysis* identifies which servers may be overutilized, which clients are hogging server resources, and what applications or protocols they are running.
- *Internetwork performance* identifies traffic rates between subnets so the network manager can find out which nodes are using WAN links to communicate. This information can be used to define typical throughput rates between interconnected devices.
- *Data correlation* allows peak network usage intervals to be selected throughout the day to determine which nodes were contributing most to the network load. Traffic source and associated destinations can be determined with seven-layer protocol identification.

Fault management. When faults occur on the network, problems must be resolved quickly to decrease the negative impact on user productivity. The following capabilities of intelligent agents can be used to gather and sort the data needed to identify quickly the cause of faults on the network:

- *Packet interrogation* isolates the actual session that is causing the network problem, allowing the network manager to get to the heart of the problem quickly.
- *Data correlation* addresses the fact that, since managers cannot always be on constant watch for network faults, it is important to have historical data

available that provides views of key network metrics at the time of the fault. Such metrics can be used to answer such questions as what was the overall error/packet rate and the types of errors that occurred, what applications were running at the time of the fault, which servers were most active, which clients were accessing these active servers, and which applications were they running?

- *Top error-generator identification* highlights the network nodes that are generating the faults and contributing to problems such as bottlenecks caused by errors and network down time.
- *Immediate fault notification* allows managers to learn instantly when a problem is occurring, before the users do. Proactive alarms help detect and solve the problem as it is happening.
- *Automated resolution procedures* configure the intelligent agents to fix a problem automatically when it occurs. The agent even can be programmed to automatically e-mail or notify help desk personnel with on-screen instructions on how to solve the problem.

Capacity planning and reporting. Capacity planning and reporting allows for the collection and evaluation of information needed to make informed decisions about how to respond to network growth. For this purpose, the following capabilities of intelligent agents are useful:

- *Baselining capability* allows the network manager to determine the true operating parameters of the network.
- *Load balancing capability* allows the network manager to compare at once internetwork service objectives from multiple sites in order to determine which subnets are over- or underutilized. It also helps the network manager discover which subnets can sustain increased growth and which require immediate attention for possible upgrade.
- *Protocol and application distribution capabilities* can help the network manager understand which applications have outgrown which domains or subnets. For example, these capabilities can find out if certain applications are continuously taking up more precious bandwidth and resources throughout the enterprise. With this kind of information, the network manager can better plan for the future.
- *Host load balancing capability* allows the network manager to obtain a list of the top network-wide servers and clients using mission-critical applications. For example, the information collected from intelligent agents might reveal if specific servers always dominate precious LAN or WAN bandwidth, or spot when a CPU is becoming overloaded. In either case, an agent on the LAN segment, WAN device, or host can initiate load balancing automatically when predefined performance thresholds are met.
- *Traffic profile optimization capability* ensures adequate service-level performance, giving network managers the valuable ability to compare actual network configurations against proposed configurations. From the information gathered and reported by intelligent agents, traffic profiles can be developed that allow what-if scenarios to be put together and tested before incurring the cost of physically redesigning the network.

For example, within the Bay Networks Optivity Enterprise family of network management products is the Optivity Design and Analysis suite of network design and optimization applications for Ethernet and Token-Ring environments. Among the tools available in this suite is DesignMan, which performs simulation activities using live traffic information gathered by embedded management agents on the network (Fig. N2).

Security management. A properly functioning and secure corporate network plays a key role in maintaining an organization's competitive advantage. Setting up security objectives related to network access must be considered before mission-critical applications are run over "untrusted" networks, particularly the In-

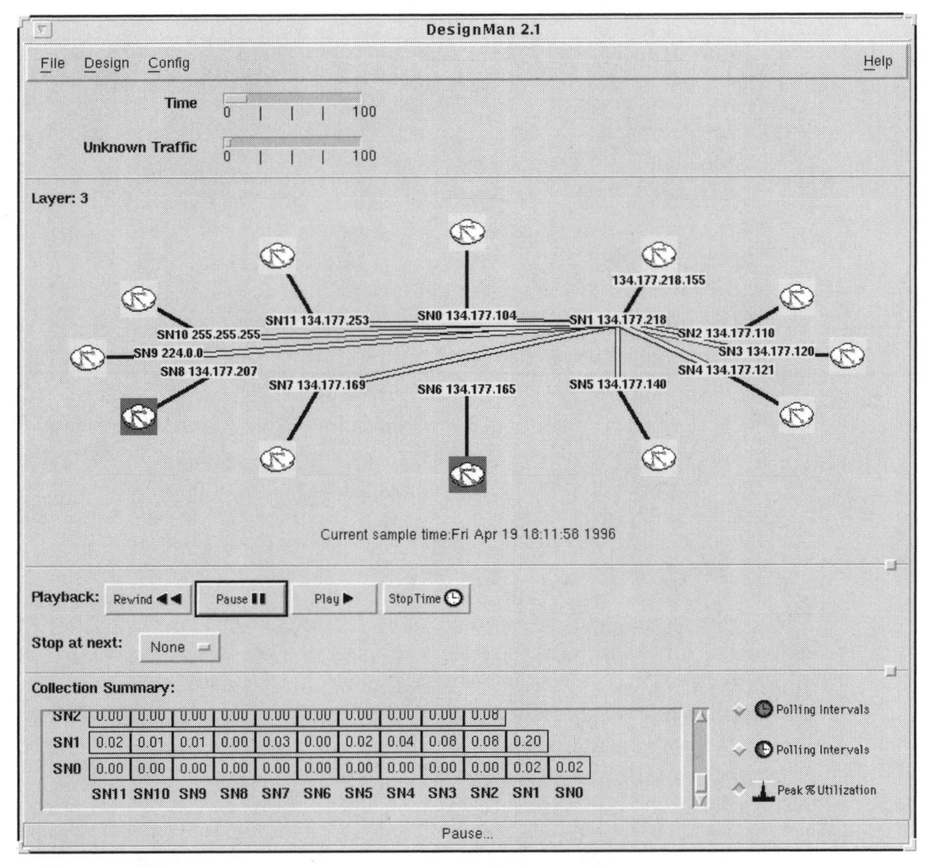

Fig. N2 *Operating at OSI layer 3 (the Network layer), the Bay Networks DesignMan application shows the traffic flow between logically connected subnets. The user can apply what-if scenarios to the traffic data collected by intelligent agents to see the effect of moving a server, for example, from one subnet to another. The application uses a VCR metaphor, allowing the user to play, pause, stop, and rewind the scenario to view its impact on the entire network.*

ternet. The following capabilities of intelligent agents can help discover holes in network security by continuously monitoring network access:

- *Monitor effects of firewall configurations:* By monitoring firewall traffic, the network manager can determine if the firewall is functioning properly. For example, if the firewall was just programmed to disallow access to a corporate host via Telnet, but the program's syntax is wrong, the intelligent agent will report this fact immediately.
- *Show access to and from secure subnets:* By monitoring access from internal and external sites to secure data centers or subnets, the network manager can set up security service-level objectives and firewall configurations based on the findings. For example, the information reported by the intelligent agent can be used to determine whether external sites (and which ones) should have access to the company's database servers or legacy hosts.
- *Trigger packet capture of network security signatures:* Intelligent agents can be set up to issue alarms and automatically capture packets upon the occurrence of external intrusions or unauthorized application access. This information can be used to track down the source of security breaches. Some intelligent agents even have the ability to initiate a trace procedure to discover a breach's point of origination.
- *Show access to secure servers and nodes with data correlation:* This capability reveals which external or internal nodes are accessing potentially secure servers or nodes and identifies which applications they are running.
- *Show applications running on secure nets with application monitoring:* This capability evaluates applications and protocol use on secure networks, or traffic components to and from secure nodes.
- *Watch protocol and application use throughout the enterprise:* This capability allows the network manager to select applications or protocols for monitoring by the intelligent agent so that the flow of information throughout the enterprise can be viewed. This information can identify who is browsing the Web, accessing database client-server applications, or running Lotus Notes, for example.

Intelligent agents can be used for a variety of other tasks, including Internet-related tasks. They can monitor information logged by servers on the World Wide Web, for example. When the log entries exceed a designated threshold, it may indicate a high demand for applications and impending congestion if the logging rate continues. An intelligent agent can act on this information to redirect traffic to another server to balance the load across the available Web servers.

The intelligent agents themselves typically are bundled with the vendor's products. Whenever a new device is placed on the network, its agent is automatically discovered by the central management system. Many vendors post new versions of their agents on the Internet or BBS-type services such as CompuServe. They then can be loaded remotely to the network devices via Telnet connections established through the network management system.

Summary. Intelligent agents have in recent years proven to be indispensable tools for providing network management assistance. In many cases, intelligent agents can implement restoral actions automatically in response to certain threshold indications. These actions can be as simple as resetting a device by turning it off and then back on. At other times, the restoral action might consist of balancing the load across multiple lines or servers to avoid impending congestion, or putting into service a backup line when the primary line degrades or fails. Problems can be identified and resolved locally by intelligent agents, rather than by harried operators at a central management console or technicians sent to remote locations, both of which are expensive and time-consuming.

See also

NETWORK MANAGEMENT SYSTEMS
SIMPLE NETWORK MANAGEMENT PROTOCOL

Network backup

Network backup is not a simple matter for most businesses. One reason is that it is difficult to find a backup system capable of supporting different network operating systems and data, especially if midrange systems and mainframes are involved. Protecting mission-critical data stored on LANs requires backup procedures that are well-defined and rigorous. These procedures include backing up data in a proper rotation, using proper media, and testing the data to ensure that it can be restored easily and quickly in an emergency. Enterprise-wide backups are especially problematic. This is because typically there are multiple servers and operating systems, as well as isolated workstations that often hold mission-critical data. Moreover, the network and client-server environments have special backup needs: Back up too often and throughput suffers. Back up too infrequently and data can be lost.

Backup procedures. Deciding which files to back up can be more complicated than picking the right storage medium. The most thorough backup is a full backup in which every file on every server is copied to one or more tapes or disks. However, the size of most databases makes this impractical to do more than once a month.

Incremental backups copy only files that have changed since the last backup. Although this is faster, it requires careful management because each tape may contain different files. Restore a system from incremental backups requires playing back (in the right order) all the incremental backups made since the last full backup.

Differential backups split the difference between full and incremental techniques. Like an incremental backup, a differential backup requires a tape with the full set of files. Each differential tape contains all the files that have changed since the last full backup, however, so restoration requires just that full set and the most recent differential.

Scheduling and automation. The scheduling of backups is determined by several factors, including the criticality of applications, network availability, and legal requirements. Network backup software with calendar-based planning features allows the system administrator to do such things as schedule the weekly archiving of all files on LAN-attached workstations. The backup can be scheduled for nonbusiness hours, both to avoid disrupting user applications and to avoid network congestion.

Some scheduling tools allow the system administrator to set precise parameters with regard to network backups. For example, the backup can target only files that have not been accessed in the preceding 60 days, with the objective of freeing at least 100 megabytes of disk space on a particular server. When the backup is complete, a report is generated listing the files that have been archived to tape, along with their file sizes and dates of last access. The total number of bytes also is provided, allowing the system administrator to confirm that at least 100 megabytes of storage has been freed on the server.

Event-based scheduling allows the system administrator to run predefined workloads when dynamic events occur in the system, such as the close of a specific file or the start or termination of a job. With regard to network backups, the administrator can decide what events to monitor and what the automated response to those events will be. For example, the administrator can decide to archive all files in a directory after the last print job, or do a database update after closing a particular spreadsheet.

Although most network backup programs can grab files from individual workstations on the LAN, there might be thousands of users with similar or identical system configurations. Instead of backing up 1000 copies of Windows 95, for example, the network backup program can be directed to copy only each user's system configuration files; if a workstation experiences a disk crash, a new copy of Windows 95 can be downloaded from the server, along with the user's applications, data, and configuration files.

With the right management tools, network backup can be automated under centralized control. Such tools can go a long way toward lowering operating and resource costs by reducing time spent on backup and recovery. These tools enhance media management by providing overwrite protection, log file analysis, media labeling, and the ability to recycle backup media. In addition, the journaling and scheduling capabilities of some tools relieve the operator of the time-consuming tasks of tracking, logging, and rescheduling network and system backups.

Another useful feature of such tools is data compression, which reduces media costs by increasing media capacity. This automated feature also increases backup performance while reducing network traffic.

When these tools are integrated with high-level management platforms (such as Hewlett-Packard's OpenView or IBM's NetView, or operating systems such as Sun's Solaris) problems or errors that occur during automated network backup are reported to the central management console. The console operator is notified of any problem or error via a color change of the respective backup application symbol on the network map. By clicking on the symbol, the operator can directly access the network backup application to determine the cause of the problem or correct the error to resume the backup operation.

Capabilities and features. Depending on the particular operating environment, some of the key areas to consider when evaluating network backup software include:

- Storage capacity and data transfer rate of the backup system.
- A fast-start capability that allows full network backups to be performed immediately, with fine-tuning of the backup parameters done later.

- The ability to back up the NetWare bindery (if applicable), security information, and file and directory attributes.
- Support for multiple file systems including NetWare, the Apple File Protocol (AFP), OS/2 High Performance File System (HPFS), Sun Microsystems Network File System (NFS), and OSI File Transfer, Access, and Management (FTAM).
- Ability to back up, monitor, and log the activities of multiple file servers simultaneously.
- Tape labeling, rotation, and script file schemes for automating the backup and recovery process.
- Reporting and audit log capability.
- Fast-search capability that allows a system administrator to easily and quickly find and retrieve files.
- File archiving and grooming methods that allow automatic file and directory storage, including the ability to delete data that has not been accessed for a specified period of time.
- Integrated network virus protection when backup and restore operations are performed.
- Security features that limit access to backups to only authorized users.

One of the latest features of network backup software is the availability of agents that enables such programs to bypass operating system constraints to store files that are still open, even if the application is accessing or updating them while the backup is in progress. This capability eliminates incomplete backups that typically result when files are not closed. It is of particular value to organizations that need around-the-clock access to information while performing complete backups.

When evaluating software for LAN backups in the mainframe environment, some of the key areas to consider include:

- Whether all or most of the platforms at the server level are supported, such as LAN Manager, NetWare, and Unix.
- Whether all or most workstation platforms are supported, such as DOS/Windows, OS/2, Unix, and Macintosh.
- Whether non-LAN PCs are supported, such as those with 3270 emulation cards with direct connections to controllers or FEPs.
- Whether other WAN connections are supported, or just TCP/IP.
- Whether users are able to set windows of availability to force backups and recoveries to take place during nonpeak hours.
- Whether the product supports options for restores to be performed by the central administrator, the LAN administrator, or individual workstation users.
- Whether the product supports both a command-line interface for expert use and a graphical user interface for end-user access.
- Whether the product supports automatic archiving of files that have not been accessed for a specified period, thus freeing up server or workstation disk space.

- Whether the product supports the skipping of redundant files in the backup process.
- Whether the product supports other features such as heterogeneous file transfers, remote command execution, and job submission from PC to host.

Summary. As LANs continue to carry increasing volumes of critical data in varying file formats, vendors continue to push the limits of backup technology. On the software side, the trend is toward increasing levels of intelligence. Backup systems must ensure not only that files are backed up, but that they are easily located and restored. Systems intelligence has already progressed to the point where the user need not know the tape, the location on the tape, or even the exact name of a lost file in order to restore it.

See also

HIERARCHICAL STORAGE MANAGEMENT
STORAGE MEDIA

Network computing

The concept of network computing originated with Oracle Corporation, whose Network Computing Architecture embraces the Internet, client-server, and legacy systems to provide a platform capable of dynamically linking all of these technologies together. Through the architecture, existing client-server applications can take advantage of Web technology with minimal change. Web applications can integrate seamlessly and take advantage of existing client-server systems without modification.

Java is expected to play a key role in this net-centric architecture. Java was designed by Sun Microsystems to provide a cleaner, simpler language that could be processed faster and more efficiently than C or C++ on nearly any microprocessor. Whereas C or C++ source code is optimized for a particular platform, Java source code is compiled into a universal format; the compiler writes for a virtual machine in the form of simple binary instructions. Compiled byte code is executed by a Java run-time interpreter, performing all the usual activities of a real processor, but within a safe, virtual environment instead of a particular computer platform.

Although Java can be used for real programming tasks such as building networked applications and creating functional user interfaces, most of the excitement surrounding Java right now centers around its capabilities for building special Web applications called *applets*. These applets range from legacy data gateways to interactive databases for electronic commerce and online banking.

The reason why Java applications development centers around the Web is that the Internet is increasingly being considered as the foundation for network computing, providing an economical way to access corporate information from remote locations and mobile computers. Under this concept, "thin" clients (i.e., network computers) rely on the Internet for running applications. A Java-enabled Web browser is used as the interface. The use of thin clients allows companies to save money by equipping their work force with less expensive computers, minimizing the number of copies of application software, and streamlining software maintenance tasks.

Role of applets. Applets are programs that are meant to be embedded in and controlled by a larger application, such as a Java-enabled Web browser. Netscape Navigator or Microsoft's Internet Explorer have built-in Java interpreters.

With a Java-enabled browser, Web users can take advantage of all the functionality offered by the applets. Through a requisition applet, for example, Web users can have easy access to a company's product catalog and order merchandise online via an electronic order form. Once the user has completed a purchase order, it is automatically routed for approval and processing through a workflow application until it reaches the shipping department, where the order is filled. A copy of the completed order is routed to the customer service database. The advantage of using Java in this case is that the Web user need not download the entire workflow application to his or her desktop, only the applet required to collect the order information.

Java is being used as the foundation for developing interactive trading, insurance, investment planning, and stock-quote applications that can be accessed over the Web. Java applets are being used for implementing online banking, allowing customers to download their account information and interactively conduct bank business. Java also is being used by transportation companies to access shipping documents (bills of lading, container manifests, shipment routings, etc.) from Web browsers.

Fat vs. thin clients. The terms "fat" and "thin" refer primarily to the amount of processing being performed at the client. Terminals are the ultimate thin clients because they rely exclusively on the server or host for applications and processing. Stand- alone PCs are the ultimate fat clients because they have the resources to run all applications locally and handle the processing themselves. Spanning the continuum from all-server processing to all-client processing is the client-server environment, where there is a distribution of work between the different processors.

Client-server once was thought to be the ideal computing solution. Despite the initial promises held out for client-server solutions, today there is much dissatisfaction with their implementation. Client-server solutions are too complex, desktops are too expensive to administer and upgrade, and the applications still are not sufficiently secure and reliable. Furthermore, client-server applications take too long to develop and deploy, and incompatible desktops prevent universal access. The network computing paradigm appears to overcome these limitations.

Role of the virtual machine. Applets are essentially small applications designed to be distributed over the Internet. As such, they are always hosted by another program such as Netscape's Navigator or Microsoft's Internet Explorer, both of which contain a *virtual machine* (VM) that runs the Java code. Because the Java code is written for the virtual machine rather than for a particular computer or operating system, all Java programs are cross-platform applications by default.

Java applications are fast because today's processors can provide efficient virtual machine execution. The performance of GUI functions and graphical applications are enhanced through Java's integral multithreading capability and just-in-time (JIT) compilation. The applications also are more secure than those running native code because the Java runtime system (part of the virtual machine) checks all code for viruses and tampering before running it. Application development is facil-

itated through code reuse, making it easier to test and faster to deploy them on the Internet or corporate intranet.

Because Java applications originate at the server, clients only get the code when they need to run the application. If there are changes to the applications, they are made at the server. Programmers and network administrators do not have to worry about distributing the changes to every client. The next time the client logs into the server and accesses the application, it automatically gets the most current code. This method of delivering applications also reduces support costs.

Accessing legacy data. Java applets running on remote or mobile computers can perform emulation, providing easy access to legacy data on mainframe and minicomputer hosts via a remote access gateway on the corporate intranet or the Internet. Although a locally installed emulator may have the same capabilities as a network-delivered, Java-based emulator, there is more work to be done in installing and configuring the local emulator than the Java-based emulator that is delivered each time a user needs it. The same local emulator takes up local disk space whether it is being used or not. The Java-based emulator, in contrast, takes no local disk space. It remains in the browser's cache, which is cleared when the emulator is no longer needed.

The Java emulator supports all standard emulation features, including:

- Menu items for common keyboard functions
- Button bar
- Configuration of terminal options based on user needs
- Font size that changes when window is resized
- Cut-and-paste options
- Color
- Online help

For added flexibility, the user can open up multiple resizable environment windows. The configuration file determines whether the Java emulator initially opens as a separate window. The user also can create separate windows with the Web browser's NEW button. The Java emulator also supports *hot spots* for commands and menu selections. In response to a user clicking on a hot spot, the Java emulator treats the action as if the user had pressed the equivalent command button or function key, or had entered the menu selection and pressed the Enter key.

Summary. With corporate intranets and the greater Internet becoming increasingly important elements in information distribution and service delivery, corporations are turning to Java for applications development and network computers for cost savings. The use of Java allows remote users and mobile computing professionals to access corporate applications and legacy data through host emulators. The applications and data are downloaded from a gateway server as needed. This not only gives users access to the information they need, but simplifies the software maintenance tasks of network administrators, since emulators and other applications reside only at designated servers.

See also

CLIENT-SERVER NETWORKS
INTERNET

Network design

To stay competitive, companies are relying on their networks as never before. Typically these networks consist of different kinds of transmission facilities, LAN technologies, protocols, and standards—all cobbled together to meet the differing needs of workgroups, departments, branch offices, divisions, subsidiaries, and, increasingly, strategic partners, suppliers, and customers. Building such networks presents special design challenges.

Fortunately, various automated design tools have become available in recent years. With built-in intelligence, these tools take an active part in the design process, from building a computerized model of the network, validating its design, and gauging its performance, to quantifying equipment requirements and exploring reliability and security issues before the purchase and installation of any network component. Even faulty equipment configurations, design flaws, and standards violations are identified in the design process.

The design process usually starts by opening a blank drawing window in the design tool, into which various vendor-specific devices (workstations, servers, hubs, routers, etc.) can be dragged from a product library and dropped into place (Fig. N3). The devices are further defined by type of component, software, and protocols, as appropriate. By drawing lines, the devices are linked to form a network, with each link assigned physical and logical attributes. Rapid prototyping is aided by the ability to copy objects(devices, LAN segments, network nodes, and subnets) from one drawing to the next, editing as necessary, until the entire network is built. Along the way, various simulations can be run to test virtually any aspect of the design.

The autodiscovery capabilities found in such management platforms as Hewlett-Packard's OpenView, IBM's NetView/6000, and Sun's Solstice SunNet Manager, which automatically detect various network elements and represent them with icons on a topology map, are often useful in accumulating the raw data for network design. Some stand-alone design tools allow designers to import this data from network management systems, which eases the task of initial data compilation. Although these network management systems offer some useful design capabilities, they are not as feature-rich as high-end, stand-alone tools that also are able to incorporate a broader range of network technologies and equipment makes and models.

Designing a large, complex network requires a multifaceted tool, ideally one that is graphical, object-oriented, and interactive. It should support the entire network life cycle, starting with the definition of end-user requirements and conceptual design, to the very detailed vendor-specific configuration of network devices, the protocols they use, and the various links among them. At each phase in the design process, the tool should be able to test different design alternatives in terms of cost, performance, and validity. When the design checks out, the tool should spew out network diagrams and a bill of materials—all this before a single equipment vendor or carrier sales rep is contacted or a single RFP is written.

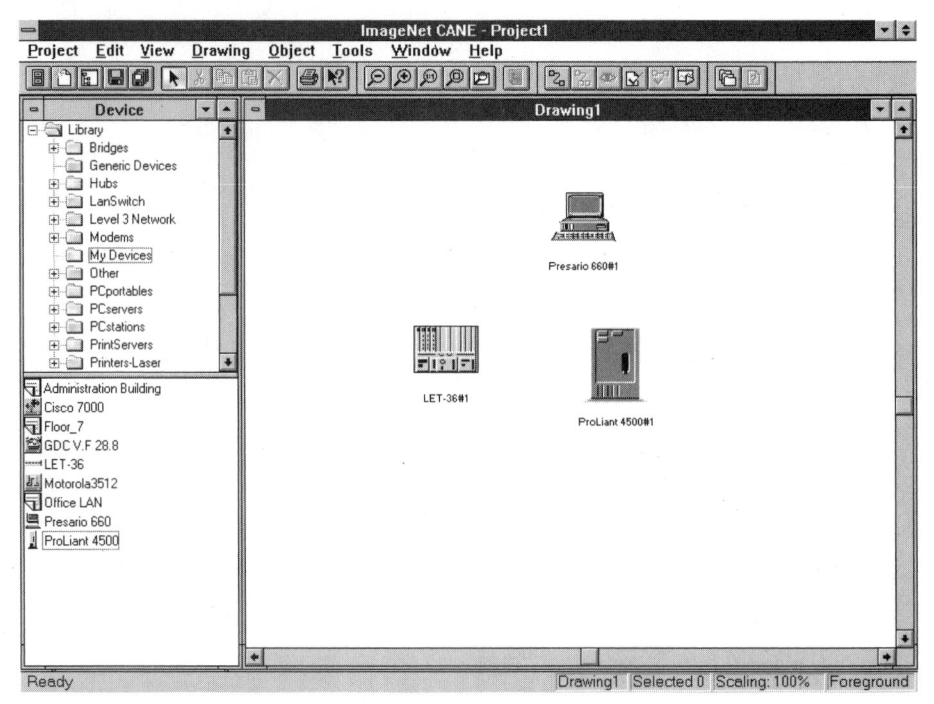

Fig. N3 *A design tool typically provides a workspace into which objects are dragged-and-dropped from device libraries to start the network design process from scratch. Source: ImageNet Ltd.*

With the right tools, modules, and device libraries, every conceivable type of network can be designed, including legacy networks such as SNA and DECnet, voice networks including ISDN, as well as T1, X.25, ATM, and TCP/IP nets. Some tools even take into account the use of satellite, microwave, and other wireless technologies.

The designer can take a top-down or bottom-up approach to building the network. In the former, the designer starts by sketching out the overall network; subsequent drawings add increasing levels of detail until every aspect of the network is eventually fleshed out. The bottom-up approach might start with a LAN on a specific floor of a specific building, with subsequent drawings linked to create the overall network structure.

As the drawing window is populated, devices can be further defined by type of component such as chassis, interface cards, and daughterboards. Even the operating system can be specified. Attributes can be added to each device taken from the library, for example to specify a device's protocol functionality. Once the devices have been configured, a simulation profile is assigned to each device, which specifies its traffic characteristics for purposes of simulating the network's load and capacity.

With each device's configuration defined, lines are drawn between them to form the network. With some design tools, the links can be validated by testing for common protocols and network functions. This prevents NetWare clients from being connected to other clients instead of servers, for example. Such online analysis also can alert the designer to undefined links, unconnected devices, insufficient available ports in a device, and incorrect addresses in IP networks. Some tools even are able to report violations of network integrity and proper network design practices.

The next step in the design process is to simulate the completed network, running it against a database that describes how the actual network devices behave under various real-world conditions. The simulator generates network events over time, based on the type of device and traffic pattern recorded in the simulation profile. This allows the designer to test the network's capacity under various what-if scenarios and fine-tune the network for optimal cost and performance. Simulators can be purchased as stand-alone programs, or may be part of the design tool itself.

Some tools are more adept at designing WANs, particularly those that are based on TDMs (time division multiplexers). With the TDM components library, for example, a designer can build an entire T1 network within specified parameters and constraints. The designer can strive for the lowest transmission cost that supports all traffic, for instance, or strive for line redundancy between all time division multiplexer nodes. By mixing and matching different operating characteristics of various TDM components, overall design objectives can be addressed, simulated, and fine-tuned. Some tools come with a tariff database to price transmission links and determine the most economical network design.

Such tools also might address clocking in the network design. Clocks are used in time division multiplexers to regulate the flow of data on the network. All clocks on the network therefore must be synchronized to ensure the uninterrupted flow of data from one node to another. The design tool automatically generates a network topology synchronization scheme, taking into account any user-defined criteria and ensuring that there are no embedded clock loops.

Summary. Today's computer systems are distributed and the networks are global. With intelligent hubs, switched LANs, sprawling router networks, broadband facilities, and advanced services like frame relay and ATM thrown into the mix, sophisticated tools that are easy to use while enforcing design integrity are required. A new generation of intelligent design tools with built-in error detection, simulation and analysis capabilities, and plug-in modules for ancillary functionality, is now available. These tools do not require managers and planners to be intimately familiar with every aspect of their networks. The essential information can be retrieved on a moment's notice—often with point-and-click ease—analyzed, queried, manipulated, and reanalyzed if necessary, with the results displayed in easy-to-understand graphical form or exported to other applications for further manipulation and study.

See also

NETWORK DRAWING
NETWORK MANAGEMENT SYSTEMS

Network drawing

Network administrators faced with managing detailed and often large quantities of information on local and worldwide corporate networks need tools that can accurately depict these complex infrastructures. While the automatic discovery capabilities of high-end network management systems can help in this regard, they are not very useful for planning and documenting new networks or adding new nodes. A variety of drawing tools have become available that can aid the network design process. They provide the five major features considered critical to network planners:

1. An easy-to-use drawing engine for general graphics,
2. An extensive library of predrawn images representing vendor-specific equipment,
3. A drill-down capability, which allows multiple drawings to be linked to show various views of the network,
4. A database capability to assign descriptive data to the device images, and
5. A high degree of embedded intelligence that makes images easy to create and update.

Most network drawing tools are Windows-based. Many of them allow network designs to be published on the corporate intranet or the public Internet, allowing any authorized user to view them with a Web browser. Some can automatically discover devices on an existing network to ease the task of drawing and documenting the network.

Equipment shapes. A library of shapes typically includes modems, telephones, hubs, PBXs, and DSU/CSUs from different manufacturers. Representations of LANs and WANs, databases, buildings and rooms, satellite dishes, microwave towers, and a variety of line connectors are included. There also are shapes that represent such generic accessories as power supplies, PCs, towers, monitors, keyboards, and switches. There are even shapes for equipment racks, shelves, patch panels, and cable runs (Fig. N4).

Typically an annual subscription provides unlimited access to the hundreds of new network devices, adapters, and accessories added to the device library. Depending on the drawing tool vendor, new objects even might be downloadable from the company's Web site.

While many drawing tools offer exact-replica hardware device images from hundreds of network equipment manufacturers, some tools have embedded intelligence into the shapes, which lets component symbols such as network cards to be snapped into equipment racks and remain in place even when the rack symbol is moved. Each shape includes product-specific attributes, including vendor, product name, description, and part number (Fig. N5). This permits users to generate detailed inventory reports for network asset management.

The shapes are even programmable so they can behave like the objects they represent. This reduces the need for manual adjustments while drawing, and ensures the accuracy of the final diagram. For example, the shape representing an equipment rack from a specific vendor can be programmed to know its own dimensions. When the user populates the drawing with multiple instances of this

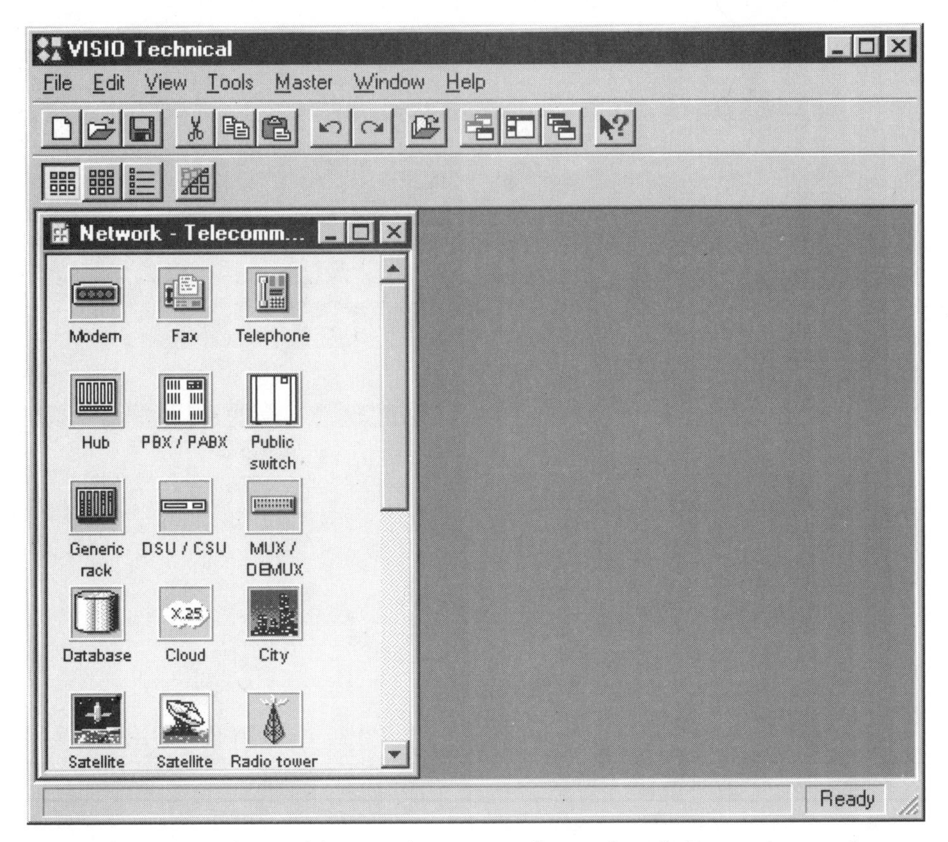

Fig. N4 *From a device library, shapes are dragged and dropped into place as needed to design a whole network or a new node. Source: Visio Corp.*

shape, it could issue an alert if there is a discrepancy between the space available on the floor plan and the space requirements of the equipment racks.

Each shape also can be embedded with detailed information. For example, the user can associate a spreadsheet with any network element—to provide cost information on a new switch node or LAN segment, for example—along with a bar chart to perhaps illustrate cost data by system component. The spreadsheet data can be manipulated until costs fit within budgetary parameters. The changes will be reflected in the bar chart the next time it is opened.

The drawing process. To start a network diagram, typically the user opens the template for the manufacturer whose equipment will be placed in the diagram. This causes a drawing page that contains rules and grid to appear. The drawing page itself can be sized to show the entire network or just a portion of it.

Various other systems and components can be added to the diagram using the drag-and-drop technique. The user has the option of having (or not having) the shapes snap into place within the drawing space so they will be precisely positioned

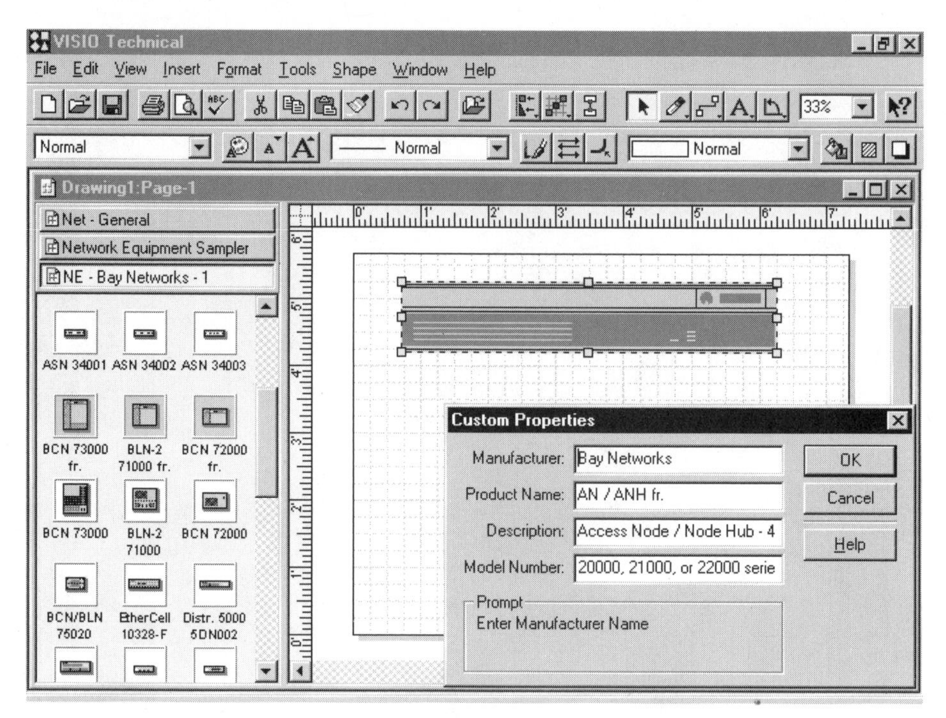

Fig. N5 *Details of various network equipment can be assigned using the Custom Properties dialog box. This information is useful in asset management. Source: Visio Corp.*

on grid lines. Once placed in the drawing space, the shapes can be moved, resized, flipped, rotated, and glued together. Expansion modules, for example, can be dragged into the chassis so that the modules' endpoints glue to the connection points on the chassis expansion slots. This allows the chassis and modules to be moved anywhere in the diagram as a single unit. Via the cut-and-paste method, the user can add as many copies of the component as desired to quickly populate the network drawing.

To show the connections between various systems and components, the user can choose shapes that represent different types of networks, including LANs, X.25, satellite, microwave, and radio. Alternatively, the user can choose to connect the shapes with simple lines that can have square or curved corners.

Each network equipment shape has properties associated with it. Custom properties can be assigned to shapes for use in tracking equipment and generating reports, such as inventories. Text can be added to any network system or component, including a Lotus Notes field specifying font, size, color, style, spacing, indent, and alignment. Text blocks can be moved and resized. Some tools even include a spell checker and a search-and-replace tool. The user can add words not in the standard dictionary that comes with the program. The user can specify a search of the entire drawing, a particular page, or selected text only.

AutoCAD files and clip art can be added to network drawings. The common file formats usually supported for importing graphics from other applications, including Encapsulated PostScript (EPS), Joint Photographic Experts Group (JPEG), Tag Image File Format (TIFF), and ZSoft's PC PaintBrush bitmap (PCX).

The various shapes used in a network drawing can be kept organized using layers. A *layer* is a named category of shapes. For example, the user can assign walls, wiring, and equipment racks to different layers in a space plan. This allows the user to:

- Show, hide, or lock shapes on specific layers so they can be edited without affecting shapes on other layers.
- Select and print shapes based on their layer assignments.
- Temporarily change the display color of all shapes on a layer to make them easier to identify.
- Assign a shape to more than one layer, as well as assign the member shapes of a group to different layers.

The user also can group shapes into customizable stencils. If the same equipment is used at each node in a network, for example, the user can create a stencil containing all the devices. All of the graphics and text associated with each device will be preserved in the newly created stencil. This saves time in drawing large-scale networks, especially those that are based on equipment from a variety of manufacturers.

At any step in the design process, the user can share the results with other network planners by sending copies via e-mail. The diagram is converted to an image file, which is displayed as an icon in the message box, and sent as an attachment. When the recipient receives the message, the attached diagram, including all embedded information, can be opened by clicking on the icon. The document then can be marked up by creating a separate layer for review comments. Using a separate layer for comments protects the original drawing and makes the comments easier to view, print, and color separately from the rest of the drawing.

Some network drawing tools provide a utility that converts network designs and device details into a series of hyperlinked HTML documents that can be accessed over the Web. These documents show device configurations, port usage, and even device photographs. Users can activate the links to navigate from device to device to trace connectivity and review device configurations (Fig. N6). In addition to supporting fault identification, the hyperlinked documents aid in planning design changes.

There are several ways that the network diagrams can be protected against inadvertent changes, especially if they are shared via e-mail or posted on the Web:

- The shapes can be locked to prevent them from being modified in specific ways.
- The attributes of a drawing file (styles, for example) can be protected against modification.
- The file can be saved as read-only, so it cannot be modified in any way.
- The shapes on specific layers can be protected against modification.

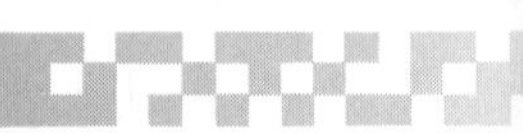

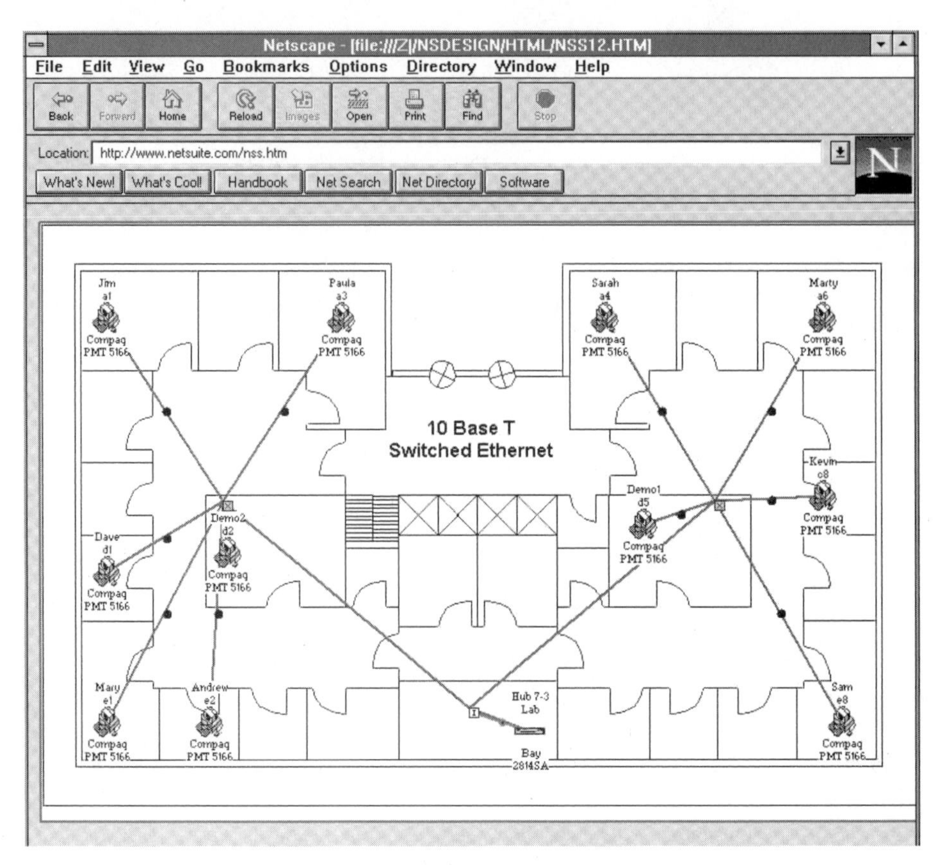

Fig. N6 *This floor plan of a 10Base-T network is a hyperlinked drawing rendered by Netscape Navigator.* *Source: NetSuite Development.*

Users can password-protect their work to prevent attributes of a drawing file from being changed. For example, a background containing standard shapes or settings can be password-protected. Users also can set a password for a drawing's styles, shapes, backgrounds, or masters. A password-protected item can be edited only if the correct password is entered.

Summary. Unlike traditional CAD programs, today's drawing tools are specifically designed for network planners. They can improve communications and productivity with their easy-to-use and easy-to-learn graphics capabilities that offer seamless integration with other applications on the Windows desktop. Their simplified graphical representations of complex projects also allows more people to understand and participate in the planning process. Despite their simplicity, this new generation of drawing tools provides a high degree of intelligence, programmability, and Web awareness that makes them well suited to the demanding needs of network designers.

See also

ASSET MANAGEMENT
INTERNET

NETWORK DESIGN
NETWORK MANAGEMENT SYSTEMS
WORLD WIDE WEB

Network integration

Distinct from systems integration, which focuses on getting different computer systems to talk to each other, network integration is concerned with getting diverse, far-flung local area networks and host systems interconnected over the wide area network. Network integration typically requires that attention be given to a plethora of different physical interfaces, protocols, frame sizes, and data transmission rates. Companies also face a bewildering array of carrier facilities and services to extend the reach of information systems globally over high-speed WANs. To get everything working properly can require a fair amount of hardware and software customization.

A network integrator brings objectivity to the task of tying together diverse products and systems to form a seamless, unified network. To do this, the network integrator draws upon its expertise in IS, office automation, LAN administration, telephony, data communications, and network management systems. Added value is provided through strong business planning, needs analysis, and project management skills, as well as accumulated experience in meeting customer requirements in a variety of industry segments and operating environments.

A qualified network integrator will have in place a stable support infrastructure capable of handling a high degree of ambiguity and complexity, as well as any technical challenge that might step in the way. In addition to financial stability, this support infrastructure includes staff representing a variety of technical and management disciplines, and strategic relationships with specialized companies such as cable installers and software firms.

Integration services. There are a number of discrete services that are provided by network integration firms, including:

- *Design and development* includes such activities as network design, facilities engineering, equipment installation and customization, and acceptance testing.
- *Consulting* includes needs analysis, business planning, systems/network architecture, technology assessment, feasibility studies, RFP development, vendor evaluation and product selection, quality assurance, security auditing, disaster recovery planning, and project management.
- *Systems implementation* includes procurement, documentation, configuration management, contract management, and program management.
- *Facilities management* deals with operations, technical support, hot-line services, move and change management, and trouble ticket administration.
- *Network management* covers network optimization, remote monitoring and diagnostics, network restoral, technician dispatch, and carrier and vendor relations.

Network integration may be performed by in-house technical staff or through an outsourcing arrangement with a computer company, local or interexchange car-

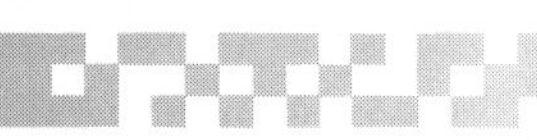

rier, management consulting firm, traditional IS-oriented system integrator, or interconnect vendor.

Each type of firm has specific strengths and weaknesses. The wrong selection can delay the implementation of new information systems and LANs, disrupt network expansion plans, and impede applications development—any of which can inflate operating costs over the long term and have adverse competitive impact.

It is therefore advisable to choose an integrator whose products and services are particularly pivotal to the application. For example, if the network integration application is such that the computer requirements are extremely well defined and no significant computer changes are expected, but a range of new communications services might be involved, a carrier would be a better choice of integrator than a computer vendor. On the other hand, if the project is narrow in scope and the needs are well understood, in-house staff might be able to handle the integration project, with as-needed assistance from the computer firm or carrier.

Summary. While some companies have the expertise required to design and install complex networks, others are turning to network integrators to oversee the process. The evaluation of various integration firms should reveal a well-organized and staffed infrastructure that is enthusiastic about helping the customer reach its networking objectives. This includes having the methodologies already in place, the planning tools already available, and the required expertise already on staff. Beyond that, the integrator must be able to show that its resources have been successfully deployed in previous projects of a similar nature and scope.

See also

OUTSOURCING

Network management systems

The task of keeping multivendor networks operating smoothly with a minimum of downtime is an ongoing challenge for most organizations. While many companies prefer to retain total control over their network resources, others rely on computer vendors and carriers to find and correct problems on their networks, or depend on third-party service firms. Wherever these responsibilities ultimately reside, the tool used for monitoring the status of the network and initiating corrective action is the network management system (NMS).

With an NMS, technicians can remotely diagnose and correct problems associated with each type of device on the network. Although today's network manager is concerned primarily with diagnosing failures, the likelihood of problems occurring can be predicted also, so that traffic can be diverted from failing lines or equipment with little or no inconvenience to users.

Network management begins with such basic hardware components as modems, data sets (CSUs/DSUs), multiplexers, and dial backup units (Fig. N7). Each component typically has the ability to monitor, self-test, and diagnose problems regarding its own operation and report problems to a central management station. The management station operator can initiate test procedures on systems at the other end of a point-to-point line. On more complex multipoint and multidrop configurations, the ability to test and diagnose problems from a central loca-

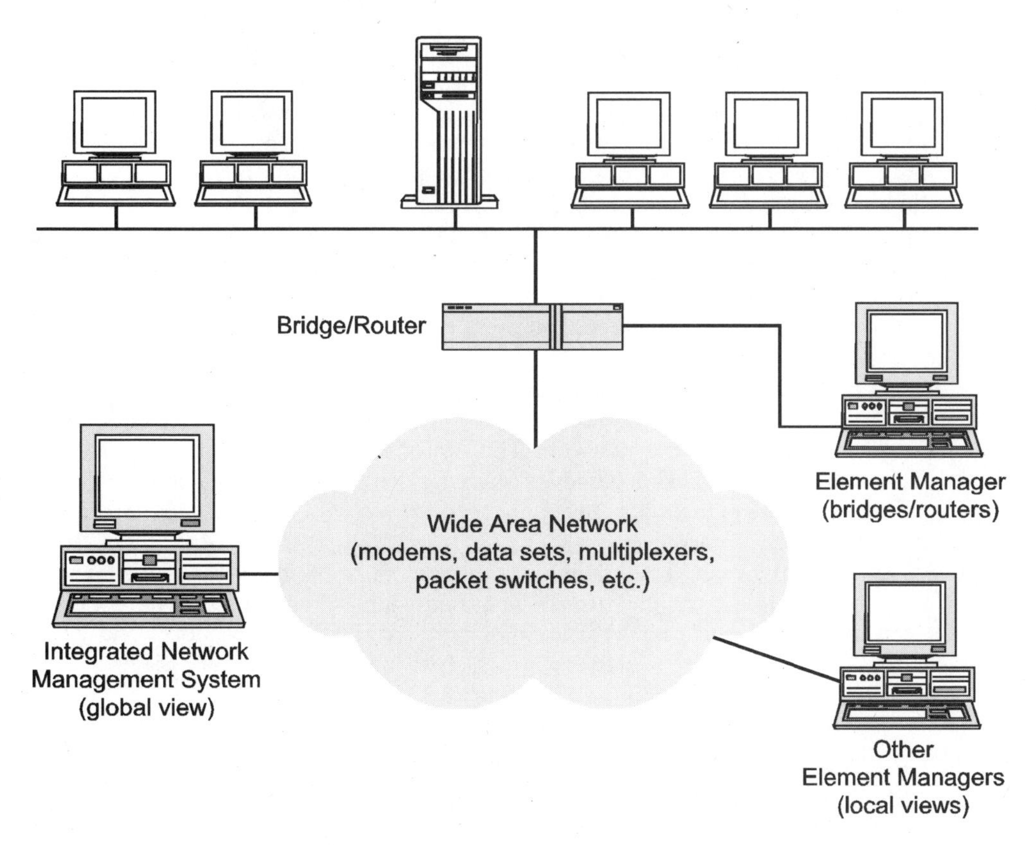

Fig. N7 *Each type of device on the network can have its own Element Management System (EMS), which reports to an Integrated Network Management System (INMS).*

tion greatly facilitates problem resolution. This capability also minimizes the need to dispatch technicians to remote locations, and reduces maintenance costs.

A minimal network management system consists of a central processing unit, system controller, operating system software, storage device, and an operator's console. The central processor can be a minicomputer or microcomputer. The system controller, the heart of the network management system, continuously monitors the network and generates status reports from data received from various network components. The system controller also isolates network faults and restores segments of the network that have failed, or which are in the process of degrading. The controller usually runs on a powerful OS platform such as Unix or Windows NT.

NMS functions. Although differing from vendor to vendor, the basic functions that are common to most network management systems include topology mapping, administration, performance measurement, control and diagnostics, configuration management, applications management, and security. Some network management systems include other functions such as network modeling, for example, which would allow the operator to simulate aggregate, node, or circuit failures to test various disaster recovery scenarios.

Topology mapping. Many network management systems have an automatic discovery capability that finds and identifies all devices or nodes connected to the network. Based on the discovered information, the NMS automatically draws the required topology maps. Nodes that cannot be discovered automatically can be represented in either of two ways: by manually adding custom or standard icons to the appropriate map views, or by using the network management system's SNMP-based application programming interfaces (APIs) for building map applications without having to modify the configuration manually to accommodate non-SNMP devices.

A network map is useful for ascertaining the relationships of various equipment and connections, for keeping accurate inventory of network components, and for isolating problems on the network. The network map is updated automatically when any device is added or removed from the network. Device status is displayed via color changes to the map. Any changes to the network map are carried through to the relevant submaps.

Administration. The administration element of an NMS allows the user to take stock of the network in terms of what hardware is deployed and where it is located. It also tells the user what facilities are serving various locations, and what lines and equipment are available with which to implement alternate routing. The vehicle for storing and using this information is the database.

For administrative tasks, multiple specialized databases that relate to one another are used. One of these databases accumulates trouble ticket information. A *trouble ticket* contains such information as the date and time the problem occurred, the specific devices and facilities involved, the vendor(s) from which they have been purchased or leased, and the service contact. It will also contain the name of the operator who responded to the alarm, any short-term actions taken to resolve the problem, and space for recording follow-up information such as visits from the vendor's service personnel, dates on which parts were returned for repair, serial numbers of spares installed, and the date of the problem's final resolution.

A trouble ticket database can be used for long-term planning. The network manager can call up reports on all outstanding trouble tickets, trouble tickets involving particular segments of the network, trouble tickets recorded or resolved within a given period, trouble tickets involving a specific type of device or vendor, and even trouble tickets over a given period not resolved within a specific time frame. The user may customize report formats to meet unique needs.

Such reports provide network managers with insight on the reliability of a given operator, the performance record of various network components, the timeliness of on-site vendor maintenance and repair services, and the propensity of certain segments of the network to fail. With information on both active and spare parts, network managers can readily support their decisions on purchasing and expansion. In some cases, cost and depreciation information on the network's components are even provided.

Performance measurement. Performance measurement concerns network response time and network availability. Many network management systems measure response time at the local end, from the time the monitoring unit receives a start-of-transmission (STX) or end-of-transmission (EOT) signal from a given unit. Other systems measure end-to-end response time at the remote unit. In either case, the network management system displays and records response time information and generates operator-specified statistics for a particular terminal, line, network segment, or the network as a whole. This information may be reported in real time, or stored for a specified time frame for future reference.

Network personnel can use this information to track down the cause of the delay. When an application exceeds its allotted response time, they can decide whether to reallocate terminals, place more restrictions on access, or install faster communications equipment to improve response time.

Availability is a measure of actual network uptime, either as a whole or by segments. This information might be reported as total hours available over time, average hours available within a specified time, and Mean Time Between Failures (MTBF).

With response time and availability statistics, calculated and formatted by the NMS, managers can establish current trends in network use, predict future trends, and plan the assignment of resources for specific present and future locations and applications.

Control and diagnostics. With control and diagnostic capabilities, the NMS operator can determine from various alarms (i.e., an audio or visual indication at the operator's terminal) what problems have occurred on the network, and pinpoint the sources of those problems so that corrective action can be taken. Alarms can be correlated to certain events and triggered when a particular event occurs. For example, an alarm can be set to go off when a line's bit error rate (BER) approaches a predefined threshold. When that event occurs and the alarm is raised, automated procedures can be launched without operator involvement. In this case, traffic can be diverted from the failing line and routed to an alternate line or service. If the problem is equipment-oriented, another device on "hot standby" can be placed into service until the faulty system can be repaired or replaced.

Configuration management. Configuration management gives the NMS operator the ability to add, remove, or rearrange nodes, lines, paths, and access devices as business circumstances change. If a T1 link degrades to the point that it can no longer handle data reliably, for example, the network management system might automatically reroute traffic to another private line or through the public network. When the quality of the failed line improves, the system could reinstate the original configuration. Some integrated network management systems, particularly those that unify the host (LAN) and carrier (WAN) environments under a single management umbrella, are even capable of rerouting data but leaving voice traffic where it is.

Voice and data traffic can even be prioritized. This NMS capability is very important because failure characteristics for voice and data are very different: voice is more delay-sensitive and data is more line error–sensitive. On networks that serve multiple business entities and on statewide networks that serve multiple government agencies, the ability to differentiate and prioritize traffic is very important.

On a statewide network, for example, state police have critical requirements 24 hours per day, 7 days per week, whereas motor vehicle branch offices use the network to conduct relatively routine administrative business only 8 hours per day,

5 days per week. Consequently, the response time objectives of each agency are different, as would be their requirements for restoral in case of an outage.

On the high-capacity network there can be two levels of service for data and another for voice. Critical data will have the highest priority in terms of response time and error thresholds, and will take precedence over other classes of traffic when it comes to restoral. Because routine data will be able to tolerate a longer response time, the point at which restoral is implemented can be prolonged. Voice is more tolerant than data with regard to error, so restoral may not be necessary at all. The ability to prioritize traffic and reroute only when necessary ensures maximum channel fills, which impacts the efficiency of the entire network and, consequently, the cost of operation.

Configuration management not only applies to the links of a network, but to equipment as well. In the WAN environment, the features and transmission speeds of software-controlled modems may be changed. If a nodal multiplexer fails, the management system can call its redundant components into action, or invoke an alternate configuration. And when nodes are added to the network, the management system can devise the best routing plan for the traffic it will handle.

Applications management. Applications management is the ability to alter circuit routing and bandwidth availability to accommodate applications that change by time of day. Voice traffic, for example, tends to diminish after normal business hours, while data traffic may change from transaction-based to wideband applications that include inventory updates and remote printing tasks.

Applications management includes having the ability to change the interface definition of a circuit so that the same circuit can alternatively support both asynchronous and synchronous data applications. It also includes having the ability to determine appropriate data rates in accordance with response time objectives, or to conserve bandwidth during periods of high demand.

Security. Network management systems have evolved to address the security concerns of users. Although voice and data can be encrypted to protect information against unauthorized access, the management system represents a single point of vulnerability to security violations. Terminals employed for network management may be password-protected to minimize disruption to the network through database tampering. Various levels of access may be used to prevent accidental damage. A senior technician, for example, might have a password that allows changes to be made to the various databases, whereas a less experienced technician's password allows read-only (no changes) database review. Other possible points of entry, such as gateways, bridges and routers, may be protected with hardware- or software-defined partitions that restrict internal access.

Individual users also might be given passwords that permit them to make use of certain network resources, but deny them access to others. A variety of methods are even available to protect networks from intruders who might try to access network resources with dial-up modems. For instance, the management system can request a password and hang up if it does not obtain one within 15 seconds, or it can hang up and call back over a preapproved number before establishing the connection. To frustrate persistent crackers, the system can limit unsuccessful call attempts before denying further attempts. All successful and unsuccessful entry attempts are automatically logged to monitor access and to aid in the investigation of possible security violations.

Summary. Today's network management systems have demonstrated their value in permitting technicians to control individual segments or the entire network remotely. In automating various capabilities, network management systems can speed up the process of diagnosing and resolving problems with equipment and lines. Combined, the capabilities of network management systems permit maximum network availability and reliability, thus enhancing the management of geographically dispersed operations, while minimizing revenue losses from missed business opportunities that could occur as a result of network downtime.

See also
SIMPLE NETWORK MANAGEMENT PROTOCOL

Network restoral

Businesses increasingly relying on communication networks to improve customer service, exploit new market opportunities, and secure strategic competitive advantages. When these networks become severely congested or fail, effective restoral solutions must be implemented as soon as possible.

Although local and long-distance carriers build reliability into their networks at the design stage and monitor performance of the network on a continuing basis, there is always the chance that unforeseen problems will occur. When problems occur, automated processes perform such functions as raise alarms, reroute traffic, activate redundant systems, perform diagnostics, isolate the cause of the problem, generate trouble tickets and work orders, dispatch repair technicians, and return primary facilities and systems to their original service configuration.

Network redundancy. Most networks are designed with a certain amount of redundancy built in. There usually is a duplicate or backup system that can be called into service immediately upon failure of the primary system.

Most central office switches are equipped with dual processors so that if one processor fails, the second one can take over automatically. The switches are designed to run self-diagnostic tests periodically to help ensure proper operation. If a problem occurs, systems often can fix themselves automatically, rebooting software, for example, or switching automatically to a backup system so the primary system can reinitialize itself. These systems also have the ability to alert technicians and network managers if the problem cannot be corrected automatically.

Redundancy also applies to Signal Transfer Points (STPs). These are the computers used to route messages over the carrier's packet-based signaling network, which is used to set up calls and create intelligent services. Each STP has two computers that operate at just under 50 percent capacity. The STP pairs are not collocated, but usually are many miles away from one another. If something happens to one STP, its mate can pick up the full load and operate until repair or replacement of the damaged STP can be accomplished. Should both halves of a mated pair of STPs fail, the switch that normally relies on them can access the signaling network through helper switches that use a different STP pair.

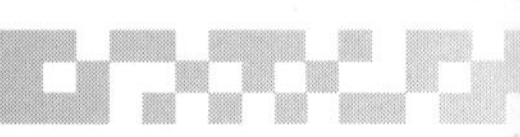

Network Control Points (NCPs), the customer databases for advanced services such as 800 or virtual private network (VPN), have not only dual processors, but also a backup NCP for the protection of all customer configuration information if the second processor should happen to fail.

Digital Interface Frames (DIFs) provide access to and from interoffice switches for processing long-distance calls. The digital interface units that actually handle this work have spares that take over immediately when a problem occurs. Guiding the overall work of the DIF are two controllers running simultaneously; if one experiences a problem, the backup controller can take over without customers even noticing that a problem has occurred. As an option, however, a customer's traffic can be sent to another DIF at another interexchange office if the primary switch encounters a problem. This ensures that the customer's calls continue to flow should the primary DIF experience a prolonged outage.

The power systems used to operate the carrier's network also have backup protection. In normal operation, the carrier's power system provides direct current from redundant rectifiers fed by commercial power. If commercial power fails, batteries (which are kept charged by the rectifiers) provide backup power. An additional level of redundancy is provided by diesel-powered generators, which can replace commercial power during prolonged outages.

Network diversity. Diversity is the concept of providing as many alternative paths as possible to ensure survival of the network when some kind of natural or manmade disaster strikes. Like redundancy, diversity is built into the network during the design stage.

One way carriers ensure diversity is to arrange transmission lines as a series of circles or loops to form an interconnecting grid. Should any particular circle be cut, such as by a backhoe operator hitting a buried cable, traffic can be sent over another facility on one or more adjacent circles. In the case of fiber-optic rings around major metropolitan areas, the use of dual fiber configurations allows carriers to offer disaster recovery services. In the event of a node failure on the fiber-optic network, traffic is automatically routed to the other ring in a matter of milliseconds.

Carriers in larger metropolitan areas often offer their business customers building diversity. In being able to reach the carrier's network from two distinct points, businesses can enhance the reliability of their mission-critical applications.

To further ensure uninterrupted service, carriers also offer *route diversity* for their signaling systems. Each pair of STPs is connected to every other pair of STPs by multiple links. To ensure that connectivity will always be available, these links are laid out over multiple, geographically separated routes. Should something happen along one route to disrupt service, the other routes remain available to keep the carrier's signaling system operational.

Optional restoral services. For most businesses, a temporary interruption of service lasting only a few minutes does not present a problem. For businesses that need a much shorter restoral period, carriers offer optional services that can be tailored to meet specific requirements. These services can range from having the carrier plan and build a complete private network to meet certain reliability and performance specifications, to selecting one or more of the following lower-cost alternatives:

- For businesses with toll-free 800 service and virtual private networks, the ability to receive calls is of primary importance. The carrier provides routes from two separate switches to the corporate location. In the event of a network disruption, calls are automatically directed to the working switch.
- For businesses with toll-free 800 service, if traffic is blocked at one location for any reason, calls are automatically sent to another corporate location.
- For businesses that use digital services, the carrier can provide a geographically separate backup facility, allowing traffic to be switched to a standby link within milliseconds of a service interruption on the primary link.
- For the access portion of a circuit, the carrier can mitigate the effects of certain network failures by automatically transferring service to a dedicated, separately routed access circuit.
- Customer-controlled reconfiguration (CCR) is a carrier-provided service that gives businesses the means to organize and manage their own circuits from an on-premises terminal that issues instructions to the carrier's digital crossconnect system (DCS). If a circuit drops to an unacceptable level of performance or fails entirely, the network manager can issue rerouting instructions to the DCS. If several circuits have failed, the network manager can upload a pretested rerouting program to the DCS to restore the affected portion of the network.
- Some carriers offer businesses a reservation service in which one or more dedicated digital facilities are brought online after the customer verbally requests it with a phone call. This restoral solution requires that the customer presubscribe to the service and that access facilities already be in place with the local carriers at each end.

Site recovery options. Some carriers offer optional site recovery options. This type of service is meant to deal with the loss of a primary data center that runs mission-critical applications. If a customer's data center suffers from a catastrophic fire or natural disaster, for example, traffic will be quickly rerouted to another comparably equipped site. This service is far more economical than having to set up and maintain another data center and links.

To offer site recovery services, the carrier typically partners with an established firm such as Comdisco Disaster Recovery Systems. When disaster strikes, the customer calls the carrier and requests activation of links to the alternate site, a process that takes about 2 hours to complete and which might entail reprogramming each router's routing tables to reflect the changes.

Summary. Not long ago, most organizations relied on their long-distance carrier for maintaining acceptable network performance. More often than not, the carriers were not up to the task. This led to the emergence of private networks in the 1980s, which allowed companies to exercise close control of leased lines with an in-house staff of network managers and technicians. In their eagerness to recapture lost market share, the carriers have made great strides in improving their response to network congestion and outages. This has gone a long way toward restoring lost confidence that once prompted companies to set up and maintain their own networks. Today, companies are once again comfortable in relying on the carriers for maintaining acceptable network performance.

See also
NETWORK MANAGEMENT SYSTEMS

Network security

Protecting vital information from unauthorized access has always been a high-priority concern among most companies. While access to distributed data networks improves productivity by making applications, processing power, and mass storage readily available to a large and growing user population, it also makes those resources more vulnerable to abuse and misuse. Among the risks are unauthorized access to mission-critical data, information theft, and malicious file tampering that can result in immediate financial loss and, in the long term, damage to competitive position. However, various protective measures can be taken to safeguard information in transit as well as information stored at various points on the network,including servers and desktop computers.

Physical security. Protecting data in distributed environments starts with securing the premises. Such precautions as locking office doors and wiring closets, restricting access to the data center, and having employees register when they enter sensitive areas can greatly reduce risk. Issuing badges to visitors, providing visitor escorts, and having a security guard station in the lobby can reduce risk even further.

Other measures, such as keyboard and disk drive locks, also are effective in deterring unauthorized access to unattended workstations. These are important security features, especially since some workstations provide management access to wiring hubs, LAN servers, bridge/routers, and other network access points. In addition, locking down workstations to desks can help protect against equipment theft.

Access controls. Access controls can prevent unauthorized local access to the network and control remote access through dial-up ports. The three minimum levels of user access usually assigned are *public, private,* and *shared access.* Public access allows all users to have read-only access to file information. Private access gives specific users read/write file access, while shared access allows all users to read and write to files.

When a company offers network access to one or more databases, it should restrict and control all user query operations. Each database should have a protective "key," or series of steps, known only to those individuals entitled to access the data. To ensure that intruders cannot duplicate the data from the system files, users should first have to sign on with passwords and then prove that they are entitled to the data requested.

Login security. Network operating systems or add-on software can offer effective login security, which requires that the user enter a login ID and password to access local or remote systems. Passwords not only can identify the user, but also associate the user with a specific workstation, as well as a designated shift, workgroup, or department. The effectiveness of these measures hinges on users' ability maintain password confidentiality.

A user ID should be suspended after a certain number of passwords have been entered to thwart trial-and-error attempts at access. Changing passwords frequently, especially when key personnel leave the company, and using a multilevel password-protection scheme can enhance security. With multilevel passwords, users can gain access to a designated security level, as well as all lower levels. With specific passwords, on the other hand, users can access only the intended level and not the others above or below. Finally, users should not be allowed to make up their own passwords; they should be assigned using a random password generator. Although such schemes entail an increased administrative burden, the effort usually is worthwhile.

The effectiveness of passwords can be enhanced by using them in combination with another control measure, such as a keyboard lock, card reader, or even a biometric device that identifies an authorized user based on such characteristics as a handprint, voice pattern, or the layout of capillary blood vessels in the retina of the eye. Of course, the choice of control measure will depend on the level of security desired.

Data encryption. Protecting data (and voice) as it traverses the network requires that it be scrambled with an encryption algorithm. One of the most effective encryption algorithms is that offered by PGP (Pretty Good Privacy), a method that uses a public key to protect computer and e-mail data. The program generates two keys that belong uniquely to the user. One PGP key is secret and stays in the user's computer. The other key is public and is given out to people with whom the user wants to communicate; the public key can be distributed as part of the message. Users often include the key as part of their signature file that is appended to e-mail, or have it sent as an automatic response to the Unix `finger command`.

PGP does more than encrypt. It has the ability to produce digital signatures, allowing the user "sign" and authenticate messages. A *digital signature* is a unique mathematical function derived from the message being sent. A message is signed by applying the secret key to it before it is sent. By checking the digital signature for a message, the recipient can make sure that the message has not been altered during transmission. The digital signature also can prove that a particular person originated message. The signature is so reliable that nobody—not even the originator—can deny creating it.

Firewalls. A *firewall* is a method of protecting one network from another, untrusted network. Though the actual mechanisms by which this is accomplished vary widely, in principle the firewall can be thought of as a pair of mechanisms: one that exists to block traffic, and the other that exists to permit traffic. Some firewalls place a greater emphasis on blocking traffic, while others emphasize permitting traffic.

One way firewalls protect networks is through packet filtering, which can be used to restrict access from or to certain machines or sites. It also can be used to limit access based on time of day or day of week, by the number of simultaneous sessions allowed, service host, destination host, or service type. This kind of firewall protection can be set up on various network routers, communications servers, or front-end processors.

Transparent proxies are also used to provide secure outbound communication to the internetwork from the internal network. The firewall software achieves this by appearing to act as the default router to the internal network. When packets hit the firewall, however, the software does not route the packets, but immediately starts a dynamic, transparent proxy. The proxy connects to a special intermediate host that actually connects to the desired service.

Proxies are often used instead of router-based traffic controls to prevent traffic from passing directly between networks. Many proxies contain extra logging or support for user authentication. Since proxies must "understand" the application protocol being used, they also can implement protocol-specific security; for example, an FTP proxy might be configurable to permit incoming FTP and block outgoing FTP.

Remote-access security. With an increasingly decentralized and mobile work force, organizations are coming to rely on remote-access arrangements that allow telecommuters, traveling executives, salespeople, and home-based offices to dial into the corporate network. This calls for appropriate security measures to prevent unauthorized access to corporate resources. One or more of the following security methods can be employed:

- *Authentication:* This involves verifying the remote caller by user ID and password, thus controlling access to the server. Security is enhanced if ID and password are encrypted before going out over the communications link.
- *Access restrictions:* This involves assigning each remote user a specific location (i.e., directory or drive) that can be accessed in the server. Access to specific servers also can be controlled.
- *Time restrictions:* This involves assigning each remote user a specific amount of connection time, after which the connection is dropped.
- *Connection restrictions:* This involves limiting the number of consecutive connection attempts or the number of times connections can be established on an hourly or daily basis.

Callback systems. Callback security systems are useful in remote-access environments. When a user dials into the corporate network, the answering modem requests the caller's identification, disconnects the call, verifies the caller's identification against a directory, and then calls back the authorized modem at the number matching the caller's identification. This scheme is an effective way to ensure that data communication occurs only between authorized devices, more so when used in combination with data encryption.

Security procedures can be implemented even before the modem handshaking sequence, rather than after it, as is usually the case. This effectively denies the access opportunity to potential intruders. This method uses a precision, high-speed, analog security sequence that is not even detectable by advanced line monitoring equipment.

While these callback techniques work well for branch offices, most callback products are not appropriate for mobile users whose locations vary on a daily basis. There are now products on the market that accept roving callback numbers. This feature allows mobile users to call into a remote access server or host computer, type in their user ID and password, and then specify a number where the server or host should call them back. The callback number is then logged and may be used to help track down security breaches.

To safeguard very sensitive information, there are third-party authentication systems that can be added to the server. These systems require a user password and also a special credit card–sized device that generates a new ID every 60 seconds, which must be matched by a similar ID number-generation process on the remote user's computer.

Link-level security. When peers at each end of a serial link support the PPP protocol suite, link-level security features can be implemented. This is because PPP can integrally support the Password Authentication Protocol (PAP) and Challenge Handshake Authentication Protocol (CHAP) to enforce link security. PPP is a versatile WAN connection standard that can be used for tying dispersed branch offices to the central backbone via dial-up serial links. It is actually an enhanced version of the older Serial Line Internet Protocol (SLIP). SLIP is limited to the IP-only environment, while PPP is used in multiprotocol environments. Since PPP is protocol-insensitive, it can be used to access both AppleTalk and TCP/IP networks, for example.

PPP framing defines how data is encapsulated before transmission over the WAN. It supports multiple Network layer protocols, including TCP/IP and IPX. PPP also offers remote protocol configuration, the ability to define the framing format over the wire, and password authentication.

PAP uses a two-way handshake for the peer to establish its identity. This handshake occurs only upon initial link establishment. An ID/password pair is repeatedly sent by the peer to the authenticator until verification is acknowledged or the connection is terminated. Passwords are sent over the circuit in text format, however, which offers no protection from playback by network intruders.

CHAP periodically verifies the identity of the peer using a three-way handshake. This technique is employed throughout the life of the connection. With CHAP, the server sends a random token to the remote workstation. The token is encrypted with the user's password and sent back to the server. Then the server does a lookup to see if it recognizes the password. If the values match, the authentication is acknowledged; otherwise, the connection is terminated. Every time remote users dial in, they are given different tokens. This provides protection against playback because the challenge value changes in every token.

Some vendors of remote-node products support both PAP and CHAP, while low-end products tend to support only PAP, which is the less robust of the two authentication protocols.

Policy-based security. With today's LAN administration tools, security goes far beyond mere password protection to include implementation of a policy-based approach characteristic of most mainframe systems. Under the policy-based approach to security, files are protected by their description in a relational database. This means that newly created files are automatically protected, not at the discretion of each creator, but consistent with the defined security needs of the organization.

Some products use a graphical calendar through which various assets can be made available to select users only during specific hours of specific days. For each asset or group of assets, a different permission type may be applied: *Permit, Deny,* and *Log. Permit* allows a user or user group to have access to a specified asset. *Deny* allows an exception to be made to a *Permit,* not allowing writes to certain files, for example. *Log* allows an asset to be accessed, but stipulates that such access will be logged.

Although the LAN administrator usually has access to a full suite of password controls and tracking features, today's advanced administration tools also provide the ability to determine whether or not a single login ID can have multiple terminal sessions on the same system. Through the console, the LAN manager can review real-time and historical violation activity online, along with other system activity.

Summary. To protect valuable information, however, companies must establish a sound security policy before an intruder has an opportunity to violate the network and do serious damage. This means identifying security risks, implementing effective security measures, and educating users on the importance of following established security procedures.

See also

FIREWALLS

Network support

Today's networks have increased in functionality and complexity, pushing support issues into the forefront of management concerns. Regardless of how problems are revealed, whether through network management tools (alarms, diagnostics, predictive methods) or through user notification, the need for timely and qualified network support services is of critical importance, especially in the WAN environment. Recognizing these concerns, the long-distance carriers now offer network support options in conjunction with their services and facilities. In some cases, these support service are very specialized, as in wide-area SNA management.

Types of services. The support concept encompasses dozens of individual activities from which the customer may select. Generally, these activities include, but are not limited to:

- Site engineering, utilities installation, cable laying, and rewiring.
- Performance monitoring of the system or network, alarm interpretation, and initiation of diagnostic activities.
- Identification and isolation of system faults and degraded facilities on the network.
- Notification of the appropriate hardware vendor or carrier for restoral action.
- The repair or replacement of the faulty system or component.
- Monitoring of the repair/replacement process and the escalation of problems.
- Testing of the restoral action to verify proper operation of the system or network.
- Trouble ticket and work order administration, inventory tracking, maintenance histories, and cost control.
- Administration of moves and changes.
- Network design, tuning, and optimization.
- Systems documentation and training.
- Preventive maintenance.

Various other types of support are also available, such as 24-hour telephone (hot line) assistance, short-term equipment rental, fast equipment exchange, guaranteed response time to trouble calls, and customized cooperative maintenance plans that qualify the organization for premium reductions if an internal help desk is established to weed out routine problems, many of which are application-related and beyond the support purview of the carrier or vendor. An increasingly popular support offering is remote diagnostics and network management from the vendor or carrier's network control center.

Levels of support. Carriers also offer multiple levels of technical support. The most basic form of technical support is toll-free telephone access to technical specialists during normal business hours. This type of service assists customers in resolving hardware or software problems. Typically there is no charge for this service and calls are handled on a first-come, first-served basis. There usually is no expiration date for this service; it is available to customers for as long as they use the carrier's services or facilities.

Extended or priority technical assistance is provided via phone 24 hours a day, 7 days a week, to assist customers in resolving hardware or software problems. As an extra-cost service, it ensures that customers are called back within 30 minutes during normal business hours and within 1 hour after normal business hours.

Some carriers offer subscription services that provide the most up-to-date technical product information on maintaining network efficiency and reliability. Written by the carrier's own engineers and field service personnel, this kind of service usually emphasizes how to more effectively operate and manage various data communications and network access products. This information can come in a variety of forms, including technical bulletins, product application notes, software release notes, user guides, and field bulletins, in print or on CD-ROM. Increasingly, the World Wide Web (WWW) is being used to distribute such information. Because access is limited to customers, a valid user ID and password is usually required.

Remote dial-in software support addresses the needs of customers operating mission-critical networks. Technical specialists remotely dial into the customer's network to resolve software problems via diagnostic testing, or by modifying a copy of the system configuration and then downloading the revised configuration file directly to the affected equipment.

Carriers also can assume single-point responsibility for remote network management, providing customers with a proactive approach to service delivery. Technical staff at a central control facility continuously monitors network performance and immediately responds to and resolves any fault resulting from hardware, software, or circuits. From the control facility, network faults are identified and alarm conditions resolved through continuous end-to-end diagnostics. Once a problem is recognized, the latest diagnostic equipment and isolation techniques are used to identify the source of the problem and provide effective resolution. Often, problems are identified and corrected before they become apparent to network users.

If the problem originates with the carrier, it assumes ownership until it is resolved. If the problem originates from a local telephone company or competitive access provider, the long-distance carrier reports the problem, makes appropriate status inquiries, and, if necessary, escalates the problem within the other company's organization.

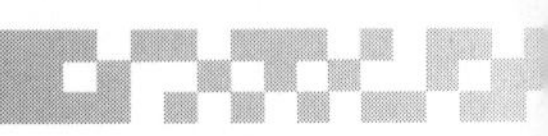

Summary. The long-distance carriers are competing with equipment vendors and third-party service firms in the provision of network support services, providing customers with a broad range of plans that encompass just about every aspect of problem identification, diagnostics, and resolution. Sometimes the support is application-related (as in the case of managing SNA over the wide area network), the cost of which is bundled with a service like frame relay.

See also

MANAGED SNA SERVICES
NETWORK INTEGRATION
OUTSOURCING

Number portability

Number portability means that a customer is able to change local service providers without having to change his or her phone number. Without this requirement, it is unlikely that customers would ever even consider switching to another provider. A local exchange carrier (LEC) must provide, to the extent technically feasible, number portability in accordance with the requirements prescribed by the FCC. In addition, it is the duty of the LEC to provide nondiscriminatory access to telephone numbers, with no unreasonable dialing delays.

Object-oriented networks

Object-oriented technology has been in practical use on the public telecommunications network in one form or another for several years. The network's object orientation not only permits carriers to administer and manage various network elements easily, but gives corporate users the means to easily upgrade, change, and customize telecommunications services without carrier involvement.

Object-oriented technology permits applications to be broken up into classes of objects that can be reused and easily modified for use elsewhere. This greatly reduces application development time, simplifies maintenance, and increases reliability. With each object viewed as a separate functional entity, reliability is improved because there is less chance that a change will produce new bugs in previously stable sections of code.

The use of objects also improves programmer productivity. Each instance of an object draws upon the same piece of error-free code, resulting in less application development time. Over time, this approach also makes it easier to maintain program integrity despite changes in personnel.

Objects. In a network management application, the functions of a switch, multiplexer, bridge, or router—indeed, any device that exists on the network—can be described in an appropriate object. This collection of objects swaps messages with the network management system, triggering events such as status and performance reports. Through messaging, the reports can be sent to other objects, such as printers or disk drives.

In the TCP/IP environment, the collection of object definitions that a given management system can work with is called the *Management Information Base* (MIB). The MIB is a repository of information necessary to manage the various devices on the network. The MIB contains a description of SNMP-compliant objects on the network and the kinds of management information they provide. The objects can be hardware, software, or logical associations such as a connection or virtual circuit. The attributes of an object might include such things as the number of packets sent, routing table entries, and protocol-specific variables for IP routing.

The messaging functions between an object and the network management system are carried out via datagrams that traverse virtual circuits. These datagrams contain commands that request various types of information from the object (such as its status) or collect performance information such as the number of packets sent. For example,

```
frCircuitSentFrames
```

is an object definition for the number of frames sent from a specified frame relay virtual circuit since its creation. Upon request, the appropriate response is sent back to the network management station.

Intelligent networks. Emerging intelligent networks offer another illustration of how object-oriented principles are being applied. Intelligent networks provide the means by which carriers can create, and uniformly introduce and support, new services and features via a common architectural platform. The creation and support of new services is accomplished through the manipulation of software objects that are accessible at intelligent nodes distributed throughout the network.

Instead of investing heavily in premises-based equipment and leasing lines to provide a high level of performance, functionality, and control via private networks, corporate users can tap the service logic of intelligent nodes embedded in the public network for services, advanced calling features, and bandwidth on demand. Users are able to build their own networks, create services, and customize features without carrier involvement simply by combining and recombining the available objects at a workstation.

The required resources, in the form of functional components, are then assembled automatically by the intelligent network in accordance with the user's design specifications. It also is possible to test the integrity of network models by simulation prior to actual implementation. It might be necessary, for example, to predict the delay performance of particular links to ensure that certain applications will not time out. Even additional bandwidth can be made available through object manipulation. In essence, companies can manage their portions of the public network as though they were private networks.

Carriers benefit from this object orientation as well. With all services and features defined in software (as objects), and the programs distributed among fewer locations (intelligent nodes) instead of at every switch, carriers can deploy new services more quickly. Once new services are developed, they can be made immediately available to customers from intelligent nodes. In accessing these nodes, customers can instantly implement a uniform set of services for maximum efficiency and economy, regardless of the location of all their business units.

In allowing users to design their own networks at management workstations and giving them the means to add or change services without outside involvement, carriers are relieved of much of the administrative burden associated with customer service. The decrease in demand for customer support reduces the carrier's staffing requirements and, consequently, the cost of network operation. Cost savings can be passed on to customers in the form of lower service rates.

Summary. The object-oriented paradigm signals a fundamental shift in the way networks, applications, databases, and operating systems are put together, as well as how they are used, upgraded, and managed. The ability to create new objects from existing objects, change them to suit specific needs, and otherwise reuse them across different applications, promises compelling new efficiencies and economies.

In addition, the object orientation can provide network managers with the means to better control services and features in ways that suit the ever-changing needs of the organization.

See also

ADVANCED INTELLIGENT NETWORK
SIMPLE NETWORK MANAGEMENT PROTOCOL

Open Management Architecture

As network-based computing in general, and network management in particular, increasingly migrate toward an object orientation, network administrators need new methods to track and use object-oriented resources. To a large extent this is achievable through the Object Management Architecture (OMA), which is overseen by the Object Management Group (OMG), an industry consortium representing over 500 software vendors, software developers, and end users. Since its establishment in 1989, the OMG's objective has been to promote the use of object technology for developing and managing distributed computing environments.

OMA is a framework that defines detailed interface specifications that lead to interoperable, reusable, portable software components based on open, standard, object-oriented interfaces. At the heart of OMA is the *Common Object Request Broker Architecture* (CORBA), which provides the basic definition for interoperable *Object Request Brokers* (ORBs), allowing end users and system managers to exploit fully their network-based resources.

ORB concepts. An ORB is the method used to transparently make requests of, and receive responses from, other objects. Fundamentally, ORBs are comparable to interprocess communications (IPC) in stand-alone computers and remote procedure calls (RPCs) in client-server environments. The fundamental theory of object orientation technology demands that objects of any type can request services of other objects—at any time—without concern for the underlying implementation details.

Under CORBA's object technology model, objects perform specific services in response to requests and return the results. The following features simplify program construction, lower software maintenance costs, and increase software product life cycles:

- A uniform *Interface Definition Language* (IDL), which specifies independent interface execution. IDL is object-oriented, allowing abstract representations such as encapsulation, polymorphic messaging, and inheritance.
- An *Interface Repository* (IR), which provides dynamic representations of available objects (or classes) in the distributed environment.
- *Dynamic invocation,* which calls extensions that discover interfaces, objects, and message requests, and provides for message handling.
- *Contextual extensions,* which pass named rather than positional parameters to control program execution.
- *Object adapters,* which shield implementation details from messaging requirements.

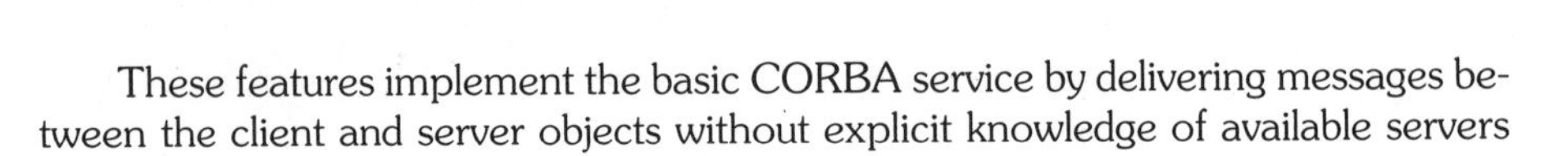

These features implement the basic CORBA service by delivering messages between the client and server objects without explicit knowledge of available servers or interfaces.

ORB structure. The CORBA structure consists of an ORB core, dynamic-invoke and Interface Definition Language (OMG IDL) stub interfaces, direct ORB interface, OMG IDL skeleton, and object adapters. Together these components permit any client object to communicate with any other object.

IDL is the OMG's standard language for defining an object's boundaries, so other objects know how to interact with the object and what it can do. IDL interfaces give developers working in C/C++, Java, and other object languages a consistent way to interact with these protocols.

Central to CORBA is that there are definitions of object interfaces accomplished by OMG IDL. This interface language defines the types of objects according to the operations that may be performed and the parameters needed by those operations. A client object makes a request through an OMG IDL compiled stub or skeleton, allowing abstraction on interface representation (encapsulation), polymorphic messaging (encapsulation), and inheritance of the interface.

The ORB is responsible for all the mechanisms required to find the "object implementation" in the network, prepare it to receive the request, and transmit the request. The client object does not care where the implementation is located, what programming language was used to implement it, or what underlying protocols, operating system, or chip technology it uses. Depending on the application, a particular ORB might support multiple interface options and communications protocols.

The dynamic-invoke, object adapters, and OMG IDL stub interfaces are the keys to initiating a request. Any ORB implementation that provides an appropriate interface is acceptable. With dynamic invocation, the interface is independent of the target object and the interaction is more unpredictable; the OMG IDL stub is specific to it. Static binding requires less messaging code and supports smaller (and simpler) objects. With static invocation, the object interface is known, as it is with C++ and Pascal, and the interaction is more predictable. The invocation method is transparent to the receiving object. Neither the originating object nor the target object knows or depends on the type of interface invoked. The CORBA standards let programmers use whatever approach is appropriate in the given circumstance, even if it is just a matter of preference.

Object adapters provide flexibility by letting the ORB use interfaces customized to target particular groups of object implementations with similar requirements. The wide range of object details, life expectancies, policies, implementation styles, and other object properties makes it difficult for ORBs to provide a single interface style that is convenient and efficient for all objects.

CORBA functionality. Once a request is initiated, the ORB calls the object implementation via an IDL skeleton. Each object implementation may choose which object adapter to use in responding to a client object's request; the adapter decision is based on what kind of services the object implementation requires.

Most object implementations provide their behavior using host system facilities. For example, although a basic object adapter provides some persistent data associated with an object, its purpose is to identify or point to the actual object information in storage, for example, in an RDBMS. With this structure, different object implementations can use storage services the same way.

An object adapter uses the ORB core and other components to perform its functions or maintain its own state. A particular adapter must provide the same interface and service for all the ORBs on which it is implemented. All object adapters do not provide the same interface or functions. Some object implementations have special requirements; for example, an object-oriented database (OODB) might wish to register implicitly its many thousands of objects without doing individual calls to the object adapter. In object adapters, as throughout the CORBA specification, flexibility is encouraged. Because of the object implementation's dependence on the adapter, however, having fewer adapters is generally better than having many adapters.

CORBA specifications. Although there are very few firm demands in CORBA, it is a specification with certain requirements. The most important requirements are:

- A particular adapter must provide the same interface and service for all the ORBs on which it is implemented.
- All ORBs must provide the same language mapping to an object reference for a particular programming language.
- In any particular computing environment, an ORB must be capable of distinguishing from others its own object references, and must be capable of passing as parameters the object references of other ORBs.

The CORBA 2.0 specification, unlike CORBA 1.*x*, attempts to achieve interoperability and portability across all compliant implementations. A major part of the specification covers the interoperability of ORBs and the integration of different object systems. The specification provides ways for nonobjects, object systems, and ORBs themselves to communicate. For example, nonobjects can be encapsulated and act as clients, and object systems can appear to be object implementations. Different types of object adapters will be created to handle the different interoperability possibilities. Logical, physical, and gateway-type connections will evolve.

Summary. Business can no longer wait to build distributed systems because the present business environment demands a distributed solution now. They also cannot wait for proprietary solutions to be retrofitted to handle large-scale, Internet-based applications. Increasingly, organizations are turning to CORBA for building their distributed systems. The advantages of CORBA are compelling: speed and cost savings in application development and implementation, better performance at run time, and interoperability across languages and platforms.

See also

OBJECT-ORIENTED NETWORKS

Open Network Architecture

Open Network Architecture (ONA) refers to the overall design of an incumbent local exchange carrier's (ILEC) network, specifically its ability to provide competitive

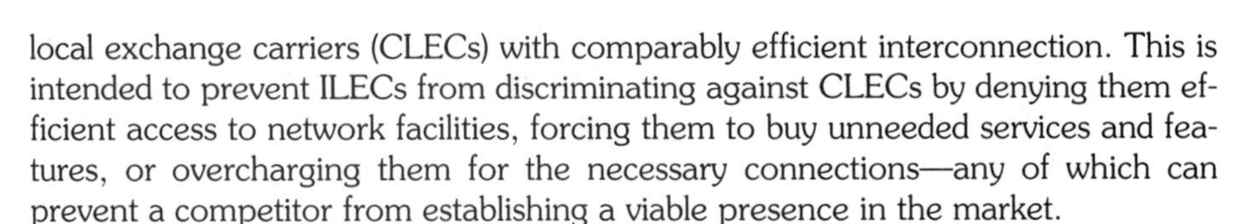

local exchange carriers (CLECs) with comparably efficient interconnection. This is intended to prevent ILECs from discriminating against CLECs by denying them efficient access to network facilities, forcing them to buy unneeded services and features, or overcharging them for the necessary connections—any of which can prevent a competitor from establishing a viable presence in the market.

Implementation of ONA requires that existing feature group access arrangements be unbundled, and that new access charge subelements, known as basic service elements (BSEs) and basic serving arrangements (BSAs), be established. The Federal Communications Commission supervises the efforts of the ILECs to open their networks in this manner, which in turn determines whether the ILECs can participate in markets that previously have been closed to them.

Regulatory history. Under Computer II, the Regional Bell Operating Companies (RBOCs) were permitted only to provide enhanced services through a structurally separate subsidiary. In its Computer III decisions, the FCC permitted the RBOCs to integrate their enhanced service and basic service offerings provided that they complied with certain nonstructural safeguards, including *comparably efficient interconnection* (CEI) requirements.

In the first stage of implementing Computer III, the FCC required the RBOCs to obtain its approval for service-specific CEI plans prior to offering individual enhanced services on an integrated basis. In these CEI plans, the FCC required the RBOCs to demonstrate how they would provide competitors with equal access to all basic underlying network services the RBOCs had used to provide their own enhanced services.

During the second stage of Computer III, the RBOCs developed and implemented Open Network Architecture (ONA) plans detailing the unbundling of basic network services. After the FCC approved these ONA plans and the RBOCs filed tariffs for ONA services, they were permitted to provide integrated enhanced services without filing service-specific CEI plans. ONA incorporates and subsumes CEI equal-access requirements and provides for the further unbundling of network service elements not limited to those associated with specific RBOC enhanced services. After the implementation of ONA, the RBOCs were still required to offer network services to competitors on a CEI equal-access basis, even though they were no longer required to file a CEI plan for each service they wished to offer.

Comparably efficient interconnection. Comparably efficient interconnection is achieved when the ILEC can demonstrate that it offers:

- Standardized interfaces to provide access to the transmission, switching, and signaling resources of the network.
- Unbundled basic services.
- Common basic service rates.
- Common basic service performance characteristics.
- Common installation, maintenance, and repair services.
- Common end-user access.
- Common knowledge of impending availability of new basic service features.
- Comparable interconnection costs for competitors.

CEI has been expanded to include access to the Operations Support Systems (OSS) of the ILECs. These are databases and information that an ILEC uses to provide telecommunications services to its customers. Among the functions of OSS are preordering, ordering, provisioning, maintenance and repair, and billing. The FCC considers access to OSS functions as necessary for meaningful competition.

ONA building blocks. As noted, implementation of ONA requires that existing feature group access arrangements be unbundled, and that new access charge subelements, known as basic service elements (BSEs) and basic serving arrangements (BSAs), be established.

Basic Serving Arrangements. The BSA specifies the access links and transport elements that comprise a basic transmission service. For example, circuit-switched trunk-side access is a BSA that provides a trunk-side access connection to the CLEC's premises. This service may be provided directly from an end office or optionally from a tandem switch to deliver one-way originating traffic to the CLEC. This service includes a 7-digit number with which users can access the alternative service.

Another BSA provides dedicated connections between end users and the CLEC so that a channel of up to 9.6 kbps may be used for such applications as the transmission of alarm signals from subscriber locations to a central alarm monitoring company.

Other examples of BSAs include X.25 and X.75 interfaces to packet switches, broadband links for video transmission, in-band signaling, and central office announcements.

Basic Service Elements. Through a series of Basic Service Elements (BSEs), a variety of network capabilities can be offered. Under CEI, the BSEs must be offered on an unbundled basis. The CLEC and ILEC decide which BSEs are appropriate to support a given service, and the CLEC pays only for those BSEs. There are four general categories of BSEs:

- *Switching* supports services that require call processing, routing, and management.
- *Signaling* supports monitoring services that require a derived channel over subscriber lines.
- *Transmission* allocates appropriate bandwidth to a customer application.
- *Network management* provides the means to monitor system performance and reallocate assigned capabilities.

The BSEs associated with circuit-switched services might include call forwarding, variable ringing, three-way calling, and automatic number identification (ANI). BSEs associated with private lines might include an out-of-band diagnostic channel, line conditioning, customer-controlled reconfiguration, and route diversity.

There is room for interpretation among ILECs in determining what elements are classified as BSAs or BSEs. For example, while multiline hunt groups are universally considered BSEs, detection of telco line breaks within 60 seconds might be considered a BSA by one ILEC and a BSE by another ILEC.

Ancillary Services. Services that provide utility to the service provider, but which are not associated with a specific network feature or function, fall under the

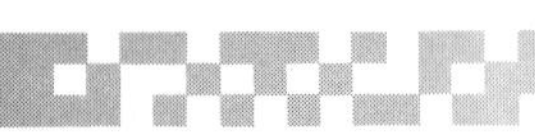

category of Ancillary Services. These services typically include maintenance and diagnostics, billing services, and the collection of traffic statistics.

Because telephone companies differ widely in their interpretation of what constitutes an "ancillary service," the FCC has directed that all regulated services must be classified as BSAs or BSEs, and that only unregulated services may be classified as ancillary services.

Complementary Network Services. Complementary Network Services (CSNs) are those features that are applied to the end user's local service to make it interact more efficiently with the service provider's BSAs or BSEs. Examples of CNS might include the multifaceted Call Forward feature:

- *Call Forward Busy Line/Don't Answer* allows user calls to a busy line or unanswered line to be forwarded to another number for call completion.
- *Call Forward Don't Answer with Variable Ring Count* allows user calls to be forwarded after a specified number of rings on a Don't Answer condition.
- *Customer Control of Call Forward Busy Line/Don't Answer* allows the service provider's operator to override the Call Forward Busy Line/Don't Answer feature on a demand basis.

Summary. ONA provides the framework for competition in the telecommunications market. Under the concept of comparably efficient interconnection, the ILECs must provide CLECs with the same economic and technical efficiencies as they use to provide telecommunications services to their own subscribers. In support of their claims for achieving CEI, the ILECs are required to submit engineering studies, time and wage studies, or other cost accounting studies to identify direct costs. They also must provide a projection of costs for a representative 12-month period, estimates of the effect of the service on the carrier's traffic and revenues (including the traffic and revenues of other services), and supporting work papers for estimates of costs, traffic, and revenues. Achieving CEI is a prerequisite for ILEC entry into other markets in which they have previously been excluded (i.e., long-distance service).

See also

OPERATIONS SUPPORT SYSTEMS

TELECOMMUNICATIONS ACT OF 1996

Open Systems Interconnection

The seven-layer Open Systems Interconnection (OSI) Reference Model was first defined in 1978 in ISO standard 7498. The lower layers (1 to 3) represent local communications, while the upper layers (4 to 7) represent end-to-end communications (Fig. O1). Each layer contributes protocol functions that are necessary to establish and maintain the error-free exchange of information between network users.

The model provides a useful framework for visualizing the communications process and comparing products in terms of standards conformance and interoperability potential. This layered structure not only aids users in visualizing the communications process, it also provides vendors with the means for segmenting and allocating various communications requirements within a workable format. This can reduce much of the confusion normally associated with the complex task of supporting successful communications.

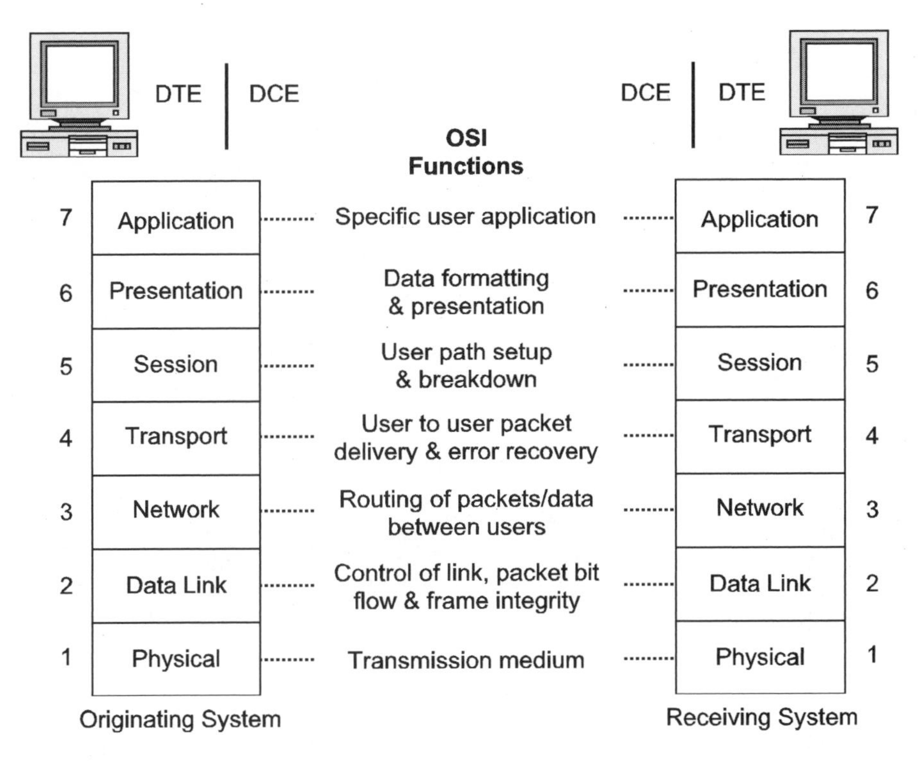

Fig. O1 *The 7-layer OSI Reference Model.*

OSI layers. Each layer of the OSI model exchanges information with a comparable layer at the other side of the connection, a process known as *peer-protocol* communications.

Application layer. The highest layer in the OSI reference model is the Application layer. It includes not only application programs, but also basic network services such as file or print services. This level applies to the actual meaning rather than the format or syntax (as in layer 6) of applications, and permits communication between users. According to the model, each type of application must employ its own layer 7 protocol; with the wide variety of available application types (including file transfer, job transfer, business data interchange, virtual terminal operation, and electronic mail), layer 7 offers definitions for each.

Presentation layer. Layer 6 deals with the format and representation of data that applications use; specifically, it controls the formats of screens and files. Layer 6 defines such things as syntax, control codes, special graphics, and character sets. Additionally, this level determines how variable alphabetic strings will be transmitted, how binary numbers will be presented, and how data will be formatted.

The Presentation layer is particularly valuable for potential carrier services such as videotext, where text and images are transported over telephone lines. Standardizing the way information is presented, irrespective of end-user terminal types, allows worldwide dissemination of information without concern for display compatibility.

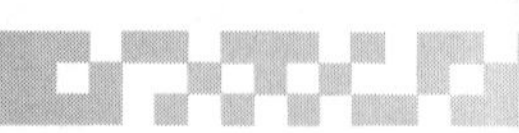

Session layer. The Session layer manages communications; for example, it sets up, maintains, and terminates virtual circuits between sending and receiving devices. It sets boundaries for the starts and ends of messages, and establishes how messages will be sent: half-duplex, with each computer taking turns sending and receiving, or full-duplex, with each computer sending and receiving at the same time. These details are negotiated during session initiation.

Transport layer. Layer 4 handles end-to-end transport. If there is a need for reliable, end-to-end sequenced delivery, then the transport layer performs this function. For example, each packet of a message might have followed a different route through the network toward its destination. The transport layer reestablishes packet order through a process called *sequencing* so that the entire message is received exactly the way it was sent. At this layer, lost data is recovered and flow control is implemented. With flow control, the rate of data transfer is adjusted to prevent excessive amounts of data from overloading network buffers.

Layer 4 may also support datagram transfers, that is, transactions that need not be sequenced. This is required for voice and video, which may tolerate loss of information but need to have low delay and low variance in transmittal time. This flexibility is the result of the protocols implemented in this layer, ranging from the five OSI protocols (TP0 to TP4) to TCP and UDP in the TCP/IP suite, and many others in proprietary suites. Some of these protocols do not perform retransmission, sequencing, checksums, and flow control.

Network layer. Layer 3 formats the data into packets, adds a header containing the packet sequence and the address of the receiving device, and specifies the services required from the network. The network does the routing to match the service requirement. Sometimes a copy of each packet is saved at the sending node until it receives confirmation that the packet has arrived at the next node undamaged, as is done in X.25 packet-switched networks. When a node receives the packet, it searches a routing table to determine the best path for that packet's destination without regard for its order in the message. In a network where not all nodes can communicate directly, this layer takes care of routing packets through the intervening nodes. Intervening nodes may reroute the message to avoid congestion or node failures.

Data Link layer. All modern communications protocols use the services defined in layer 2. The Data Link layer provides the lowest level of error control. It detects errors and requests the sending node to retransmit the data. This layer has assumed a greater role as communications lines have become less noisy through the replacement of analog lines with digital lines, while end stations have become more intelligent through the use of more powerful processors and high-capacity memory. Combined, these factors have lessened the need for high-level information protection mechanisms in the network, moving them to the end systems. Layer 2 does not know what the information or packets it encapsulates mean or where they are headed. Networks that can tolerate this lack of information are rewarded by low transmission delays.

Physical layer. The lowest OSI layer is the Physical layer. This layer represents the actual interface, electrical and mechanical, that connects a device to a transmission medium. Because the physical interface has become so standardized, it is usually taken for granted in discussions of OSI connections. Yet physical connections—cables and connectors, with their pinouts and transmission characteristics—still can be a problem in designing a reliable network if they do not conform to a common model.

Conformance vs. interoperability. There are 12 laboratories accredited by the National Institute of Standards and Technology (NIST) to run a suite of tests that certify vendor products for conformance to the OSI reference model. While the products of different vendors may conform to the OSI model, however, this does not necessarily mean that they are interoperable.

Conformance testing is the process of comparing a vendor's protocol implementation against a model of the protocol. Conformance test results are sent to NIST for approval. Approved results for each product are then entered into NIST's registry of OSI-conformant products.

Conformance by itself does not guarantee that the product of one vendor will work with the product of another vendor, however, even though both products have passed the same conformance test. OSI product conformance testing only increases the probability of successful interoperability in a customer's multivendor OSI network. To ensure that the products of both vendors do indeed work together on the network, they must be specifically tested for interoperability at the highest level of OSI, the Application layer. This involves running both vendors' protocol implementations of FTAM or X.400, for example, to see if they work properly across their respective products.

Summary. Throughout the 1980s, the prediction was often made that OSI would replace TCP/IP as the preferred technique for interconnecting multivendor networks. This has yet to happen. There are several reasons for this, including the slow pace of OSI standards progress in the 1980s, as well as the expense of implementing complex OSI software and having products certified for OSI interoperability. Furthermore, TCP/IP was already widely available and doing an acceptable job interconnecting multivendor networks. The situation is different in Europe, where OSI compliance was mandated early on by the regulatory authorities in many countries.

See also
SIMPLE NETWORK MANAGEMENT PROTOCOL
TCP/IP

Operations support systems

Operations support systems (OSS) comprised databases and information that a local exchange carrier (LEC) uses to provide telecommunications services to its customers. Among the functions of OSS are preordering, ordering, provisioning, maintenance and repair, and billing.

The FCC has determined that OSS functions fall within the definition of a network element, and that it is technically feasible for incumbent LECs to provide access to OSS functions on an unbundled basis to requesting competitive local exchange carriers (CLECs). The FCC considers access to OSS functions as necessary for meaningful competition, and that failing to provide such access could impair the ability of requesting telecommunications carriers to provide competitive service.

The FCC had set a January 1, 1997, deadline for the incumbent LECs to provide access to OSS functions on a parity basis. There are no standards for OSS, however, and as of year-end 1997, these systems were still not fully automated. Complaints of faulty OSSs continued from many CLECs, who claim that the ILECs are not giving priority attention to this matter.

See also

TELECOMMUNICATIONS ACT OF 1996

Operator language translation services

Language can be a significant problem for those wishing to place international calls. Callers may find it hard to use services because they are confronted with operators who do not speak their language and have trouble processing the call. Because of the limited number of international operators, it is often difficult for callers to get through to place a collect call.

To overcome these problems, the large global telecommunications companies such as AT&T, MCI, and Sprint are starting to offer language translation services in conjunction with collect or credit card calls. AT&T's Direct In-Language Service offering, for example, allows callers access to a bilingual AT&T operator who completes the collect call to the United States. The call is processed the same as a regular call except that the operator completes the call in the preferred language of the caller. AT&T currently provides language translation service from more than 20 countries.

To place a AT&T Direct In-Language Service call, a caller dials the appropriate access code and is connected to an AT&T operator in the U.S. who speaks the caller's language. The operator takes the name and number of the person being called and completes the call. If the call is completed in a language other than English, the operator remains on the line with the caller through completion of the call. Because calls are placed in queue while ringing, a different operator might complete the call than the one who originally answered it. When the call is completed, it is billed to the called party.

This type of operator service also is being provided by telecommunications carriers in other countries, making translation services available to their citizens who are traveling in the United States. Available to more than 70 countries, callers can access direct service numbers from both the U.S. mainland and Hawaii. The PTTs and AT&T are handling the service together. Calls must be collect or billed to a PTT credit card. For example, a German traveler in the United States who wants to call back to Germany can call the German Direct Service and be connected to a German-speaking operator in Germany, who places the collect or PTT credit card call.

MCI provides this kind of service through its WorldPhone offering. To use WorldPhone, callers dial the toll-free access number of the country from which they are calling and an operator who speaks the person's language will connect the call. Fifteen languages are available for translation.

WorldPhone offers several other travel-related features. For example, callers can speak with an MCI Traveler's Assist Specialist who will give emergency local medical, legal, translation, restaurant, and entertainment referrals.

Summary. As more companies in different countries participate in the global economy and more people travel to international locations for both business and vacation, service providers see an increasing need to support their communications offerings with multilingual operators who can facilitate call completion. Early in the next century, operators might even be dispensed with entirely for this task as language conversion systems are added to the network.

Outsourcing

In today's down economy, it makes sense to outsource tasks that tend to consume a disproportionate share of corporate resources. Running information systems and communication networks involves a commitment of time and money that is becoming increasingly difficult to justify in the face of other pressing concerns. Consequently, many companies are turning to service firms that specialize in such things as running data centers, managing networks, integrating diverse computer systems, and developing software. The arrangement is called *outsourcing*.

Given the increasing complexity of today's communications networks and the need for companies to focus more on core business to succeed in the global economy, many are seeking ways to offload communications management responsibilities to those with more knowledge, experience, and hands-on expertise than they alone can afford.

Outsourcing firms typically provide an analysis of an organization's business objectives, application requirements, and current and future communications needs. The resulting network design may incorporate the interexchange facilities of any carrier and include equipment from multiple vendors. Acting as the client's agent, the outsourcing firm coordinates the activities of equipment vendors and carriers to ensure efficient and timely installation and service turn-up.

Typical services. In a typical outsourcing arrangement, an integrated control center (located at the outsourcing firm's premises or that of its client) serves as a single point of service support where technicians are available 24 hours per day, 365 days per year, to monitor network performance, contact the appropriate carrier or dispatch field service as needed, perform network reconfigurations, and take care of any necessary administrative chores.

Integration. Today's communications networks consist of number of different elements: legacy hosts, clients and servers, LANs and cable hubs, bridges and routers, PBXs and key systems, and wide area network facilities. The selection, installation, integration, and maintenance of these elements requires a broad range of expertise that is not usually found within a single organization. In consequence, many companies are increasingly turning to the services of outsourcing firms.

Briefly, the integration function is concerned with unifying disparate computer systems and transport facilities into a coherent, manageable utility. This typically involves the reconciliation of different physical connections and protocols. The outsourcing firm also ties in additional features and services offered through the public switched network. The objective is to provide compatibility and interoperability among different products and services, making access transparent to end users.

Project management. Project management entails the coordination of many discrete activities, starting with the development of a customized project plan based on the client's organizational needs. For each ongoing task, critical requirements are identified, lines of responsibility are drawn, and problem escalation procedures are defined.

Line and equipment ordering may also be included in project management. Acting as the client's agent, the outsourcing firm interfaces with multiple suppliers and carriers to economically upgrade or expand the network without sacrificing predefined performance requirements. Before new systems are installed at client locations, the outsourcing firm performs site survey coordination and preparation, ensuring that all power requirements, air conditioning, ventilation, and fire protection systems are properly installed and in working order.

When an entire node must be added to the network or a new host must be brought into the data center, the outsourcing firm will stage all equipment for acceptance testing before bringing it online, thus minimizing potential disruption to normal business operations. When new lines are ordered from various carriers, the outsourcing firm will conduct the necessary performance testing before making them available to user traffic.

Trouble ticket administration. In assuming responsibility for daily network operations, a key service performed by the outsourcing firm is trouble ticket processing, which typically is automated. The sequence of events is as follows:

1. An alarm indication is received at the network control center operated by the outsourcing firm.
2. The outsourcing firm uses various diagnostic tools to isolate and identify the cause of the problem.
3. Restoral mechanisms are initiated (manually or automatically) to bypass the affected equipment, network node, or transmission line until the faulty component can be brought back into service.
4. A trouble ticket is opened. If the problem is with hardware, a technician is dispatched to swap out the appropriate board; if the problem is with software, analysis may be performed remotely; if the problem is with a particular line, the appropriate carrier is notified.
5. The client's help desk is kept informed of the problem's status so that the help desk operator can assist local users.
6. Before closing out the trouble ticket, the repair is verified with an end-to-end test by the outsourcing firm.
7. Upon successful end-to-end testing, the primary CPE or facility is turned back over to user traffic and the trouble ticket is closed.

Vendor-carrier relations. Another benefit of the outsourcing arrangement comes in the form of improved vendor-carrier relations. Instead of having to manage multiple relationships, the client only needs to manage one: the outsourcing firm. Dealing with only one firm has several advantages in that it:

- Improves response time to trouble calls/alarms,
- Eliminates delays caused by vendor/carrier finger-pointing,

- Expedites order processing,
- Reduces time spent in invoice reconciliation,
- Frees staff time for applications development and planning, and
- Reduces cost of network ownership.

Maintenance/repair/replacement. Some outsourcing arrangements include maintenance, repair and replacement services. Not only does this arrangement eliminate the need for ongoing technical training, the company is also buffered from the effects of technical staff turnover. Repair and replacement services can increase the availability of systems and networks, while eliminating the cost of maintaining a spare parts inventory, test equipment, and asset tracking system.

Disaster recovery. Disaster recovery includes numerous services that may be customized to ensure maximum network availability and performance:

- Disaster impact assessment
- Network recovery objectives
- Evaluation of equipment redundancy and dial backup
- Network inventory and design, including circuit allocation
- Vital records recovery
- Procedure for initiating the recovery process
- Location of "hot site," if necessary
- Installation responsibilities
- Test run guidelines
- Escalation procedures
- Recommendations to prevent network loss

Long-term planning support. An outsourcing firm can provide many services that can assist the client with strategic planning. Specifically, the outsourcing firm can assist the client in determining the impact of:

- Emerging services and products
- Industry and technology trends
- International developments in technology and services

With experience drawn from a broad customer base, as well as its daily interactions with hardware vendors and carriers, the right outsourcing firm can have much to contribute to clients in the way of strategic planning assistance.

Training. Outsourcing firms can fulfill the varied training requirements of users, including:

- Basic communications concepts
- Product-specific training
- Resource management
- Help desk operator training

The last type of training is particularly important, since 80 percent of reported problems are applications-oriented and can be solved without the outsourcing firm's

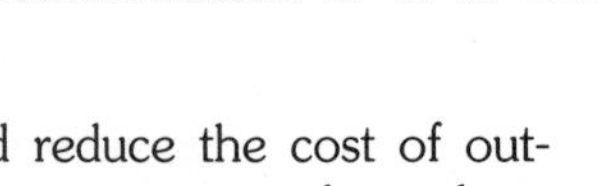

involvement. This can speed up problem resolution and reduce the cost of outsourcing. For this to be effective, however, the help desk operator must know how to differentiate between applications problems, system problems, and network problems. Basic knowledge may be gained by training and improved with experience.

Equipment leasing. An outsourcing arrangement can include equipment leasing. There are a number of financial reasons for including leasing in the outsourcing agreement, depending on the organization's financial situation. Leasing can improve a company's cash position, because costs are spread over a period of years. Leasing can free up capital for other uses, and even cost-justify technology acquisitions that normally would prove too expensive to purchase. With new technology becoming available every 12 to 18 months, leasing can prevent the organization from becoming saddled with obsolete gear.

Standards. Recognizing that today's networks involve complex, high-maintenance technologies, the Networking Technical Support Alliance (NTSA) was formed in 1994 to address the growing need for better service and support among users of multivendor networks.

An industry consortium composed of leading computer and data communications equipment manufacturers, NTSA concerns itself with networking infrastructure, specifically support and interoperability for multivendor local and wide area networks. The NTSA has put into place a process for resolving problems that are beyond the normal ability of typical front-line end users or service providers to address, and that require manufacturer involvement.

NTSA members provide information and cooperation among their service organizations to the extent necessary to resolve multivendor service problems experienced by mutual customers. As a worldwide, open vendor alliance, NTSA provides for more efficient service of multivendor networks through the development and implementation of voluntary, standardized, cooperative technical problem resolution procedures and measurements relating to multivendor service problems.

Customer problems are resolved through multivendor problem resolution procedures adopted by the NTSA Management Committee. These procedures define a methodology for identifying a multivendor problem, requesting technical assistance from other manufacturers, escalating to the appropriate level within each member organization any disagreements about appropriate service responses, sharing information necessary for the resolution of the problem, and measuring problem response activities.

These procedures are voluntary, except that members who consistently fail to adhere to them will be subject to suspension or termination from the NTSA. Requests for use of the procedures may be initiated by members or by others, including customers and associate members.

NTSA's multivendor problem resolution process (MVPRP) is invoked when mutual customers have problems involving the interoperation of member companies' products.

Let's say that a customer has a mix of IBM and Unisys equipment. The problem resolution process starts when the customer calls one of the vendors for assistance. If that vendor needs assistance, it establishes contact with the other vendor and references NTSA.

If the requesting vendor is IBM, the company provides the customer's name to Unisys. Both companies work together until the root cause of the problem is identified. Since IBM originated the call to Unisys, IBM owns the problem unless/until all parties, including the customer, mutually agree that ownership should be transferred.

During the problem-resolution process, the vendors share technical information but must respect each others' proprietary information by not disclosing it to a third party.

Persistent problems are escalated to management within the vendors' respective service organizations. The trouble call is closed when the issue is resolved. All of this assumes, of course, that the customer is entitled to service under both vendors' warranty or service terms and conditions. Although NTSA members have agreed to work together in good faith to resolve issues, ultimately customers must be entitled to service and vendors can decide when to provide such service.

Of note is that the alliance does not imply unlimited service entitlement at no charge; each vendor determines its own customer service entitlement criteria. This includes the products covered, operating hours, and response times. And absent a warranty or service contract, customers may be charged for such service.

A problem management record for each NTSA incident is retained by IBM at its facilities in Raleigh and Atlanta. Each NTSA incident is thoroughly documented, listing:

- Receiving vendor
- Call ID or incident number
- Requesting vendor
- Open date
- Close date
- Time spent
- Product(s) involved
- Summary problem description

This information is summarized and circulated among NTSA members, providing feedback on the overall value of their efforts and a yardstick for improvement. As members increasingly implement NTSA procedures, customers will experience faster multivendor problem resolution.

Participation in NTSA is intended to demonstrate to prospective buyers that the vendor has a sincere interest in resolving customer problems and is willing to own the problem until it can be resolved in cooperation with other vendors. With the complexity of today's internetworks, anything less is unacceptable. Therefore, prospective buyers are advised to query vendors about their membership in NTSA and other such alliances and make membership a condition of the sale whenever possible. At the same time, if a vendor's membership in NTSA has been suspended or terminated, this should be a red flag, indicating that the vendor's track record in providing service and support is seriously flawed.

Summary. While outsourcing promises numerous benefits, determining whether such an arrangement makes sense is a difficult process that requires the organiza-

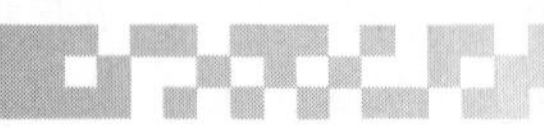

tion to consider a range of factors. Besides calculating the baseline cost of managing one's own information systems and communications network and determining their strategic value, the decision to outsource often hinges on the company's business direction, its present systems and network architecture, the internal political situation, and the company's readiness to deal with the culture shock that inevitably occurs when two firms must work closely together on a daily basis.

See also

HELP DESKS
NETWORK INTEGRATION
NETWORK SUPPORT

Packet-switched networks

Packet-switched networks are based on the feature-rich X.25 protocol, a worldwide method of data transport over analog lines standardized by the CCITT (now the ITU). The X.25 protocol was developed in the early 1970s in a networking environment dominated by copper lines and electromechanical switches, which are subject to a variety of impairments that made the transmission of data difficult. To deal with this environment, packet switches were deployed using the X.25 protocol. Among the many features of this protocol is error correction, which allows any node on the network to request a retransmission of errored data from the node that sent it, thus overcoming the poor performance of analog lines and equipment. For some applications, this makes X.25-based packet networks of value today even with the availability of high-speed digital networks.

Applications. Error correction in X.25 entails an overhead burden that limits network throughput. This in turn limits X.25 to niche applications, such as terminal-to-host interactive services like point-of-sale transaction processing, where the reliable transmission of credit card numbers and other financial information overrides speed as the primary concern.

Architecture. The X.25 standard defines three protocol layers that are used to interface various data terminal equipment (DTE) at the customer premises with data communications equipment (DCE) on service provider's network.

Physical layer. Layer 1 defines the physical, electrical, functional, and procedural characteristics required to establish the communications link between two devices. X.25 specifies the use of several standards for the physical connection of equipment to an X.25-based network. These standards include X.21, X.21*bis,* and V.24; the latter two are virtually identical to the EIA-232 standard. The Physical layer operates as a full-duplex, point-to-point synchronous circuit.

Data Link layer. X.25's Data Link layer corresponds to the second layer of the OSI model. At this layer, Link Access Procedure-Balanced (LAPB) is used to provide efficient and timely data transfer, synchronize the data link signals between the transmitter and receiver (flow control), perform error checking and error recovery, and identify and report procedural errors to higher levels of the system architecture. LAPB ensures the accurate transmission of packets that are delivered by the Network layer and contained in HDLC information frames between the DTE and the network.

This layer also defines the unit of data transfer: the frame (Fig. P1). The specific Data Link protocol determines the organization and interpretation of each field in the frame. The general definitions of each field are as follows:

- *Opening Flag* (8 bits): This field delimits or marks the beginning (opening flag).
- *Address Field* (8 bits): As a portion of the header, the Address Field identifies the destination of the frame.
- *Control Field* (8 bits): Also part of the header, the Control Field specifies the type of message (i.e., command or response), the frame sequence number, and other control information. The frame sequence number prevents a duplicate frame from being received unintentionally.
- *Information Field* (variable length): This field contains the Format Identifier, Logical Channel Number, Sequence Number, and User Data.
- *Frame Check Sequence* (16 bits): Transmitted after the data bits are sent, the FCS provides error checking using the cyclic redundancy check (CRC). Frames that are received with errors are retransmitted.
- *Closing Flag* (8 bits): This field delimits or marks end of a frame (closing flag). In some applications, the Closing Flag also acts as the Opening Flag for the next frame.

Network layer. The Network layer is the highest-level protocol stipulated in X.25. This layer provides access to services available on a public packet-switched

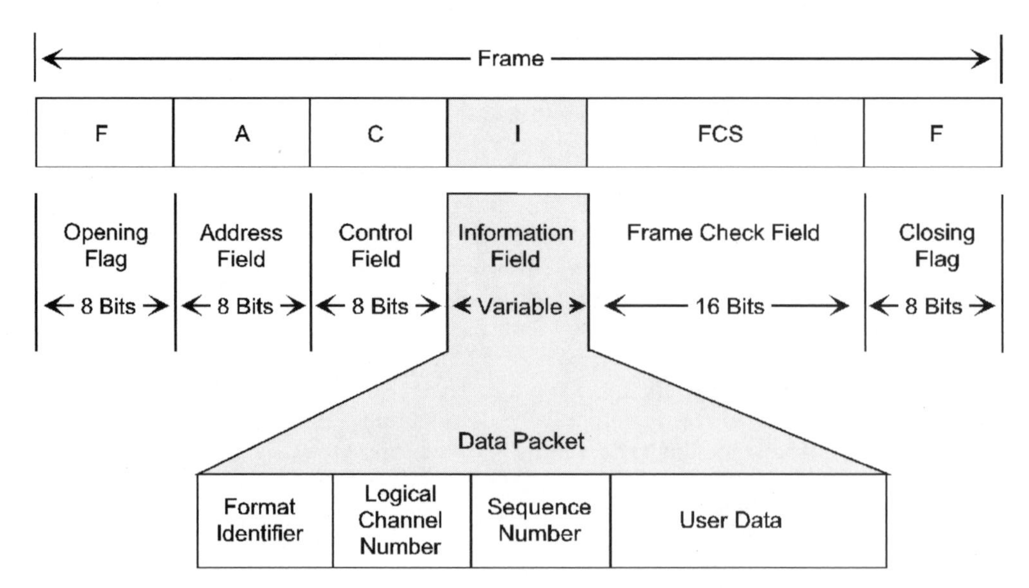

Fig. P1 *Structure of an X.25 frame.*

network. When users subscribe to an X.25 service, the packet data network (PDN) provides configuration parameters. These parameters include the following:

- *Gateway Address:* Can be dedicated or nondedicated and uses an adapter card that plugs into a communications server on a LAN or a dedicated gateway device.
- *Frame Size:* Specifies the maximum number of characters that can be sent on the line at one time.
- *Window Size:* The maximum number of packets that can be transmitted without an acknowledgment from the destination host or LAN.
- *Logical Channel Number:* Identifies either a switched or permanent virtual circuit.

Types of connections. X.25 specifies three types of connections for routing information over the network:

- *Permanent virtual circuit* (PVC): Resembles a leased line in that it is a permanent path between through a network that never changes unless manually reconfigured.
- *Switched virtual circuit* (SVC): A temporary path through a network that is maintained only for the duration of a data transmission. Switched virtual circuits are set up on request, maintained for the duration of the "call," and then released on request.
- *Datagram:* A simple delivery service, which operates on a best-effort basis, depending on bandwidth availability. Each message or packet contains enough information for it to be routed to the appropriate destination DTE without requiring that a call be established.

Packet transmission. X.25 specifies the means by which DTEs establish, maintain, and clear virtual circuits. The Network layer uses packet-interleaved statistical multiplexing to allow a DTE to set up concurrent virtual circuits with multiple DTEs. This multiplexing technique makes some basic assumptions:

- Typical virtual circuits do not always carry data.
- Frames of data are interleaved or mixed together to form full packets.
- Packet size varies by network.

X.25 assigns a logical channel number, which corresponds to a switched or permanent virtual circuit, to each packet. This channel number applies to both transmission directions. Logical channels are the equivalent of dial-in ports in a conventional timeshare network; they are a conceptual rather than physical path between the DTE and the network. When idle, logical channels are free to handle new calls requested by a local or remote DTE. After a call is established, the logical channel remains busy until the call is released.

A single logical channel number is used for a DTE that supports only one virtual circuit. If multiple virtual circuits are involved, the service company assigns a range of channel numbers to the DTE user. If a subscriber uses both permanent and switched virtual circuits, the X.25 service company statistically assigns individual permanent virtual circuits within a range of numbers, beginning with 1, while vir-

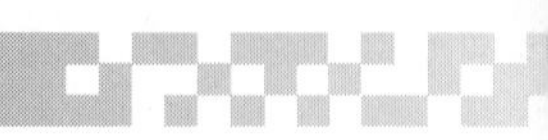

tual calls are assigned a second range of numbers above the first. Logical channel numbers for switched virtual circuits are dynamically assigned during call establishment and identify all of the packets sent while that call is in progress.

Virtual circuits. X.25 supports both permanent and switched virtual circuits. A permanent virtual circuit, generally suited for a fixed host-to-host connection, is used on point-to-point interfaces that do not require higher-level dynamic session-control features and the use of a data switching network. Host-to-host and host-to-terminal communications through a switched network generally use a switched virtual circuit.

A permanent virtual circuit performs the following functions:

- Assembles data into one logical channel using asynchronous time-division multiplexing.
- Packetizes the data stream into one logical channel.
- Performs packet-level error control using packet sequence numbers.
- Performs logical-channel flow control.
- Permits end-to-end confirmation of data delivery.

A switched virtual circuit has all of the characteristics of a permanent virtual circuit, plus the following:

- Provides a means to request the dynamic establishment of virtual circuits.
- Allows the host to accept or reject virtual-call requests from other hosts.
- Allows the host to take down a virtual call when data transfer is complete.

Delivery Confirmation. X.25's Delivery Confirmation procedure allows either the network or the DTE to select the maximum number of data packets on a virtual circuit. The network limits the number of packets based on network performance criteria, including throughput and resource availability. The DTE controls the maximum number of packets on the network with a higher-level DTE-to-DTE error-control protocol. If a DTE wants to receive end-to-end acknowledgment of the data it sends, the DTE sets the Delivery Confirmation bit in the packet's header to 1. The packet-receive sequence numbers, embedded in frames sent by the receiver, acknowledge data receipt. When a DTE activates Delivery Confirmation, the DTEs determine the maximum number of packets on the network. Delivery Confirmation limits the amount of unconfirmed data on a network and thereby facilitates error recovery. If the Delivery Confirmation bit is set to 0, acknowledgments have only local significance between DTE and DCE. In this case, the network determines the maximum number of packets within the throughput limits of the DTEs.

Throughput. A virtual circuit's maximum throughput varies according to allocated switch resources and the statistical multiplexing of data transmission. A virtual circuit's throughput is further limited by access line characteristics, including line speed, flow control parameters, and other call traffic at both the local and remote DTE/network boundaries. Use of the Delivery Confirmation procedure also affects throughput; the packet transfer rate is affected by the packet delivery confirmation rate from the receiving DTE. In addition, different types of national and international calls can vary the throughput limit.

On the other hand, the X.25 network maximizes throughput when the DTE access links at both ends of the virtual circuit are properly engineered, the receiving DTE does not control flow from the DCE, or the transmitting DTE sends full data packets.

Extended packet length. An X.25-based network can accommodate extended-length packets. The X.25 network can logically chain together data packets to convey a large block of related information. This improves throughput and minimizes delay by requiring fewer acknowledgments. This procedure is implemented when the packets have the Delivery Confirmation (D) bit set to 0 and the More-Data-To-Follow indicator bit (M) set to 1 (active)—until the last packet in the chain, where D is set to 1 and M is set to 0. This mechanism is also significant for flow control.

After the network establishes the virtual circuit, data packets can be sent across the logical channel. X.25 numbers each data packet and limits to seven the maximum number of packets that can be sent without additional authorization from the receiving DTE, DCE, or network; the default value is usually set to two. The actual limit is either set at subscription time or during call setup. (There is an extended mode of operation within LAPB that supports up to 127 packets.)

Flow control. Data packets carry a packet-receive sequence number that aids flow control. This sequence number authorizes the maximum number of unconfirmed packets that the logical channel can transmit. Either a DTE or the network can authorize transmission of one or more packets by sending a Receive Ready packet to the calling DTE. The packet-receive sequence numbers ensure that no error-free frames are lost or interpreted out of order.

When the Delivery Confirmation (D) bit is set to 0, the packet-receive sequence number provides local flow control information (i.e., packet acknowledgment has only local significance). When the Delivery Confirmation bit is set to 1, the packet-receive sequence number provides delivery confirmation information between the sending and receiving DTEs. Two communicating DTEs can operate at their locally determined packet size if the user includes the More-Data-To-Follow (M) indicator either in a full packet or in any packet that has the Delivery Confirmation bit set to 1. This indicator then informs the network and receiver that there is a logical continuation of data in the next packet on a particular logical channel.

The DTE may transmit interrupt packets even when the data packets are flow-controlled. These packets do not contain either send or receive sequence numbers. To maintain packet integrity, a network therefore can contain only one unconfirmed interrupt packet at a time between sending and receiving DTEs.

Error recovery. A typical data communications network performs error detection and recovery on various levels, some of which overlap:

- X.25 specifies several error checking levels.
- The network might provide some level of error control.
- The DTE/DCE software might contain error control mechanisms.

X.25 provides the following guidelines for handling packet-level errors:

- Procedural errors that occur during call establishment and clearing are reported to the calling DTE with a diagnostic packet that clears the call.

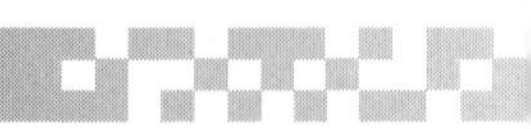

- Procedural errors that occur during the data transfer phase (such as loss of synchronization) are reported to the sending DTE with a diagnostic packet that resets the sequence counters of both the DTE and DCE.
- A diagnostic field, included in the packet, provides additional information to the DTE and to the network.
- Time-outs that resolve some deadlock conditions are defined for two major areas: the length of time the DTE has to respond to an incoming call (the minimum typically 3 minutes), and the amount of time the DCE has to wait for confirmation of a reset, clear, or restart packet. To avoid looping conditions, the DCE takes an appropriate action for the indication packet and continues operation.
- Misalignments of subscription options between the DTE and the DCE can cause DTE procedural errors.
- Error tables, which define the actions to be taken by the DCE on receipt of various packet types in various stages of the interface and the state to which the DCE enters, define the diagnostic code generated for each error condition.

X.25 also identifies a number of special error cases (such as a packet received on an unassigned logical channel) that cause a diagnostic packet to be sent to the DTE rather than resetting or clearing the logical channel. A diagnostic packet includes a diagnostic code and the logical channel number on which the error occurred. There are diagnostic codes for reset, clear, and restart packets. Because the diagnostic packet is nonprocedural, it does not affect the normal meanings of call progress signals, nor is a DTE required to take action on receipt of a diagnostic packet. The DTE logs diagnostic packets for troubleshooting information.

The transmitting DTE, the receiving DTE, and the network can detect errors in transferred data packets. If an error is detected by a DTE, it informs the other DTE and requests that the affected packets be resent. If the network detects an error, it informs both DTEs by sending a reset call-progress signal. These signals include remote DTE out-of-order (permanent virtual circuit only), procedural error at the remote DTE/network boundary, network congestion, or the inability of the remote DTE to support a particular function.

Data generated before and after an error-caused reset occurs is handled in one of two ways. If a reset occurs before data reaches its destination, that data either continues to its destination or, more likely, is discarded by the network. Data generated after both local and remote ends recover from the reset continues to its destination. Data generated by a remote DTE before it receives the error indication from the local DTE either continues to its destination or, again more likely, the network discards it. In this case, the appropriate DTE resends discarded packets. The assigned resources for a given virtual circuit and the network end-to-end transmission delay and throughput characteristics determine the maximum number of packets that may be discarded.

Optional user facilities. The various optional features that apply to the subscriber's network are determined at the time of subscription or as requested specifically as part of the call establishment procedure. The X.25 user facilities described in the next several paragraphs may be activated within the call request packet.

Closed user group facility. As an alternative to having a private data network for manageability and security needs, companies can establish closed user groups on the PDN between a group of users and the network administrator for a specified length of time.

Flow control parameter selection. A network administrator can restrict access at the data level by using specific packets and window sizes to prevent unauthorized users from communicating with the X.25 gateway. With this option, any network user without the correct configuration is denied access. Specified either at the time of subscription or during call establishment, flow control parameters include packet size and window size. (Window size determines the maximum number of packets on a network without additional authorization from the receiving DTE.) X.25 supports the following packet sizes: 16, 32, 64, 128, 256, 512, 1024, 2048, and 4096 bytes. The maximum window size is seven, with two being the most commonly used.

Throughput class negotiation. *Throughput class* is the measure of the throughput that is not normally exceeded on a virtual circuit. It is a characteristic of virtual circuits and is a function of the amount of network resources allocated to the circuit. The X.25 network and the user DTE decide default values for the maximum throughput class associated with a virtual circuit, but these values may not always be attained because of overall link utilization, network congestion, and host processing.

One-way outgoing logical channel. This optional feature restricts the use of a range of logical channels to outgoing calls only. This restraint does not affect the full-duplex data transfer process.

Incoming or outgoing call barring. X.25 provides two call-barring service options. The first bars the presentation of incoming calls to the DTE, although the DTE can initiate outgoing calls. The second, outgoing call barring, prevents the DCE from accepting calls from a DTE; however, the DTE can receive incoming calls.

Fast select facility. The fast select facility, a variation on switched virtual circuit service, is designed to satisfy short, low-volume, transaction-based applications such as point-of-sale, funds transfer, credit checks, and meter reading. Fast select allows for the inclusion of up to 128 bytes of data in the call establishment and clearing procedures for a switched virtual circuit.

Dial X.25 (X.32). Dial X.25, or X.32, allows users to dial synchronously into a PDN over public telephone lines. This service option is designed for companies that use a PDN only occasionally or are just beginning to use an X.25 service. X.32 saves a company the cost of leasing a dedicated packet-switched line, and it allows users to access the network from unsupported locations while gaining complete error-detection facilities.

Other protocols. There are other standards that govern various aspects of X.25 packet-switched networks. Some of the most commonly used are:

- *X.3* defines the functions of the packet assembler/disassembler (PAD), which is used to communicate with a remote X.25 device connected to the PDN.
- *X.28* defines the procedures used by an asynchronous terminal to connect with a PAD.

- *X.29* defines the procedures that allow a packet mode device to control the operation of a PAD.
- *X.75* defines the gateway procedures for interconnecting X.25 PDNs, giving end hosts the appearance of a single X.25 network.

The relationships of these packet network standards are shown in Fig. P2.

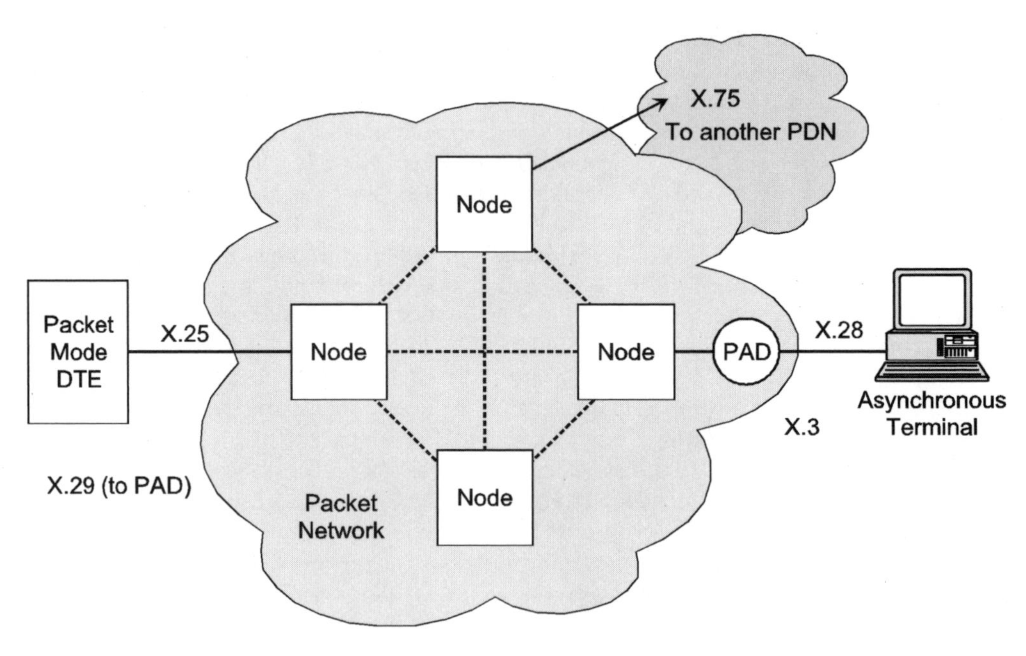

Fig. P2 *Common packet network standards.*

Summary. X.25 offers error-free communications and guaranteed delivery, making it the best choice for financial transactions and for companies that must establish international networks in countries that still have analog-based communications infrastructures in place. X.25 provides connectivity with legacy mainframes, minicomputers, and LANs. Despite the emergence of higher-speed cell- and frame-switched services that operate over more reliable digital links, such as ATM and frame relay, there is still strong demand for X.25 products and services.

See also
ASYNCHRONOUS TRANSFER MODE
FRAME RELAY

Paging

A paging system provides one- or two-way wireless messaging to give mobile users continuous accessibility to family, friends, and business colleagues while they are away from telephones. The mobile user typically carries a palm-sized device (the pager or some other portable device with a paging capability) that has a unique identification number. The calling party inputs this number, usually through the public telephone network, to the paging system, which then signals the pager to alert the called party.

Alternatively, callback numbers and short text messages can be sent to pagers via messaging software installed on a PC, or input into forms accessed on the World Wide Web for delivery via an Internet gateway. Regardless of delivery method, the called party receives an audio or visual notification of the call, which includes a display of the phone number to call back. If the pager has an alphanumeric capability, messages can be displayed on the pager's screen.

Paging applications. There are many applications for paging. Among the most popular are:

- *Mobile messaging* allows messages to be sent to mobile workers. They can respond with confirmation or additional instructions.
- *Data dispatch* allows managers to schedule work appointments for mobile workers. Upon activating their pagers each morning, their itineraries will be waiting for them.
- *Single-key callback* allows the user to read a message and respond instantly with a predefined stored message that is selected with a single key.

Some message paging services are compatible with text messaging software programs, allowing users to send messages from their desktop or notebook computers to individuals or groups. This kind of software also keeps a log of all messaging activity. This method also offers privacy because messages do not have to go through an operator before being delivered to a recipient.

Types of paging services. There are several types of paging services available.

Selective operator-assisted voice paging. Early paging systems were nonselective and operator-assisted. Operators at a central control facility received voice input messages, which were taped as they came in. After an interval of typically 15 minutes or so, these messages were then broadcast and received by all the paging system subscribers. This meant that subscribers had to tune in at appointed times and listen to all messages broadcast to see if there were any messages for them. Not only did it waste air time, the system was inconvenient, labor-intensive, and offered no privacy.

These disadvantages were overcome with the introduction of address encoders at the central control facility and associated decoders in the pagers. Each pager was given a unique address code. Messages intended for a particular called party were input to the system, preceded by this address. In this way, only the party addressed was alerted to switch on his or her pager to retrieve messages.

With selective paging, tone-only alert paging became possible. The called party was alerted by a beep tone to call the operator or a prearranged home or office number to have the message read back.

Automatic paging. Traditionally, an operator was always needed either to send the paging signal or to play back or relay messages for the called party. With automatic paging, a telephone number is assigned to each pager and the paging terminal can automatically signal for voice input (if any) from the calling party, after which it will automatically page the called party with the address code and relay the input voice message.

Tone and numeric paging. Voice messages take up a lot of air time and, as the paging market expands, frequency overcrowding becomes a potentially serious problem. Tone-only alert paging saves on-air time usage but has the disadvantage that the alerted subscriber knows only that he or she has to call certain prearranged numbers based on the kind of alert tone received.

With the introduction of numeric display pagers in the mid-1980s, the alert tone is followed by a display of a telephone number to call back, or a coded message. This method resulted in great savings in air time usage because it was no longer necessary to add a voice message after the alert tone. This is still the most popular form of paging.

Alphanumeric paging. Alphanumeric pagers display text or numeric messages entered by the calling party or operator using a modem-equipped computer or a custom page-entry device designed to enter short text messages. Although alphanumeric pagers have captured a relatively small market thus far, this could change with the introduction of value-added services that include news, stock quotes, sports scores, traffic bulletins, and other specialized information services.

Ideographic paging. Pagers capable of displaying different ideographic language—Chinese, Japanese, and others—also are available. The particular language supported is determined by the firmware (computer program) installed in the pager and in the page-entry device. The pager is similar to that used in alphanumeric display paging.

Paging system components. The key components of a paging system include an input source, the existing wireline telephone network, the paging encoding and transmitter control equipment, and the pager itself.

Input source. A page can be entered from a phone, computer with modem or other type of desktop page-entry device, a PDA, or through an operator who takes a phone-in message and enters it on behalf of the caller. Various forms posted on the World Wide Web also can be used to input messages to pagers (Fig. P3 and Fig. P4).

Telephone network. After the message is input, the page is sent through the public switched telephone network (PSTN) to the paging terminal for encoding and transmission through the wireless paging system. Typically the encoder accepts the incoming page, checks the validity of the pager number, looks up the directory or database for the subscriber's pager address, and converts the address and message into the appropriate paging signaling protocol. The encoded paging signal is then sent to the transmitters (base stations), through the paging transmission control systems, and broadcast across the coverage area on the specified frequency.

Encoder. Encoding devices convert pager numbers into pager codes that can be transmitted. There are two ways in which encoding devices accept pager numbers, manual and automatic. In manual encoding, a paging system operator enters pager numbers and messages via a keypad connected to the encoder. In automatic encoding, a caller dials up an automatic paging terminal and uses the phone keypads to enter pager numbers. Regardless of the method used, the encoding device then generates the paging code for the numbers entered and sends the code to the paging base station for wireless transmission.

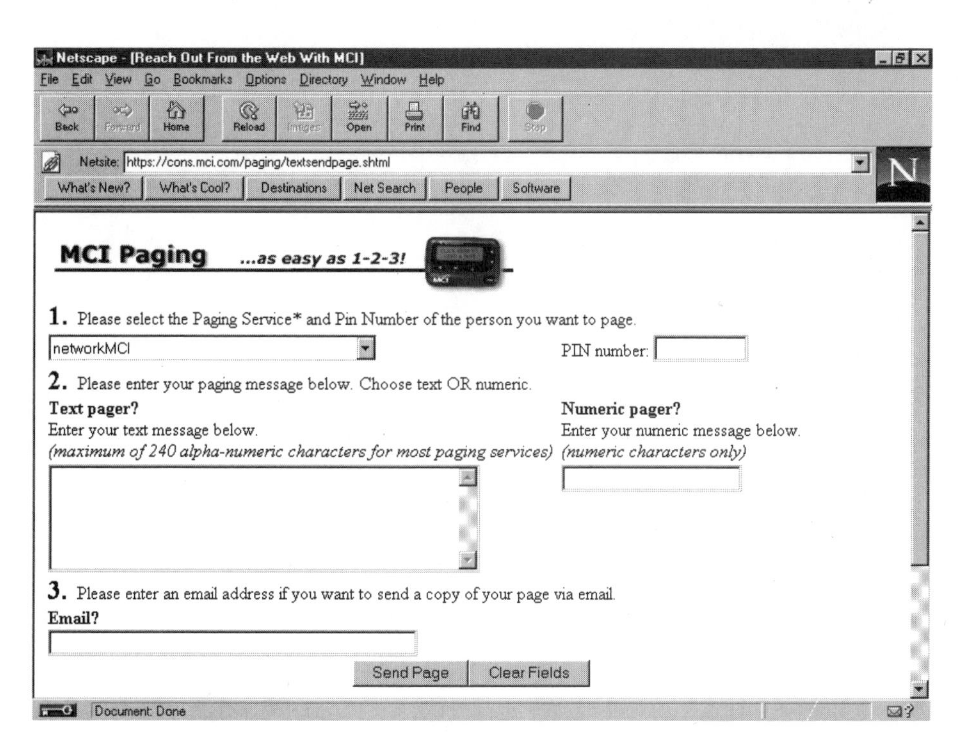

Fig. P3 *MCI offers a Web pager that lets anyone send a page to anyone who has a networkMCI or SkyTel one-way pager. All the user needs is the pager's PIN to send a 240-character message to alphanumeric pagers or 10 characters to a numeric-only pager.*

Base station transmitters. The base station transmitters send page codes on an assigned radio frequency. Most base stations are specifically designed for paging, but those designed for two-way voice can be used as well.

Pagers. Pagers are essentially FM receivers tuned to the same RF frequency as the paging base station. A decoder unit built into each pager recognizes the unique code assigned to the pager and rejects all other codes for selective alerting. Pagers can be assigned the same code for group paging, however. There also are pagers that can be assigned multiple page codes, typically up to a maximum of four, allowing the same pager to be used for a mix of individual and group paging functions.

Signaling protocols. The paging terminal in a paging system, after accepting an incoming page and validating it, will encode the pager address and message into the appropriate paging signaling protocol. The signaling protocol allows individual pagers to be uniquely identified/alerted and to be provided with the additional voice message or display message if any.

Various signaling protocols are used for the different paging service types, such as tone-only, tone and voice, etc. Most paging networks are able to support many different paging formats over a single frequency. Many paging formats are manufacturer-specific and often proprietary, but there are public-domain protocols, such

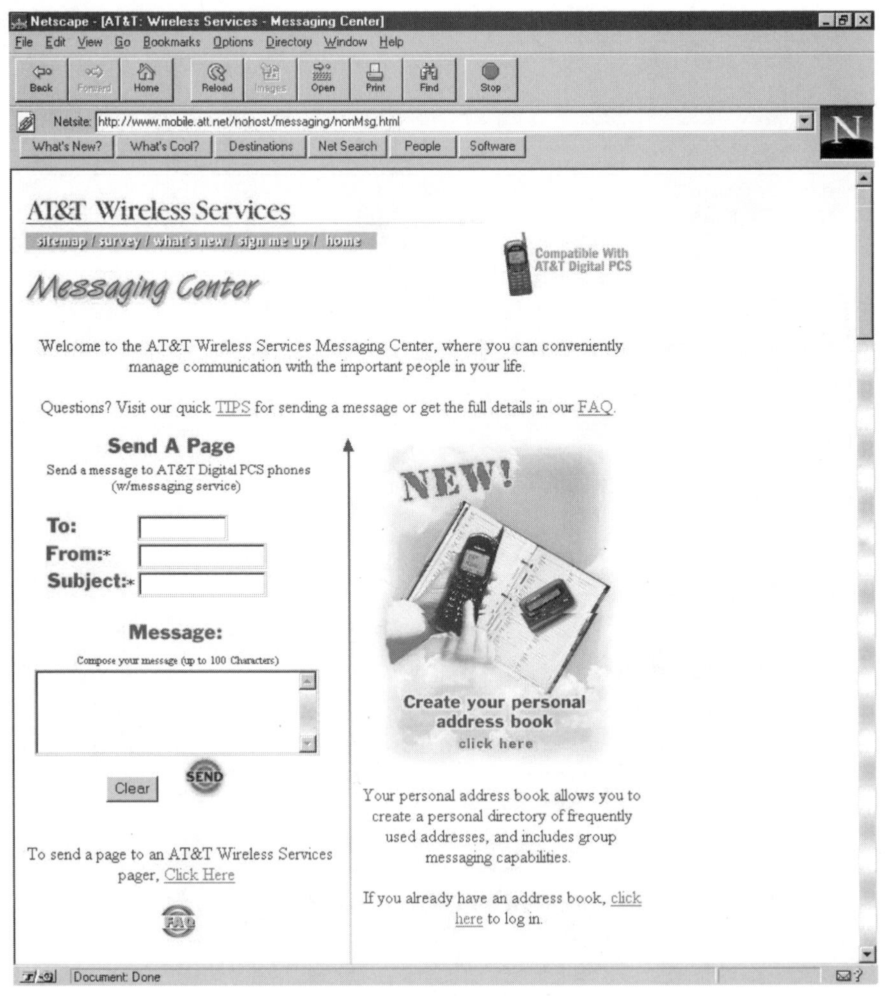

Fig. P4 *AT&T Wireless offers a Web pager that lets anyone send a message to people who carry AT&T PCS phones, alphanumeric pagers, and CDPD-compliant PocketNet phones.*

as the Post Office Code Standardization Advisory Group (POCSAG), that allow different manufacturers to produce compatible pagers.

POCSAG is a public-domain digital paging format adopted by many manufacturers around the world. It can accommodate two million codes (pagers), each capable of supporting up to four addresses for such paging functions as tone-only, tone and voice, and numeric display. POCSAG operates at data rates of up to 2400 bps. At this rate, sending a single, tone-only page requires only 13 milliseconds. This is about 100 times faster than two-tone paging.

With the explosion of wireless technology and dramatic growth in the paging industry in many markets, existing networks are becoming more and more overcrowded. In addition, RF spectrum is not readily available because of demands by other wireless applications. In response to this problem, Motorola has developed a one-way messaging protocol called Flex (Feature-rich Long-life Environment for eXecuting), which is intended to transform and broaden paging from traditional low-end, numeric services into a range of PCS/PCN and other wireless applications.

Relative to POCSAG, Flex can transmit messages at up to 6400 bps and permit up to 600,000 numeric pagers on a single frequency, compared to POCSAG's 2400 bps transmission rate and 300,000 users per frequency. In addition, Flex provides enhanced bit error correction and much higher protection against signal fades common in FM simulcast paging systems. The combination of increased bit error correction and improved fade protection increases the probability of receiving a message intact, especially longer alphanumeric messages and data files that will be sent over PCS/PCN. Motorola has also developed ReFlex, a two-way protocol that will allow users to reply to messages, and InFlexion, a protocol that will enable high-speed voice messaging and data services at up to 112 kbps.

Summary. The computer hardware and software used in radio paging systems have also evolved from simple operator-assisted systems to terminals that are fully computerized, with such features as message handling, future delivery, user-friendly prompts to guide callers to a variety of functions, and automatic reception of messages. Paging's low cost, ease of use, and numerous practical benefits make it one of the fastest-growing communications services. Today there are more than 65 million paging subscribers worldwide, of which more than half—some 35 million—are Americans. Global demand is expected to grow by 25 to 30 percent a year during the next few years.

See also

ELECTRONIC MAIL
PERSONAL COMMUNICATION SERVICES
PERSONAL DIGITAL ASSISTANTS

Pay-per-call services

Pay-per-call services, also known as "audiotext" or "900" services, provide telephone users with a variety of recorded and interactive information programs for which they are charged rates different from, and usually higher than, the normal transmission rates for ordinary telephone calls.

In 1991 the FCC adopted regulations governing interstate pay-per-call services to address complaints from consumers of widespread abusive practices involving 900 services. Among other protective measures, the FCC:

- Required that pay-per-call programs begin with a preamble disclosing the cost of the services and affording the caller an opportunity to hang up before incurring charges.
- Required local exchange carriers (LECs), where technically feasible, to offer telephone subscribers the option of blocking access to 900 numbers.
- Prohibited common carriers from disconnecting basic telephone service for failure to pay pay-per-call charges.

To expand upon this regulatory framework, Congress enacted legislation in 1992 requiring both the FCC and the Federal Trade Commission (FTC) to adopt rules intended to increase consumers' protection from fraudulent and deceptive practices and promote the development of legitimate pay-per-call services. In response to complaints from consumers, businesses, and organizations alleging that they had been billed for calls made from their phones to toll-free numbers, this legislation also mandated explicit restrictions on the use of 800 and other toll-free numbers to provide information services.

In mid-1993, the FCC amended its pay-per-call regulations to be consistent with the Congressional mandate. The new rules required that all interstate pay-per-call services be provided through 900 numbers. In other words, use of 800 numbers, or any other number advertised or widely understood to be toll-free, cannot be used to charge callers for information services.

Summary. Even with these safeguards, carriers and information providers are still free to use 800 numbers to provide a wide variety of information services. For example, information services charged on a per-call basis may be made using 800 numbers when they are charged to a credit card or provided under a written pre-subscription arrangement. The safeguards simply recognize the significant governmental interest in shielding consumers from deceptive practices associated with a service that the public widely perceives as free.

PCS 1900

PCS 1900 is the American National Standards Institute (ANSI) radio standard for 1.9 GHz personal communications service (PCS) in the United States. As such, it is compatible with the Global System for Mobile (GSM) telecommunications, an international standard adopted by 160 operators supporting 30 million subscribers in 86 countries. GSM-based networks are expected to be established in more than 100 countries by the end of the decade, with 230 operators and a combined subscriber base of more than 100 million.

PCS 1900 can be implemented with either TDMA or CDMA technology. TDMA-based technology enjoys an initial cost advantage over rival CDMA equipment because suppliers making TDMA infrastructure equipment and handsets have already reached economies of scale. In contrast, CDMA equipment is still in its first generation and, therefore, is generally more expensive.

At present, the CDMA (IS-95) standard has been chosen by about half of all the PCS licensees in the U.S., giving it the lead in the total number of potential subscribers. The first operational PCS networks have been using PCS 1900 as their standard, however, mainly because of the maturity of the GSM-based technology employed.

Although similar in appearance to analog cellular service, PCS 1900 is based on digital technology. As such, PCS 1900 provides better voice quality, broader coverage, and a richer feature set. Not only is voice quality improved, but fax and data transmissions are more reliable. Laptop computer users can connect to the handset with a PCMCIA card and send fax and data transmissions at higher speeds with less chance of error.

Architecture. The PCS 1900 system architecture consists of four major components:

- *Switching System:* Controls call processing and subscriber-related functions.
- *Base Station:* Performs radio-related functions.
- *Operation and Support System* (OSS): Supports the operation and maintenance activities of the network.
- *Mobile Station:* Is the end-user device that supports voice and data communications as well as short message services.

The Switching System contains five main functional elements (Fig. P5):

- *Mobile Switching Center* (MSC): Performs the telephony switching functions for the network. It controls calls to and from other telephone and data communications networks such as public switched telephone networks (PSTN), Integrated Services Digital Network (ISDN), Public Land Mobile Radio Services (PLMRS) networks, public data networks (PDN), and various private networks.
- *Visitor Location Register* (VLR) database: Contains all temporary subscriber information needed by the MSC to serve visiting subscribers.
- *Home Location Register* (HLR) database: Stores and manages subscriptions. It contains all permanent subscriber information including the subscriber's service profile, location information, and activity status.
- *Authentication Center* (AC): Provides authentication and encryption parameters that verify the user's identity and ensure the confidentiality of each call. This functionality protects network operators from common types of fraud found in the cellular industry today.
- *Message Center* (MC): Supports numerous types of messaging services, for example voice mail, facsimile, and e-mail.

Advanced services and features. Like GSM, PCS 1900's digital orientation makes possible several advanced services and features that are not efficiently and economically supported in analog cellular networks. Among them are:

- *Short Message Service:* This allows alphanumeric messages up to 160 characters to be sent to and from PCS 1900–compatible handsets. Short Message Service applications include two-way point-to-point messaging, confirmed message delivery, cell-based messaging, and voice mail alert. These messaging and paging capabilities create a broad array of potential new revenue-generating opportunities for carriers.
- *Voice Mail:* The PCS 1900 network provides one central voice mailbox for both wired and wireless service. In addition, the voice mail alert feature ensures that subscribers do not miss important messages.
- *Personal Call Management:* This offers subscribers a single telephone number for all their physical telecommunication devices. For example, a single number can be assigned for home and mobile use, or office and mobile use. This allows subscribers to receive all calls regardless of their physical location.

➢ *Data Applications:* Wireless data applications that can be supported by PCS 1900 networks include Internet access, electronic commerce, and fax transmission.

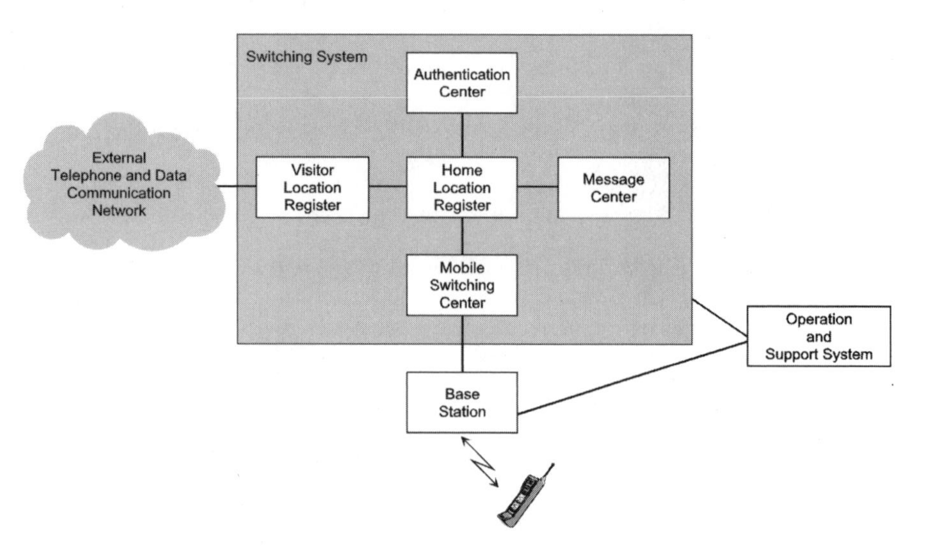

Fig. P5 *PCS 1900 switching system architecture.*

Smart cards. The PCS 1900 standard supports the Smart Card, which provides similar features as GSM's Subscriber Identity Module (SIM). The size of a credit card, Smart Cards contain embedded computer chips with user profile information. By removing the Smart Card from one PCS 1900 phone and inserting it into another PCS 1900 phone, the user is able to receive calls at that phone, make calls from that phone, or receive other subscribed services such as wireless Internet access. The handsets cannot be used to place calls (except 911 emergency calls) until the subscriber inserts the Smart Card and enters a personal identification number (PIN).

The profile information stored in the Smart Card also enables international roaming. When traveling in the United States, international GSM customers will be able to rent handsets, insert their SIMs, and access their services as if they were back home. By the same token, when U.S. subscribers travel internationally to cities with compatible networks and mutual roaming agreements, they need only take their Smart Card with them to access the services they subscribed to back home via the local GSM network.

Like SIMs, Smart Cards also provide storage for features such as frequently called numbers and short messages. Smart Cards also include the AT modem command set extensions, which integrate computing applications with cellular data communications. In the future, Smart Cards and PCS 1900 technology also will link subscribers to applications in electronic commerce, banking, and health care.

Summary. PCS 1900 is a frequency-adapted version of GSM, which is made necessary because the FCC assigned the 1.9 GHz frequency for broadband PCS. Otherwise, PCS 1900 and GSM are similar in all respects, including the network architecture and types of services supported. An advantage U.S. carriers have in supporting the PCS 1900 standard is that it is interoperable with the worldwide GSM standard, which mean customers can roam globally, since most countries support GSM.

See also

GLOBAL SYSTEM FOR MOBILE (GSM) TELECOMMUNICATIONS
PERSONAL COMMUNICATIONS SERVICES

PCS-over-cable

Emerging digital personal communications services (PCS) are designed to offer high voice quality and support an array of feature-rich services not available over analog cellular networks. The lightweight phones are small enough to fit in a pocket and battery life greatly surpasses that of conventional cellular communications devices. Many carriers are using Code Division Multiple Access (CDMA), a spread-spectrum technology, to implement PCS as well as to provide a digital overlay to their analog networks.

Implementing PCS with the usual tower-in-a-cell configuration is both expensive and time-consuming. This has led to the development of an alternative architecture called *PCS-over-cable,* which implements CDMA over existing cable television (CATV) or emerging hybrid fiber/coax (HFC) networks. CDMA is more compatible with cable infrastructures than Time Division Multiple Access (TDMA), a competing technology, and requires less frequency planning over cable's extensive terrestrial network.

Cox Communications—one of the nation's largest cable television operators, serving about 3.2 million customers—pioneered the development of PCS-over-cable and demonstrated the nation's first PCS phone call through a cable television system in 1992. In December 1996, Cox announced the commercial availability of its PCS-over-cable service in San Diego. Cox chose CDMA as the underlying technology because it provides the best long-term solution for a seamless network. Since then, other service providers have adopted CDMA for their PCS networks, including Sprint PCS, which hopes to have a nationwide PCS service fully operational by year-end 1997.

Approaches to PCS-over-cable. Many cable television operators have purchased PCS licenses through the FCC auction process. Because companies paid premium prices for these licenses, they have every incentive to introduce PCS services as soon as possible to recover their huge investments in radio spectrum. Because of their higher frequency and lower base station output power, PCS networks require the installation of considerably more sites than standard cellular networks.

Acquiring locations for base stations is becoming difficult and expensive, in large part because providers increasingly face community opposition to proposed site locations. Leveraging the extensive physical plant of cable television operators can overcome these obstacles, however. Two PCS-over-cable methods are available: the *cable microcell integrator/head-end interface converter* (CMI/HIC) and *fiber microcells.* Either method offers CATV operators a viable option for quickly providing PCS coverage in their service areas and generating an additional revenue stream.

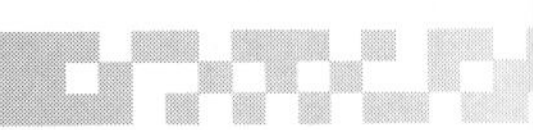

CMI/HIC. The first method of leveraging existing cable is known as cable microcell integrator/head-end interface converter (CMI/HIC), sometimes known as the remote antenna driver/remote antenna signal processor (RAD/RASP). The CMI is a small box that is similar in size to a standard cable amplifier. Where aerial facilities are available, the CMI is stand-mounted, with three small antennas mounted alongside; if the plant is underground, the CMI can be buried and connected to the antennas mounted above ground.

The subscriber's signal is received by the antennas and converted from 1900 MHz to the cable's frequency (5 to 40 MHz), amplified by the CMI and sent to the head end. There the HIC sends the PCS signal to the associated base station. For a call to the subscriber, the process is reversed, but the PCS signal is converted to a frequency between 450 and 550 MHz. Each CMI covers about a 2000-ft radius, much smaller than a standard base station. Several CMIs can simulcast the same frequency, however, to create an area similar in size to a base station's.

Fiber microcells. Fiber-optic microcells allow CATV/PCS operators to deploy a network rapidly by centrally locating base station equipment. The microcell consists of a hub unit and remote units. The hub, which is the microcell's interface to the base station, usually serves up to four remote units. Small, remote fiber-optic microcells can be placed in the desired locations to distribute RF signals, reducing the number of standard sites. Because fiber has a low attenuation factor, remote units can be placed up to 20 km away from the hub equipment. Initially, a number of fiber-optic microcells can be connected to a single base station. As capacity demands increase, capacity can be added at the distribution point, so each microcell acts as its own cell site.

This microcell is not limited to a specific antenna; directional and omni antennas can be used to control where the signal is sent, which also can facilitate frequency planning. Because fiber-optic microcells operate at higher power than CMI antennas, fewer microcells need to be installed. With an appropriate power amplifier, a fiber microcell can serve nearly the same coverage area as a conventional base station.

Summary. With locations for new base stations and RF antennas becoming increasingly more difficult to obtain, CATV/PCS operators must look for alternative methods for implementing PCS services. Fiber microcells and CMI/HIC provide economical and quickly deployable solutions. Both support CDMA, which offers the added advantages of better voice quality and higher call capacity than competing TDMA. Support for CDMA also provides better integration between terrestrial CATV networks and wireless PCS networks, which will become increasingly important in the future.

See also

CABLE TELEVISION NETWORKS
PERSONAL COMMUNICATIONS SERVICES

Personal Access Communications Systems

PACS (Personal Access Communications Systems) is a standard adopted by ANSI for personal communications services (PCS). Adopted in June, 1995, PACS provides an approach for implementing PCS in North America that is fully compatible with the local-exchange telephone network and interoperable with existing cellular systems. Based on the Personal Handyphone System (PHS) developed in Japan and the Wireless Access Communications System (WACS) developed by Bellcore, PACS is designed to support mobile and fixed applications in the 1.9 GHz frequency range. It promises low installation and operating costs, while providing very high quality for voice and data services. In the United States, limited trials of PACS equipment began in 1995 and equipment rollout began in 1996.

Most of the standards (including upbanded versions of CDMA, TDMA, and GSM) look like cellular systems in that they have high transmit powers and receivers designed for the large delay spreads of the macrocell environment, and typically use low bit-rate voice coders (vocoders). PACS fills the niche between these classes of systems, providing high-quality services, high data capability, and high user density in indoor and outdoor environments. PACS equipment is simpler and less costly than macrocell systems, yet more robust than indoor systems.

PACS capabilities include pedestrian and vehicular-speed mobility, data services, licensed and unlicensed systems, simplified network provisioning, maintenance, and administration. Key features of PACS include:

- Voice and data services comparable in quality, reliability, and security with wireline alternatives.
- Optimized service to the in-building, pedestrian, and city-traffic operating environments.
- High cost-effectiveness to serve high-density traffic areas.
- Small, inexpensive, line-powered radio ports for unobtrusive pole or wall mounting.
- Signal processing with low complexity per circuit.
- Low transmit power and efficient sleep mode requiring only small batteries to power portable subscriber units for hours of talk time and multiple days of standby time.

Like PHS, PACS uses 32 kbps ADPCM waveform encoding, which provides virtual landline voice quality. ADPCM has demonstrated a high degree of tolerance to the cascading of vocoders, as experienced when a mobile subscriber calls a voice mail system, and the mailbox owner retrieves the message from a mobile phone. Under other mobile technologies, the message becomes unintelligible, but with PACS it is very clear. Similarly, the compounding of delays in mobile-to-PCS through satellite calls (a routine situation in Alaska and many developing countries) can be troublesome. PACS provides extremely low delay.

The low complexity and transmit power of PACS yield limited cell sizes, which makes it well-suited for urban and suburban applications where user density is high. Antennas can be installed inconspicuously, piggybacking on existing structures. This avoids the high costs and delays associated with obtaining permits for the construction of high towers.

Applications. Wireless local loops, pedestrian venues, commuting routes, and indoor wireless venues are typical PACS applications. Additionally, PACS is designed to offer high capacity, superior voice quality, and ISDN data services. Inter-

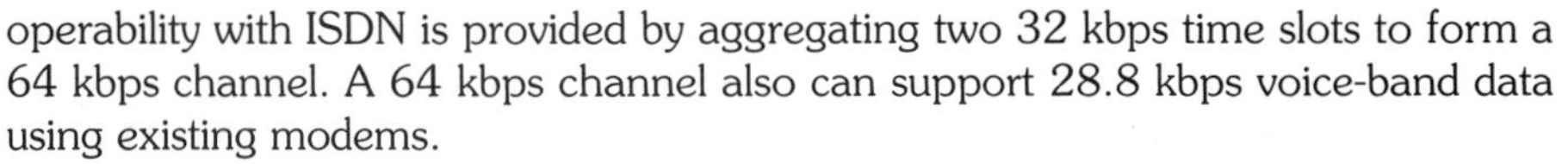

operability with ISDN is provided by aggregating two 32 kbps time slots to form a 64 kbps channel. A 64 kbps channel also can support 28.8 kbps voice-band data using existing modems.

PACS also can be used for providing wireless access to the Internet. The packet data communications capabilities defined in the PACS standards, together with the ability to aggregate 32 kbps channels, makes it possible for users to access the Internet from their personal computers equipped with suitable wireless modems at speeds of up to 200 kbps. When using the packet mode of PACS for Internet traffic, radio channels are not dedicated to users while they are on active Internet sessions, which can be very long. Rather, users use radio resources only when they are sending or receiving data, resulting in very efficient operation and minimally impacting the capacity of the PACS network to support voice communications.

Bellcore has designed PACS to support the full range of AIN services, including custom calling features, terminal and personal mobility, etc. As new AIN features are developed, the Bellcore standards defining this technology will evolve to facilitate incorporation of the new services into the PCS technology.

Summary. The market for PCS will be very competitive. PCS already is exerting downward price pressure on cellular services where the two compete side by side. PACS will permit PCS operators to differentiate their offerings through digital voice clarity, high-bit-rate data communications, and advanced intelligent network services—all in a lightweight handset. Moreover, the cost savings and ease-of-use associated with PACS will make it very economical for residential and business environments compared to competing high-powered wide area systems.

See also

ADVANCED INTELLIGENT NETWORK
PERSONAL HANDYPHONE SYSTEM
PERSONAL COMMUNICATIONS SERVICES
WIRELESS COMMUNICATIONS SERVICE

Personal communications services

PCS is a set of wireless communication services personalized to the individual. Subscribers can tailor their service packages to include only the services that they want, which might include stock quotes, sports scores, headline news, voice mail, e-mail and fax notification, and caller ID. Telephone numbers used in PCS handsets are tied specifically to individuals and will belong to them for as long as they wish, regardless of whether they move to other locations. These handsets have full roaming capability, allowing anywhere-to-anywhere communication.

Unlike the existing cellular network, PCS is completely digital. The digital nature of PCS allows antennas, receivers, and transmitters to be smaller. It also allows for the simultaneous transmission and reception of data and voice with no performance penalty. PCS eventually will overtake cellular technology as the preferred method of wireless communication.

Several technologies are being used to implement PCS. In Europe the underlying digital technology for PCS is GSM. Personal communications networks (PCN) in Europe operate at 1800 MHz. GSM has been adapted for operation at 1900

MHz for PCS in the United States (i.e., PCS 1900). Other versions of GSM are employed to provide PCS services in other countries, such as the Personal Handyphone System (PHS) in Japan. Many U.S. service providers have standardized their PCS networks on CDMA.

The PCS network. A typical PCS network operates around a system of *microcells* (smaller versions of a cellular network's cell sites), each equipped with a base station transceiver. The microcell transceivers require less power to operate, but cover a more limited range. The base stations used in the microcells can even be placed indoors, allowing seamless coverage as the subscriber walks into and out of buildings. Smaller versions of these base station transceivers are even available for residential installation. This not only allows subscribers to take full advantage of their PCS services at home, but in certain cases it allows them to bypass the local phone company.

Similar to packet radio networks today, terminal devices stay connected to the network even when not in use, allowing the network to locate an individual within the network via the nearest microcell, and routing calls and messages directly to the subscriber's location. For a "follow me" service that incorporates more than one device, a subscriber may be required to turn on a pager, for example, to receive messages on that device. If the subscriber receives a phone call while the pager is on, the network may store the call, take a message, or send the call to a personal voice mail system, and simultaneously page the subscriber. In this way, PCS allows the concept of universal messaging to be fully realized.

Cellular switching systems currently operate separately from the public switched network. When cellular subscribers call a landline phone (and vice versa), the two systems are interconnected to complete the communications circuit. In the future, many PCS services will be switched on the same switches that service the wireline network so that the only distinction between a wireless and wireline call will be the medium used to complete the last leg of the communication at each end of the circuit. In some places, wireless PCS and cable television (CATV) already are integrated in a unified CDMA-based architecture called *PCS-over-cable.* The use of CATV allows PCS to reach more potential subscribers with a lower start-up cost for service providers.

Broadband and narrowband. There are two technically distinct types of PCS, narrowband and broadband, each of which operates on a specific part of the radio spectrum and has unique characteristics.

Narrowband PCS is intended for two-way paging and other types of communications that handle small bursts of data. These services have been assigned to the 900 MHz frequency range, specifically 901 to 912, 930 to 931, and 940 to 941 MHz.

Broadband PCS is intended for more sophisticated data services. These types of services have been assigned a frequency range of 1850 to 1990 MHz, the same spectrum previously reserved for microwave users. The FCC has opened up this range for broadband PCS with the understanding that microwave users must eventually migrate out of that area of the spectrum. The FCC has set up rules to facilitate this process, which is supposed to be completed on a voluntary basis in 1998.

Both narrowband PCS and broadband PCS license ownership has been determined by public auction. PCS service areas are divided into 51 regional service areas, which are subdivided into a total of 492 metropolitan areas. There will be competition in each service area by at least two service providers. There will be 10 national service providers, plus 6 regional providers in each of 5 multistate regions called *major trading areas.*

An unlicensed portion of the PCS spectrum has been allocated from 1830 to 1930 MHz. This service is designed to allow unlicensed operation of short-distance (typically for indoor or campus-oriented environments) voice and data services provided by wireless LANs and wireless PBXs.

Summary. PCS is likely to bring about a variety of new mobile and portable devices such as small, lightweight telephone handsets that work at home, in the office, or on the street; portable wireless facsimile machines; advanced "smart" paging devices; and wireless electronic mail services. At this writing, PCS services are becoming available in all regions of the United States. Eventually all of the smaller PCS networks will be interconnected to the nationwide PCS networks.

See also

CELLULAR COMMUNICATIONS
CODE DIVISION MULTIPLE ACCESS
GLOBAL SYSTEM FOR MOBILE (GSM) TELECOMMUNICATIONS
PCS 1900
PCS-OVER-CABLE
PERSONAL HANDYPHONE SYSTEM
UNIVERSAL MESSAGING

Personal digital assistants

Personal digital assistants (PDAs) are lightweight, handheld computers that are equipped with an operating system, applications software, and communications capabilities for short-text messaging, e-mail, fax, news updates, and voice mail. They are intended for mobile users who require instant access to information, regardless of their location at any given time.

The first PDA made its market debut in 1993. Introduced by Apple Computer, the company's Newton technology was trumpeted as a major milestone of the information age. Apple's MessagePad, the first handheld device based on Newton technology, was soon joined by similar products from such companies as Hewlett-Packard, Motorola, Sharp, and Sony.

The earliest models were hampered by poor performance, excessive weight, and unstable software. Without a wireless communications infrastructure, there was no compelling advantage to owning a PDA. With the performance limitations largely corrected and the emergence of new wireless personal communications services (PCS)—plus continuing advances in operating systems, connectivity options, and battery technology—PDAs may yet fulfill their potential.

Applications. Real estate agents, medical professionals, field service technicians, and delivery people are just a few of the people using PDAs. Real estate agents can use PDAs to conveniently browse through property listings at client locations. Health care professionals can use PDAs to improve their ability to access, collect, and record patient information at the point of care. Numerous retailers and distributors can collect inventory data on the store and warehouse floor, and later export it into a spreadsheet on a PC. Insurance agents, auditors, and inspectors can use PDAs to record data in the field, and then instantly transfer that data to personal computers and databases at the home office. Even mortgage loan officers can use PDAs at temporary mall stations to collect credit histories of potential loan applicants.

PDA components. Aside from the case, PDA components include a screen, keyboard or other type of input device, an operating system, memory, and battery. Some PDAs can be outfitted with fax/modem cards and a docking station to facilitate direct connection to a PC or LAN for data transfers and file synchronization.

Screen. The biggest limitation of PDAs is the size of the screen. Visibility is greatly improved through the use of nonglare screens and backlighting, which aid viewing and entering information in any lighting condition. In a dim indoor environment, backlighting is a virtual necessity, but it drains the battery faster. Some PDAs offer user-controllable backlighting, while others let the user set a timer that shuts off the screen automatically after the unit has been idle for a specified period of time. Both features greatly extend battery life.

A new screen technology developed jointly by Motorola and Polaroid improves screen brightness and readability. The Optimax technology illuminates the PDA screen like backlighting, but without drawing any power from the battery. The liquid crystal display screen, which is enhanced by a holographic reflector, appears as a soft green glow to provide sufficient light for easy viewing.

Keyboard. Keyboards have always been a problem for PDAs. The problem with PDA keyboards is that they are too small to permit touch-typing. Pen devices were intended to overcome this limitation. Pens speed up and simplify navigation. They also make drag-and-drop task selection and text editing much easier. Of course, pens can be used for handwritten notes, but the recognition technology still needs improvement.

A promising scheme developed by Palm Computing employs a small 2×1-in writing surface and a special character set for data input. The character set uses the same symbol for both upper- and lowercase letters. Pen gestures alert the recognizer that the user is about to enter numbers, letters, or symbols. Limiting the character set, as well as the way to draw the characters, greatly improves handwriting recognition. Once the character set is learned, entry of up to 15 words per minute is possible.

Operating systems. A PDA's operating system provides the foundation upon which applications run. The operating system might offer handwriting recognition, for example, and include solutions for organizing and communicating information via fax or electronic mail, as well as the ability to integrate with Windows and MacOS-based computers in enterprise environments. The operating system might also include built-in support for a range of landline modems and third-party paging and cellular communication solutions. Because memory is limited in a PDA, the operating system and the applications that run on it must be compact.

Some operating systems come with useful utilities. There are utilities that set up direct connections between the PDA and desktop applications to transfer files

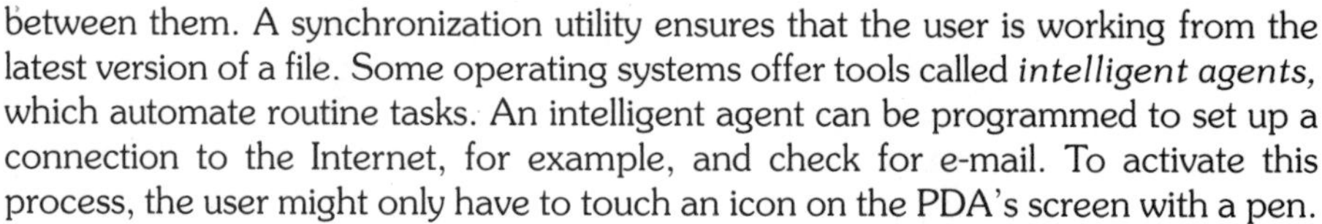

between them. A synchronization utility ensures that the user is working from the latest version of a file. Some operating systems offer tools called *intelligent agents,* which automate routine tasks. An intelligent agent can be programmed to set up a connection to the Internet, for example, and check for e-mail. To activate this process, the user might only have to touch an icon on the PDA's screen with a pen.

Memory. Although PDAs come with a base of applications built into ROM (usually a file manager, word processor, and scheduler), users can install other applications as well. New applications and data are stored in RAM. The combined limit in many of these devices is only 1 MB, while others offer 2 MB configurations. However, some PDAs have a PC Card (formerly PCMCIA) slot that can accommodate storage cards that are purchased separately.

Battery. Some PDA makers claim a battery life of 45 hours when users search for data 5 minutes out of every hour the unit is turned on. Of course, using the backlight will drain the batteries much more quickly. Using the backlight will reduce battery life by about 22 percent.

Fax/modems. Some PDAs come with an external fax/modems to support basic message needs. Others offer a PC Card slot (formerly PCMCIA) that not only can accept fax/modems, but storage cards as well. With fax/modems, a PDA user can receive a fax from the office, annotate it, and fax it back with comments written on it in "electronic ink."

Docking station. A noteworthy new peripheral is a docking unit that permits the PDA to be used as a companion to a desktop PC. While many PDAs include PC connectivity, others are designed specifically as a PC peripheral. The user simply drops the PDA into the docking unit, presses a button, and the desktop software automatically synchronizes with the PDA.

A standard for data exchange between portable devices and desktop computers and peripherals was finalized in 1993 by the Infrared Data Association. Before that, companies used proprietary protocols and wireless technologies for exchanging data between devices. Adoption of the first IrDA standard paved the way for interoperability among the wireless devices of different manufacturers, allowing users to send documents to printers or to and from wireless modems.

The IrDA physical-layer standard specifies point-to-point standards of operation at up to 1 m at 115.2 kbps maximum data rate over a ±15-degree minimum and a ±30-degree maximum viewing angle. Unlike many wireless networking plans that are constrained by the domestic and international regulations that control radio frequency transmissions, infrared connections are free of this encumbrance.

In April 1995, IrDA adopted a standard for 4 Mbps transmission, which ensures compatibility with products equipped with the original 115.2 kbps transmission capability. The 4 Mbps and 115.2 kbps standard allow users to move into more data-intensive applications, such as networking and transferring large amounts of data, including color graphics files, to printers.

Hybrid cell phones. A potential competitor to the PDA is the hybrid cell phone, which can be used for fax, e-mail, and short-text messaging, as well as for voice conversations. The key advantage of the hybrid cell phone is that it eliminates the need for users to carry around separate devices, one for voice communication and another for data communication. These hybrids, however, only work over GSM networks, which are widely available in Europe and Asia but only just catching on in the

United States. The availability of hybrid cell phones in the United States depends upon the future introduction and penetration of GSM cellular services.

Summary. Improvements in technology and the availability of wireless communications services, including PCS, overcome many of the limitations of early products, making today's PDAs very attractive to mobile professionals. In the process, PDAs may find acceptance beyond vertical markets and finally become popular among consumers, particularly those looking for an alternative to notebook computers.

See also

GLOBAL SYSTEM FOR MOBILE (GSM) TELECOMMUNICATIONS
PERSONAL COMMUNICATIONS SERVICES
PERSONAL HANDYPHONE SYSTEM

Personal Handyphone System

The Personal Handyphone System (PHS) offers high quality, low-cost mobile telephone services using a fully digital system operating in the 1.9 GHz spectrum. Originally developed by NTT, the Japanese telecommunications giant, PHS is based on GSM technology. PHS made its debut in Japan in July 1995, where service was initially offered in metropolitan Tokyo and Sapporo. Although PHS was developed in Japan, it is now considered a pan-Asian standard.

Advantages over cellular. PHS phones (Handyphones) operate at 1.9 GHz, whereas cellular phones operate at 800 MHz. To achieve superior wireline voice quality, PHS uses a portion of its capacity advantage to support a high-performance voice encoding algorithm called Adaptive Differential Pulse Code Modulation (ADPCM). With this algorithm, PHS can support a much higher data throughput (32 kbps) than a cellular-based system, allowing PHS to support fax and voice mail services and emerging multimedia applications such as high-speed Internet access and photo and video transmission.

PHS supports the handover of calls from one microcell to the next during roaming. PHS goes a step farther than cellular, however, by giving users the flexibility to make calls at home (just like conventional cordless phones), at school or in the office, while riding the subway, or while roaming through the streets.

Not only are PHS handsets extremely small and lightweight—almost half the size and weight of cellular handsets—the battery life of PHS handsets is superior to that offered by cellular handsets. PHS phones output 10 to 20 mW, whereas cellular phones output between 1 and 5 W. Whereas the typical cellular handset has a battery life of 3 hours talk time, the typical PHS handset has a battery life of 6 hours talk time. Whereas the typical cellular handset has a battery life of 50 hours in standby mode, the typical PHS handset has a battery life of 200 hours in standby mode (more than a week). The low-power operation of PHS handsets is achieved through strict built-in power management and "sleep" functions in individual circuits.

PHS gives subscribers more security and complete privacy. And, unlike cellular phones, PHS phones cannot be cloned for fraudulent use.

Another key advantage of PHS over cellular is cost: PHS can provide mobile communications more economically that cellular. Through its efficient microcellular architecture and use of the public network, start-up and expansion costs are minimized. As a result, total per-subscriber costs tend to be much lower than with traditional cellular networks. Because PHS is a "low-tier" microcellular wireless network, it offers far greater capacity per dollar of infrastructure than existing cellular networks, which results in lower calling rates. In Japan, the cost of a 3-minute call using a PHS handset is only about ¥10 (about 10 cents) more than making that same call on a public phone.

Applications. There are a number of applications for PHS technology. In the area of mobile telecommunications, users can establish communications by accessing public cell stations that are installed throughout an area. PHS phones can also be used with a home base station as a residential cordless telephone at the public switched telephone network (PSTN) tariff.

When used in the local loop, PHS provides the means to access the PSTN in areas where conventional local loops consisting of copper wire, optical fiber, or coaxial cable are impractical or not available. Wireless technologies are fast being adopted throughout the Asia/Pacific region because eliminating the costly and slow process of laying copper wire and optical fiber means faster installation for network customers, and faster returns on investment for service providers. And in being able to charge customers lower rates, there is also increased cash flow, since customers typically pay their bills sooner. In addition, relocation or expansion of the network is as simple as rewriting the database. A PHS Wireless Local Loop can be easily expanded as its customer base increases, or can be quickly and efficiently scaled back if the customer base decreases.

PHS also can be adopted as a digital cordless PBX for office use, providing readily expandable, seamless communications throughout a large office building or campus. Users carry PHS handsets with them and are no longer chained to their desks by their communications systems. As a digital system, PHS provides a level of voice quality not normally associated with a cordless telephone. In addition, the digital signal employed by PHS provides security for corporate communications, and the system's microcell architecture can be reconfigured easily to accommodate increases or decreases in the number of users. Another benefit of PHS is the ease and minimal expense with which the entire network can be dismantled and set up again in another facility if, for example, a business decides to relocate its offices.

For personal use as a cordless telephone at home, PHS is a low-cost mobile solution that allows the customer to use a single handset at home and outdoors, with a digital signal that provides improved voice quality for a cordless phone, and enough capacity for data and fax transmissions that are increasingly a part of users' home communications.

Network architecture. The PHS radio interface has a four-channel time division multiple access capability with time division duplexing (TDMA/TDD), which provides one control channel and three traffic channels for each cell station.

The base station allocates channels dynamically and is not constrained by a frequency reuse scheme, thus deriving the maximum advantage of carrier-switched TDMA. This means PHS handsets communicating to a base station may all be on different carrier frequencies.

The PHS system uses a microcell configuration that creates a radio zone with a diameter of 100–300 m. The base stations themselves are spaced at a maximum of 500 m apart. In urban areas, the microcell configuration is capable of supporting several million subscribers. This configuration also makes possible smaller and lighter handsets, as well as the more efficient reuse of radio spectrum to conserve frequency bands. In turn, this permits very low transmitter power consumption and, as a result, much longer handset talk times and standby times than are possible with cellular handsets.

A drawback of the lower operating power level is the smaller radius that a PHS base station can cover: only 100–300 m, versus at least 1500 m for cellular base stations. The extra power of cellular systems improves penetration of the signals into buildings, whereas PHS may require an extra base station inside some buildings. Another drawback of PHS is that the quality of reception can diminish significantly when mobile users are traveling at a rate greater than 15 MPH.

Because PHS uses the public network, rather than dedicated facilities, between microcells, the only service startup requirements are handsets, cell stations, PHS server, and a database of services to support PHS network operation. With no separate transmission network needed for connecting cell stations and for call routing, carriers can introduce PHS service with little initial capital investment.

Service features. PHS is a feature-rich service, giving Handyphone users access to a variety of call handling options, including:

- *Call forwarding* is available to a fixed line, to another PHS phone, or to a voice mailbox.
- *Call waiting* alerts the subscriber of an incoming call.
- *Call hold* allows the subscriber to alternate between two calls.
- *Call barring* restricts any incoming local or international call.
- *Calling line identification* (CLI) displays the number of the incoming call, informing the subscriber of the caller's identity.
- *Voice mail* allows the subscriber receives recorded messages, even when the phone is busy or turned off.
- *Text messaging* allows the subscriber to send and receive text messages through the PHS phone.
- *International roaming* allows subscribers to use their PHS phones in another country and be billed by the service provider in their home country.

Depending on the implementation progress of the service provider, the following value-added services also might be available to Handyphone users:

- *Virtual fax* allows subscribers to retrieve fax messages anywhere, have fax messages sent to a Handyphone, or have it redirected to any fax machine.
- *Fax* allows subscribers to send and receive faxes anywhere by attaching the Handyphone to a laptop or desktop computer.
- *E-mail/Internet access* allows subscribers to retrieve e-mail from the Internet through the Handyphone.
- *Conference call* allows subscribers to talk to as many as four other parties at the same time.

Data communications. Full-fledged PHS data communications services are set to start in the near future. Among the services currently undergoing testing include:

- *E-mail service:* E-mail is seen as being a basic service of the PHS multimedia communications menu. This service will make it possible to send and receive memos, schedules, daily reports, and other information from locations around town, from the office, and from home. PHS users want various types of service features, such as a function for screening mail by the sender or title and downloading only essential messages.
- *Fax service:* Uses envisioned for this service include sending handwritten memos or faxing data stored in a PDA. It is expected that faxes will be used frequently as a simple and convenient communication alternative to the telephone.
- *Internet access:* Internet access services have mushroomed in recent years; there are many people who want to obtain information from around the world via the Web. This trend has also influenced PHS in that many users want to be able to obtain necessary information in a timely manner when they are away from the office. It is also projected that PHS will be used extensively to form intranets for in-house communications by facilitating the expansion of office LAN access points.
- *Photograph transmission service:* This service can be realized by transmitting the signals of a digital still camera directly or through the medium of a personal computer. As such, it can be regarded as another variation of data transmission service. Since it involves transmitting large volumes of data and has a wide range of potential business applications, this service is can be expected to become a major multimedia offering of PHS with its high-speed capability.
- *Mobile office service:* The widespread use of groupware entails groups of users sharing common databases in carrying out or supporting the execution of collaborative work. There are demands to extend this collaborative environment even to outside locations through the use of mobile communications. The quantities of information being handled have also increased considerably to include the latest product information and business information.

Many other types of services also are being considered. Together with the ongoing advancement of telephone services, new consumer demands will likely be created, which suggests that users will be looking increasingly to the further evolution of PHS. For example, there is a strong possibility that real-time news, sports scores, and stock quotes will be added to the services currently being considered. There also is the possibility of using PHS terminals as an electronic wallet that can be used to access bank balances, move funds from one account to another, and debit a subscriber's bank account when making purchases at retail stores.

Another possible service is videophone communication. Although PHS is limited to speech today, extensions for data and multimedia service are in the works. These extensions have to take into account that PHS channels are only 32 kbps. Prototype videophones use H.261 compression to squeeze video signals down to size, as well as other technologies to overcome the inherent transmission errors common to wireless links. These devices support transmission of three to seven frames per second, however, which is consistent with the frame rate that users experience when videoconferencing over the Internet. By comparison, television-quality video operates at 30 frames per second.

Summary. Although PHS cannot be used while traveling at high speeds (such as in a train or a car), at walking speeds it provides seamless connections at home, outdoors, and in the office. This seamless feature will become even more important in future applications, such as when it becomes possible to use PHS terminals as an electronic wallet.

While 32 kbps transmission is now available in Japan, research is now under way to achieve a transmission rate of 64 kbps through the use of two channels. With this much capacity, PHS can be extended to a variety of other services in the future, including full-motion video.

In combination with a small, lightweight, portable data terminal, PHS might also be used to realize Oracle Corporation's concept of "network computing," whereby users would access application software stored on the Internet for use when needed. With the limited memory and disk storage capacity of such network computers, the applications and associated programs would stay on the Internet, preventing the PHS devices from becoming overwhelmed.

See also

CELLULAR COMMUNICATIONS

GLOBAL SYSTEM FOR MOBILE (GSM) TELECOMMUNICATIONS

PERSONAL ACCESS COMMUNICATIONS SYSTEMS

PERSONAL COMMUNICATIONS SERVICES

PERSONAL DIGITAL ASSISTANTS

Point-to-Point Protocol

Point-to-Point Protocol (PPP) provides the means to transfer data across any full-duplex (i.e., two-way) circuit, including dial-up links to the Internet via modems, ISDN, and high-speed SONET over fiber-optic lines. PPP is an enhanced version of the older Serial Line Internet Protocol (SLIP). While SLIP is typically used in an IP-only environment, PPP is more versatile in that it can be used in multiprotocol environments. In addition, PPP allows traffic for several protocols to be multiplexed across the link, including IP, IPX, DECnet, ISO, and others. PPP also carries bridged data over complex internetworks.

Features. PPP supports authentication, link configuration, and link monitoring capabilities via several subprotocols, including:

- *Link Control Protocol* (LCP) negotiates details and desired options for establishing and testing the overall serial link.
- *Password Authentication Protocol* (PAP) uses a two-way handshake for the peer to establish its identity at the time of link establishment.
- *Challenge Handshake Authentication Protocol* (CHAP) periodically verifies the identity of the peer using a three-way handshake, which is employed throughout the life of the connection. Of the two authentication protocols, CHAP is the more robust.

- *Network control protocols* are used to dynamically configure different network layer protocols, such as IP and IPX. For each type of network layer protocol there is a network control protocol that is used to initialize, configure, and terminate its use.
- *Link quality monitoring* provides a standardized way of delivering link quality reports on the quality/accuracy of a serial link.

During TCP sessions, PPP's compression feature can reduce the typical 40-byte TCP header to only 3–5 bytes, providing a significant savings in transmission time. This is accomplished by sending only the changes in a PPP frame's header values. Since most header information does not change from one frame to the next, the savings can be substantial.

Multilink PPP. A relatively new protocol is Multilink PPP (MPPP), which combines multiple B channels of an ISDN link into a single, higher-speed channel. Although extra channels can added to an established ISDN connection, MPPP does not offer dynamic control. This is remedied by the *Bandwidth Allocation Control Protocol* (BACP), which works in conjunction with MPPP. With BACP, ISDN channels can be added as needed and dropped when no longer required to support the application.

BACP allows bandwidth to change on demand through a standard set of rules, while minimizing the need for the end user to be involved in complex connection configuration issues. BACP can even interact with the resource ReSerVation Protocol (RSVP) to provide enhanced functionality. For example, if a bandwidth reservation is queued for lack of bandwidth somewhere on the network, this could trigger the creation of additional channels to support the application. If the application is a videoconference, for example, when the router senses that network load has gone down because participants are dropping out of the session, it starts terminating B channels. This minimizes the usage charges associated with ISDN.

Summary. While SLIP typically is used to connect computers to an IP network via a dial-up link, it has severe limitations in that it cannot support any other protocol and does not perform error-checking. PPP is more versatile in terms of the protocols it can handle and is more functional, particularly with regard to authentication, link configuration, and link monitoring. PPP is gradually displacing SLIP. Other protocols (such as MPPP, BACP, and RSVP) are used with PPP to support more sophisticated applications such as videoconferencing.

See also

INTEGRATED SERVICES DIGITAL NETWORK

MULTIMEDIA NETWORKING

Private branch exchange

Over the years, the private branch exchange (PBX) has emerged to become the undisputed cornerstone of today's corporate voice network. In the simplest terms, the PBX is a circuit switch which, through control signaling, performs three basic functions:

- In response to a call request, establishes end-to-end connectivity among its subscribers (on-net) and from its own subscribers to remote subscribers (off-net) through intermediate nodes, which may consist of other PBXs or central office switches on the public telephone network. The connected path is dedicated to the user for the duration of the call.

- Supervises the circuit to detect call request, answer, signaling, busy, and disconnect (hang up).
- "Tears down" the path upon call termination (disconnect) so that another user can access the resources available over that circuit.

These functions closely parallel those of the central office switch. In fact, the PBX evolved from the operator-controlled switchboards that were used on the public telephone network. The first of these simple devices was installed in 1878 by The Bell Telephone Company to serve 21 subscribers in New Haven, Connecticut. The operator had full responsibility for answering call requests, setting up the appropriate connections, supervising for answer and disconnect, and tearing down the path upon call completion. Interconnectivity among subscribers was accomplished via cable connections at a patch panel.

Today's PBXs are much more complicated, of course, but they provide the same basic functionality as the first generation of circuit switching devices. The difference is that the process of receiving call requests, setting up the appropriate connections, and tearing down the paths upon call completion is now entirely automated. Because the intelligence necessary to do all this resides on the user's premises, the PBX allows organizations to exercise more control over internal operations and incorporate communications planning into their long-term business strategies.

In a little over 100 years, PBXs have evolved from simple patch panels to sophisticated systems capable of integrating voice and data. The five generations characterizing the development of PBXs may be summarized as follows:

- First-generation PBXs consisted of the operator-controlled patch panels.
- Second-generation PBXs evolved from the electromechanical central office switches: step-by-step (Strowger) and crossbar. Automatic dialing and space division switching are the capabilities that differentiated second- from first-generation switches.
- Third-generation PBXs include the attributes of second-generation PBXs. Instead of electromechanical control, third-generation PBXs use electronic componentry under stored program control, making possible the addition of many new features as well as distributed architectures.
- Fourth-generation devices are computer-based to permit the integration of capabilities that previously had to be added on with external components, such as automatic call distribution and voice mail. Fourth-generation PBXs also use time division switching, which permits the integration of voice and data over T1 trunks and ISDN services.
- Fifth-generation PBXs are more data-oriented, adding LAN and Internet connectivity, support for ATM, and management via SNMP.

Basic features. Many PBX features are under direct user control and can be implemented right from the telephone keypad, including:

- *Add-on conference* allows the user to establish another connection while having a call already in progress.

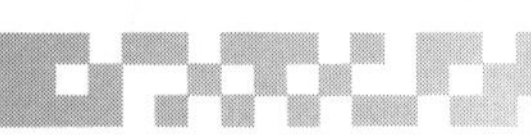

- *Call forwarding* allows a station to forward incoming calls to another station. This includes forwarding calls when the station is busy or unattended, or as needed.
- *Call hold* allows the user to put the first party on hold so that an incoming call can be answered.
- *Call waiting* lets the user know that an incoming call is waiting. While a call is in progress, the user will hear a special tone that indicates another call has come through.
- *Camp-on* allows the user to wait for a busy line to become idle, at which time a ring signal notifies both parties that the connection has been made.
- *Last number redial* allows users to press one or two buttons on the keypad to activate dialing of the previously dialed number.
- *Message waiting* allows the user to signal an unattended station that a call has been placed. Upon returning to the station, an indicator tells the person that a message is waiting.
- *Speed dialing* allows the user to implement calls with an abbreviated number. This feature also allows users to enter a specified number of speed-dial numbers into the main database. These numbers may be private or shared among all users. Entering and storing additional speed-dial numbers is accomplished via the telephone keypad.

There are also PBX capabilities that operate in the background, transparent to the user. The most common of these system capabilities include:

- *Automated attendant* allows the system to answer incoming calls and prompt the caller to dial an extension or leave a voice message without going through the operator.
- *Automatic call distribution* (ACD) allows sharing of incoming calls among a number of stations so that the calls are served in order of their arrival. This is usually an optional capability, but it may be integral to the PBX, or purchased separately as a stand-alone device.
- *Automatic least-cost routing* ensures that calls are completed over the most economic route available. This feature may be programmed so that mail room staff always get the cheapest carrier, while executives get to choose whatever carrier they want.
- *Call detail recording* (CDR) allows the PBX to record information about selected types of calls for management and cost control.
- *Call pickup* allows incoming calls made to an unattended station to be picked up by any other station in the same trunk group.
- *Class of service restrictions* control access to certain services or shared resources. Access to long-distance services, for example, may be restricted by area code or exchange. Access to the modem pool for transmission over analog lines may be similarly controlled.
- *Database redundancy* allows the instructions stored on one circuit card may be dumped to another card as a protection against loss.

- *Direct inward dialing* (DID) allows incoming calls to bypass the attendant and ring directly on a specific station.
- *Direct outward dialing* (DOD) allows outgoing calls to bypass the attendant for completion anywhere over the public telephone network.
- *Hunting* is a capability that routes calls automatically to an alternate station when the called station is busy.
- *Music on hold* indicates to callers that the connection is active while the call waits in queue for the next available station operator.
- *Power-fail transfer* permits the continuance of communication paths to the external network during a power failure. This capability works in conjunction with an uninterruptible power supply (UPS), which kicks in within a few milliseconds after detecting a power outage.
- *System redundancy* allows sharing of the switching load, so that in the event of failure, another processor can take over all system functions.

PBX components. Aside from the line and trunk interfaces, there are three major elements that comprise the typical PBX: processor, memory, and switching matrix. Together these elements provide all of the intelligence necessary to place calls anywhere on the public or private network without the need for human intervention.

Processor. The processor is responsible for controlling the various operations of the PBX. This includes monitoring all lines and trunks that provide connectivity, establishing line-to-line and line-to-trunk paths through the switching matrix, and tearing down connections upon call completion. The processor even controls such optional capabilities as voice mail and the recording of billing and traffic information. Because the processor is programmable, features and services can be added or changed at a management terminal.

Many PBXs may be optioned for two or more processors, which adds to the reliability of the system in that if one fails, another takes over. Programs and configuration information are automatically downloaded from the main processor to the standby processor to ensure uninterrupted service.

Memory. The processor uses the memory element to implement the sophisticated functions of the PBX. Those functions are defined and implemented in software instead of hardware. There are two types of memory: nonvolatile and volatile. The former is fixed, whereas the latter may be changed as needed. Nonvolatile memory (various forms of ROM) contains the operating instructions and stores system configuration information. Volatile memory, or random access memory (RAM) is used for temporary storage of frequently used programs, or for workspace.

In the event that a system failure destroys the contents of nonvolatile memory, a reserve program, also stored in nonvolatile memory, is put into operation automatically. (Some systems, however, still require that a spare program be loaded manually via disk or tape cartridge after a catastrophic failure.) As the term implies, nonvolatile memory can withstand a power outage, retaining its contents. Information stored in nonvolatile EEROM (electronically erasable read-only memory) is automatically dumped into RAM for access by the redundant processor. Although EEROM is nonvolatile, it can be changed by a programmer using special equipment.

Switch matrix. The switch matrix, under control of the processor, interconnects lines and trunks.[1] This may be accomplished through space division switching or time division switching. Space division switching originated in the analog environment. As its name implies, a space division switch sets up signal paths that are physically separate from one another, or divided by space. Each connection establishes a physical point-to-point circuit through the switch that is dedicated entirely to the transfer of signals between the two end points. The basic building block of the space division switch is a metallic crosspoint (relay contact) or semiconductor gate that can be enabled and disabled by the processor or control unit. Thus, physical interconnection is achieved between any two lines by enabling the appropriate crosspoint.

Although the single-stage space-division switch is virtually nonblocking,[2] it has several limitations, the most serious of which is the number of crosspoints that are required as the number of input lines and output lines grows. This not only is costly, but results in ever-greater inefficiencies in the utilization of available crosspoints. These limitations are mitigated through the use of multistage crosspoint matrices. Although requiring a more complex control scheme, the use of multiple stages reduces the number of crosspoints, while increasing overall reliability.

Summary. Despite humble beginnings more than 100 years ago, there is still plenty of room for innovation in PBX systems. Today's PBXs emphasize their integration into enterprise networking infrastructures, thereby addressing the applications that will be most in demand for the rest of the 1990s and beyond: LAN interconnectivity, videoconferencing, and multimedia. Some vendors offer the means to take voice traffic off the PBX through its T1/E1 interface and feed it to an ATM switch, where it can be combined with data for more economical transport over private networks. There also are PBX add-ons that support in-building mobile communications using wireless technology. PBXs can also be connected to Internet gateways, allowing users to send e-mail and run multimedia applications such as voice calls and videoconferences more economically.

See also

CENTRAL OFFICE SWITCHES
CENTREX
COMMUNICATIONS SERVICES MANAGEMENT
KEY TELEPHONE SYSTEMS
TELECOMMUNICATIONS MANAGEMENT SYSTEMS

1 Although the terms "line" and "trunk" are often used interchangeably, there is an important difference. A *line* refers to the link between each station (telephone set) and the switch, whereas a *trunk* refers to the link between switches. The term "tie line," then, is really a misnomer. It should really be called a *tie trunk*.

2 Blocking refers to the inaccessibility of the switch due to the unavailability of crosspoints, which establish the connections between various end points. In theory, all switches can experience blocking no matter how they are designed. As a practical matter, however, some switch designs are less prone to blocking than others. Blocking can also be the result of having fewer trunks than lines, which is a normal condition. The assumption behind high line-to-trunk ratios is that not all users will try to access the switch at once. But if 21 lines out of 200 try to access 20 trunks, the 21st will be blocked.

Private land mobile radio services

Since the 1920s, the private land mobile radio services (PLMRS) have been meeting the internal communication needs of private companies, state and local governments, and other organizations. These services provide voice and data communications that allow users to control their business operations and production processes, protect worker and public safety, and respond quickly in times of natural disaster or other emergencies.

In 1934, shortly after the agency was established, the Federal Communications Commission (FCC) identified four private land mobile services: Emergency Service, Geophysical Service, Mobile Press Service, and Temporary Service. (The latter applied to frequencies used by the motion picture industry.) Over the years the FCC refined these categories. Today, PLMRS consists of 21 services spread among six service categories: Public Safety, Special Emergency, Industrial, Land Transportation, Radiolocation, and Transportation Infrastructure.

PLMRS are used by organizations that are engaged in a wide variety of activities. Police, fire, ambulance, and emergency relief organizations such as the Red Cross use private wireless systems to dispatch help when emergency calls come in or disaster strikes. Utility companies, railroad and other transportation providers, and other infrastructure-related companies use their systems to provide vital day-to-day control of their systems (including monitoring and control and routine maintenance and repair), and also to respond to emergencies and disasters, often working with public safety agencies. A wide variety of businesses, including package delivery companies, plumbers, airlines, taxis, manufacturers, and even the American Automobile Association (AAA) rely on private wireless systems to monitor, control, and coordinate their production processes, personnel, and vehicles.

Although commercial services can serve some of the needs of these organizations, private users generally believe that their own systems provide them with capabilities, features, and efficiencies that commercial systems cannot. Some of the requirements and features that PLMRS users believe make their systems unique include:

- Immediate access to a radio channel (no dialing required).
- Coverage in areas where commercial systems cannot provide service.
- Peak usage patterns that could overwhelm commercial systems.
- High reliability.
- Priority access, especially in emergencies.
- Specialized equipment required by the job or federal regulations.

The FCC has continually allocated more frequencies for PLMRS in several different spectrum bands to overcome congestion and keep pace with advances in technology that make possible the use of higher frequencies. Currently, over 1 million licensed stations are authorized to operate over 12 million transmitters in support of private land mobile radio services. This represents an investment of over US$25 billion.

Public Safety Radio Services. The first uses of mobile radio date to the 1920s, when police departments, marine fireboats, and industries that worked in remote locations were the primary users. When the FCC was established in 1934, these uses were classified into four Emergency Services: marine fire, municipal police, state police, and special emergency. Since then they have evolved into six different public safety radio services: Local Government, Police, Fire, Highway Maintenance, Forestry-Conservation, and Emergency Medical.

Special Emergency Radio Service. The Special Emergency Radio Service was once categorized under the public safety services, but now is governed separately, even though it serves similar needs. the term *special emergency* originally applied to telegraph stations used by power companies when regular communications were disrupted by storms or other emergencies. In 1946, however, most power companies switched to a new service for public utilities and the term became restricted to matters directly related to public safety and the protection of life and property. Special Emergency is now classified as a service for the communications needs of hospitals and clinics, ambulance and rescue services, veterinarians, handicapped persons, disaster relief organizations, school buses, beach patrols, persons or organizations in isolated areas, and emergency standby and repair facilities for telephone and telegraph systems.

Industrial Radio Services. The FCC created an "industrial" classification of radio services in 1949, after most of the industrial uses of radio had already been established under "miscellaneous" or "experimental" categories. By that time, such diverse industries as oil exploration, mining, news reporting, and motion picture production had been using radio for about 20 years. Like other early users, these industries first relied on radio mainly for the safety of work crews in remote locations, but they quickly learned the value of mobile radio as an economical tool for carrying company instructions to remote operations, for dispatching and diverting work vehicles, and for coordinating the activities of workers and machines on location. Today there are nine Industrial Radio Services: Power, Petroleum, Forest Products, Film and Video Production, Relay Press, Special Industrial, Business, Manufacturers, and Telephone Maintenance.

Land Transportation Radio Services. Land Transportation Radio Services also became a special classification in 1949. Before that date, most transportation industries shared two frequency bands in the experimental General Mobile Radio Service. One band served vehicles operating over highways, such as intercity buses and large trucks, and the other band served urban vehicles such as taxicabs, delivery vans, and tow trucks. As the various transportation industries grew, the FCC allocated more frequencies and created exclusive radio services for different types of transportation: Railroad, Urban Transit, Taxicab, Intercity Bus, Highway Truck, and Automobile Emergency. Today, Railroad, Taxicab, and Automobile Emergency remain separate categories, but Urban Transit, Intercity Bus, and Highway Truck are now classified together under the Motor Carrier Radio Service. Land transportation radio stations may not be used for passenger communications.

Radiolocation Service. *Radiolocation* is the use of radio waves to determine an object's distance, direction, speed, or position for any purpose except navigation. The Radiolocation Service authorizes persons engaged in commercial, industrial, scientific, educational, or government activities to use radiolocation devices in connection with those activities. Various types of radar (such as police radar and weather radar) are examples of radiolocation applications.

Transportation Infrastructure Radio Service. The Transportation Infrastructure Radio Services category was created in 1995 to integrate radio-based technologies into the nation's infrastructure and to develop and implement the nation's intelligent transportation systems. It includes the Location and Monitoring Service (LMS). LMS systems are used to determine the location and status of vehicles and equipment. The railroad industry, for example, operates an extensive automatic equipment identification system that allows companies to track, identify, and monitor the movement and location of over 1.3 million railroad cars and equipment throughout the United States.

Other private radio services. There are other private radio services, such as Private Land Mobile Paging and Private Operational Fixed Microwave Systems.

A private paging service is one that is not-for-profit and serves the licensee's internal communications needs. Private paging systems in general provide the same applications offered by commercial paging services: tone, tone-voice, numeric, or alphanumeric. Shared-use, cost-sharing, or cooperative arrangements, multiple licensed systems that use third-party managers, or users combining resources to meet compatible needs for specialized internal communications facilities are presumed to be private paging services.

In addition to their mobile operations, many private companies, public utilities, and state and local governments also make extensive use of Private Operational-Fixed Microwave Systems. These systems connect specific locations in either a point-to-point or point-to-multipoint configuration, and can carry or relay voice, data (including teletype, telemetry, facsimile, and other digital communications), and video communications. Such fixed links often connect mobile radio base stations or far-flung offices, but also are used for a variety of other purposes, including: the remote control of unattended equipment, valves, and switches; the recording of data such as pressure, temperature, or speed of machines; the monitoring of voltage and current in power lines; and other control or monitoring functions, such as would be necessary for rivers, railroads, and highways.

Summary. Private radio systems serve a great variety of communication needs that common carriers and other commercial service providers traditionally have not been able or willing to fulfill. Companies large and small use their private systems to support their business operations, safety, and emergency needs. The one characteristic that all these uses share—and which differentiates private wireless use from commercial use—is that private wireless licensees use radio as a tool to accomplish their missions in the most effective and efficient ways possible. Private radio users employ wireless communications as they would any other tool or machine; radio contributes to their production of some other good or service.

For commercial wireless service providers, by contrast, the services offered over the radio system *are* the end products. Cellular, PCS, and SMR providers sell service or capacity on wireless systems, permitting a wide range of mobile and portable communications that extend the national communications infrastructure.

This difference in purpose is significant because historically it has been the foundation of the different regulatory treatments afforded to the different communities.

See also

CELLULAR COMMUNICATIONS
FEDERAL COMMUNICATIONS COMMISSION
PERSONAL COMMUNICATIONS SERVICES
SPECIALIZED MOBILE RADIO
TELEMETRY

Protocol analyzers

A category of test equipment known as the *protocol analyzer* is used to monitor and diagnose performance problems on LANs and WANs by decoding upper-layer protocols. There are protocol analyzers for all types of communications circuits, including frame relay, X.25, T1, and ISDN. There also are protocol analyzers that offer full seven-layer decodes of NetBIOS, SNA, SMB, TCP/IP, DEC LAT, XNS/MS-NET, NetWare, and VINES, as well as the various LAN cabling, signaling, and protocol architectures, including those for AppleTalk, ARCnet, Ethernet, StarLAN, and Token-Ring.

In the case of a LAN, the protocol analyzer connects directly to the cable as if it were just another node, or to the test port of a communication device (DTE or DCE) where trouble is suspected (Fig. P6).

Troubleshooting features. Many protocol analyzers add features, such as data capture to RAM or disk, automatic configuration, counters, timers, traps, masks, and statistics. These features can dramatically shorten the time it takes to isolate a problem. Some sets also incorporate sophisticated programmability and simulation features.

Monitoring and simulation. Protocol analyzers are generally used either in a passive monitoring application or a simulation application. In the monitoring application, the analyzer sits passively on the network and monitors both the integrity of the cabling and the level of data traffic, logging such things as excessive packet collisions and damaged packets that can tie up an Ethernet LAN, for example. The information on troublesome nodes and cabling are compiled for the network manager. In a monitoring application, the protocol analyzer merely displays the protocol activity and user data (packets) that are passed over the cable, providing a window into the message exchange between network nodes.

In the simulation application, the protocol analyzer is programmed to exhibit the behavior of a network node, such as a gateway, communications controller, or front-end processor (FEP). This makes it possible to replace a suspect device on the network with a simulator that is running a program to simulate proper operation. This also enhances the ability to do fault isolation. For example, a dual-port protocol analyzer can monitor a gateway while running a simulation. More sophisticated protocol analyzers can run simulations designed to stress-test individual nodes to verify their conformity to standards. Protocol simulation is most often used to verify the integrity of a new installation.

Trapping. The trapping function allows the troubleshooter to command the protocol analyzer to start recording data into its buffer or onto disk when a specific

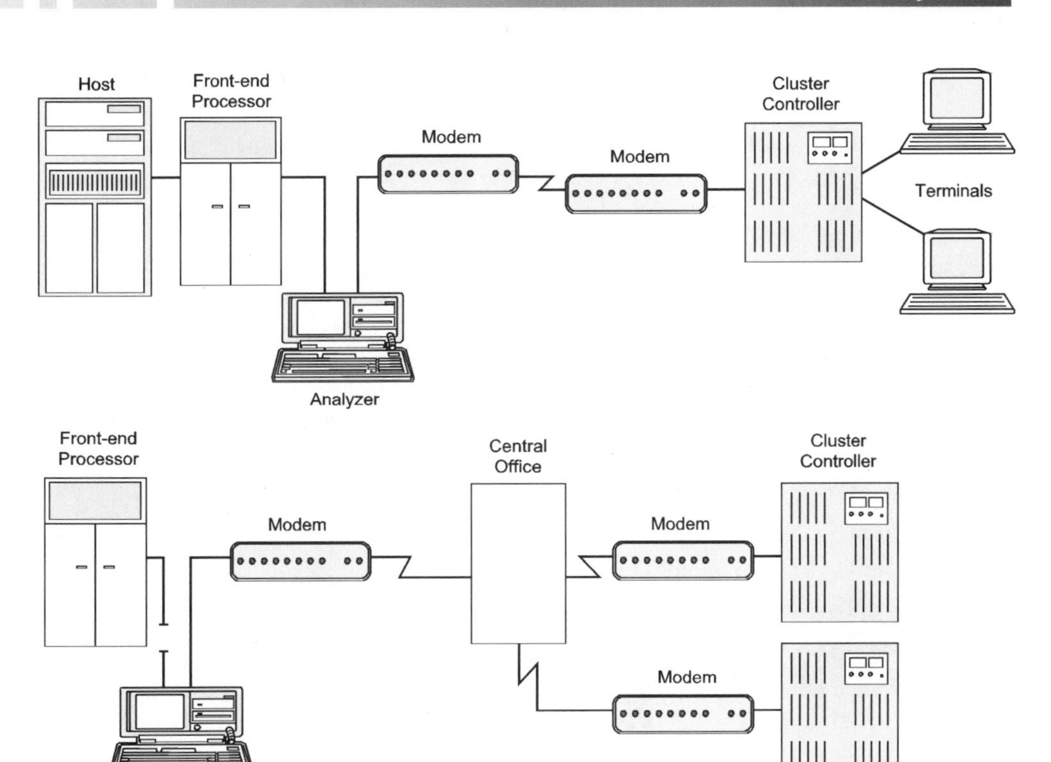

Fig. P6 *Protocol analyzer connections to the network. The protocol analyzer in monitor mode (top) allows the user to check events taking place between a front-end processor and local modem over a 9.6 kbps synchronous link. The protocol analyzer in simulation mode (bottom) allows the user to run a program that exhibits proper operation of the suspect front-end processor over a 9.6 kbps synchronous multidrop link, which includes two cluster controllers.*

event occurs. For example, the protocol analyzer could be set to trap the first errored frame it receives. This feature permits the capture of only essential information. Some protocol analyzers allow the user to set performance thresholds according to the type of traffic on the network. When these performance thresholds are exceeded, an alarm message is triggered, indicating that there is a problem.

Filtering. With the protocol analyzer's filtering capability, the user can exclude certain types of information from capture or analysis. For example, the technician might suspect that errors are being generated at the Data Link layer, so Network layer packets can be excluded until the problem is located. At the Data Link layer, the analyzer will track information such as where the data was generated and whether it contains errors. If no problems are found, the user can set the filter to include only Network layer packets. At this layer, the protocol analyzer tracks information such as where the data is destined and the type of application under which it was generated. If the troubleshooter has no idea where to start looking for

problems, then all of the packets may be captured and written to disk. The filters may be applied later for selective viewing.

Bit error rate testing. Bit error rate testing is used to determine whether data are being passed reliably over a carrier-provided communications link. This is accomplished by sending and receiving various bit patterns and data characters to compare what is transmitted with what is received. The bit error rate is calculated as a ratio of the total number of bit errors divided by the total number of bits received. Any difference between the two is indicated and displayed as an error. Additional information that may be presented includes sync losses, sync loss seconds, errored seconds, error-free seconds, time unavailable, elapsed time, frame errors, and parity errors.

Packet generation. In being able to generate packets, the protocol analyzer allows the user to test the impact of additional traffic on the network. Using a set of configuration screens, the technician can set the following parameters of the packets:

- The source address from which the packets will be sent,
- The destination address to which the packets will be sent,
- The maximum and minimum frame size of the packets,
- The spacing between the packets, expressed in μs, and
- The number of packets sent out with each burst.

The technician also can customize the contents of the data field section of the packets to simulate real or potential applications. When the packets are generated, the real-time impact of the additional traffic on the network can be observed on the monitor of the protocol analyzer. Packets also can be generated to force a suspected problem to recur, thereby expediting the troubleshooting process.

Load generation. A related capability is load generation, whereby varying traffic rates on the network may be created. By loading the network from 1 to 98 percent, network components such as repeaters, bridges, and transceivers can be stressed for the purpose of identifying any weak links on the network before they become serious problems later.

Timing. Protocol analyzers contain timers that measure the time interval between events. By setting up two traps, one for the transmit path and one for the receive path, the troubleshooter can verify if a handshake procedure has exceeded its maximum time interval, for example. Although this type of problem is most often handled better from the central network control point or host location, a protocol analyzer that supports simple timing measurements between data events can often solve the problem without invoking these resources.

Terminal emulation. Asynchronous terminal emulation is a feature found in some analyzers. If there is a requirement to communicate with intelligent network devices for configuration management, or to access remote databases, this feature can save the cost and inconvenience associated with carrying a separate terminal. Implementations of terminal emulation can vary considerably among analyzers, with some supporting complete 24×80 characters on a single display, and others requiring windowing to access an entire page.

Cable testing. Cable problems are the source of more than 50 percent of LAN network failures. Many protocol analyzers include the ability to test for cable breaks and improperly terminated connections, using a technique called *time do-*

main reflectometry (TDR). This test procedure involves sending a signal down the cable and then receiving and interpreting its echo. The status of the cable and connections may be reported simply as *no fault detected, no carrier sense, open on coax,* or *short on coax.* The distance to the problem is also reported. These devices are capable of pinpointing such problems as shorts, crimps, and water faults.

Mapping. The mapping capability of some protocol analyzers can save network managers hours of work. In automatically documenting the physical location of LAN nodes, many hours of work can be eliminated in rearranging the network map when devices are added, deleted, or moved. The mapping software allows the network manager to name nodes. Appropriate icons for servers and workstations are included. The icon for each station also provides information about the type of adapter used, as well as the node's location along the cable. When problems arise on the network, the network manager can quickly locate the problem by referring to the visual map. Some protocol analyzers can depict network configurations according to the usage of network nodes, arranging them in order of highest to lowest traffic.

Programmability. The various tasks of a protocol analyzer may be programmed, allowing performance information to be collected automatically. While some analyzers require the use of programming languages, others employ a setup screen, allowing the operator to define a sequence of tests to be performed. Once preset thresholds are met, a sequence of appropriate tests is initiated automatically. This capability is especially useful for tracking down intermittent problems. An alternative to programming or defining analyzer operation is to use off-the-shelf software that can be plugged into the data analyzer in support of various test scenarios.

Automatic configuration. Some protocol analyzers have an autoconfigure capability, which allows the device to automatically configure itself to the protocol characteristics of the line under test. This eliminates the need to go through several manually established screens for setup, which can save a lot of time and frustration.

English translation. Some protocol analyzers are unique in their ability to decode packets and display their contents in English-like notation, in addition to the hexadecimal or binary code. Further details about a specific protocol may be revealed through the analyzer's zoom-in capability, which allows the troubleshooter to display each bit field, along with a brief explanation of its status. This feature may be applied to any protocol, such as SNA, X.25, or TCP/IP.

Editing. Some analyzer software packages include text editors that can be used in conjunction with captured data. This allows the user to delete unimportant data, enter comments, print reports, and even create files in common database formats.

See also

ANALOG LINE IMPAIRMENT TESTING

Public telephone service

Public telephone service is either outgoing only or two-way service and can be coin or coinless. There are at least four public telephone service options:

- *Coin-operated payphone service* (classic or "dumb" payphone): This service relies on an operating company's central office controls to collect and return coins at the payphone set. End users have access to local, toll, and operator network services. The payment options available to end users are cash or billing as a collect call, a charge to a third-party number, or a calling card.

- *Coin-operated payphone service* ("smart" payphone): This service offering uses payphone sets that are capable of rating local calls, collecting and returning coins, rating and routing toll calls, and providing station-based operator services. The payment options available to end users are cash or billing as collect, third-party, or calling card.
- *Card and coin payphone service* (advanced payphone): This service offering is similar to the coin-operated payphone service but provides the additional payment options of using commercial credit cards or cash cards.
- *Coinless payphone service:* This service offering uses payphone sets that rely on central office and operator systems to rate and bill for calls placed from the sets. It has a unique screening feature that prohibits coin-paid, direct-dialed-call billing. End users have access to local and toll services. The payment options available to end users are limited to billing as collect, third-party, or calling card.

Some carriers offer *semipublic telephone service* and *shared payphone service*. These services utilize a "dumb" payphone set that relies on central office control to collect and return coins. The end users have access to local, toll, and operator network services. The payment options available to end users are cash or billing as collect, third-party, or calling card. The semipublic telephone service and shared payphone service recover a portion of the carrier's costs through separate monthly service charges billed to the location site provider.

There is also *inmate calling service,* which utilizes a payphone set that relies on central office and operator or premises-based call management systems for the control of rating and billing. It has unique screening features that prohibit direct-dialed calls, calling card, and third-number billing. This service option provides end users access to local and toll services on a collect billing arrangement.

Summary. The Federal Communications Commission (FCC) regulates payphone service to the extent that the carrier is not permitted to subsidize its payphone service directly or indirectly from its telephone exchange or exchange access service operations. In addition, the incumbent local exchange carrier (ILEC) cannot prefer or discriminate in favor of its payphone service.

See also

CALLING CARD

Public utility commissions

In the United States, each state has a commission charged with the responsibility for regulating public utilities and intrastate service rates. Utilities include privately owned corporations providing electric, natural gas, water, sewer, telephone, and radio service to the public and may also include railroads, motor buses, truck lines, ferry, and other transportation companies. The public utility commissions (PUCs) supervise and regulate utilities and public transportation in the state to ensure that service and facilities are made available at rates that are reasonable and just.

PUC policies benefit ratepayers through lower rates, new and improved utility products and services; they also protect consumers where competition otherwise does not. The PUC uses competitive markets to accomplish these goals where possible and appropriate. Generally, regulated utilities must seek approval from the PUC to change rates or services.

The PUCs are independent, quasi-judicial regulatory bodies whose jurisdiction, powers, and duties are delegated by the state legislature. The PUCs have quasi-legislative and quasi-judicial authority in that they establish and enforce administrative regulations and, like a court, may take testimony, subpoena witnesses and records, and issue decisions and orders. Usually the PUCs conduct public hearings on applications, petitions, and complaints.

In its decision-making, the PUC seeks to balance the public interest and need for reliable, safe utility services at reasonable rates with the need to assure that utilities operate efficiently, remain financially viable, and provide stockholders with an opportunity to earn a fair return on investment. The PUC encourages ratepayers, utilities, and consumer and industry organizations to participate in its proceedings and seeks their assistance in resolving complex issues.

The commissioners usually are appointed by the governor, with the approval of the state legislature, for a term of 6 years. The commissioners make all final policy, procedural, and other decisions. Their terms may be staggered to assure that the PUC always has the benefit of experienced members.

Summary. In cases where state and federal jurisdictions might overlap, a federal-state Joint Board is established to arrive at a mutually agreeable course of action. Under the provisions of the Communications Act, each Joint Board comprises three FCC commissioners and four state commissioners, and is moderated by the chairman of the FCC.

The issue of universal service is a case where federal and state jurisdictions overlap. A Federal-State Joint Board on Universal Service was established to make recommendations to the FCC with respect to the implementation of the universal service provisions of the Telecommunications Act of 1996. The public meetings included panel discussions held to address issues raised in the universal service proceeding. The panels discussed competition and universal service, and universal service for consumers in rural, insular, and high-cost areas and for low-income consumers, as well as the provision of advanced telecommunications services to schools, classrooms, libraries and health care providers.

See also

FEDERAL COMMUNICATIONS COMMISSION
UNIVERSAL SERVICE

Radio communication interception

When it comes to the interception of radio communications, the Federal Communications Commission (FCC) has the authority to interpret Section 705 of the Communications Act, 47 U.S.C. Section 605, "Unauthorized Publication of Communications."

Although the act of intercepting radio communications may violate federal or state statutes, this provision generally does not prohibit the mere interception of radio communications. For example, if someone happens to overhear a neighbor's cordless telephone, this is not a violation of the Communications Act. Similarly, if someone listens to radio transmissions on a scanner, such as emergency service reports, this is not a violation of Section 705.

A violation of Section 705 would occur, however, if someone were to divulge or publish what they hear or use it for their own or someone else's benefit. An example of using an intercepted call for a beneficial use in violation of Section 705 would be someone listening to accident reports on a police channel and then sending his or her tow truck to the reported accident scene in order to obtain business.

The Communications Act does allow for the divulgence of certain types of radio transmissions. The statute specifies that there are no restrictions on the divulgence or use of radio communications that have been transmitted for the use of the general public, such as transmissions of a local radio or television broadcast station. Likewise, there are no restrictions on divulging or using radio transmissions originating from ships, aircraft, vehicles, or persons in distress. Transmissions by amateur or citizens band radio operators are also exempt from interception restrictions.

In addition, courts have held that the act of viewing a transmission (such as pay television signal) that the viewer was not authorized to receive is a "publication" violating Section 705. Section 705 also has special provisions governing the interception of satellite television programming that is being transmitted to cable operators. The section prohibits the interception of satellite cable programming for private home viewing if the programming is either scrambled, or is not scrambled but is sold through a marketing system. In these circumstances, authorization must be obtained from the programming provider to legally intercept the transmission.

The Act also contains provisions that affect the manufacture of equipment used for listening or receiving radio transmissions, such as scanners. Section 302(d) of the Communications Act, 47 U.S.C. Section 302(d), prohibits the FCC from authorizing scanning equipment that is capable of receiving transmissions in the frequencies allocated to domestic cellular services, that is capable of readily being altered by the user to intercept cellular communications, or that may be equipped with decoders that convert digital transmissions to analog voice audio. Effective April 26, 1994 (47 CFR 15.121), such receivers may not be manufactured in the United States or imported for use therein. FCC regulations also prohibit the sale or lease of scanning equipment not authorized by the FCC (47 CFR 2.803).

Summary. The FCC receives many inquiries regarding the interception and recording of telephone conversations. To the extent these conversations are radio transmissions, there would be no violation of Section 705 if there is no divulgence or beneficial use of the conversation. Again, however, the mere interception of some telephone-related radio transmissions—whether cellular, cordless, or landline conversations—may constitute a criminal violation of other federal or state statutes.

See also

TELEPHONE FRAUD

Redundant array of inexpensive disks

An increasingly popular method of protecting data is the *redundant array of inexpensive disks* (RAID). Instead of risking all data on one high-capacity disk, this solution distributes the data across multiple smaller disks. RAID products usually are grouped into the following categories:

- *RAID Level 0:* These products are technically not RAID products at all because they do not offer parity or error-correction data to provide redundancy in the event of system failure. Although data striping is performed, it is accomplished without fault tolerance. Data is simply striped block-by-block across all the drives in the array. This is a high-performance data storage solution.
- *RAID Level 1:* These products duplicate data that is stored on separate disk drives. Also called *mirroring,* this approach ensures that critical files will be available in case of individual disk drive failures. Each disk in the array has a corresponding mirror disk and the pairs run in parallel. Blocks of data are sent to both disks at the same time. While highly reliable, RAID Level 1 is costly because every drive requires its own mirror drive, which doubles the hardware cost of the system.
- *RAID Level 2:* These products distribute the code used for error detection and correction across additional disk drives. The controller includes an error-correction algorithm, which enables the array to reconstruct lost data if a single disk fails. As a result, no expensive mirroring is required. But the code requires that multiple disks be set aside to do the error-correction function. Data is sent to the array one disk at a time.
- *RAID Level 3:* These products store user data in parallel across multiple disks. The entire array functions as one large logical drive. Its parallel operation is ideally suited to supporting imaging applications that require high data transfer rates when reading and writing large files. RAID Level 3 is configured with one *parity* (error-correction) drive. The controller determines which disk has failed by using additional check information recorded

at the end of each sector. Because the drives do not operate independently, however, every time an image file must be retrieved, all of the drives in the array are used to fulfill that request. Other users are put into a queue.

- *RAID Level 4:* These products store and retrieve data using independent writes and reads to several drives. Error correction data is stored on a dedicated parity drive. In RAID Level 4, data striping is accomplished in sectors, not bytes or blocks. Sector-striping offers parallel operation in that reads can be performed simultaneously on independent drives, which allows multiple users to retrieve image files at the same time. While multiple reads are possible, multiple writes are not because the parity drive must be read and written to for each write operation.
- *RAID Level 5:* These products interleave user data and parity data, which are then distributed across several disks. Because data and parity codes are striped across all the drives, there is no need for a dedicated parity drive. This configuration is suited for applications that require a high number of I/O operations per second, such as transaction-processing tasks that involve writing and reading large numbers of small data blocks at random disk locations. Multiple writes to each disk group are possible because write operations do not have to access a single common parity drive.
- *RAID Level 6:* These products improve reliability by implementing drive mirroring at the block level so that data is mirrored on two drives instead of just one. This means that up to two drives in the five-drive disk array can fail without loss of data. If a drive in the array fails with RAID 5, for instance, data must be rebuilt from the parity information spanned across the drives. With RAID 6, however, the data is simply read from the mirrored copy of the blocks found on the various striped drives. No rebuilding is required. Although this results in a slight performance advantage, it requires at least 50 percent more disk capacity to implement.
- *RAID Level 10:* Some vendors offer products that combine the performance advantages of RAID 0 with the data availability and consistent high performance of RAID 1, also referred to as *striping over a set of mirrors.* This method offers high performance and high availability for mission-critical data.

Summary. Businesses today have multiple data storage requirements. Depending on the application, performance might be valued more than availability; other times, the reverse could be true. Today it is even common to have different data structures in different parts of the same application. Until recently, the choice among specific RAID solutions involved trade-offs between cost, performance, and availability; once installed, they cannot be changed to take into account the different storage needs of applications that might arise in the future.

Vendors have responded with storage solutions that support a mix of RAID levels (or non-RAID technologies) simultaneously. Individual disk drives or groups of drives now can be configured via a PC-based resource manager for high performance, high availability, or as an optimized combination of both. This solves a classic data storage dilemma: meeting the exacting requirements of current applications, while staying flexible enough to adapt to long-term needs.

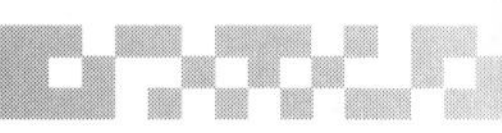

See also

HIERARCHICAL STORAGE MANAGEMENT
STORAGE MEDIA

Remote control

Remote control is a software-based solution for remotely accessing another computer. There are two types of computers in a typical scenario, the remote system and the host system. The remote system can be a branch office PC, a PC located at home, or a portable computer whose location varies on a daily basis. The host system can be any computer that a remote user wishes to access, including LAN-attached workstations and stand-alone PCs equipped with modems. Both the remote and host computers must be equipped with the same remote control software, and the user must be authorized to access a particular host.

With remote control, the remote computer user takes full control of a host system physically attached to the corporate network. All the user's keystrokes and mouse movements are sent to the host system; the image on that screen is sent back to the remote PC for display as if the user were sitting at the host system. Among other ways, remote control differs from remote node (discussed later) in that traffic along the remote connection consists only of screen and keyboard information; in remote node, the traffic is normal network packets to and from a client that happens to reside on the remote machine.

Applications. There are many applications for remote control software. It allows a mobile user to check e-mail at the office, query a corporate contact database from a hotel room, or access an important file stored on a server. Help desk operators and technicians can use remote control software to troubleshoot problems with specific systems. Trainers can periodically monitor the performance of users to determine their need for additional training.

With remote control, security can become an important concern. This is because the host system's monitor displays all the information that is being manipulated remotely, allowing any casual observer to view all screen activity, including electronic mail, financial data, and confidential documents. A possible solution would be to turn off the monitor and leave the computer running, but this is not always possible.

Features. The leading remote control programs are feature-rich, providing such useful capabilities as:

- *Callback* enhances security by having a remote host call back to a guest. With fixed callback, the guest is automatically called back at a number previously specified. With a roving callback, the guest has the opportunity to enter the callback number.
- *Chat* allows two connected users to type messages to one another. Some products allow messaging during a remote control or file transfer session.
- *Clipboard* allows the user to highlight and copy material between the remote computer and host.
- *Compression* condenses transferred data by a factor of at least 2-to-1 to cut dial-up line costs.

- *Directory synchronization* ensures that directories on the host and remote machines contain exactly the same directory trees.
- *Drag-and-drop* allows the user to copy or move files from one machine to the other with the drag-and-drop technique.
- *Drive mapping* makes the disk drives on both the remote and local machines seamlessly accessible to users on both ends. This feature would allow a user on a remote host to open a document stored on a local drive from within a word processor. Without remote drive mapping, the user would have to transfer the document to the remote host before being able to open it.
- *Emulation* allows the remote control software also to be used as a conventional dial-up telecommunications program. Some vendors offer limited modem support, supplying only TTY terminal emulation and only ASCII and Xmodem file transfer protocols. Others provide extensive emulation support, even allowing users to log into commercial services such as CompuServe.
- *Encryption* encrypts the data stream during transmission so that remote sessions cannot be monitored.
- *File management* allows files to be displayed and sorted by name, size, modification date and time, and attribute flags. The directory can be filtered to display only specified types of files.
- *File synchronization* ensures that the remote machine contains exactly the same files as the host. File transfers can be expedited by a feature that copies only the parts of files that have changed.
- *Printer redirection* is the ability to reroute printer output from the host to the local printer.
- *Screen data caching* stores locally the screen data from the host so that only changes have to be transmitted to the remote. The cache can be configured to retain data between sessions.
- *Scripting facilities* allow unattended file transfer operations, and various tasks to be automated. A script can even automate login procedures. Multiple tasks can be automated in a single script.
- *Transfer restart* allows the remote-control program to restart a file transfer where it left off if the connection is dropped in midstream.
- *Virus scanning* automatically scans files for viruses as they are being transferred.
- *Voice-to-data switching* establishes a remote-control connection from a voice call without requiring hanging up and redialing. This feature is especially convenient for technical support applications.

Summary. Remote control software allows users to dial up a specific modem-equipped or LAN-attached PC to access its files or troubleshoot a problem. As long as the software is installed on both ends, the remote computer assumes the capabilities of the host. Everything on the host's screen is mirrored on the remote computer's screen. Today's increasing distributed workforce makes this kind of connectivity a near-necessity.

See also
BRANCH OFFICE ROUTING
REMOTE NODE

Remote monitoring

The common platform from which to monitor multivendor networks is SNMP's Remote Monitoring (RMON) MIB. Although a variety of SNMP MIBs collect performance statistics to provide a snapshot of events, RMON enhances this monitoring capability by keeping a record of past events that can be used for fault diagnosis, performance tuning, and network planning.

Hardware- or software-based RMON-compliant devices (i.e., probes) placed on each network segment monitor all data packets sent and received. The probes view every packet and produce summary information on various types of packets, such as undersized packets, and on events such as packet collisions. The probes also can capture packets according to predefined criteria set by the network manager or test technician. At any time, the RMON probe can be queried for this information by a network management application or an SNMP-based management console so that detailed analysis can be performed, in an effort to pinpoint where and why an error occurred.

The original Remote Network Monitoring MIB, as described in the Internet Engineering Task Force (IETF) request for comment (RFC) 1271, defined a framework for remote monitoring functions implemented on a network probe. (As of February, 1995, RFC 1271 has been superseded by RFC 1757.) The RMON MIB defines objects broken down into nine functional groups. Some of those functional groups, such the statistics and history groups, have a view of the Data Link layer that is specific to the media type and requires specific objects to be defined for each media type. RFC 1271 defined those specific objects necessary for Ethernet.

RFC 1513 defines those specific objects necessary for Token-Ring LANs. In addition, RFC 1513 defines some additional monitoring functions specifically for Token-Ring. These are defined in the Ring Station Group, the Ring Station Order Group, the Ring Station Configuration Group, and the Source Routing Statistics Group. A map of the RMON MIB for Ethernet and Token-Ring is shown in Fig. R1. The RMON MIB has been extended further to include FDDI.

RMON applications. A management application that views the internetwork, for example, gathers data from RMON agents running on each segment in the network. The data is integrated and correlated to produce various internetwork views that give end-to-end visibility of network traffic, both LAN and WAN. The operator can switch among a variety of views.

For example, the operator can switch between a Media Access Control (MAC) view (which shows traffic going through routers and gateways), a network view (which shows end-to-end traffic), or apply filters to see only traffic of a given protocol or suite of protocols. These traffic matrices provide the information necessary to configure or partition the internetwork to optimize LAN and WAN utilization.

In selecting the MAC-level view, for example, the network map shows each node of each segment separately, indicating intrasegment node-to-node data traffic. It also shows total intersegment data traffic from routers and gateways. This

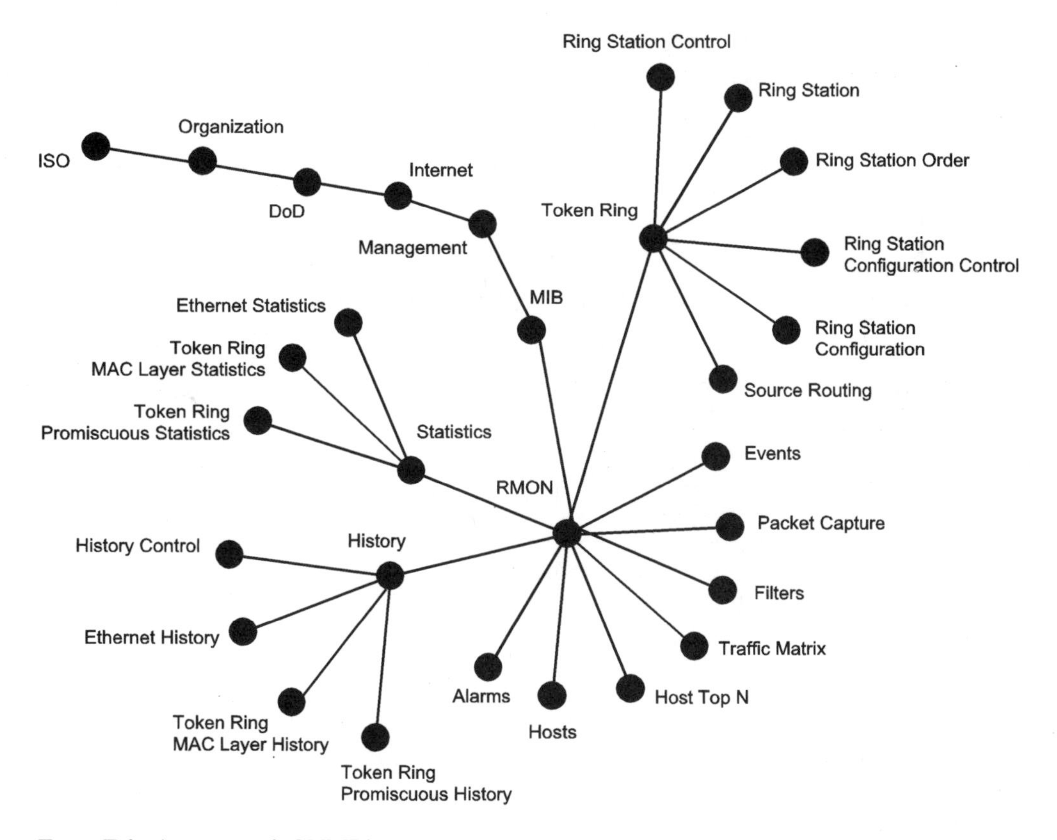

Fig. R1 *A map of SNMP's remote monitoring management information base, RMON MIB.*

combination allows the operator to see consolidated internetwork traffic and how each end node contributes to it.

In selecting the network-level view, the network map shows end-to-end data traffic between nodes, across segments. By connecting source and ultimate destination, without clouding the view with routers and gateways, the operator can immediately identify specific areas contributing to an unbalanced traffic load.

Another type of application allows the network manager to consolidate and present multiple-segment information, configure RMON alarms, provide complete Token-Ring RMON information, as well as perform baseline measurements and long-term reporting. Alarms can be set on any RMON variable. Notification via traps can be sent to multiple management stations. Baseline statistics allow long-term trend analysis of network traffic patterns, which can be used to plan for network growth.

Ethernet object groups. The RMON specification consists of two RFCs. RFC 1757 contains nine Ethernet and Ethernet/Token-Ring groups; RFC 1513 defines ten specific Token-Ring RMON extensions (Fig. R2).

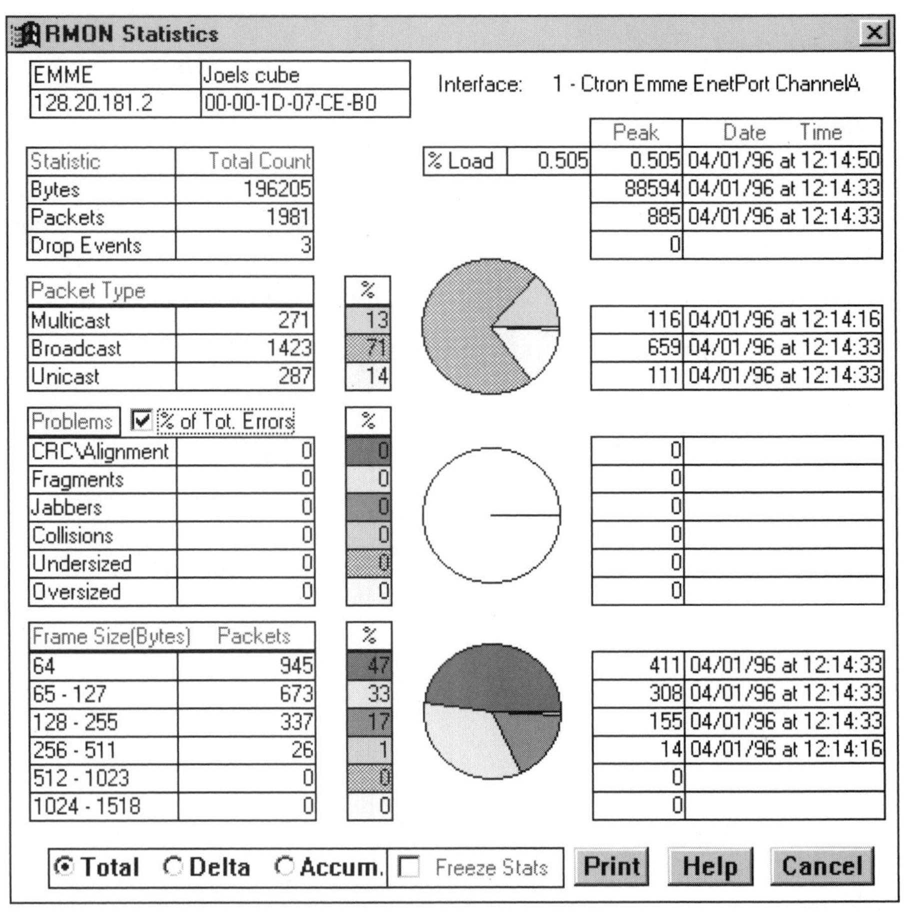

Fig. R2 *The Ethernet Statistics Window accessed from the Cabletron Systems Spectrum Element Manager for Windows. The column on the left side of the window displays the statistic name, total count, and percentage; the column on the right displays the peak value for each statistic, and the date and time that value occurred. Note that peak values are always delta values.*

Ethernet Statistics Group. The Statistics Group provides segment-level statistics. These statistics show packets, octets (or bytes), broadcasts, multicasts, and collisions on the local segment, as well as the number of occurrences of dropped packets by the agent. Each statistic is maintained in its own 32-bit cumulative counter. Real-time packet size distribution is also provided.

Ethernet History Group. With the exception of packet size distribution, which is provided only on a real-time basis, the History Group provides historical views of the statistics provided in the Statistics Group. The History Group can respond to user- defined sampling intervals and bucket counters, allowing for some customization in trend analysis.

The RMON MIB comes with two defaults for trend analysis. The first provides for 50 buckets (or samples) of 30-second sampling intervals over a period of 25 minutes. The second provides for 50 buckets of 30-minute sampling intervals over a period of 25 hours. Users can modify either of these or add additional intervals to meet specific requirements for historical analysis. The sampling interval can range from 1 second to 1 hour.

Host Table Group. The RMON MIB specifies a host table that includes node traffic statistics: packets sent/received and octets sent/received, as well as broadcasts, multicasts, and errored packets sent. In the host table, the classification "errors sent" is the combination of undersizes, fragments, CRC/alignment errors, collisions, and oversizes sent by each node.

The RMON MIB also includes a host timetable that shows the relative order in which each host was discovered by the agent. This feature not only is useful for network management purposes, but also assists in uploading those nodes to the management station of which it is not yet aware. This reduces unnecessary SNMP traffic on the network.

Host Top N Group. The Host Top N Group extends the host table by providing sorted host statistics such as the top 10 nodes sending packets, or an ordered list of all nodes according to the errors sent over the last 24 hours. Both the data selected and the duration of the study are defined by the user at the network management station, and the number of studies is limited only by the resources of the monitoring device.

When a set of statistics is selected for study, only the selected statistics are maintained in the Host Top N counters; other statistics over the same time intervals are not available for later study. This processing, performed remotely in the RMON MIB agent, reduces SNMP traffic on the network and the processing load on the management station, which otherwise would need to use SNMP to retrieve the entire host table for local processing.

Alarms Group. The Alarms Group provides a general mechanism for setting thresholds and sampling intervals to generate events on any counter or integer maintained by the agent, such as segment statistics, node traffic statistics defined in the host table, or any user-defined packet match counter defined in the Filters Group. Both rising and falling thresholds can be set, each of which can indicate network faults. Thresholds can be established on both the absolute value of a statistic or its delta value, so the manager is notified of rapid spikes or drops in a monitored value.

Filters Group. The Filters Group provides a generic filtering engine that implements all packet capture functions and events. The packet capture buffer is filled with only those packets that match the user-specified filtering criteria. Filtering conditions can be combined using the boolean parameters AND or NOT. Multiple filters are combined with the boolean OR parameter.

Packet Capture Group. The type of packets collected is dependent upon the Filter Group. The Packet Capture Group allows the user to create multiple capture buffers and to control whether the trace buffers will wrap (overwrite) when full or stop capturing. The user can expand or contract the size of the buffer to fit immediate needs for packet capturing, rather than permanently commit memory that will not always be needed.

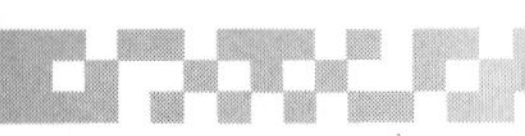

Notifications (Events) Group. In a distributed management environment, traps can be delivered by the RMON MIB agent to multiple management stations that share a single community name destination specified for the trap. In addition to the three traps already mentioned—rising threshold and falling threshold (see Alarms Group) and packet match (see Packet Capture Group)—there are seven additional traps that can be specified:

- *coldStart:* This trap indicates that the sending protocol entity is reinitializing itself such that the agent's configuration or the protocol entity implementation may be altered.
- *warmStart:* This trap indicates that the sending protocol entity is reinitializing itself such that neither the agent configuration nor the protocol entity implementation is altered.
- *linkDown:* This trap indicates that the sending protocol entity recognizes a failure in one of the communication links represented in the agent's configuration.
- *linkUp:* This trap indicates that the sending protocol entity recognizes that one of the communication links represented in the agent's configuration has come up.
- *authenticationFailure:* This trap indicates that the sending protocol entity is the addressee of a protocol message that is not properly authenticated. While implementations of the SNMP must be capable of generating this trap, they also must be capable of suppressing the emission of such traps via an implementation-specific mechanism.
- *egpNeighborLoss:* This trap indicates that an EGP neighbor for whom the sending protocol entity was an EGP peer has been marked down and the peer relationship is no longer valid.
- *enterpriseSpecific:* This trap indicates that the sending protocol entity recognizes that some enterprise-specific event has occurred.

The Notifications (Events) Group allows users to specify the number of events that can be sent to the monitor log. From the log, any specified event can be sent to the management station. The log includes the time of day for each event and a description of the event written by the vendor of the monitor. The log overwrites when full, so events may be lost if not uploaded to the management station periodically.

Traffic Matrix Group. The RMON MIB includes a traffic matrix at the Media Access Control (MAC) layer. A traffic matrix shows the amount of traffic and number of errors between pairs of nodes: one source and one destination address per pair. For each pair, the RMON MIB maintains counters for the number of packets, number of octets, and error packets between the nodes. Users can sort this information by source or destination address.

Applying remote monitoring and statistics-gathering capabilities to the Ethernet environment offers a number of benefits. The availability of critical networks is maximized, because remote capabilities allow for more timely problem resolution. With the ability to resolve problems remotely, operations staff can avoid costly travel to troubleshoot problems on-site. With the ability to analyze data collected at specific intervals over a long period of time, they can track down intermittent problems that normally would go undetected and unresolved.

Token-Ring extensions. Initially, RMON defined media-specific objects for Ethernet only. With RFC 1513, media-specific objects for Token-Ring became available.

Token-Ring MAC-Layer Statistics. This extension tracks statistics, diagnostics, and event notification associated with MAC traffic on the local ring. Statistics include the number of beacon, purge, and 803.5 MAC management packets and events; MAC packets; MAC octets; and ring soft error totals.

Token Ring Promiscuous Statistics. This extension collects utilization statistics of user data (non-MAC) traffic on the local ring. Statistics include the number of data packets and octets, broadcast and multicast packets, and data frame size distribution. "Promiscuous" refers to all user data traffic (non-MAC).

Token Ring MAC-Layer History. This extension offers historical views of MAC-layer statistics based on user-defined sample intervals, which can be set from 1 second to 1 hour to allow short-term or long-term historical analysis.

Token Ring Promiscuous History. This extension offers historical views of promiscuous statistics based on user-defined sample intervals, which can be set from 1 second to 1 hour to allow short-term or long-term historical analysis.

Ring Station Control Table. This extension lists status information for each ring being monitored. Statistics include ring state, active monitor, hard error beacon fault domain, and number of active stations.

Ring Station Table. This extension provides diagnostics and status information for each station on the ring. The types of information collected include station MAC address, status, and isolating and nonisolating soft error diagnostics.

Source Routing Statistics. The extension for source routing statistics is used for monitoring the efficiency of source-routing processes by keeping track of the number of data packets routed into, out of, and through each ring segment. Traffic distribution by hop count provides an indication of how much bandwidth is being consumed by traffic-routing functions.

Ring Station Configuration Control. The extension for station configuration control provides a description of the network's physical configuration. A media fault is reported as a *fault domain,* an area that isolates the problem to two adjacent nodes and the wiring between them. The network administrator can discover the exact location of the problem—the fault domain—by referring to the network map. Faults that result from changes to the physical ring, including each time a station inserts or removes itself from the network, are discovered by comparing the start of symptoms with the timing of physical changes.

The RMON MIB not only keeps track of the status of each station, it also reports the condition of each ring being monitored by a RMON agent. On large Token-Ring networks with several rings, the health of each ring segment and the number of active and inactive stations on each ring can be monitored simultaneously. Network administrators can be alerted to the location of the fault domain should any ring go into a beaconing (fault) condition. Network managers also can be alerted to any changes in backbone ring configuration, which could indicate loss

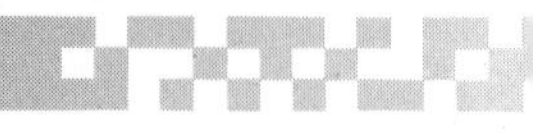

of connectivity to an interconnect device (such as a bridge) or to a shared resource (such as a server).

Ring Station Configuration. The ring station group collects Token-Ring–specific errors. Statistics are kept on all significant MAC-level events to assist in fault isolation, including ring purges, beacons, claim tokens, and such error conditions as burst errors, lost frames, congestion errors, frame copied errors, and soft errors.

Ring Station Order. Each station can be placed on the network map in a specified order relative to the other stations on the ring. This extension provides a list of stations attached to the ring in logical ring order. It lists only stations that comply with 802.5 active monitoring ring poll or IBM trace tool present advertisement conventions.

RMON II. The RMON MIB is basically a MAC-level standard. Its visibility does not extend beyond the router port, meaning that it cannot see beyond individual LAN segments. As such, it does not provide visibility into conversations across the network or connectivity between the various network segments. Given the trends toward remote access and distributed workgroups, which generate a lot of intersegment traffic, visibility across the enterprise is an important capability to have.

RMON II extends the packet capture and decoding capabilities of the original RMON MIB to layers 3 through 7 of the OSI Reference Model. This will allow traffic to be monitored via Network layer addresses—which lets RMON "see" beyond the router to the internetwork—and distinguish between applications.

Analysis tools that support the Network layer view can sort traffic by protocol, rather than just report on aggregate traffic. This means that network managers will be able to determine, for example, the percent of IP versus IPX traffic traversing the network. In addition, these higher-level monitoring tools can map end-to-end traffic, giving network managers the ability to trace communications between two hosts or nodes, even if the two are located on different LAN segments. RMON II functions that will allow this level of visibility include:

- *Protocol directory table* provides a list of all the different protocols a RMON II probe can interpret.
- *Protocol distribution table* permits tracking of the number of bytes and packets on any given segment that have been sent from each of the protocols supported. This information is useful for displaying traffic types by percentage in graphical form.
- *Address mapping* permits identification of traffic-generating nodes or hosts by Ethernet or Token-Ring address, in addition to MAC address. It also discovers switch or hub ports to which the hosts are attached. This is helpful in node discovery and network topology applications for pinpointing the specific paths of network traffic.
- *Network layer host table* permits tracking of bytes, packets, and errors by host according to individual Network layer protocol.
- *Network layer matrix table* permits tracking, by Network layer address, of the number of packets sent between pairs of hosts.

➢ *Application layer host table* permits tracking of bytes, packets, and errors by host and according to application.

➢ *Application layer matrix table* permits tracking of conversations between pairs of hosts by application.

➢ *History group* permits filtering and storing of statistics according to user-defined parameters and time intervals.

➢ *Configuration group* defines standard configuration parameters for probes that include such items as network address, serial line information, and SNMP trap destination information.

RMON II is focused more on helping network managers understand traffic flow for the purpose of capacity planning, rather than for the purpose of physical troubleshooting. The ability to identify traffic levels and statistics by application has the potential to reduce greatly the time it takes to troubleshoot certain problems. Without tools that can pinpoint which software application is gobbling up a disproportionate share of the available bandwidth, for example, network managers can only guess. Often it is easier just to upgrade a server or a buy more bandwidth, which inflates operating costs and shrinks budgets.

Despite the advantages of RMON II, it falls short in terms of monitoring switched networks. There is still no uniform way to discover where traffic is coming from in the switched environment. RMON II helps somewhat, however, by providing an address- mapping function that identifies the switch port to which a given host is attached. What is still needed are standards for analyzing the next generation of switched networks, specifically ATM.

Summary. Applying remote monitoring and statistics-gathering capabilities to the Ethernet and Token-Ring environments via the RMON MIB offers a number of benefits. The availability of critical networks is maximized, since remote capabilities allow for more timely problem resolution. With the ability to resolve problems remotely, operations staff can avoid costly travel to troubleshoot problems on-site. With the ability to analyze data collected at specific intervals over a long period of time, they can track down intermittent problems that would normally go undetected and unresolved. Now, with RMON II, these abilities are enhanced and extended across the enterprise.

See also

NETWORK AGENTS
OPEN SYSTEMS INTERCONNECTION
SIMPLE NETWORK MANAGEMENT PROTOCOL

Remote node

Remote node is a method of remote access that permits users to dial into the corporate network and perform tasks as if they were attached locally to the LAN. (In effect, they *are* attached locally.) With remote node, the remote system performs client functions, while the office-based host systems perform true server functions. This allows remote users to take advantage of a host's processing capabilities. Only the results are transmitted over the connection to the remote computer.

With remote node, the user's modem-equipped PC dials into the LAN and behaves as if it were a local LAN node. Instead of keystrokes and screen updates, the traffic on the remote node's dial-up line is essentially normal network packet traffic. The remote PC does not control another PC, as in remote control; rather, it runs regular applications as if it were directly attached to the LAN.

Remote node usually relies on a remote access server that is set up and maintained at a central location. This allows many remote users, all at different locations, to share the same resources. The server usually has Ethernet and/or Token-Ring ports and built-in modems for dial-up access at up to 56 kbps. Instead of modems, some vendors offer high-speed asynchronous ports. Depending on vendor, there might be optional support for higher-speed connections over such services as ISDN or frame relay.

Remote node offers many of the same features as remote control and, in some cases, surpasses them in functionality. This is especially true in such key areas as management, event reporting, and security.

Management. Most vendors allow the remote access server's routing software to be configured from a local management console, a remote Telnet session, or SNMP management station. SNMP support facilitates ongoing management and integrates the remote access server into the management environment commonly used by network administrators.

Management utilities allow for status monitoring of individual ports, the collection of service statistics, and the viewing of audit trails on port access and usage. Other statistics include port address, traffic type, connect time/break connection, and connect time exceeded. Filters can be applied to customize management reports with only the desired type of information.

The SNMP management station, in conjunction with the vendor-supplied management information base (MIB), can then display alerts about the operation of the server. The network administrator is notified when, for example, an application processor has been automatically reset due to a time-out, or when there is a hardware failure on a processor that triggers an antilocking mechanism reset (a type of reset that ensures that the entire system and all other users are not locked by the failing processor).

Event reporting. Remote node products typically include management software that is installed at the server to provide usage statistics, including packets sent/received and transmission errors, as well as who is logged on, how long they have been connected, and what types of modems are attached to the device.

Support for SNMP allows the server to trap a variety of meaningful events. Special drivers pass these traps to any SNMP-based network management platform, such as Hewlett-Packard's OpenView, IBM's NetView/6000, or Sun's Solstice SunNet Manager.

Security. Remote node products generally offer more levels of security than remote control products. Depending on the size of the network and the sensitivity of the information that can be remotely accessed, one or more of the following security methods can be employed:

- *Authentication:* This involves verifying the remote caller by user ID and password, thus controlling access to the server. Security is enhanced if ID and password are encrypted before going out over the communications link.
- *Access restrictions:* This involves assigning each remote user a specific location (i.e., directory or drive) that can be accessed in the server. Access to specific servers also can be controlled.
- *Time restrictions:* This involves assigning each remote user a specific amount of connection time, after which the connection is dropped.
- *Connection restrictions:* This involves limiting the number of consecutive connection attempts and/or the number of times connections can be established on an hourly or daily basis.
- *Protocol restrictions:* This involves limiting users to a specific protocol for remote access.

There are other options for securing the LAN, such as callback and cryptography. With callback, the remote client's call is accepted, the line is disconnected, and the server calls back after checking that the phone number is valid. While this works well for branch offices, most callback products are not appropriate for mobile users whose locations vary on a daily basis. There are products on the market that accept roving callback numbers, however. This feature allows mobile users to call into a remote access server or host computer, type in their user IDs and passwords, and then specify a number where the server or host should call them back. The callback number is then logged and may be used to help track down security breaches.

To safeguard very sensitive information, there are third-party authentication systems that can be added to the server. These systems require a user password and also a special, credit card-sized device that generates a new ID every 60 seconds, which must be matched by a similar ID number-generation process on the remote user's computer.

In addition to callback and encryption, security can be enforced via IP filtering and login passwords for the system console and for Telnet- and FTP-server programs. Many remote-node products also enforce security at the Link level using the Point- to-Point Protocol (PPP) with the Challenge Handshake Authentication Protocol (CHAP) and Password Authentication Protocol (PAP).

Summary. Companies are being driven to provide effective and reliable remote access solutions—remote control as well as remote node—to meet the productivity needs of the today's increasingly decentralized workforce. While remote control usually relies on dial-up connections established by modems, remote node connections to the remote access server also can be established by dial-up bridge/routers. Dial-up bridge/routers serve branch offices, while modems are typically used by mobile workers and telecommuters.

See also

BRANCH OFFICE ROUTING
NETWORK SECURITY
REMOTE CONTROL

Repeaters

A repeater is a device that extends the inherent distance limitations of various networks by regenerating the signals. As such, the repeater operates at the lowest level of the OSI reference model, the Physical layer (Fig. R3).

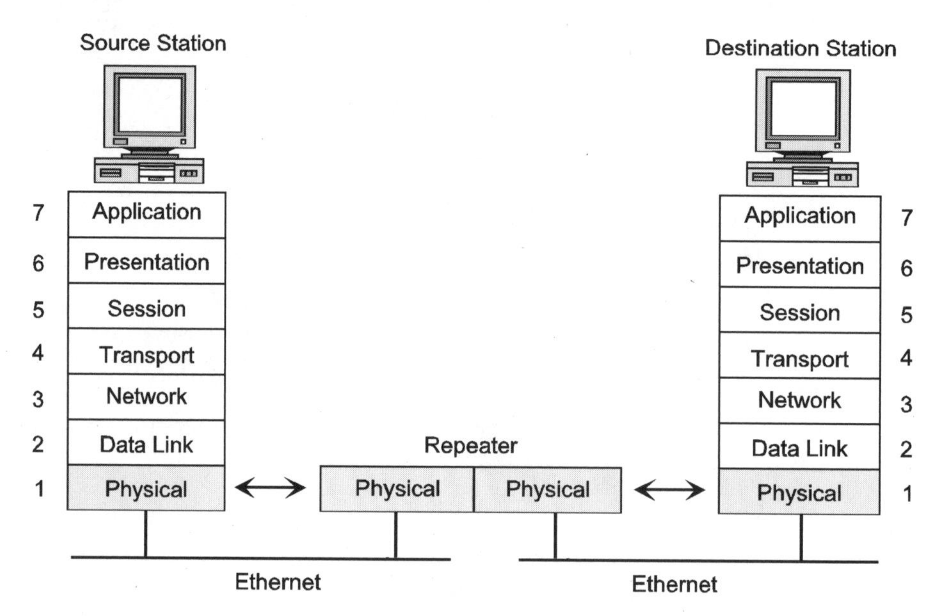

Fig. R3 *Repeaters operate at the Physical layer of the OSI Reference Model.*

Signal regeneration is necessary because signal strength weakens with distance: The longer the path a signal must travel, the weaker it gets. This condition is known as *signal attenuation.* On a telephone call, a weak signal will cause low volume, interfering with the parties' ability to hear one another. In cellular networks, when a mobile user moves beyond the range of a cell site, the signal fades to the point of disconnecting the call. In the LAN environment, a weak signal can result in corrupt data, which can substantially reduce throughput by forcing retransmissions when errors are detected. When the signal level drops low enough, the chances of interference from external noise increase, rendering the signal unusable.

In the LAN environment, the protocols limit the use of repeaters. With Ethernet, for example, repeaters are usually required every 500 feet. The IEEE 802.3 design guidelines for Ethernet LANs specify that the span between the two users farthest apart, including the cable connecting each user to the LAN, cannot exceed 500 m (about 1500 ft). Even with repeaters, the typical Ethernet LAN application requires that the entire path not exceed 1500 m end-to-end. This is the limit established for the proper operation of Ethernet's media access control mechanism, known as Carrier Sense Multiple Access with Collision Detection (CSMA/CD).

Aside from signal regeneration, repeaters also can be used to link different types of network media, e.g., fiber to coaxial cable. Often LANs are interconnected in a campus environment by means of repeaters that form the LANs into connected network segments. The segments may employ different transmission media: thick or thin coaxial cable, twisted-pair wiring, or optical fiber.

See also
BRIDGES
GATEWAYS
ROUTERS

Request for Proposal

The request for proposal (RFP) is a formal, technical document that completely describes an organization's requirements for new systems or networks. After the RFP is written, it is distributed to qualified vendors or service providers, who must respond by a specified date with a proposal that describes how they plan to meet the organization's requirements and at what cost. A goal of the RFP process is to provide the organization with the following:

- A consistent set of vendor responses, which are narrow in scope for easy comparison.
- A formal statement of requirements from which a purchase contract can be written, and against which vendor performance can be benchmarked.
- A mechanism within which vendors, fostered by the implied competition of a general solicitation for bids, assure the terms and conditions.

A properly written RFP and carefully managed evaluation process can accomplish these goals. The RFP should not be so rigid that it locks out vendor-recommended alternative solutions, however, that might be more efficient and economical. At the same time, an overly broad RFP that invites vendors to propose whatever is the optimum solution every step of the way is essentially no RFP at all. Not only does such an open-ended approach produce responses that are difficult to compare, but it leaves too much room for generalities and obfuscation on the part of vendors.

An overly interpretive RFP also can open the door to challenge from the losing bidders, which can tie up corporate resources and delay installation. To avoid this situation, it is best to know as much as possible about the business objectives and feasible solutions, and describe them clearly and concisely, so all vendor responses will be focused enough for easy comparison and fair evaluation.

RFP alternatives. The purpose of the RFP is bid solicitation. Other types of documents are used when different forms of assistance are required. For example, the *Request for Quotation* (RFQ) is used is used when planning the bulk purchase of commodity products such as PCs, printers, modems, and applications software. The RFQ is used when the most cost-effective solution is the overriding concern.

The *Request for Information* (RFI) is used when the organization is looking for the latest information on a particular technology, but has no immediate need. The purpose of the RFI is merely to get briefed on new technologies, how vendors plan to employ a particular technology in the future, or to get vendors' perspectives

on the feasibility of using or integrating a particular technology in a current network. Compared to the RFP and RFQ, the RFI is a very informal document. Vendor responses tend to be brief, and they may or may not include information on product pricing and availability. Nevertheless, the RFI responses can be useful for planning purposes and for deciding which vendors might qualify for a future RFP.

Summary. A technically knowledgeable team usually develops the RFP with input from all parts of the organization. The RFP is packaged in a professional manner; it is organized simply and logically, thereby making it easy for vendors to follow and helping them to develop a timely response that addresses all of the important issues. A good RFP also is written in a style that invites input from the vendors or service providers bidding on the project. A team manager plays the leading role in overseeing the development of the RFP, evaluating vendor proposals, and conducting meetings with vendors, sometimes with the aid of a consultant.

See also

BUSINESS PROCESS REENGINEERING

Routers

A router is a general-purpose device that is used to segment a network, with the goals of limiting broadcast traffic and providing security, control, and redundant paths. A router operates at a higher layer of the OSI Reference Model, distinguishing among network layer protocols and making intelligent packet-forwarding decisions. A router also provides firewall service and economical WAN access.

A router is similar to a bridge in that both provide filtering and bridging functions across the network. While bridges operate at the Physical and Data Link layers of the OSI reference model, however, routers join LANs at the Network layer (Fig. R4). Routers convert LAN protocols into wide area network protocols and perform the reverse process at the remote location. They may be deployed in mesh as well as point-to-point networks and, in certain situations, can be used in combination with bridges.

Although routers include the functionality of bridges, they differ from bridges in that they generally offer more embedded intelligence and, consequently, more sophisticated network management and traffic control capabilities than bridges. Perhaps the most significant distinction between a router and a bridge is that a bridge delivers packets of data on a best-effort basis, which can result in lost data unless the host computer protocol provides protection. By contrast, a router has the potential for flow control and more comprehensive error protection.

Types of routing. There are two types of routing, static and dynamic. In *static routing,* the network manager configures the routing table to set fixed paths between two routers. Unless reconfigured, the paths on the network never change. Although a static router will recognize that a link has gone down and will issue an alarm, it will not automatically reroute traffic.

A *dynamic router,* on the other hand, reconfigures the routing table automatically and recalculates the most efficient path in terms of load, line delay, or bandwidth. Some routers even balance the traffic load across multiple links. This allows the various links to better handle peak traffic conditions.

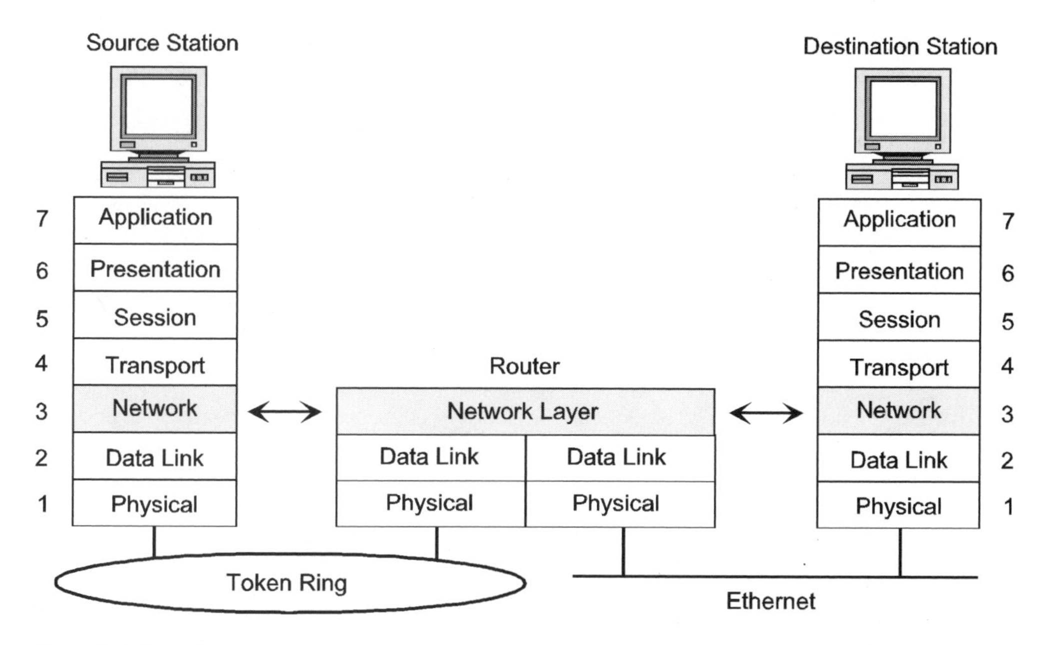

Fig. R4 *Routers operate at the Network layer of the OSI Reference Model.*

Routing protocols. Each router on the network keeps a routing table and moves data along the network from one router to the next using such routing protocols as Open Shortest Path First (OSPF), Intra-autonomous System to Intra-autonomous System (IS-IS), External Gateway Protocol (EGP), Border Gateway Protocol (BGP), Inter-Domain Policy Routing (IDPR), and Routing Information Protocol (RIP).

Although still supported by many vendors, RIP does not perform well in today's increasingly complex networks. As the network expands, routing updates grow larger under RIP and consume more bandwidth to route the information. When a link fails, the RIP update procedure slows route discovery, increases network traffic and bandwidth usage, and may cause temporary looping of data traffic. RIP also cannot calculate routes based on such factors as delay and bandwidth, and its line selection facility is capable of choosing only one path to each destination.

A newer routing standard, OSPF, overcomes the limitations of RIP and even provides capabilities not found in RIP. The update procedure of OSPF requires that each router on the network transmit a packet with a description of its local links to all other routers. On receiving such a packet, each router acknowledges it and, in the process, distributed routing tables are built from the collected descriptions. Because these description packets are relatively small, they produce a minimum of overhead. When a link fails, updated information floods the network, allowing all the routers to simultaneously calculate new tables.

Types of routers. Multiprotocol nodal, or hub, routers are used for building highly meshed wide area internets. In addition to allowing several protocols to share the same logical network, these devices pick the shortest path to the end node, balance the load across multiple physical links, reroute traffic around points of failure or congestion, and implement flow control in conjunction with the end nodes. They also provide the means to tie remote branch offices into the corporate backbone, which might use such WAN services as TCP/IP, T1, ISDN, and ATM. Some vendors also provide an interface for SMDS.

Access routers typically are used at branch offices. Available in Ethernet and Token-Ring versions, these usually are fixed-configuration devices that support a limited number of protocols and physical interfaces. They provide connectivity to high-end multiprotocol routers, allowing large and small nodes to be managed as a single logical enterprise network. Although low-cost, plug-and-play bridges can meet the need for branch office connectivity, low-end routers can offer more intelligence and configuration flexibility at comparable cost.

Midrange routers provide network connectivity between corporate locations, supporting workgroups or the corporate intranet, for example. These routers can be stand-alone devices or packaged as modules that occupy slots in an intelligent wiring hub or LAN switch. In fact, this type of router often is used to provide connectivity between multiple wiring hubs or LAN switches over high-speed LAN backbones such as ATM, FDDI, and Fast Ethernet.

Summary. Routers fulfill a vital role in implementing complex mesh networks and have become an economical means of tying branch offices into the enterprise network. Like other interconnection devices, routers are manageable via SNMP, as well as vendor-proprietary management systems. Just as bridging and routing functions made their way into a single device, routing and switching functions are being combined in the same way.

See also

BRANCH OFFICE ROUTING
BRIDGES
GATEWAYS
REPEATERS

Rural Radiotelephone Service

The Rural Radiotelephone Service allows common carriers to use wireless technology to provide telephone service to the homes of subscribers that live in extremely remote rural areas, where it is not feasible to provide telephone service by wire or other means. Conventional Rural Radiotelephone Service stations employ standard duplex analog technology to provide telephone service to the subscribers' homes.

The quality of conventional rural radiotelephone service is similar to that of precellular mobile telephone service. Several subscribers may have to share a radio channel pair (similar to party-line service), each waiting until the channel pair is not in use by the others before making or receiving a call.

Conventional Rural Radiotelephone Service generally is considered by state regulators to be a separate service that is interconnected to the public switched telephone network. This service has been available to rural subscribers for more than 25 years.

Satellite communications

The first satellites for communications use, Echo 1 and Echo 2, were launched by the United States in the early 1960s. They were little more than metallic balloons that simply reflected microwave signals from point A to point B. Although highly reliable, these passive satellites could not amplify the signals. Reception was often poor and the range of transmission limited. Ground stations had to track them across the sky and communication between two ground stations was only possible for a few hours a day when both could "see" the satellite at the same time. Geosynchronous (or geostationary) satellites overcame this problem.

Satellites are categorized by type of orbit and area of coverage as follows:

- *Geostationary–earth-orbit* (GEO) satellites orbit the equator in a fixed position about 22,000 miles above the earth. Three GEO satellites can cover most of the planet, with each unit capable of handling 20,000 voice channels. Because of their large coverage "footprint," these satellites are ideal for radio and television broadcasting and long-distance domestic and international communications.
- *Middle–earth-orbit* (MEO) satellites circle the earth at about 6100 miles up. It takes about 12 satellites to provide global coverage. The lower orbit reduces power requirements and transmission delays that can affect signal quality and service interaction.
- *Low–earth-orbit* (LEO) satellites circle the earth only 600 miles up (Fig. S1). As many as 200 satellites might be required to provide global coverage. Because their low altitude means that they have nonstationary orbits and they pass over a stationary caller rather quickly, calls must be handed off from one satellite to the next to keep the session alive. The omnidirectional antennas of the earthbound devices do not have to be pointed at a specific satellite. There also is very little propagation delay. The low altitude of these satellites also means that earthbound transceivers can be packaged as low-powered, handheld devices that cost less money.

The International Telecommunications Union (ITU) is responsible for all frequency and orbit assignments. The Federal Communications Commission (FCC) regulates all service rates, competition among carriers, and interstate and international telecommunications traffic in the United States, ensuring that U.S. satellite

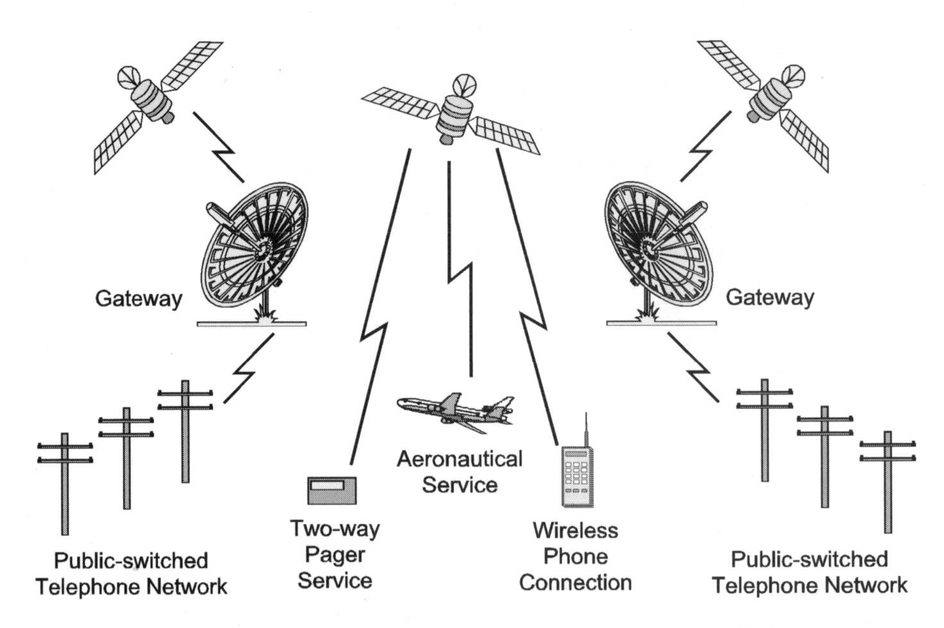

Fig. S1 *Low-earth-orbiting satellites hold out the promise of ubiquitous personal communications services, including telephone, pager, and two-way messaging services worldwide.*

operators conform to ITU frequency and orbit assignments. The FCC also issues licenses to domestic satellite service providers.

Satellite technology. Each satellite carries *transponders,* devices that receive radio signals at one frequency and convert them to another for transmission. The uplink and downlink frequencies are separated to minimize interference between transmitted and received signals.

Satellite channels allow one sending station to broadcast transmissions to one or more receiving stations simultaneously. In a typical scenario, the communications channel starts at a host computer, which is connected through a traditional telephone company medium to the central office (i.e., master earth station or hub) of the satellite communications vendor. The data from this and other local loops is multiplexed into a fiber-optic or microwave signal and sent to the satellite vendor's earth station. This signal becomes part of a composite transmission that is sent by the earth station to the satellite (uplink) and then transmitted by the satellite to the receiving earth stations or VSATs (downlink). At the receiving earth station, the data is transferred by a fiber-optic or microwave link to the satellite vendor's central office. The composite signal is then separated into individual communications channels and dispatched over local telephone company facilities to their destinations.

Satellite communication is very reliable. The bit error rate (BER) for a typical terrestrial link is one error in every 100,000 bits transmitted. A satellite channel can achieve a BER in the range of one error in one billion bits transmitted.

A potential problem with satellite communication, however, is delay. Round-trip satellite transmission takes approximately 500 ms, which can hamper voice

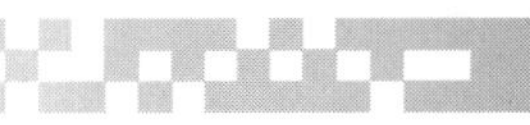

communications and create significant problems for real-time, interactive data transmissions. Satellite delay compensation units can nullify the effects of this lag in most bisynchronous protocol applications. For voice communications, digital echo cancelers can correct voice echo problems caused by the transmission delay.

Very Small Aperture Terminals. VSAT networks have evolved to become mainstream communication networking solutions that are affordable to both large and small companies. Today's VSAT is a flexible, software-intensive system built around standard communications protocols. Using a satellite as the serving office and using radio frequency (RF) electronics instead of copper or fiber cables, these systems can be truly considered packet-switching systems in the sky.

VSATs can be configured for broadcast (one-way) or interactive (two-way) communications. The typical star topology provides a flexible and economical means of communications with multiple remote or mobile sites. Applications include broadcasting database information, insurance agent support, reservations systems, retail point-of-sale credit checking, and interactive inventory data sharing.

Today's VSATs are used for supporting high-speed message broadcasting, image delivery, integrated data and voice, and mobile communications. VSATs are increasingly used for supporting LAN-to-LAN connectivity and LAN-to-WAN crossbridging, as well as in providing route and media diversity for disaster recovery.

To make VSAT technology more affordable, VSAT providers offer compact hubs and sub-meter antennas (i.e., less than 1 m across) that provide additional functionality at approximately 33 percent less than the cost of full-size systems. Newer sub-meter antennas are even supporting direct digital TV broadcasts to the home.

Network management. The performance of the VSAT network is monitored continuously at the hub location by the network control system. A failure anywhere on the network automatically alerts the network control operator, who can reconfigure capacity among individual VSATs. In the case of signal fade due to adverse weather conditions, for example, the hub detects the weakness or absence of a signal and alerts the network operations staff so that corrective action can be taken.

Today's network management systems indicate whether power failures are local or remote. They also can locate the source of communications problems and determine whether the trouble is with the software or hardware. Such capabilities often eliminate the expense of dispatching technicians to remote locations. When technicians must be dispatched, the diagnostic capabilities of the network management system can ensure that service personnel have with them the appropriate replacement parts, test gear, software patch, and documentation to solve the problem in a single service call.

Overall link performance is determined by the bit error rate, network availability, and response time. Because of the huge amount of information transmitted by the hub station, uplink performance requirements are more stringent. A combination of uplink and downlink availability, coupled with BER and response time, provides the network control operator with overall network performance information on a continuous basis.

The VSAT's management system offers a full range of accounting, maintenance, and data flow statistics, including those for inbound versus outbound data flow, peak periods, and total traffic volume by node. Also provided are capabilities for identifying fault conditions, performing diagnostics, and initiating restoral procedures. The VSAT's management system provides an interface to the major enterprise management platforms for single-point monitoring and control.

Communications protocols. Within the VSAT network there are three categories of protocols: those associated with the backbone network, those of the host computer, and those concerned with transponder access. The scope and functionality of protocol handling differ markedly among VSAT network providers.

The *backbone network protocol* is responsible for flow control, retransmissions of packets delivered with errors, and concurrent multiple sessions. The backbone network protocol could be associated with either the host or the communications link. The *host protocol* is related to the user application interface, which provides a compatible translation between the backbone protocol and the host communications protocol. Several host protocols are used in VSAT networks, including SNA/SDLC, 3270 BSC, Poll Select, and HASP. Multiple protocols can be used at the same VSAT location.

Transponder access protocols are used to assign transponder resources to various VSATs on the network. The three key transponder access protocols used on VSAT networks include Frequency Division Multiple Access (FDMA), Time Division Multiple Access (TDMA), and Code Division Multiple Access (CDMA).

Frequency Division Multiple Access. With FDMA, the radio frequency is partitioned so that bandwidth can be allocated to each VSAT on the network. This permits multiple VSATs to use their portions of the frequency spectrum simultaneously.

Time Division Multiple Access. With TDMA, each VSAT accesses the hub (via the satellite) by the bursting of digital information onto its assigned radio frequency carrier. Each VSAT bursts at its assigned time relative to the other VSATs on the network. Dividing access in this way—by time slots—is inherently wasteful because bandwidth is available to the VSAT in fixed increments whether or not it is needed. To improve the efficiency of TDMA, other techniques are applied to ensure that all the available bandwidth is used, regardless of whether the application contains bursty or streaming data. A reservation technique even can be applied to ensure that bandwidth is available for priority applications.

Code Division Multiple Access. With CDMA, all VSATs share the assigned frequency spectrum and also can transmit simultaneously. This is possible through the use of spread-spectrum technology, which employs a wideband channel, as opposed to the narrowband channels employed by other multiple access techniques such as FDMA and TDMA. Over the wideband channel, each transmission is assigned a unique code: a long row of numbers resembling a combination to a lock. The outbound data streams are coded so that they can be identified and received only by the station(s) having that code. This technique also is used in mobile communications as a means of cutting down interference and increasing available channel capacity by as much as 20 times.

Mobile satellite communications. Mobile satellite communications are used by the airline, maritime, and shipping industries. In fact, maritime mobile satellite services are quite advanced, primarily because of the efforts of the International Maritime Satellite Organization (INMARSAT), formed in the late 1970s.

INMARSAT's original, primary goal was to design and construct a worldwide maritime communications system. As a result of its efforts, today more than 7000 ships, offshore drilling rigs, and other vessels and maritime structures are equipped with satellite communications capabilities, including telex, voice, data, and video transmission. Currently, COMSAT (a private U.S. company formed to establish communications via satellites) is the only authorized organization in the United States that can directly access the INMARSAT system.

There are two land-based mobile satellite services currently being implemented in the United States: radio determination satellite service (RDSS) and mobile satellite service (MSS).

RDSS is a set of radio telecommunications and computational techniques that helps users determine their precise geographical location so that they can relay their position and other digital information to other subscribers, or to a central control center. The system constantly maintains and processes the information flow between a control center, satellites, and the mobile terminals.

MSS technology provides both voice and data communications to mobile users and offers wider geographic coverage than RDSS by utilizing relay satellites. The communications link identifies a transmitting vehicle's position between the mobile units and the satellite using Loran C navigational technology. MSS's uplink operates in the L band, while the downlink operates in the K band.

Summary. Satellites provide a reliable, economical way of providing communications to remote locations or supporting mobile telecommunications. When taking into account the large number of satellites that can be employed, along with their corresponding radio frequency assignments, it is clear that satellite communications systems offer ample room for expansion. Conversion of satellite transmissions from analog to digital, and use of more sophisticated multiplexing techniques, will further increase satellite transmission capacity. Other technological advances are focusing on the higher frequency bands, applying them in ways that decrease signal degradation.

See also
GLOBAL POSITIONING SYSTEM
TRANSMISSION FACILITIES

Signaling System 7

To place a telephone call successfully requires the use of a signaling protocol to convey information about the call, at minimum the call's origin and destination. Early signaling systems used dial pulses or tone pulses to represent the called number. Over the years these in-band signaling systems were expanded to generate and detect tone pulses representing the status of the called number (i.e., subscriber free, busy, etc.) or the calling number for automatic number identification (ANI). The number of signal types is limited, however, by the number of available tones in the signaling system; moreover, call setup time is extended each time more features are added.

Common channel signaling (CCS) overcomes the limitations of in-band signaling systems by using a separate signaling network to convey call information. Separating the signaling and voice networks means not only that signaling can occur

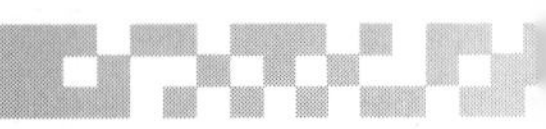

during a call without affecting voice traffic, but also that more sophisticated services and features can be supported. Compared to in-band signaling, common channel signaling allows:

- ➢ Faster call setup times,
- ➢ More efficient use of voice circuits,
- ➢ Support for services that require signaling during a call, and
- ➢ Better control over fraudulent network usage.

With the emergence of ISDN and the increased use of 800 and credit card services in the 1970s, the existing Common Channel Signaling System 6 (CCSS6) network architecture had to be revamped to handle not only circuit-related messages, but database access messages as well. These database access messages convey information between toll centers and centralized databases to permit real-time access to billing-related information and other sophisticated services. Since high-speed digital facilities were being employed to support these new services, a new end-to-end advanced signaling arrangement was required. This new system became known as Common Channel Signaling System 7, or just SS7. This signaling system is an ITU standard and is in use by all carriers in the United States and most other industrialized countries.

The SS7 network comprises packet-switching elements that are tied together with digital transmission links (Fig. S2):

- ➢ *Signal Transfer Point (STP):* A packet switch that routes signaling messages within the network.
- ➢ *Service Control Point (SCP):* A network element that interfaces with the STP and contains the network configuration and call completion database.
- ➢ *Service Switching Point (SSP):* Usually an end office (local switch) or access tandem office (long-distance switch) that contains the network signaling protocols and can access the SCP.

These service elements also are key building blocks of the Advanced Intelligent Network (AIN). In fact, SS7 is what makes the intelligent network "intelligent."

Message types. Two types of signaling messages are conveyed via SS7: circuit-related messages and database access messages. Circuit-related messages are used to establish and disconnect calls between two signaling points (i.e., SSPs). The STP conveys these messages over the appropriate circuit from the originating signaling point to the terminating signaling point. The information contained in these messages includes the identity of the circuit that connects both signaling points, the called number, answer indication, release indication, and release completion indication.

Database access messages retrieve information stored in the Service Control Point (SCP). An inquiry message requesting the necessary data needed to complete the call is transmitted from the originating signaling point via the STP to the SCP. The SCP sends a response message containing the requested data over the SS7 network to the originating signaling point, indicating that the network resources are available to complete the call.

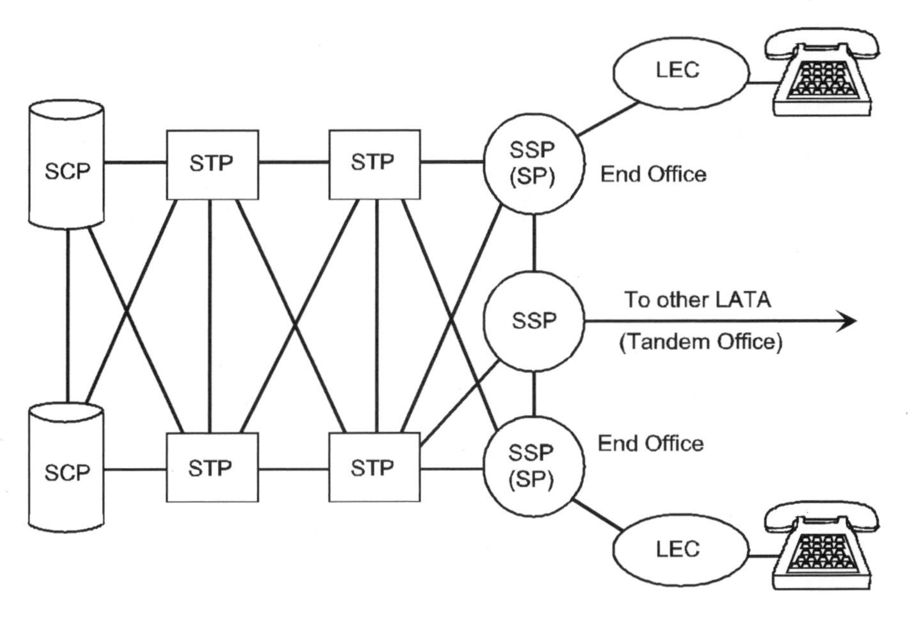

Fig. S2 *Key elements of an SS7 network.*

SS7 and OSI. The hardware and software functions of the SS7 protocol are divided into functional abstractions called *levels*. These levels are somewhat analogous to the seven-layer Open Systems Interconnect (OSI) Reference Model (Fig. S3) defined by the International Standards Organization (ISO).

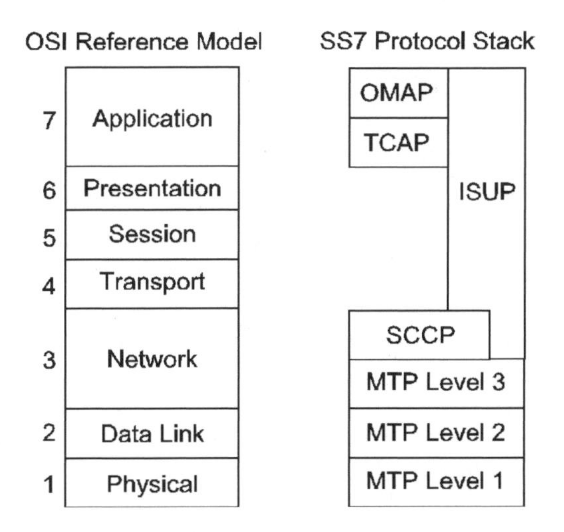

Fig. S3 *Relationship of SS7 to the OSI Reference Model.*

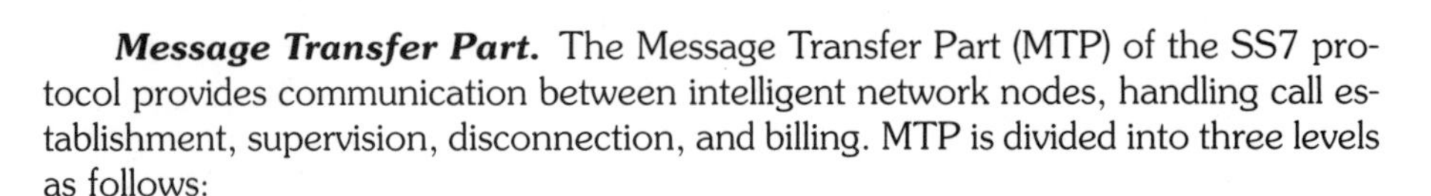

Message Transfer Part. The Message Transfer Part (MTP) of the SS7 protocol provides communication between intelligent network nodes, handling call establishment, supervision, disconnection, and billing. MTP is divided into three levels as follows:

- *MTP Level 1:* The lowest level (Level 1) is equivalent to the OSI Physical layer. MTP Level 1 defines the physical, electrical, and functional characteristics of the digital signaling link. Physical interfaces defined include E1 (2.048 Mbps), DS1 (1.544 Mbps), V.35 (64 kbps), DS0 (64 kbps), and DS0-A (56 kbps).
- *MTP Level 2:* MTP Level 2 ensures accurate end-to-end transmission of a message across the signaling link. When an error occurs on the link, the message (or set of messages) is retransmitted. SS7 signal units (messages) are formatted as follows (Fig. S4):

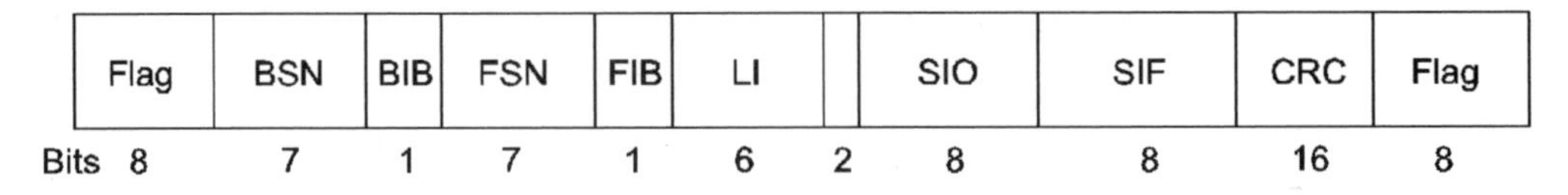

Fig. S4 *SS7 message signal unit format.*

- *Flag* indicates the beginning and end of a signal unit.
- *Backward Sequence Number (BSN)* is used to acknowledge the receipt of signal units by the remote signaling point.
- *Backward Indicator Bit (BIB)* indicates a negative acknowledgment by the remote signaling point.
- *Forward Sequence Number (FSN)* contains the sequence number of the signal unit.
- *Forward Indicator Bit (FIB)* is used for error recovery.
- *Length Indicator (LI)* indicates the number of octets that follow the LI and precede the CRC.
- *Service Information Octet (SIO)* contains the network indicator (e.g., national or international) and the message priority.
- *Signaling Information Field (SIF)* contains the routing label and signaling information.
- *Cyclic Redundancy Check (CRC)* is calculated and appended to the forward message. Upon receiving the message, the remote signaling point checks the CRC and copies the value of the FSN into the BSN of the next available message scheduled for transmission back to the initiating signaling point. If the CRC is correct, the backward message is transmitted. If the CRC is incorrect, the remote signaling point indicates negative acknowledgment by toggling the BIB prior to sending the backward message. When the originating signaling point receives a negative acknowledgment, it retransmits all forward messages, beginning with the corrupted message, with the FIB toggled.

- *MTP Level 3:* MTP Level 3 provides message routing and network management functions. Messages are routed based on the routing label in the signaling information field (SIF) of message signal units. MTP Level 3 also provides network management and maintenance functions to reconfigure the SS7 network in the event of link failures, and to control signaling traffic when congestion occurs.

Signaling Connection Control Part. The Signaling Connection Control Part (SCCP) of SS7 is used as the transport layer for TCAP-based services including 800/888, 900, and calling card services. In support of such services, SCCP provides connectionless and connection- oriented network services and global title translation capabilities. A *global title* is an address (a dialed 800 or 888 number, for example) that is translated by SCCP into a destination point code and subsystem number. Each service has a unique subsystem number that identifies the SCCP user at the destination signaling point. (For example, the 800 service SSN is 254). SCCP messages are contained within the signaling information field (SIF) of a message signal unit. The SIF contains the routing label followed by the SCCP message contents.

Transaction Capabilities Applications Part. TCAP enables the deployment of advanced intelligent network services by supporting non–circuit-related information exchange between signaling points using the SCCP connectionless service. An SSP uses TCAP to query an SCP to determine the routing number(s) associated with a dialed 800, 888, or 900 number. The SCP uses TCAP to return a response containing the routing number(s)—or an error or reject component—back to the SSP.

In the wireless environment, TCAP query and response messages make possible:

- Roamer Registration/Validation
- PCS Home Database
- Roamer Service Profile Delivery
- PCS Routing
- Roamer Call Delivery
- PCS Controller Access

These transactions are performed in real time. Calling card calls are also validated using TCAP query and response messages.

ISDN User Part. ISUP messages are used to control the setup and release of trunk circuits that carry voice and data between the calling party and called party at different terminating line exchanges. (Calls that originate and terminate at the same switch do not use ISUP signaling.) ISUP information is carried in the Signaling Information Field (SIF) of a message signal unit.

Various types of ISUP messages are exchanged between intelligent nodes and signal transfer points (STPs) within the intelligent network; among them are:

- *Forward address message,* which is used to set up a circuit between end points.

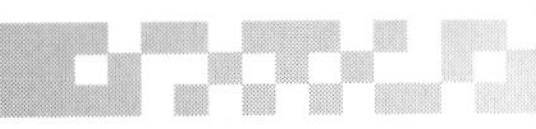

- *General setup message* conveys additional information required during call setup and provide the means to check that a circuit crossing multiple ISDNs maintains the desired transmission characteristics across all networks.
- *Backward setup message* supports call setup and initiates appropriate call accounting and charging procedures.
- *Call supervision message* supports call establishment with additional information, including whether the call was answered or not, and provides for manual operator intervention on ISDN calls between national boundaries.
- *Circuit supervision message* supports three functions on a pre-established circuit: release (which terminates the call), suspend and resume, and outgoing call blocking (which also permits incoming calls on an established but inactive circuit).
- *Circuit group supervision message* performs the same functions as the preceding message type, but on a group of circuits treated as a single unit for purposes of control.
- *In-call modification message* is used to alter the characteristics or associated network facilities of an active call.
- *Node-to-node message* allows the management and control of closed user groups so that incoming or outgoing (or both) types of calls are permitted only between members of the group.

Operations, Maintenance, Administration and Provisioning. Operations, Maintenance, Administration and Provisioning (OMAP) specifies network management functions and messages related to the common operations and maintenance procedures of the carriers.

Summary. The limitations of the old signaling systems prompted the development of common channel signaling, which evolved into SS7. In its simplest form, SS7 is a common channel, out-of-band signaling system that evolved from and became the replacement for multifrequency in-band telephony signaling systems. SS7 is much more powerful and flexible than the earlier signaling systems, however, because it is a message-based system designed to operate on separate digital facilities. This capability has led SS7 to grow beyond the initial support of voice services; it now supports enhanced services for the mass market, such as credit and debit card validation, follow-me services, 800 number routing, wireless applications, and automatic call distribution capabilities. The robustness of the SS7 architecture also will serve the needs of the emerging personal communications services (PCS). Work continues within the ITU to evolve and improve SS7.

See also

ADVANCED INTELLIGENT NETWORK
OPEN SYSTEMS INTERCONNECTION

Simple Network Management Protocol

Since 1988, the Simple Network Management Protocol (SNMP) has been the de facto standard for managing multivendor TCP/IP-based networks. SNMP is an industry-standard protocol that specifies a structure for formatting messages and for transmitting information between reporting devices and data-collection programs on the network. The SNMP-compliant devices on the network are polled for performance-related information, which is passed to a network management console. Alarms also are passed to the console. There, the gathered information can be viewed to pinpoint problems on the network or stored for later analysis.

SNMP runs on top of TCP/IP's datagram protocol (the User Datagram Protocol, or UDP), a transport protocol that offers a connectionless-mode service. This means that a session need not be established before network management information can be passed to the central control point. Although SNMP messages can be exchanged across any protocol, UDP is well suited to the brief request/response message exchanges characteristic of network management communications.

SNMP is a very flexible network management protocol that can be used to manage virtually any object, even OSI (Open System Interconnection) objects. An *object* refers to hardware, software, or a logical association such as a connection or virtual circuit. An object's definition is written by its vendor. The definitions are held in a management information base, or MIB, which is often thought of as a database. In reality, a MIB is a list (of switch settings, hardware counters, in-memory variables, or files) that is used by the network management system to determine the alarm and reporting characteristics of each device on the network, including those connected over Ethernet, Fast Ethernet, Token-Ring, and FDDI.

SNMP is basically a request/response protocol. The management system retrieves information from the agents through SNMP's GET and GET-NEXT commands. The GET request retrieves the values of specific objects from the MIB. The MIB lists the network objects for which an agent can return values. These values may include the number of input packets, the number of input errors, and routing information. The GET-NEXT request permits navigation of the MIB, allowing the next MIB object to be retrieved, relative to its current position. A SET request is used to ask a logically remote agent to alter the values of variables.

In addition to these message types, there are TRAP messages, which are unsolicited messages conveyed from management agent to management stations. Other commands are available that allow the network manager to take specific actions to control the network. Some of these commands look like SNMP commands, but are really vendor-specific implementations. For example, some vendors use a STAT command to determine the status of network connections.

All of the major network management platforms support SNMP, including Hewlett-Packard's OpenView, IBM's NetView/6000, and Sun's Solstice SunNet Manager. In addition, many of the third-party systems and network management applications that plug into these platforms support SNMP. The advantage of using such products is that they take advantage of SNMP's capabilities, while providing a graphical user interface (GUI) to make SNMP easier to use. Even MIBs can be selected for display and navigation through the GUI.

Another advantage of commercial products is that they can use SNMP to provide additional functionality. For example, OpenView, NetView, and SunNet Manager are used to manage network devices that are IP-addressable and run SNMP. Their automatic discovery capability finds and identifies all IP nodes on the network, including those of other vendors that support SNMP. Based on discovered information, the management system automatically draws the required topology maps. Nodes that cannot be discovered automatically can be represented in either of two ways: by manually adding custom or standard icons to the appropriate map views, or using SNMP-based APIs for building map applications without having to modify the configuration to accommodate non-SNMP devices manually.

Architectural components. SNMP is one of three components comprising a total network management system (Fig. S5). The other two are the management information base (MIB) and the network manager (NM). The MIB defines the controls embedded in network components, while the NM contains the tools that allow network administrators to comprehend the state of the network from the gathered information.

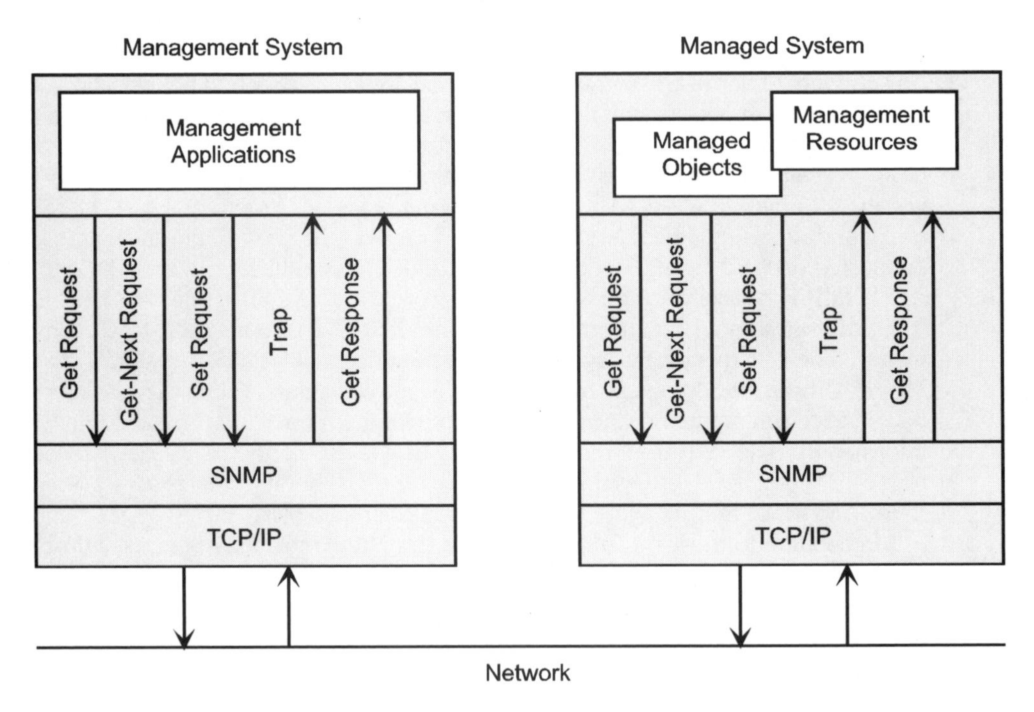

Fig. S5 *The SNMP architecture.*

Network manager. The network manager is a program that may run on one host or more than one host, each of which manages a particular subnet. SNMP communicates network management data to a single site, called a *network management station* (NMS). Under SNMP, each network segment must have a device, called an *agent,* that can monitor devices, called *objects,* on that segment and report the information to the NMS. The agent could be a passive monitoring device the sole purpose of which is to read the network, or it could be an active device that performs other functions as well, such as bridging, routing, and switching. Devices that are not SNMP-compliant must be linked to the NMS by means of a *proxy agent.*

The NMS provides the information display, communication with agents, information filtering, and control capabilities. The agents and their appropriate information are displayed in a graphical format, often against a network map. Network technicians and administrators can query the agents and read the responses on the NMS display. The NMS also periodically polls the agents, searching for anomalies. Detection of an anomaly results in an alarm at the NMS.

Management information base. The MIB is a listing of information necessary to manage the various devices on the network. The MIB contains a description of SNMP-compliant objects on the network and the kinds of management information they provide. An *object* refers to hardware, software, or a logical association such as a connection or virtual circuit. The attributes of an object might include such things as the number of packets sent, routing table entries, and protocol-specific variables for IP routing.

The first MIB was concerned primarily with IP routing variables used for interconnecting different networks. There are over 110 objects that form the core of the standard SNMP MIB. The latest generation MIB, known as MIB II, defines more than 160 objects. It extends SNMP capabilities to a variety of media and network devices, marking a shift from Ethernets and TCP/IP wide area networks to all media types used on LANs and WANs. Many vendors want to add value to their products by making them more manageable, so they create private extensions to the standard MIB, which can include 200 or more additional objects.

Many vendors of SNMP-compliant products include MIB tool kits that generally include two types of utilities. One, a MIB compiler, acts as a translator that converts ASCII text files of MIBs for use by an SNMP management station. The second type of MIB tool converts the translator's output into a format that can be used by the management station's applications or graphics. These output handlers, also known as *MIB editors* or *MIB walkers,* let users view the MIB and select the variables to be included in the management system. Some vendors of SNMP management stations do not offer MIB tool kits, but rather an optional service whereby they will integrate into the management system any MIB a user requires for a given network. This service includes debugging and technical support.

Summary. SNMP's popularity stems from the fact that it works, it is reliable, and it is widely supported. The protocol itself is in the public domain. SNMP capabilities have been integrated into just about every conceivable device that is used on today's LANs and WANs, including intelligent hubs and carrier services such as frame relay.

See also

NETWORK MANAGEMENT SYSTEMS
OPEN SYSTEMS INTERCONNECTION
REMOTE MONITORING

Slamming

Since the start of long-distance competition in the telephone industry in the early 1980s, when consumers could for the first time switch from AT&T to another long-distance carrier, telecommunications companies have been using various marketing techniques to win over new customers. Some of these (media advertising, telemar-

keting, frequent flyer bonuses, and offers of lower rates) are legitimate. Other techniques, like slamming, are not.

Slamming is the illegal practice of having a long-distance or other telecommunications service switched without the subscriber's knowledge or permission. According to the FCC, slamming is one of the top three complaints the agency receives from consumers.

Federal law requires a customer's approval before his or her long-distance carrier can be changed. However, some carriers may inform the local telephone company that a customer has given it authorization to carry his or her long-distance call, when in fact no such authorization was provided. Some typical ways that slamming can occur include:

- A telemarketer might call with an offer to save the subscriber money if he or she switches long-distance carriers. Even if the subscriber says no, they may wind up being "slammed" if the telemarketer still reports that the subscriber has agreed to switch.
- The subscriber might receive a check in the mail that states in small print that by signing and cashing the check he or she agrees to switch to a new carrier.
- The subscriber might fill out what he or she believes to be a contest entry blank that actually is also a consent form for switching.

In September, 1995, the FCC issued rules designed to strengthen protections for consumers while preserving their right to choose or change long-distance carriers. Signatures on contest and sweepstakes forms and other such gimmicks no longer may be used as authorizations to switch companies. An endorsement on a bonus check will be accepted as an authorization to change a long-distance provider, but only if the authorization form to switch carriers is separate from promotional material and if it is written in plain language and in print size comparable to accompanying advertising copy. In addition, the form must clearly state that its purpose is to change one's long-distance carrier.

Summary. Despite state and federal laws against the practice, slamming is still a problem. As recently as May, 1997, the FCC slapped an $80,000 fine on Long Distance Services Inc. (LDSI), a Washington, D.C.–based interexchange carrier, for slamming two of its competitors' customers.

To prevent becoming a slamming victim, customers can call their local phone company's business office to request a "freeze" of their long-distance phone carrier. In that way, the long-distance carrier cannot be switched unless the local company receives the customer's authorization, including a personal ID code, verbally and in writing.

See also
TELEPHONE FRAUD

Specialized Mobile Radio

Specialized Mobile Radio service was created by the FCC in 1974 to let carriers provide two-way radio dispatch service for the public safety, construction, and transportation industries. SMR systems generally provide dispatch services using "push to talk" technology for companies with multiple vehicles.

SMR services include voice dispatch, data broadcast, and mobile telephone service; SMR currently has limited roaming capabilities, however. An SMR subscriber can interconnect with the public telephone network much like a cellular subscriber. Both operate over different assigned frequencies within the range of 800–900 MHz. Cellular services are assigned to bands between 824–849 MHz and 869–894 MHz.

SMR networks traditionally used one large transmitter to cover a wide geographic area. This limited the number of subscribers because only one subscriber could talk on one frequency at any given moment. The number of frequencies allocated to SMR is smaller than for cellular and there have been several operators in each market. Because dispatch messages are short, SMR services were able to work reasonably well.

In April, 1990, Nextel asked the FCC for permission to build enhanced SMR systems in several of its key markets. These systems would consist of multiple, lower-power transmitters that would allow the same frequencies to be reused some distance away. Nextel also planned its new systems to be digital in order to further expand calling capacity. Since ESMR systems integrate voice, messaging, paging, and dispatch capabilities over the same network, the resulting cellular-like network would open new consumer and business communications markets to the SMR industry.

In February, 1991, the FCC approved Nextel's request, and its first digital mobile network came on-line in Los Angeles in August, 1993. Other SMR carriers soon followed with enhanced systems. The growth of digital is expected to drive the SMR industry to over 4 million total subscribers by the year 2000.

See also

CELLULAR COMMUNICATIONS

Spread spectrum radio

Spread spectrum is a digital coding technique in which the signal is taken apart or "spread" so that it sounds more like noise to the casual listener. The coding operation increases the number of bits transmitted and expands the bandwidth used. With the signal's power spread over a larger band of frequencies, the result is a more robust signal that is less susceptible to impairment from electromechanical noise and other sources of interference.

Using the same spreading code as the transmitter, the receiver correlates and collapses the spread signal back down to its original form. Spread spectrum is used for wireless Ethernet LANs and is the basis for other advanced wireless transmission techniques such as Code Division Multiple Access (CDMA), which is being used to support a variety of services, including emerging personal communications services (PCS).

Technology. Spread spectrum uses the industrial, scientific, and medical (ISM) bands of the electromagnetic spectrum. The ISM bands include the frequency ranges at 902–928 MHz and 2.4–2.484 GHz, which do not require an FCC site license.

Spread spectrum is a highly robust wireless data transmission technology that offers substantial performance advantages over conventional narrowband radio systems. As noted, the digital coding technique used in spread spectrum takes the signal apart and spreads it over the available bandwidth, making it appear as random noise. The coding operation increases the number of bits transmitted and expands the bandwidth used. Noise has a flat, uniform spectrum with no coherent peaks and generally can be removed by filtering. The spread signal has a much lower power density, but the same total power.

This low power density, spread over the expanded transmitter bandwidth, provides resistance to a variety of conditions that can plague narrowband radio systems, including:

- *Interference,* a condition in which a transmission is being disrupted by external sources, such as the noise emitted by various electromechanical devices, or internal sources such as crosstalk.
- *Jamming,* a condition in which a stronger signal overwhelms a weaker signal, causing a disruption to data communications.
- *Multipath,* a condition in which the original signal is distorted after being reflected off a solid object.
- *Interception,* a condition in which unauthorized users capture signals in an attempt to determine its content.

Non–spread spectrum narrowband radio systems transmit and receive on a specific frequency that is just wide enough to pass the information, whether voice or data. By assigning users different channel frequencies, confining the signals to specified bandwidth limits, and restricting the power that can be used to modulate the signals, undesirable crosstalk (interference between different users) can be avoided. These rules are necessary because any increase in the modulation rate widens the radio signal bandwidth, which increases the chance for crosstalk.

The main advantage of spread spectrum radio waves is that the signals can be manipulated to propagate fairly well through the air, despite electromagnetic interference, to virtually eliminate crosstalk. In spread-spectrum modulation, a signal's power is spread over a larger band of frequencies. This results in a more robust signal that is less susceptible to interference from similar radio-based systems, since they too are spreading their signals, but with different spreading algorithms.

There are two spreading techniques in common use today: direct sequence and frequency hopping.

Direct sequence. In direct sequence spreading, the most common implementation of spread spectrum technology, the radio energy is spread across a larger portion of the band than is actually necessary for the data. This is done by breaking each data bit into multiple sub-bits called "chips" to create a higher modulation rate. The higher modulation rate is achieved by multiplying the digital signal with a chip sequence. If the chip sequence is ten, for example, and it is applied to a signal carrying data at 300 kbps, then the resulting bandwidth will be ten times wider. The amount of spreading is dependent upon the ratio of chips to each bit of information.

Because data modulation widens the radio carrier to increasingly larger bandwidths as the data rate increases, this chip rate of 10 times the data rate spreads the radio carrier to 10 times wider than it otherwise would be for data alone.

The rationale behind this technique is that a spread-spectrum signal with a unique spread code cannot create the exact spectral characteristics as another spread-coded signal. Using the same code as the transmitter, the receiver can correlate and collapse the spread signal back down to its original form, while other receivers using different codes cannot.

This feature of spread spectrum makes it possible to build and operate multiple networks in the same location. By assigning each one its own unique spreading code, all transmissions can use the same frequency band, yet remain independent of each other. The transmissions of one network appear to the other as random noise and are filtered out because the spreading codes do not match.

This spreading technique would appear to result in a weaker signal-to-noise ratio, because the spreading process lowers the signal power at any one frequency. Normally, a low signal-to-noise ratio would result in damaged data packets that would require retransmission. The processing gain of the despreading correlator recovers the loss in power when the signal is collapsed back down to the original data bandwidth, however, but is not strengthened beyond what would have been received had the signal not been spread.

The FCC has set rules for direct sequence transmitters. Each signal must have ten or more chips. This rule limits the practical raw data throughput of transmitters to 2 Mbps in the 902 MHz band and 8 Mbps in the 2.4 GHz band. The number of chips is directly related to a signal's immunity to interference. In an area with a lot of radio interference, users will have to give up throughput to limit interference successfully.

Frequency hopping. Frequency hopping entails the transmitter jumping from one frequency to the next at a specific hopping rate in accordance with a pseudorandom code sequence. The order of frequencies selected by the transmitter is taken from a predetermined set as dictated by the code sequence. For example, the transmitter may have a hopping pattern of going from channel 3 to channel 12 to channel 6 to channel 11 to channel 5, and so on (Fig. S6). The receiver tracks these changes. Since only the intended receiver is aware of the transmitter's hopping pattern, then only that receiver can make sense of the data being transmitted.

Other frequency hopping transmitters will be using different hopping patterns that usually will be on noninterfering frequencies. Should different transmitters coincidentally attempt to use the same frequency and the data of one or both become garbled at that point, retransmission of the affected data packets is required. Those data packets will be sent again on the next hopping frequency of each transmitter. Most LAN protocols have an integral error-detection capability. When the protocol's error-checking mechanism recognizes incoming packets that are bad or determines that there are missing packets, the receiving station requests a retransmission of only those packets. When the new packets arrive to rendezvous with those held in queue, the protocol's sequencing capability puts them in the correct order.

The FCC mandates that frequency-hopped systems must not spend more than 0.4 seconds on any one channel in each 20 seconds (or 30 seconds in the 2.4 GHz band). Furthermore, they must hop through at least 50 channels in the 900 MHz band, or 75 channels in the 2.4 GHz band. These rules reduce the chance of repeated packet collisions in areas with multiple transmitters.

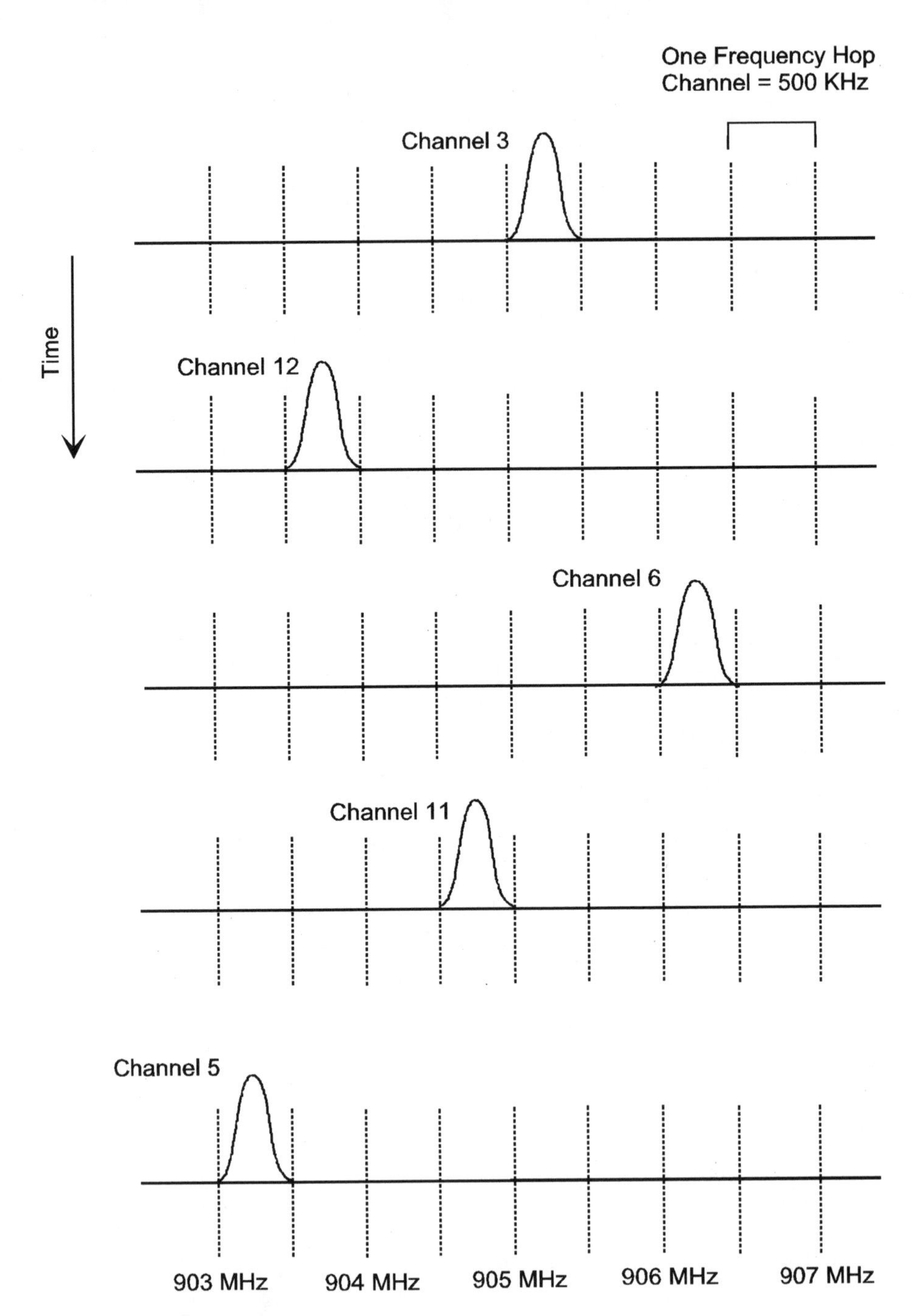

Fig. S6 *Frequency-hopping spread spectrum.*

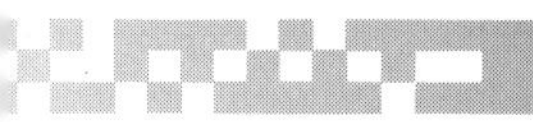

Summary. Direct-sequence spread spectrum offers better performance, but frequency-hopping spread spectrum is more resistant to interference and is preferable in environments with electromechanical noise. Direct sequence is more expensive than frequency hopping and uses more power. Although spread spectrum provides more secure data transmission than conventional narrowband radio systems, this does not mean the transmissions are immune from interception and decoding by knowledgeable intruders with sophisticated tapping equipment. For this reason, many vendors provide optional encryption for added security.

See also

INFRARED NETWORKING

StarLAN

AT&T originally developed StarLAN to satisfy the need for a low-cost, easy-to-install local area network that would offer more configuration flexibility than Token-Ring and more availability than Ethernet. The hub-based StarLAN was offered in two versions, 1 Mbps and 10 Mbps. Because it was based on IEEE 802.3 standards, StarLAN offered interoperability with Ethernet and Token-Ring through driver software.

In 1991, AT&T and NCR merged under the name AT&T Global Information Solutions, where responsibility for StarLAN resided until 1996. That year, AT&T Global Information Solutions changed its name back to NCR in anticipation of being spun off to AT&T shareholders as an independent, publicly traded company. Around that time, NCR discontinued the StarLAN product line.

See also

ARCNET
ETHERNET
TOKEN-RING

Storage media

A variety of storage media are available for backing up mission-critical data over the network, including multigigabyte hard drives, optical disks, and tape. A common backup strategy employs a backup server that grabs files from other servers at high speeds and writes them to high-capacity disk(s). When data does not have to be available for instant retrieval, it can be archived on more economical media such as optical disk and tape. An effective network backup strategy would be to use all three media types—hard disk, optical disk and tape—in a hierarchical arrangement that is based on how readily data needs to be available.

Backup media. There are several backup media to choose from, depending on the frequency of access and the age of the data.

Magnetic disk. When hard disks are used for network backup, they are usually housed in a dedicated server. Today's hard disks may hold several gigabytes (GB) of data on several platters, support a transfer rate of 15 to 40 megabytes (MB) or more per second (depending on the speed of the interface) and a seek time of less than 12 ms. The cost per megabyte is less than 15 cents. Disk capacity can be doubled using software that performs compression/decompression in real time, bringing the cost per megabyte down to 7.5 cents.

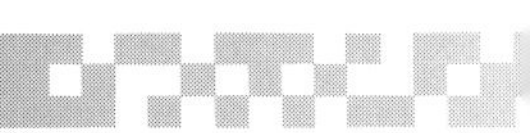

There are removable disk cartridges available to back up data on stand alone PCs. One is a 100 MB cartridge that costs about 20 cents per megabyte and the other is a 1 GB cartridge that costs 10 cents per megabyte. Both do not rely on compression and deliver performance that approaches that of many internal drives. In addition to low cost, they have the advantage of providing high-capacity data portability.

Magnetic tape. Magnetic tape is the low-cost choice for long-term data storage. The popular QIC (Quarter-Inch Cartridge) format varies in storage capacity, from 400 MB without compression to about one GB with compression (and the use of extended- length tape); it is used mainly for backing up the files of a single PC or workstation. There are other formats based on QIC, such as Travan and Taumat, which offer slightly higher storage capacity.

Two other tape standards are 4 mm digital audio tape (DAT) and 8 mm helical scan tape (HST). DAT can store up to 8 GB using 2:1 compression. The HST systems offer 16 MB of storage, and generally cost more than DAT systems. Some vendors use proprietary HST formats to achieve a native capacity of up 25 GB per cartridge and an average data transfer rate of 3 megabytes per second.

Another popular tape format uses half-inch tape available in 9-, 18- and 36-track versions. These are most often used in the distributed Unix, midrange, and mainframe environments. These tape formats are also packaged in cartridge form.

All of these tapes can be used with autochangers or jukeboxes that hold a dozen to a few hundred tapes. Special software ensures that tapes are changed when necessary, labeled properly, and stored in the right place and in the right order. Higher- capacity library storage facilities are available, which are complete installations potentially consisting of thousands of individual tape cartridges that are retrieved by robotic mechanisms.

Optical media. Many businesses are finding that conventional fixed disk drives are running out of storage capacity too quickly. Optical disk technology offers an economical solution.

Compact Disk-Read Only Memory. Commercially available since 1985, CD-ROM is similar to the familiar CD disks used for music, except that they have been formatted to store data and must be used on drives that also read data. As its name implies, CD-ROM is a read-only medium. The disks themselves can contain about 700 MB of uncompressed data, which equates to 120,000 printed pages. CD-ROM is not only inexpensive to manufacture and duplicate, it is also one of the most economical (in cost per megabyte) long-term storage solutions available.

To create a master disk, a laser beam burns pits into the track where a binary 1 is to be stored and leaves the reflective surface intact when a 0 is required. When reading data on the disk, a laser beam in the CD-ROM drive is focused on the track and a light detector calculates the amount of light being reflected back from the surface. When the reflected light is of similar intensity to the original laser beam, a bit value of 0 is assumed. When the laser beam hits a pit on the disk, the light scatters, causing the detector to see a very small amount of light. When the reflected light is of lesser intensity to the original laser beam, a bit value of 1 is assumed.

This reading process combines with the slow rotation speed of the disk (between 200 and 530 rpm) to deliver average access times of about 300 ms. This is very slow compared to Winchester-type hard disks, which deliver average access times of well under 15 ms. However, continued innovations in drive technology (i.e., 2X through 24X speeds) have greatly improved the performance of CD-ROMs.

CD-ROM is an ideal medium for distributing reference databases that do not change frequently. Hundreds of references are now offered on CD-ROM, including encyclopedias, dictionaries, and technical periodicals. Many companies routinely issue software and product documentation on CD-ROM. While CD-ROM is useful for specialized applications, its inflexibility makes it a poor choice for real-time network backup.

Compact Disk-Recordable. CD-ROM would be a much more useful technology for network backup if users could store their own data on CDs, as well as read data from them. This is the promise behind CD-Recordable (CD-R), which represents an economical and standards-based alternative to WORM. Once a CD is created in the standard ISO 9660 format, it can be read by any CD-ROM drive connected to any platform. This is not true of WORM or magneto-optical media, for which there are almost as many standards as there are products.

CD-Rewritable. While CD-R allows users to write to a CD, it does not allow data to be erased so that more data can be recorded; the user just keeps writing to the CD until it becomes full. With CD-Rewritable (formerly known as CD-Erasable), the CDs are not only recordable, but also erasable and rewritable. The user can add, erase, or replace files by dragging and dropping them. When a file is deleted, the directory reflects the change, but the file remains on the CD. The drawback to CD- RW is that the user must erase the entire disk to reuse it.

Write Once Read Many. WORM operates similarly to CD-ROM in that a mark on the disk represents 1 and no mark is 0. WORM technology uses a laser beam to heat a small spot on the track, creating a pit. During playback, the laser light scatters upon hitting the heat-treated spot. The drive circuitry interprets the lower level of reflection as 1.

Once the laser beam creates a pit on the disk, the change is permanent. Although data can be continuously added to the WORM disk until it runs out of room, previously input data cannot be written over. What appears to be a technological weakness, however, is actually a strength for some applications. While WORM is not suitable for applications that require frequent changes, it is useful for long-term archival storage of such critical information as legal documents, tax records, and regulatory filings, since it provides assurance that the data cannot be tampered with.

WORM disks come in several sizes: 5.25-inch disks that hold about 400 MB, 12-inch disks that hold 3.5 GB, and 14-inch disks that hold about 5 GB. With continual improvements in technology, higher storage capacities are possible.

Magneto-optical. Magneto-optical (MO) drives marry magnetic and optical technologies. The optical component reads data by bouncing a laser beam off the spinning disk and uses optical sensors to examine the reflected beam. The magnetic component records information by altering the polarity of bits on the film that coats the disk surface, which in turn alters the nature of the reflective (optical) coating above it. This complex interaction of the laser beam and the magnetic recording head on a disk permits the data to be overwritten, offering a true rewritable medium.

The first wave of MO drives were limited by slow (2400 rpm) spin rates and bulky head mechanisms that provided an average access time of 70 ms versus the

15 ms or less of hard disks, relegating them to archival storage uses. The latest MO drives, however, offer average access times of 28 ms.

Summary. Network managers have many choices of storage media for backing up corporate data. With the right tools, backup can be automated under centralized control. These tools enhance media management by providing overwrite protection, log file analysis, media labeling, and the ability to recycle backup media. In addition, the journaling and scheduling capabilities of some tools relieve the operator of the time-consuming tasks of tracking, logging, and rescheduling network and system backups. Adding data compression reduces media costs by increasing media capacity, while reducing network traffic.

See also

HIERARCHICAL STORAGE MANAGEMENT
REDUNDANT ARRAY OF INEXPENSIVE DISKS

Switched digital video

Switched digital video (SDV), often referred to as *fiber-to-the-curb,* is a broadband network technology that is intended to deliver multimedia services to homes and businesses. In the SDV model, a phone company installs fiber from a central office (CO) to a curbside unit, then passes digital signals over ordinary twisted-pair wiring to the premises for a distance of up to a few hundred feet. SDV uses a method of encoding known as *common Amplitude modulation/phase modulation* (CAP), which is less expensive than discrete-multitone (DMT) encoding used in Asymmetrical Digital Subscriber Line (ADSL), a competing local-loop technology. Both increase the transmission capacity of existing twisted-pair wiring, so video and other high-bit-rate signals can reach multimedia workstations and television sets without expensive rewiring.

The CAP technology greatly increases the data transmission capacity of unshielded twisted-pair (UTP) wiring. Digital bits of audio, video, or data can travel at a rate of up to 51.84 Mbps over UTP wiring to the home or office, and up to 1.6 Mbps back into the network. By contrast, ADSL supports up to 6.144 Mbps downstream (CO to customer) and 16 kbps to 640 kbps upstream (customer to CO). Because of its lower speed, links can be up to 22,000 feet from customer to CO, whereas SDV's higher speed reduces the distance to only 1000 feet. This limitation necessitates a different network architecture.

SDV also has advantages over hybrid fiber/coax (HFC) networks that rely on cable modems at the customer premises. SDV is dedicated, both downstream and upstream, and it relies on a point-to-point connection, rather than shared connections among subscribers. This does have security implications: With shared connections, there is always the chance of a neighbor stealing the signal, whereas this is not possible with dedicated connections. And because SDV is essentially an Asynchronous Transfer Mode (ATM) link, the bandwidth can be allocated appropriately to deliver symmetrical services such as videoconferencing.

Applications. The applications for SDV include Internet and remote LAN access, video on demand, CD-ROM links, and a multitude of multimedia services such as video real estate listings, distance learning, videoconferencing, interactive advertising, and interactive shopping. A home receiving SDV-based services could simultaneously be running several movies stored in MPEG1 format and transmitted at 1.5 Mbps on separate televisions. At the same time, there can be a videoconference operating at 384 kbps and files being downloaded from a server via a 128 kbps ISDN link. Incoming telephone calls could be answered without disturbing the other activities.

Given the demands of multimedia services, several network requirements have become clear. The broadband multimedia network must provide the following capabilities:

- High bandwidth downstream to carry video and other high-speed data.
- Variable bandwidth upstream from low-speed, VCR-like control of video-on-demand, and up to 1 Mbps for telecommuters who require remote LAN access.
- Dedicated bandwidth to each home to ensure that capacity is there when consumers want it.
- Low latency, so that consumers can get quick responses to their requests.
- Security, so that signal pirates cannot obtain valuable services without paying for them.
- Service-independence to allow various services to be carried over the network simultaneously with minimal changes to the network infrastructure.

Other technologies such as HFC, wireless cable (i.e., MMDS), and digital satellite promise to meet some of the requirements for emerging broadband multimedia networks. But in each case there are limitations in the fundamental network architecture that limit its ability to meet all of the requirements.

For example, HFC systems rely on a bus architecture in which all signals are on the same coaxial cable, raising congestion and security issues. Wireless cable has limited bandwidth downstream and no upstream bandwidth. Digital satellite service does not have an upstream data link. An Internet user, for example, still must dial into an Internet service provider (ISP) and issue commands, even while receiving data at 400 kbps via satellite. All of these technologies have their place and will be deployed in the next few years, but their applications ultimately will be limited.

The network. SDV networks meet all of these requirements. The SDV network is a local-loop network based on telephony standards for the next-generation digital loop carrier (NGDLC) and for fiber-to-the-curb (FTTC). Using an FTTC system as a base, the enhanced SDV system provides a fiber link from the CO to a point within 1000 feet of the home. The unit that terminates the fiber is called an *optical network unit* (ONU).

The Host Digital Terminal (HDT) is installed in the CO. It provides interfaces to the telephony network, to the DS1 network, and to the digital video network via 16 OC-3c ATM/SONET data streams. The interactive signaling is also carried in these data streams. A pair of single-mode fibers extends from the HDT to each of up to 160 ONUs. Each fiber pair may be split as many as four times via a passive optical splitter. The ONU provides the optical terminations and interfaces to the homes.

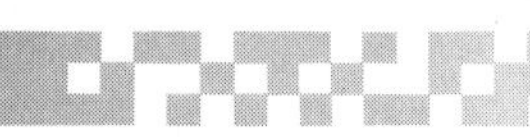

For telephony, standard channel units are plugged in at the ONU and twisted-pair wire extends to the premises. For DS1, specialized channel units are plugged in at the ONU. For digital video service, broadband transceivers are plugged in at the ONU. Twisted-pair wires extend from the ONU, carrying digital video to the home. It is possible to combine the digital video and the analog video near the premises so that the incoming coax cable carries both.

An SDV network offers the following features and benefits:

- Bandwidth of 51 Mbps downstream and up to 1.62 Mbps upstream at every home, in addition to the telephony and DS1 service that also is available. Bandwidth assignment is all software-provisionable. The bandwidth is dedicated from the HDT to the home via the star topology employed by the system.
- Low latency (response of under 250 ms) is provided via the software in the HDT. For requests that must be sent upstream, latency is reduced because the HDT does not perform any protocol conversions.
- The switching fabric architecture provides effective security. The switches in the HDT are controlled from commands issued from the upstream video providers. These switches only respond to requests from the viewer if the services have been preauthorized from the video providers. A data stream never leaves the HDT if the customer is not authorized to have it. Because the SDV architecture relies on the star topology, no customer ever has the opportunity to access to programs being received by another customer.
- The network provides service-independence by delivering a standardized ATM packetized signal (i.e., cells) to the premises. ATM is an open standard for transporting any type of digital data. Service provider can easily adapt their specialized premises-based terminals to accept an ATM input using currently available chip sets. The availability of the chip sets and the standard-compliant interface will speed the availability of services. The result will be that customers will be able to purchase different types of services more quickly.

Summary. For mass deployment of interactive multimedia services to become a reality, today's local loops must deliver greater throughput, which neither conventional cable nor telephone networks can provide. SDV is one of several architectures that are available to telephone companies and other network operators to meet the challenge of providing multimedia services to the premises. For the foreseeable future, the choice of technology will hinge on which one best meets the requirements for service and costs in a given territory.

See also

ASYNCHRONOUS TRANSFER MODE
DIGITAL SUBSCRIBER LINE TECHNOLOGIES
HYBRID FIBER/COAX
PCS-OVER-CABLE

Switched Multimegabit Data Services

Initially offered in December of 1991, SMDS is a carrier-provided, connectionless, cell-switched service developed by Bellcore and standardized by the IEEE. As such,

SMDS provides organizations with the flexibility they need for distributed computing and supporting bandwidth-intensive applications.

As a connectionless service, SMDS eliminates the need for carrier switches to establish a call path between two points before data transmission can begin. Instead, SMDS access devices pass 53-byte cells to a carrier switch. SMDS cells are a fixed size of 53 octets (bytes), the same type of cell used in ATM. The switch reads addresses and forwards cells one-by-one over any available path to the desired end point. SMDS addresses ensure that the cells arrive in the right order. The benefit of this connectionless "any-to-any" service is that it puts an end to the need for precise traffic-flow predictions and for dedicated connections between locations. With no need for a predefined path between devices, data can travel over the least congested route in an SMDS network, providing faster transmission, increased security and greater flexibility to add or drop network sites.

Applications. SMDS is useful for a number of broadband applications, including:

- Channel-attached IT services such as "dark" data centers, data vaults, and print factories.
- Disaster recovery using mirroring, shadowing, and off-site magnetic tape storage.
- Networked computer centers.
- Distributed supercomputer applications such as crash and process simulation.
- Distribution of expensive, bit-hungry resources such as document image and CAD/CAM files.
- Bulk file transfer for program testing, verification, and interchange.
- Remote access to high-resolution satellite images, x-ray images, and CAT scans.

Although offered by one carrier (MCI) as a nationwide service, SMDS is primarily used for metropolitan or regional LAN internetworking. As a technology-independent service, SMDS offers several major benefits/features:

- *Simplicity:* Virtual connections are made as needed; there are no permanent, fixed connections between sites.
- *E.164 addressing:* SMDS addresses are like standard telephone numbers. If a user knows the SMDS address of another user, he or she can call up and begin sending and receiving data.
- *Call control:* SMDS supports call blocking, validation, and screening for the secure interconnection of LANs and distributed client-server applications.
- *Multicasting:* SMDS supports group addressing.
- *Multiprotocol support:* SMDS supports the key protocols used in local and wide area networking, including TCP/IP, Novell IPX/SPX, DECNet, AppleTalk, SNA, and OSI.
- *Management:* SMDS is easier to manage than other services such as frame relay, which can have a multitude of virtual circuits that makes setup, reconfiguration, and testing much more difficult.

- *Reconfiguration:* Sites can be connected and disconnected easily and inexpensively—usually within 30 minutes—without impacting other network equipment. This makes SMDS ideal for external corporate relationships that change frequently.
- *Security:* SMDS includes built-in security features that allow intracompany transmission of confidential data.
- *Scalability:* SMDS is easily and cost-effectively scaled as organizations grow or application requirements change.
- *Migration path:* Customers have a well-defined migration path to ATM, since both services use the same 53-byte frame structure.
- *Cost:* Because SMDS is a switched service, users pay only for the service when it is used, which can make it less expensive than other services such as frame relay.

Because SMDS is able to coexist with dedicated facilities, it allows customers to create hybrid public/private networks. SMDS also allows for the easy expansion of existing networks, since new sites can be quickly added to an SMDS net without totally reconfiguring the network. Additions to an SMDS network only require a simple update to a screening database on the SMDS switch.

Architecture. SMDS defines a three-tiered architecture (Fig. S7):

- A *switching infrastructure* comprising SMDS-compatible switches that may or may not be cell-based.
- A *delivery system* made up of T1, T3, and lower-speed circuits called Subscriber Network Interfaces (SNIs).
- An *access control system* for users to connect to the switching infrastructure without having to become a part of it.

LANs provide the connectivity for end users on the customer premises. The LAN is attached to the SMDS network via a bridge or router, with an SMDS-capable CSU/DSU at the front end of the connection.

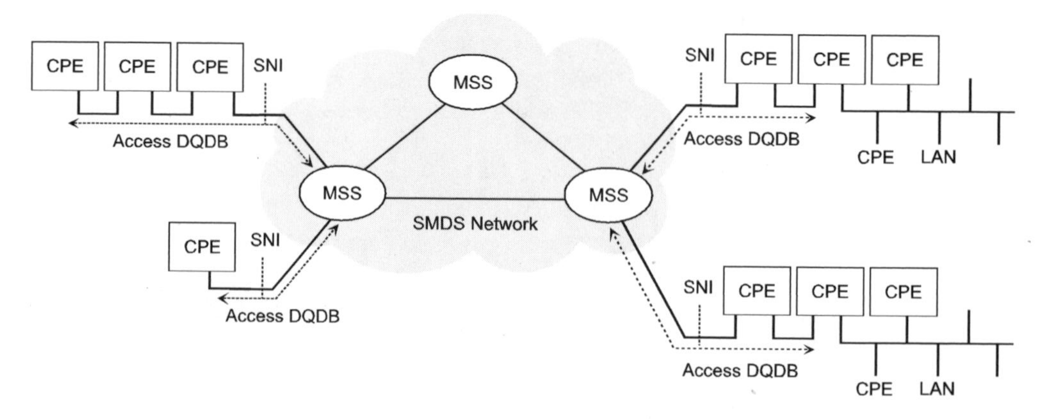

Fig. S7 *The SMDS architecture.*

T1 SNIs are used to access 1.17-Mbps SMDS offerings, while T3 SNIs are used to tap into offerings at 4, 10, 16, 25, or 34 Mbps. A fractional T3 circuit can be used to access intermediate-speed SMDS offerings. Some carriers offer low-speed SMDS access at 56 kbps, 64 kbps and in increments of 56/64 kbps. This allows smaller companies, large companies that have some small sites, and current users of frame relay technology to also take advantage of SMDS.

Each subscriber has a private SNI, and may connect multiple user devices (CPE) to it. At this interface point, the CPE attaches to a dedicated access facility that connects to an SMDS switch. Security is enforced, since only data originating from or destined for that subscriber will be transported across that SNI.

The SMDS Interface Protocol (SIP), operating across the SNI, is based on the IEEE 802.6 Distributed Queue Dual Bus (DQDB) media access control (MAC) scheme. The SIP consists of three protocol layers that describe the network services and how these services are accessed by the user. The SIP defines the frame structure, addressing, error control, and data transport across the SNI.

The SMDS network itself is a collection of SMDS Switching Systems (SS). The SS is a high-speed packet switch (most likely an ATM switching platform) providing the SMDS service interface. An SS typically will be located in a service provider's central office. Interconnecting several SS locations forms the foundation for a metropolitan or regional network.

The Inter–Switching Systems Interface (ISSI) provides communications between different switching vendors within the same network, while the Interexchange Carrier Interface (ICI) allows local telephone companies and interexchange carriers to interconnect SMDS networks.

Summary. SMDS provides users with the cost-effectiveness of a public switched network; the benefits of fully meshed, wide-area interconnection; and the privacy and control of dedicated, private networks. The key benefits subscribers can realize with SMDS include widespread current availability and increased LAN performance. It provides data management features, flexibility, bandwidth on demand, network security and privacy, multiprotocol support, and technology compatibility. SMDS is offered on a regional basis from most of the incumbent local exchange carriers (ILECs); MCI currently is the only long-distance carrier that offers SMDS. Despite the many advantages of SMDS, the service lags in popularity behind frame relay and ATM. Many industry watchers believe that the ILECs have simply done a poor job of marketing SMDS.

See also
ASYNCHRONOUS TRANSFER MODE

Synchronous Optical Network

The Synchronous Optical Network (SONET) is an industry standard for high-speed transmission over optical fiber. The SONET standard was developed by the Alliance for Telecommunications Industry Solutions (ATIS), formerly known as the Exchange Carriers Standards Association (ECSA), with input from Bellcore, formerly the research and development arm of the seven RBOCs. The standard was published and distributed by the American National Standards Institute (ANSI).

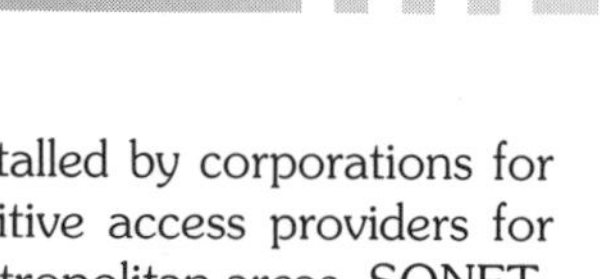

SONET-compliant fiber-optic facilities are being installed by corporations for backbone networks, as well as by carriers and competitive access providers for long-haul routes and fault-tolerant rings around major metropolitan areas. SONET-based services have a performance objective of 99.9975 percent error-free seconds and an availability rate of at least 99.999 percent. SONET combines bandwidth and multiplexing capabilities, allowing users to fully integrate voice, data, and video over a single fiber-optic facility.

Advantages. SONET is expected to provide the transport infrastructure for the next three to four decades, in much the same way that T1 (1.544 Mbps) and its T3 extension (44.736 Mbps) have provided the transmission infrastructure of the past two. SONET offers numerous benefits in many areas, some of which are described in subsequent paragraphs.

Bandwidth. The enormous amounts of bandwidth available with SONET and its inherently flexible management capability permit carriers to create global intelligent networks capable of supporting the next generation of services. Any information that can be represented electronically is likely to be transported over SONET-based information superhighways, including:

- Bit-intensive, three-dimensional computer-aided design (CAD)
- Medical imaging
- Collaborative computing
- Interactive virtual reality programs
- Multipoint videoconferencing
- New consumer services such as video-on-demand, interactive entertainment, and multiuser games.

Under SONET, bandwidth is scalable from 51.84 Mbps to 13 Gbps and higher.

Standardization. Because T3 implementation is unique to each equipment vendor, carriers are severely limited in terms of product selection and configuration flexibility. SONET standards make possible seamless interconnectivity among compliant equipment, eliminating the need to deploy equipment from the same vendor end-to-end and making it much easier to interconnect compliant networks, even among international locations.

In eliminating proprietary optics, carriers are freed from the necessity of dealing with a single vendor for all of their equipment needs. Instead, they can buy equipment based on price and performance, and mix and match hardware from multiple vendors. SONET also gives corporate users flexibility in choosing customer premises equipment instead of being locked into the carriers' vendors.

Bandwidth management. A major advantage of SONET is its ability to manage huge amounts of bandwidth easily. Within the SONET infrastructure, carriers can tailor the width of information highways in a standard way. Carriers can parcel out specific amounts of this bandwidth to meet the needs of a broad and diverse array of user applications. Such parceling can be accomplished without adding equipment to the network or manually reconnecting cables. SONET eliminates central office reliance on metallic DSX technology with its cumbersome manual cabling and jumpers, and replaces it with remotely configurable optical crossconnects. SONET provides more efficient switching and transport by eliminating the need for such midlevel network elements as back-to-back M13 multiplexers, for example, that normally crossconnect T1 facilities.

With remotely configurable SONET equipment, carriers can more expeditiously support the connectivity requirements of their customers. For example, a DS3 signal with a rate of 44.736 Mbps can be mapped directly into an STS-1 at 51.84 Mbps. The remaining STS-1 bytes are used for overhead and stuffing. Or two 51.84-Mbps channels can be combined to support LAN traffic between Fiber Distributed Data Interface (FDDI) backbones operating at 100 Mbps.

Real-time monitoring. SONET permits sophisticated self-diagnostics and fault analysis to be performed in real time, making it possible to identify problems before they disrupt service. Intelligent network elements, specifically the SONET add-drop multiplexer (ADM), can automatically restore service in the event of failure via a variety of restoral mechanisms.

SONET's embedded control channels allow tracking end-to-end performance and identifying elements that cause errors. With this capability, carriers can guarantee transmission performance, and users can readily verify it without having to go offline to implement various test procedures. For network managers, these capabilities allow earlier problem identification—before any disruption in service. Along with the self-healing capabilities of ADMs, these diagnostics ensure that properly configured SONET-compliant networks will experience virtually no down time.

Survivable networking. SONET offers multiple ways to recover from network failures, including:

- *Automatic protection switching* is the ability of a transmission system to detect a failure on a working facility and to switch to a standby facility to recover the traffic. One-to-one and one-to-*n* protection switching are provided.
- *Bidirectional line switching* requires two fiber pairs between each recoverable node. A given signal is transmitted across one pair of fibers. In response to a fiber facility failure, the node preceding the break loops the signal back toward the originating node, where the data traverses a different fiber pair to its destination.
- *Unidirectional path switching* requires one fiber pair between each recoverable node. A given signal is transmitted in two different paths around the ring. At the receiving end, the network determines and uses the best path. In response to a fiber facility failure, the destination node switches traffic to the alternate receive path.

Universal connectivity. As the foundation for future high-capacity backbone networks, SONET carries a variety of current and emerging traffic types, including Asynchronous Transfer Mode (ATM), frame relay, Switched Multimegabit Data Services (SMDS), and broadband ISDN (B-ISDN). To ensure universal connectivity down to the component level, equipment vendors are beginning to support common product specifications, including transmitters and receivers that are pin-for-pin compatible. Uniformity and compatibility are achieved by using common specifications for module pin-out, footprint, logic interface, optical performance parameters, and power supplies.

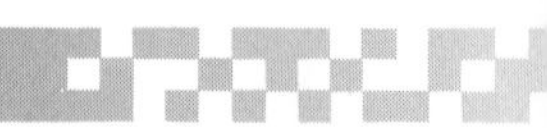

Transmission rate. The SONET standard specifies a hierarchy of rates and formats for optical transmission, ranging from 51.84 Mbps to more than 13 Gbps, far surpassing the DS3 rate of 44.736 Mbps. Table S1 summarizes the available and emerging SONET transmission rates.

Table S1. SONET Transmission Rates.

STS Level (electrical)	OC Level (optical)	Line Rate
STS-1	OC-1	51.84 Mbps
STS-3	OC-3	155.520 Mbps
STS-9	OC-9	466.560 Mbps
STS-12	OC-12	622.080 Mbps
STS-18	OC-18	933.120 Mbps
STS-24	OC-24	1.244 Gbps
STS-36	OC-36	1.866 Gbps
STS-48	OC-48	2.488 Gbps
STS-96	OC-96	4.976 Gbps
STS-192	OC-192	9.953 Gbps
STS-256	OC-256	13.271 Gbps

The basic building blocks used in the SONET signaling hierarchy are STS-1/OC-1 (51.84 Mbps) groups that are multiplexed to higher-rate signals. The STS-1/OC-1 frame, from which all larger frames are constructed, has a 9×90-byte format, which permits efficient packing of both US and European data rates in a payload of 783 bytes, plus 27 bytes for transport overhead, for a total of 810 bytes. This results in a usable payload of 48.384 Mbps (Fig. S7), which may contain asynchronous DS3, DS1, or other types of signals. The overhead bytes are used for real-time error monitoring, self-diagnostics, and fault analysis. The signal is transmitted byte-by-byte beginning with byte one, scanning left to right from row one to row nine. The entire frame is transmitted in 125 μs. Higher-level signals (STS-N) are integer multiples of the base electrical rate, which are interleaved and converted to optical signals (OC-N).

Protocol layers. The SONET transmission protocol consists of four layers:

- *Photonic layer:* This is the electrical and optical physical interface for the transport of information bits across the physical medium. Its primary function is to convert the STS-*n* electrical signals into OC-*n* optical signals. This layer performs functions associated with the bit rate, optical pulse shape, power, and wavelength; it uses no overhead.
- *Section layer:* This layer deals with the transport of the STS-*n* frame across the optical cable, performing a function similar to that of the Data Link layer (layer 2) in bit-oriented protocols such as High-level Data Link Control (HDLC) or Synchronous Data Link Control (SDLC). This layer establishes frame synchronization and the maintenance signal; functions performed in the Section layer include framing, scrambling, error monitoring, and order wire communications.

- *Line Layer:* This layer provides the synchronization, multiplexing, and automatic protection switching (APS) for the Path layer. It is primarily concerned with the reliable transport of the Path layer payload (voice, data, or video) and overhead, and allows (via APS) automatic switching to another circuit if the quality of the primary circuit drops below a specified threshold. Line overhead associated with this layer includes line error monitoring, line maintenance, protection switching, and express order wire.[1] Future uses of the embedded overhead channel (EOC) will provide for an even more sophisticated network management system.
- *Path layer:* This layer maps services such as DS3, FDDI, and ATM into the SONET payload format. This layer provides end-to-end communications, signal labeling, path maintenance, and control, and it is accessible only by equipment that terminates this layer. A SONET add-drop multiplexer (ADM) that terminates and demultiplexes a SONET OC-3 into three DS3 signals, for example, accesses the Path layer overhead. Conversely, a SONET crossconnect system would perform Section and Line layer processing but would not require access to the Path layer overhead.

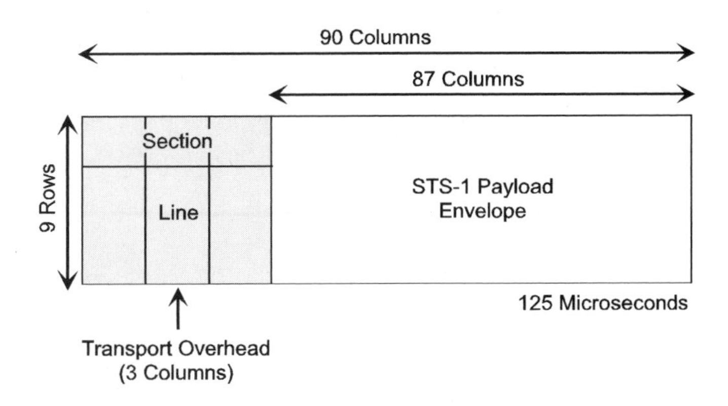

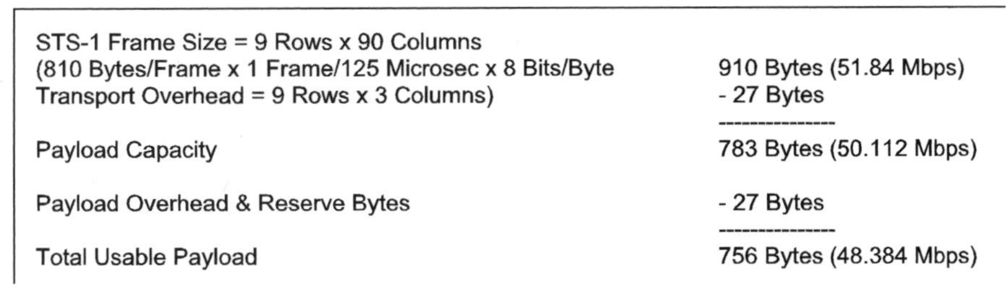

STS-1 Frame Size = 9 Rows x 90 Columns (810 Bytes/Frame x 1 Frame/125 Microsec x 8 Bits/Byte Transport Overhead = 9 Rows x 3 Columns)	910 Bytes (51.84 Mbps) - 27 Bytes
Payload Capacity	783 Bytes (50.112 Mbps)
Payload Overhead & Reserve Bytes	- 27 Bytes
Total Usable Payload	756 Bytes (48.384 Mbps)

Fig. S7 *A SONET STS-1/OC-1 frame.*

1 The term *order wire* originated in the early days of telephony when requests for new telephone service were fulfilled by manually configuring a patch panel. Today order wire is software-controlled, allowing services of all types to be installed and terminated via instructions issued at a computer keyboard. *Express order wire* refers to the installation or termination of service on a priority basis.

Virtual tributaries. A key feature of SONET framing is its ability to accommodate existing synchronous and asynchronous signal formats. The SONET payload can be subdivided into smaller "envelopes" called *virtual tributaries* (VTs) to transport signals using less than DS3 capacity. Because VTs can be placed anywhere on higher-speed SONET payloads, they provide effective transport for existing North American and international formats.

Each virtual tributary has its own overhead bits and functions as a separate container within the STS-1/OC-1 signal. Because SONET is an international standard, it defines VT mappings for the most common North American tributary (DS1 at 1.544 Mbps) and the most common international tributary (E1 at 2.048 Mbps). Less common tributaries, such as DS1C and DS2 are also represented; although DS3 is not defined as a VT, SONET does provide a special mapping for the transport of DS3 signals within a payload. The SONET standard is quite flexible, even accommodating tributaries for 10-Mbps Ethernet and 16-Mbps Token-Ring LANs.

Network elements. SONET network infrastructures consist of various types of specialized equipment known as *network elements.* Examples of SONET-compliant network elements are described in the next several paragraphs.

Add-drop multiplexer (ADM). The add-drop multiplexer provides interfaces between the different network signals and SONET signals. It is a single-stage multiplexer/demultiplexer that converts one or more DS-*n* signals into/from an OC-*n* signal. It can be used in terminal sites and intermediate (add-drop) sites. At an add-drop site, it can drop lower-rate signals to be transported on different facilities, or it can add lower-rate signals into the higher-rate OC-*n* signal. The rest of the traffic simply passes through the ADM.

Broadband digital crossconnect (BDCS). The BDCS interfaces various SONET signals and legacy DS3s. It accesses the STS-1 signals, and switches at this level. It is the synchronous equivalent of the DS3 digital crossconnect, except that the BDCS accepts optical signals and allows overhead to be maintained for integrated Operations, Administration, Maintenance, and Provisioning (OAMP). Most asynchronous systems prevent overhead from being integrally passed from optical signal to signal. The BDCS can make two-way crossconnections at the DS3, STS-1, and STS-c (concatenated) levels. It is typically used as a SONET hub that grooms STS-1s for broadband restoration purposes, or for routing traffic.

Wideband digital crossconnect (WDCS). The WDCS is a digital crossconnect that terminates SONET and DS3 signals, and maintains the basic functionality of VT- or DS1-level crossconnections. The optical equivalent to the DS3/DS1 digital crossconnect, it accepts optical carrier signals as well as DS1s and DS3s. In a WDCS, the switching is done at either the VT, DS1, or DS0 level. Because SONET is synchronous, the low-speed tributaries are visible in VT-based systems and directly accessible within the STS-1 signal. This allows the required tributaries to be extracted and inserted without demultiplexing, which is not possible with existing digital crossconnects. Also, the WDCS crossconnects the constituent DS1s between DS3 terminations, and between DS3 and DS1 terminations.

Digital loop carrier. Similar to the DS1 digital loop carrier, this SONET network element can accept and distribute SONET optical-level signals as well. This allows the network to transport many of the new services requiring large amounts of bandwidth. The integrated overhead capability of the SONET digital loop carrier allows surveillance, control, and provisioning from the central office.

Regenerator. A SONET regenerator drives a transmitter with the output from a receiver, stretching transmission distances far beyond what is normally possible over a single length of fiber.

SONET CPE. SONET customer-provided equipment provides an interface to carrier-provided SONET services.

Summary. The continued deployment of SONET-compliant equipment on the public network will have a significant impact on telephone companies, interexchange carriers, and corporate users. SONET offers virtually unlimited bandwidth, integral fault recovery and network management, interoperability between public services and corporate enterprise-wide networks, and multivendor equipment interoperability. As SONET becomes more widely available and the promised efficiencies and cost savings materialize, it will change the way networks are designed, operated, and managed. For carriers, timely and effective deployment of SONET will determine the types of broadband services that can be offered to subscribers, since SONET comprises the physical layer upon which broadband services are built. Although noteworthy progress has been made in SONET deployment, especially fault-tolerant rings in major metropolitan areas, it is still not widely available.

See also

ASYNCHRONOUS TRANSFER MODE
FIBER OPTICS TECHNOLOGY
SYNTRAN
WAVELENGTH DIVISION MULTIPLEXING

SYNTRAN

SYNTRAN (Synchronous Transmission) is a restructured DS3 signal format for synchronous transmission at the DS3 (44.736 Mbps) level of the North American Digital Transmission Hierarchy. The SYNTRAN format is an ANSI standard published in 1987 as ANSI T1.103-1987. The format specifies two modes of operation: bit-synchronous and byte-synchronous. The bit-synchronous mode provides for the transport of DS1 signals as a complete bundle. This mode is used when access to the individual DS0 channels is not required. The byte-synchronous mode provides DS0 access and allows for digital signal processing at the DS0 level. Functions such as an integrated time slot interchange (TSI), the basic functionality provided by digital crossconnects, can be directly implemented in this mode.

The SYNTRAN format provides three functions that are distinct from the older DS3 format that standards advocates sought to replace. First, the SYNTRAN DS3 signal provides immediate identification and direct access to all DS0 and DS1 signals. This promotes efficient and economical add-drop and DS0/DS1 time slot interchange.

Second, the SYNTRAN format frees up bandwidth used for the bit stuffing process required in the asynchronous DS3 signal. The freed up bandwidth, which amounts to 1.2 Mbps in capacity, is used to provide the operations channels that supply high- performance maintenance and test functionality to each SYNTRAN network element from a centralized remote location; remote provisioning of DS0 and DS1 signals throughout a SYNTRAN network; and improved DS3 performance monitoring via a cyclic redundancy check (CRC-9), which is used to assure the integrity and quality of the DS3 transmission signal throughout the network. Similar to CRC-6 used in DS1 ESF, CRC-9 provides fast and highly reliable bit error rate and burst error detection.

Finally, the SYNTRAN format provides a one-step multiplexing scheme that eliminates the need for the intermediate DS2 multiplexing stage used with asynchronous DS3.

Summary. SYNTRAN never became widely accepted. Among its shortcomings were that SYNTRAN lacked an optical standard and did not support transmission rates beyond DS3. What really was needed was a long-term solution that provided a standard, synchronous, high-speed transport vehicle flexible enough to integrate a menu of multimegabit and subrate DS3 services while providing management, real-time bandwidth provisioning, drop-and-insert routing, and nonintrusive fault location and testing. These goals were achieved with the introduction of the more advanced Synchronous Optical Network (SONET).

See also

SYNCHRONOUS OPTICAL NETWORK

T-carrier facilities

T-carrier is a type of digital transmission system employed over copper, optical fiber, or microwave to achieve various channel capacities for the support of voice and data. The most popular T-carrier facility is T1, which is implemented by a system of copper-wire cables, signal regenerators, and switches that provides a transmission rate of up to 1.544 Mbps using digital signal level 1 (DS1). In Europe, the United Kingdom, Mexico, and other countries that abide by ITU standards, the equivalent facility is E1, which provides a transmission rate of 2.048 Mbps.

T-carrier had its origins in the 1960s. It was used first by telephone companies as the means of aggregating multiple voice channels into a single high-speed digital backbone facility between central office switches. The most widely deployed T-carrier facility is T1, which has been commercially available since 1983.

Digital signal hierarchy. To achieve the DS1 transmission rate, selected cable pairs with digital signal regenerators (repeaters) are spaced approximately 6000 feet apart. This combination yields a transmission rate of 1.544 Mbps. By halving the distance between the span line repeaters, the transmission rate can be doubled to 3.152 Mbps, which is called DS1C. Adding more sophisticated electronics and/or multiplexing steps makes higher transmission rates possible, creating a range of digital signal levels, as shown in Table T1.

For example, a DS3 signal is achieved in a two-step multiplexing process (Fig. T1) whereby DS2 signals are created from multiple DS1 signals in an intermediate step. DS1C is uncommon except in highly customized private networks where the distances between repeaters is very short, such as between floors of an office building or between buildings in a campus environment. Some channel banks and multiplexers support DS2 by performing multiplexing to achieve 96 voice channels over a single T-carrier facility. DS4 is used mostly by carriers for trunking between central offices.

Quality objectives. The quality of T-carrier facilities is determined by two criteria: performance and service availability. The performance objective refers to the percentage of seconds per day when there are no bit errors on a circuit. The service availability objective refers to the percentage of time a circuit is functioning at full capability during a three-month period. If these objectives are not met, the carrier issues credits to the user. Each carrier has its own quality objectives for T-carrier services, which are based on circuit length.

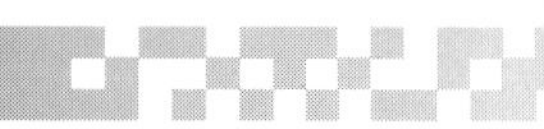

Table T1. Digital Signal Hierarchy.

Signal Level	Bit Rate	Channels	Carrier System	Medium
North America				
DS0	64 kbps	1	—	Copper wire
DS1	1.544 Mbps	24	T1	Copper wire
DS1C	3.152 Mbps	48	T1C	Copper wire
DS2	6.312 Mbps	96	T2	Copper wire/ microwave
DS3	44.736 Mbps	672	T3	Microwave/fiber
DS4	274.176 Mbps	4032	T4	Microwave/fiber
International (ITU)				
0	64 kbps	1	—	Copper wire
1	2.048 Mbps	30	E1	Copper wire
2	8.448 Mbps	120	E2	Copper wire
3	34.368 Mbps	480	E3	Microwave/fiber
4	44.736 Mbps	672	E4	Microwave/fiber
5	565.148 Mbps	7680	E5	Microwave/fiber

The quality objectives for AT&T's Fractional T1 service, for example, is 9 errored seconds per day, which translates into 99.99 percent error-free seconds per day, 4 severely errored seconds per day, and 99.96 percent service availability per year. According to AT&T, *severely errored* means that 96 percent of all frames transmitted in a second have at least one error.

A related measure of performance is *failed seconds,* which is defined by AT&T as the time starting after 10 consecutive severely errored seconds and ending when there have been 10 consecutive seconds that are not severely errored. Channel Service Unit/Data Service Unit (CSU/DSU) at each end of the circuits collect and issue reports on this type of information.

Summary. T-carrier underlies just about every type of carrier facility available today, including T1 and T3, and their fractional derivatives—dedicated or switched. These facilities, in turn, support such services as frame relay and ISDN, and provide access to SMDS, ATM, and virtual private networks. Through multiplexing techniques, companies can subdivide T-carrier facilities to achieve greater bandwidth efficiency and cost savings.

See also

CHANNEL BANKS

MICROWAVE COMMUNICATIONS

FIBER OPTICS TECHNOLOGY

T1 LINES

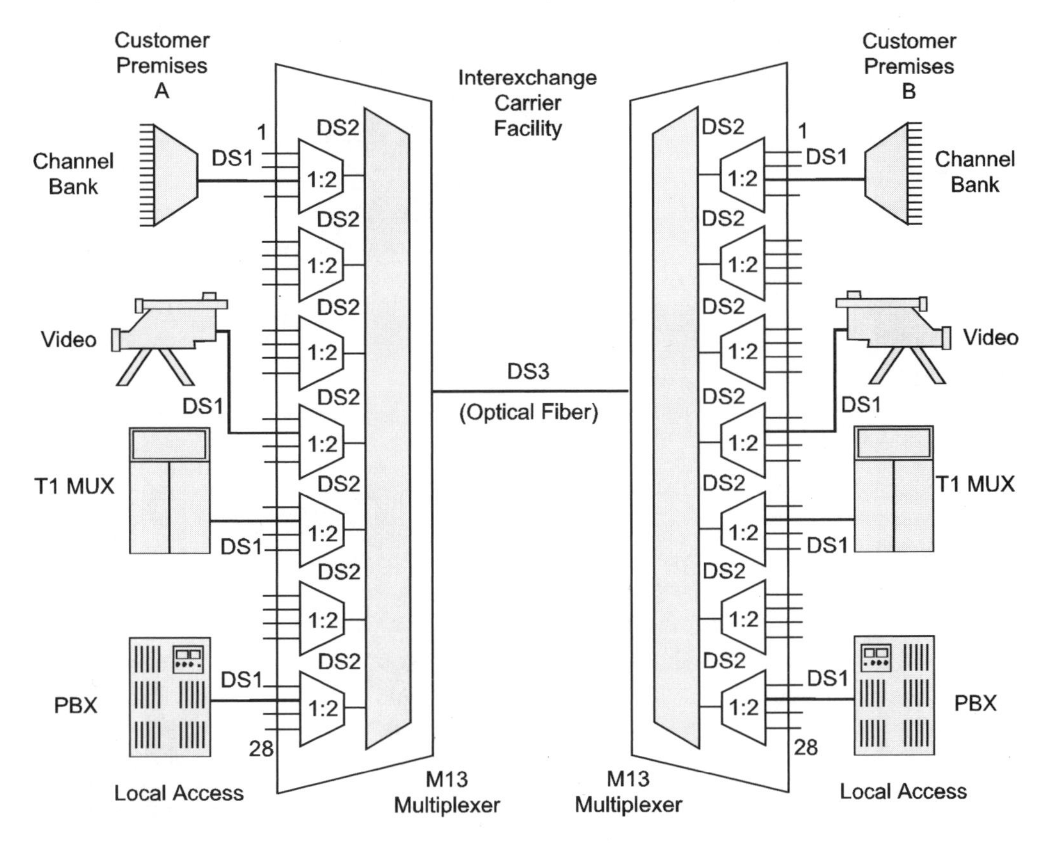

Fig. T1 *DS1 multiplexing to DS3 requires the intermediate step to DS2.*

T1 lines

T1 lines are digital facilities that provide a transmission rate of up to 1.544 Mbps using Digital Signal level 1 (DS1). The available bandwidth is divided into 24 channels operating at 64 kbps each, plus an 8 kbps channel for basic supervision and control. Voice is sampled and digitized via pulse code modulation (PCM). T1 lines are used for more economical and efficient transport over the wide area network.

Economy is achieved by consolidating multiple lower-speed voice and data channels over the higher-speed T1 line. This is more cost-effective than dedicating a separate lower-speed line to each terminal device. The economics are such that only 5 to 8 analog lines are needed to cost-justify the move to T1.

Efficiency is obtained by compressing voice and data to make room for even more channels over the available bandwidth. Individual channels also can be dropped or inserted at various destinations along the line's route. Network management information can be embedded in each channel for enhanced levels of supervision and control.

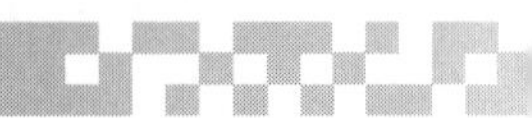

Usually a T1 multiplexer provides the means for companies to realize the full benefits of T1 lines, but channel banks offer a low-cost alternative. The difference between the two devices is that T1 multiplexers offer higher line capacity, support more types of interfaces, and provide more network management features than channel banks.

D4 framing. T1 multiplexers and channel banks transmit voice and data in frames that are called *D4 frames*. Frames are bounded by framing bits that perform two functions: They identify the beginning of each frame and help locate the channel carrying the signaling information. For voice, this bit is carried in the eighth bit position of frames 6 and 12 (Fig. T2).

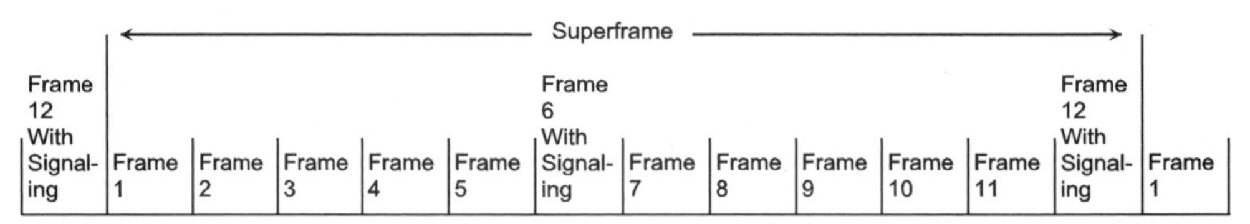

Fig. T2 *D4 framing (i.e., superframe).*

D4 frames consist of 193 bits, which equates to 24 channels of 8 bits each, plus a single framing bit. Each frame contains framing bits or signaling bits in the 193rd position, which permits the management of the DS1 facility itself. This is done by robbing the least significant bit from the data stream, which alternatively carries information or signaling data. Another bit is used to mark the start of a frame. Twelve D4 frames comprise a *superframe*.

Extended Superframe Format. Extended Superframe Format (ESF) is an enhancement to T-carrier that specifies methods for error monitoring, reporting, and diagnostics. The use of ESF allows technicians to maintain and test the T1 line while it is in service, and often fix minor troubles before they become service-affecting. ESF extends the normal 12-frame superframe structure of the D4 format to 24 frames. By doubling the number of bits available, more diagnostic functionality also becomes available (Fig. T3).

Of the 8 kbps bandwidth (repetition rate of 193rd bit or framing bit) allocated for basic supervision and control, 2 kbps are used for framing, 2 kbps for Cyclic Redundancy Checking (CRC-6), and 4 kbps for the Facilities Data Link (FDL). With CRC, the entire circuit may be segmented so that it can be monitored for errors without disrupting normal data traffic. In this manner, performance statistics can be generated to monitor T1 circuit quality. Via FDL, performance report messages are relayed to the customer's equipment, usually a Channel Service Unit (CSU), at 1-second intervals. Alarms also can use the FDL, but performance report messages always have priority.

ESF diagnostic information is collected by the CSU at each end of the T1 line for both carrier and user access. CSUs gather statistics on such things as clock synchronization errors and framing errors, as well as errored seconds, severely errored

Frame Number	D4 Format: Value of 193rd Bit	D4 Format: Use	ESF Format: Value of 193rd Bit	ESF Format: Use
1	1	F_T	X	FDL
2	0	F_S	X	CRC
3	0	F_T	X	FDL
4	0	F_S	0	F_S
5	1	F_T	X	FDL
6	1	F_S	X	CRC
7	0	F_T	X	FDL
8	1	F_S	0	F_S
9	1	F_T	X	FDL
10	1	F_S	X	CRC
11	0	F_S	X	FDL
12	0	F_T	1	F_S
13	First 12 Frames Repeated		X	FDL
14			X	CRC
15			X	FDL
16			0	F_S
17			X	FDL
18			X	CRC
19			X	FDL
20			1	F_S
21			X	FDL
22			X	CRC
23			X	FDL
24			1	F_S

Frames 1–12: 1 Superframe (D4). Frames 1–24: 1 Extended Superframe (ESF).

Notes:

- F_T Terminal framing bit } (F bit)
- F_S Multiframe alignment bit } (F bit)
- FDL 4 Kbps data link bit (M bit)
- CRC Cyclic redundancy check bit (C bit)
- X Data dependent

Fig. T3 *A comparison of the 193rd bit in D4 and ESF formats.*

seconds, failed seconds, and bipolar violations. A supervisory terminal connected to the CSU displays this information, furnishing a record of circuit performance.

Originally, the CSU compiled performance statistics every 15 minutes. This information would be kept updated for a full 24 hours so that a complete one-day history could be accessed by the carrier. The carrier would have to poll each CSU to retrieve the collected data and clear its storage register. By equipping the CSU with dual registers (one for the carrier and one for the user), both carrier and user alike have full access to the performance history.

Today the CSU is not required to store performance data for 24 hours. Also, the CSU no longer responds to polled requests from carriers, but simply transmits ESF performance messages every second.

ESF also allows end-to-end performance data and sectionalized alarms to be collected in real time. This allows the customer to narrow down problems between carrier access points and on interoffice channels, and to find out in which direction the error is occurring.

E1 frame format. In today's increasingly global economy, more and more companies are expanding their private networks beyond the United States and Canada to European locations. In doing so, the first thing they notice is that the primary bit rate is E1, not T1. Whereas T1 has a maximum bit rate of 1.544 Mbps, E1 has a maximum bit rate of 2.048 Mbps. Between the two, there is only one common characteristic: the 64 kbps channel or DS0. A T1 line carries 24 DS0s, while an E1 carries 32 DS0s. Despite this commonality, the DS0s of a T1 line and the DS0s of an E1 line are not compatible.

Although each uses PCM to derive a 64 kbps voice channel, the form of PCM encoding differs. T1 is based on mu-law, while E1 is based on A-law companding. This difference is not as great as it may seem; most multiplexers and carrier switches have the integral capability to convert between them. Conversion includes both the signaling format and companding method.

In E1, as in T1, there is the need to identify the DS0s to the receiver. The E1 format uses framing for this function, as does T1; that is, there are 8000 frames per second, with each frame containing one sample from each time slot, numbered 0 to 31.

The frame synchronization in E1 uses half of time slot 0. Signaling occupies time slot 16 in the 0 to 31 sequence, resulting in 30 channels left for user information. However, it is not possible for all 30 channels to signal within the 8 bits available in time slot 16. The channels must, therefore, take turns using time slot 16. Two channels send their signaling bits in each frame. The 30 user channels then take 15 frames to cycle through all the signaling bits. One additional frame is used to synchronize the receiver to the signaling channel, so the full multiframe ends up having 16 frames. This multiframe corresponds to the T1 superframe.

See also

CHANNEL BANKS
CHANNEL SERVICE UNIT
DATA SERVICE UNIT
MULTIPLEXERS
T-CARRIER FACILITIES
VOICE COMPRESSION (for a discussion of PCM)

Tag switching

The growth of the Internet demands higher bandwidth to accommodate more users and emerging multimedia applications. Demand for higher bandwidth in turn requires higher forwarding performance (packets per second) by routers, for both multicast and unicast traffic. Tag switching uses Network layer packet forwarding to provide an efficient solution to these challenges. It seeks to improve forwarding performance while adding routing functionality to support multicast, allowing more flexible control over how traffic is routed and providing the ability to build a hierarchy of routing knowledge. This approach, developed by Cisco systems, is under standards consideration by the Multiprotocol Integrated Switch-Routing (MISR) working group of the Internet Engineering Task Force (IETF), along with alternative proposals from IBM and Toshiba.

Tag switching components. Tag switching consists of two components, forwarding and control. The forwarding component uses the tag information (tags) carried by IP packets and the tag forwarding information maintained by a tag switch to perform packet forwarding without the need for further processing. The control component is responsible for maintaining correct tag forwarding information among a group of interconnected tag switches.

Forwarding component. When an IP packet with a tag is received by a switch, it is added to the switch's *tag information base* (TIB). Each entry in the TIB consists of an incoming tag, and one or more associated subentries containing outgoing tag, outgoing interface, and outgoing link-level information.

When the switch discovers an IP packet whose incoming tag is equal to that stored in the TIB, the switch replaces the tag in the packet with the outgoing tag, replaces the link-level information (e.g., MAC address) in the packet with the outgoing link-level information, and forwards the packet over the outgoing interface. In this way, tag switching streamlines packet processing at intermediate nodes.

Since the tag forwarding component is Network layer–independent, use of control component(s) specific to a particular Network layer protocol allows the use of tag switching with different Network layer protocols, including point-to-point links, multiaccess links, and ATM.

Control component. In tag switching there is binding between a tag and the routes. At one extreme, a tag could be associated (bound) to a group of routes. At the other extreme, a tag could be bound to an individual application flow (e.g., an RSVP flow). A tag also could be bound to a multicast tree. The control component is responsible for creating tag bindings and then distributing the tag binding information among the tag switches. The control component is organized as a collection of modules, each designed to support a particular routing function. To support new routing functions, new modules can be added.

Basic operation. Once a TIB entry is populated with both incoming and outgoing tags, the tag switch can forward packets for routes bound to the tags by using the tag switching forwarding algorithm. When a tag switch creates a binding between an outgoing tag and a route, the switch, in addition to populating its TIB, also updates its *forwarding information base* (FIB) with the binding information. This allows the switch to add tags to previously untagged packets.

In general, a tag switch will try to populate its TIB with incoming and outgoing tags for all routes to which it has reachability, so that all packets can be forwarded by simple label swapping. Tag allocation is thus driven by topology (routing), not traffic: It is the existence of a FIB entry that causes tag allocations, not the arrival of data packets.

Use of tags associated with routes, rather than flows, also means that there is no need to perform flow classification procedures for all the flows to determine whether to assign a tag to a flow. That, in turn, simplifies the overall scheme and makes it more robust and stable in the presence of changing traffic patterns.

Tag switching does not completely eliminate the need to perform normal Network layer forwarding. To add a tag to a previously untagged packet requires normal Network layer forwarding. This function could be performed by the first hop router, or by the first router on the path that is able to participate in tag switching. In addition, whenever a tag switch aggregates a set of routes (e.g., by using the technique of hierarchical routing) into a single tag, and the routes do not share a common next hop, the switch needs to perform Network layer forwarding for packets carrying that tag.

Summary. Tag switching is not constrained to a particular Network layer protocol; it is a multiprotocol solution. The forwarding component of tag switching is simple enough to facilitate high-performance forwarding; it may be implemented on high-performance forwarding hardware such as ATM switches. The control component is flexible enough to support a wide variety of routing functions, such as destination-based routing, multicast routing, hierarchy of routing knowledge, and explicitly defined routes. In allowing a wide range of forwarding granularities that could be associated with a tag, this method of switching provides both scalable and functionally rich routing.

See also

INTERNET
MULTIMEDIA NETWORKING
MULTIPROTOCOL LABEL SWITCHING
TCP/IP

TCP/IP

Transmission Control Protocol/Internet Protocol (TCP/IP) is a suite of networking protocols that is valued for its ability to interconnect diverse computer platforms, ranging from PCs, Macintoshes, and Unix systems, to mainframes and supercomputers such as the Cray. The development of TCP/IP originated from research funded by the U.S. government's Advanced Research Projects Agency (ARPA) in the 1970s. The protocol suite was developed so that research networks around the world could be joined to form a virtual network known as an *internetwork*. The original Internet was formed by converting over to TCP/IP an existing conglomeration of networks known as ARPAnet. This evolved to become the backbone of today's Internet. The Internet Engineering Task Force (IETF) oversees the development of the TCP/IP protocol suite.

Several factors have driven the acceptance of TCP/IP for mainstream business and consumer use. These include the technology's ability to support both local and wide area connections, its open architecture, and a set of specifications that are freely available in the public domain. Although not the most functional or robust transport available, TCP/IP offers a mature, dependable environment for corporate users who need a common denominator for their diverse and sprawling networks.

The key protocols in the suite include the Transmission Control Protocol (TCP), the Internet Protocol (IP), and the User Datagram Protocol (UDP). There also are application services that include the Telnet protocol, providing virtual ter-

minal service; the File Transfer Protocol (FTP); and the Simple Mail Transfer Protocol (SMTP). Management is provided by the Simple Network Management Protocol (SNMP).

Transmission Control Protocol. TCP forwards data delivered by IP to the appropriate process at the receiving host. Among other things, TCP defines the procedures for breaking up the data stream into packets and reassembling them in the proper order to reconstruct the original data stream at the receiving end. Because the packets typically take different routes to their destination, they arrive at different times and out of sequence. All packets are temporarily stored until the missing packets arrive so they can be put in the correct order. If a packet arrives damaged, it is simply discarded and another one resent.

To accomplish these and other tasks, TCP breaks the messages or data stream down into a manageable size and adds a header to form a packet. The packet's header (Fig. T4) consists of:

- *Source Port/Destination Port Address* (16 bits each): The source and destination ports correspond to the calling and called TCP applications. The port number is usually assigned by TCP whenever an application makes a connection. There are well-known ports associated with standard services such as Telnet, FTP, and SMTP.

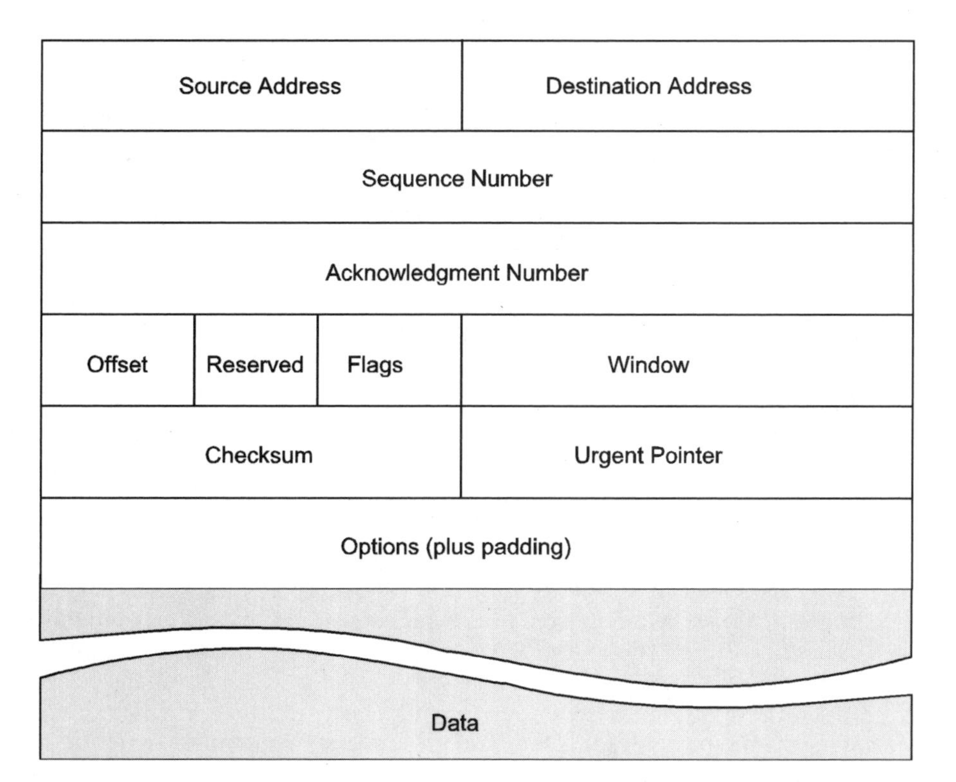

Fig. T4 *TCP packet header.*

- *Sequence Number* (32 bits): Each packet is assigned a unique sequence number that lets the receiving device reassemble the packets in sequence to form the original data stream.
- *Acknowledgment Number* (32 bits): The acknowledgment number indicates the identifier or sequence number of the next expected packet. Its value is used to acknowledge all packets transmitted in the data stream up to that point. If a packet is lost or corrupted, the receiver will not acknowledge that particular packet. This negative acknowledgment triggers a retransmission of the missing or corrupted packet.
- *Offset* (4 bits): The Offset field indicates the number of 32-bit words in the TCP header. This is required because the TCP header may vary in length according to the options that are selected.
- *Reserved* (6 bits): This field currently is not used, but may accommodate some future enhancement of TCP.
- *Flags* (6 bits): The Flags field serves to indicate the initiation or termination of a TCP session, reset a TCP connection, or to indicate the desired type of service.
- *Window* (16 bits): The Window field, also called the *receive window size,* indicates the number of 8-bit bytes that the host is prepared to receive on a TCP connection. This provides precise flow control.
- *Checksum* (16 bits): The checksum is used to determine whether the received packet has been corrupted in any way during transmission.
- *Urgent Pointer* (16 bits): The Urgent Pointer indicates the location in the TCP byte stream where urgent data ends.
- *Options* (0 or more 32-bit words): The Options field typically is used by TCP software at one host to communicate with TCP software at the other end of the connection. It passes such information as the maximum TCP segment size that the remote machine is willing to receive.

The bandwidth and delay of the underlying network impose limits on throughput. Poor transmission quality causes packets to be discarded, which in turn results in retransmissions. Too many retransmissions results in effective bandwidth being cut.

Internet Protocol. An *internet* comprises a series of *autonomous systems* or *subnetworks,* each of which is locally administered and managed. The subnetworks may consist of Ethernet LANs, X.25 packet networks, ISDN, or frame relay networks, for example. IP delivers data between these different networks through routers that process packets from one autonomous system (AS) to another.

Each node in the AS has a unique IP address. The Internet Protocol adds its own header and checksum to make sure the data is properly routed (Fig. T5). This process is aided by the presence of routing update messages that keep the address tables in each router current. Several different types of update messages are used, depending on the collection of subnets involved in a management domain. The routing tables list the various nodes on the subnets as well as the paths between the nodes. If the data packet is too large for the destination node to accept, it will be segmented into smaller packets.

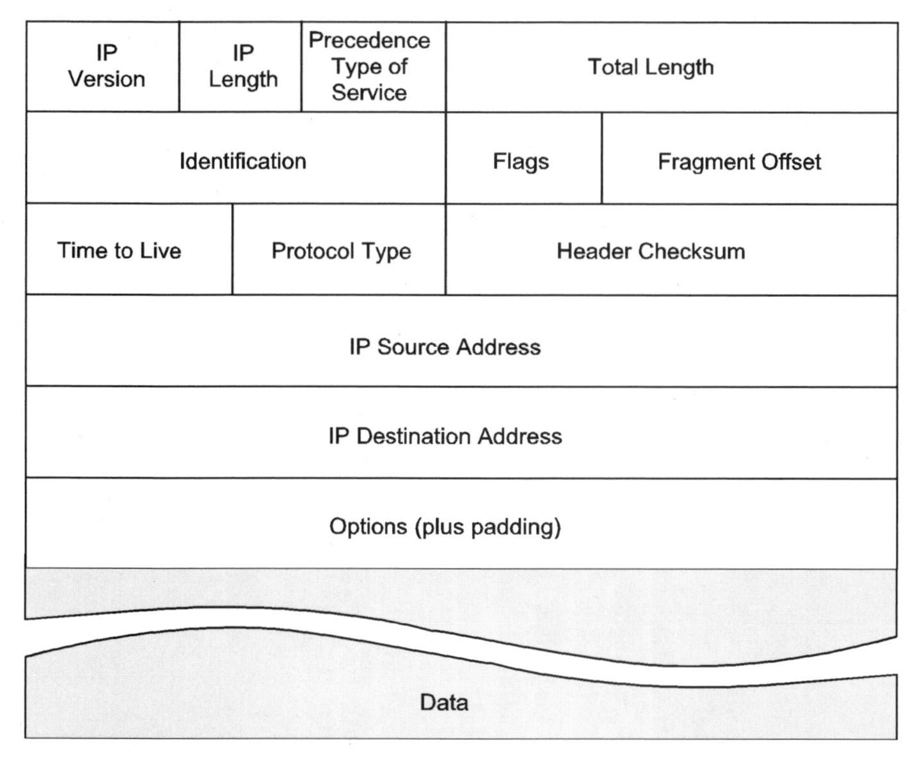

Fig. T5 *IP packet header.*

The IP header consists of the following fields:

- *IP version* (4 bits): The current version of IP is 4. The next generation of IP is 6.
- *IP header length* (4 bits): This indicates header length; if options are included, the header may have to be padded with extra 0s so it can end at a 32-bit word boundary. This is necessary because header length is measured in 32-bit words.
- *Precedence and type of service* (8 bits): Precedence indicates the priority of data packet delivery, which may range from 0 (lowest priority) for normal data to 7 (highest priority) for time-critical data (i.e., multimedia applications). Type of service contains quality of service (QoS) information that determines how the packet is handled over the network. Packets can be assigned values that maximize throughput, reliability, or security, and minimize monetary cost or delay. This field will play a larger role in the future as internets evolve to handle more multimedia applications.
- *Total packet length* (16 bits): This is the total length of the header plus the total length of the data portions of the packet.
- *Identification* (16 bits): This is a unique ID for a message that is used by the destination host to recognize packet fragments that belong together.

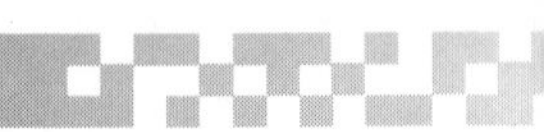

- *Flags* (3 bits): This field indicates whether or not the packets can be fragmented for delivery. If a packet cannot be delivered without being fragmented, it will be discarded and an error message will be returned to the sender.
- *Fragmentation offset* (13 bits): If fragmentation is allowed, this field indicates how IP packets are to be fragmented. Each fragment has the same ID. Flags are used to indicate that more fragments are to follow, as well as indicate the last fragment in the series.
- *Time to live* (8 bits): This field indicates how long the packet is allowed to exist on the network in its undelivered state. The hop counter in each host or gateway that receives this packet decrements the value of the time-to-live field by one. If a gateway receives a packet with the hop count decremented to 0, it will be discarded. This prevents the network from becoming congested by undeliverable packets.
- *Protocol type* (8 bits): This specifies the appropriate service to which IP delivers the packets, such as TCP or UDP.
- *Header checksum* (16 bits): This field is used to determine whether the received packet has been corrupted in any way during transmission. The checksum is updated as the packet is forwarded because the time-to-live field changes at each router.
- *IP source address* (32 bits): This indicates the address of the source host (e.g., 130.132.9.55).
- *IP destination address* (32 bits): This is the address of the destination host (e.g., 128.34.6.87).
- *Options* (up to 40 bytes): Although seldom used for routine data, this field allows one or more options to be specified. Option four, for example, timestamps all stops that the packet made on the way to its destination. This allows measurement of overall network performance in terms of average delay and nodal processing time.

Internet performance is dependent on the resources available at the various hosts and routers (transmission bandwidth, buffer memory, and processor speed) and how efficiently these resources are used. Although each type of resource is manageable, there are always tradeoffs between cost and performance.

User Datagram Protocol. The protocols themselves offer another example of the cost-performance tradeoff. While TCP offers assured delivery, it does so at the price of more overhead. UDP, on the other hand, functions with minimum overhead (Fig. T6); It merely passes individual messages to IP for transmission. Since IP is not reliable, there is no guarantee of delivery.

Nevertheless, UDP is very useful for certain types of communications, such as quick database lookups. For example, the Domain Name System (DNS) consists of a set of distributed databases that provide a service that translates between system names and their IP addresses. For simple messaging between applications and these network resources, UDP does the job.

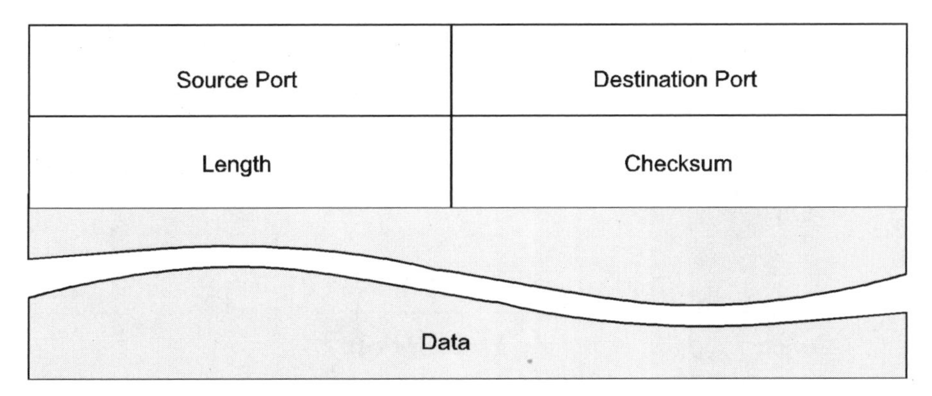

Fig. T6 *The UDP header.*

The UDP header consists of the following fields:

- *Source port* (16 bits): This field identifies the source port number.
- *Destination port* (16 bits): This field identifies the destination port number.
- *Length* (16 bits): This indicates the total length of the UDP header and data portion of the message.
- *Checksum* (16 bits): This validates the contents of a UDP message. Use of this field is optional. If it is not computed for the request, it still can be included in the response.

Applications using UDP communicate through a specified numbered port that can support multiple virtual connections, which are called *sockets.* A socket is an IP address and port, and a pair of sockets (source and destination) forms a TCP connection. One socket can be involved in multiple connections (Fig. T7).

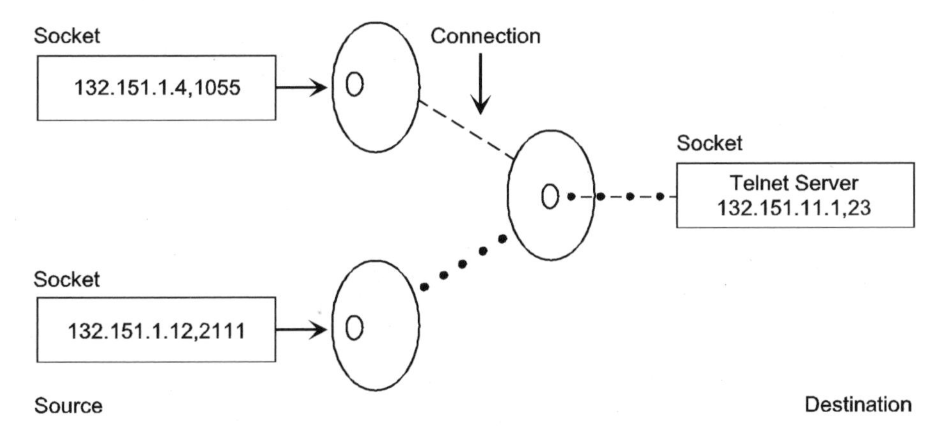

Fig. T7 *UDP socket application.*

Some ports are registered ("well-known") and can be found on many TCP/IP implementations. Well-known ports are numbered from 0 to 1023. Telnet, for example, always uses port 23 for communications, while FTP uses port 21. The well-known ports are assigned by the Internet Assigned Numbers Authority (IANA) and on most systems can be used only by system (or root) processes or by programs executed by privileged users. Other examples of UDP well-known ports are listed in Table T2.

Table T2. Some UDP Well-Known Ports.

Service	Port	Description
Users	11	Shows all users on a remote system.
Quote	17	Returns a "quote of the day"
Mail	25	Used for electronic mail via SMTP.
Domain Name Server	53	Translates system names and their IP addresses
BOOTpc	68	Client port used to receive configuration information
TFTP	69	Trivial File Transfer Protocol used for initializing diskless workstations
World Wide Web	80	Provides access to the Web via the HyperText Transfer Protocol (HTTP)
snagas	108	Provides access to an SNA Gateway Access Server
nntp	119	Provides access to a newsgroup via the Network News Transfer Protocol (NNTP)
SNMP	161	Used to receive network management queries via the Simple Network Management Protocol

In addition to the well-known ports, there are also registered ports numbered from 1024 to 49151, and private ports numbered from 49152 to 65535.

High-level TCP/IP services. The TCP/IP model includes three simple types of services for file transfers, electronic mail, and virtual terminal sessions.

File Transfer Protocol (FTP). FTP is a protocol used for the bulk transfer of data from one remote device to another. Usually implemented as applications-level programs, FTP uses the Telnet and TCP protocols. Most FTP offerings have options to support the unique aspects of each vendor's file structures. Data in the FTP environment consists of a stream of data followed by an end-of-file marker, allowing only entire files to be transferred, not selected records within files. Sending a file using FTP to a user on another TCP/IP network requires a valid user ID and password for a host on that network. To facilitate retrieving files, however, many administrators allow selective FTP access to their systems under the user ID "anonymous" and a password consisting of the user's true address.

Simple Mail Transfer Protocol (SMTP). SMTP is protocol for exchanging mail messages between systems, without regard for the type of user interface or the functionality that is available locally. SMTP sessions consist of a series of commands, starting with both ends exchanging handshake messages to identify themselves. This is followed by a series of commands indicating that a message is to be sent and receipts are needed, and by commands that actually transfer the data. Separating the data message from the address field allows a single message to be delivered to multiple users and to verify that there is at least one deliverable addressee before sending the contents. SMTP modifies every message that it receives by adding a time stamp and a reverse path indicator into each message. This means that a mail message in the SMTP environment usually consists of a fairly long header with information from each node that handled the message. Many user interfaces are able to automatically filter out this kind of information, however.

Telnet Virtual Terminal Service. The Telnet protocol defines a network-independent virtual terminal through which a user can log into remote TCP/IP hosts. The user goes through the standard login procedure on the remote TCP/IP host and must know the characteristics of the remote operating system to execute host-resident commands. Telnet allows remote terminals to access different hosts by fooling the operating system into thinking that a remote terminal is locally connected. Most Telnets operate in the full-duplex mode, meaning that they are capable of sending and receiving at the same time. There is a half-duplex mode to accommodate IBM hosts, however. In this case, a turnaround signal switches the sending of data to the other side of the connection.

Summary. Developed as a military standard, TCP/IP was once considered of interest only to research institutions, academia, and defense contractors. Today, corporations have embraced TCP/IP as a platform that can meet their needs for multivendor, multinetwork connectivity. Because it was developed with government funding, TCP/IP code is in the public domain; this availability has encouraged its use by hundreds of vendors, who apply it to support nearly all types of computers. Because of its flexibility, comprehensiveness, and nonproprietary nature, TCP/IP has captured a considerable and growing share of the commercial internetworking market.

See also

INTERNET
OPEN SYSTEMS INTERCONNECTION
SIMPLE NETWORK MANAGEMENT PROTOCOL

Telecommunications Act of 1996

The Telecommunications Act of 1996, which became law in the United States on February 8, 1996, establishes a procompetitive, deregulatory framework for telecommunications. This law makes sweeping changes affecting all consumers and telecommunications service providers. The intent of this law is to accelerate private sector deployment of advanced telecommunications and information technologies and services to all Americans by opening all telecommunications markets to competition.

For years, competing service providers have wanted to offer local telecommunications services, a market dominated by the traditional monopolistic telephone companies. At the same time, the local telephone companies have wanted to expand into long-distance services. Among the key provisions of the Telecommunications Act of 1996 is that competition in both areas is allowed for the first time.

State regulators and the Federal Communications Commission (FCC) must review Bell Operating Company (BOC) compliance with a comprehensive checklist, however, before the BOCs are allowed to provide in-region interLATA long-distance telephone service. This is meant to ensure that the BOCs do not impose burdensome restrictions on would-be competitors, even as they themselves branch out into new markets. Section 271 of the Telecommunications Act provides a 14-point checklist that is meant to ensure a truly competitive market for local and long-distance services. The following sections excerpt the relevant portions of the checklist and explain the significant points.

Checklist item 1. The act requires "... interconnection in accordance with the requirements of sections 251(c)(2) and 252(d)(1)." This means that:

- All incumbent local exchange carriers (LECs) must allow interconnection to their networks: (1) for exchange service and exchange access; (2) at any technically feasible point; (3) that is at least equal in quality to what the local exchange carrier gives itself, its affiliates, or anyone else; and (4) on rates terms and conditions that are just, reasonable, and nondiscriminatory. *[251(c)(2)]*
- Any interconnection, service, or network element provided under an approved agreement shall be made available to any other requesting telecommunications carrier upon the same terms and conditions as those provided in the agreement. (252(i)) Prices for interconnection shall be based on cost (without reference to any rate-based proceeding) and be nondiscriminatory, and may include a reasonable profit. *[252(d)(1)]*

Checklist item 2. The act requires "... nondiscriminatory access to network elements in accordance with the requirements of sections 251(c)(3) and 252(d)(1)." This means that:

- All incumbent local exchange carriers must provide, to any requesting telecommunications carrier for the provision of a telecommunications service, nondiscriminatory access to network elements on an unbundled basis at any technically feasible point on rates, terms, and conditions that are just, reasonable, and nondiscriminatory. These unbundled network elements will be provided in a manner that allows carriers to combine the elements in order to provide the telecommunications service. *[251(c)(3)]*
- A network element is a facility or equipment used in the provision of a telecommunication service, including features, functions, and capabilities such as subscriber numbers, databases, signaling systems, and information sufficient for billing and collection, or used in transmission, routing, or provision of a telecommunications service. *[3(a)(45)]*
- In determining which network elements will be made available, the FCC shall consider, at a minimum, whether (a) access to network elements that are proprietary is necessary, and (b) whether failure to provide access to these network elements would impair the ability of a carrier to provide the services it wishes. *[251(d)(2)]*

- Prices shall be based on cost (without reference to any rate-based proceeding) and be nondiscriminatory, and may include a reasonable profit. *[252 (d)(1)]*
- As part of their competitive checklist, BOCs are required to unbundle loop transmission, trunk side local transport, and local switching. *[271(c)(2) B)(iv)–(vi)]*

Checklist item 3. The act requires "... nondiscriminatory access to the poles, ducts, conduits, and rights-of-way owned or controlled by the Bell operating company at just and reasonable rates in accordance with the requirements of section 224." This means that:

- Each local exchange carrier is to afford nondiscriminatory access to the poles, ducts, conduits, and rights-of-way to competing providers of telecommunications services, but they may deny access for reasons of safety, reliability, and generally applicable engineering purposes. *[251(b)(4), 224(f)]*
- Within two years, the FCC must prescribe regulations for charges for pole attachments used by telecommunications carriers (not incumbent local exchange carriers) to provide telecommunications services, when the parties fail to agree. Charges must be just, reasonable, and nondiscriminatory. *[224(a)(5), (e)(1)]*
- Pole attachment charges shall include costs of usable space and other space. *[224(d)(1)–(3), (e)(2)]*
- Duct and conduit charges shall be no greater than the average cost of duct or conduit space. *[224(d)(1)]*
- A utility must impute and charge affiliates its pole attachment rates. *[224(g)]*

Checklist item 4. The act requires "... local loop transmission from the central office to the customer's premises, unbundled from local switching or other services." This means that:

- BOCs must unbundle loop transmission. *[271(c)(2)(B)(iv)]*
- This is to be provided at any technically feasible point and in a way that is nondiscriminatory, including rates, terms, and conditions that are just, reasonable, and nondiscriminatory. Unbundled network elements will be provided in a manner that allows carriers to combine the elements in order to provide the telecommunications service. *[251(c)(3)]*
- In determining which network elements will be made available, the FCC shall consider, at a minimum, whether (a) access to network elements that are proprietary is necessary, and (b) whether failure to provide access to these network elements would impair the ability of a carrier to provide the services it wishes. *[251(d)(2)]*
- Prices shall be based on cost (without reference to any rate-based proceeding) and be nondiscriminatory, and may include a reasonable profit. *[252 (d)]*

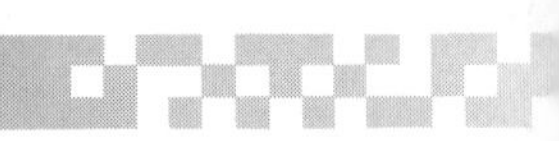

Checklist item 5. The act requires "... local transport from the trunk side of a wireline local exchange carrier switch unbundled from switching or other services." This means that:

- BOCs must unbundle trunk side local transport. *[271(c)(2)(B)(v)]*
- This is to be provided at any technically feasible point and in a way that is nondiscriminatory, including rates, terms, and conditions that are just, reasonable, and nondiscriminatory. Unbundled network elements will be provided in a manner that allows carriers to combine the elements in order to provide the telecommunications service. *[251(c)(3)]*
- In determining which network elements will be made available, the FCC shall consider, at a minimum, whether (a) access to network elements that are proprietary is necessary, and (b) whether failure to provide access to these network elements would impair the ability of a carrier to provide the services it wishes. *[251(d)(2)]*
- Prices shall be based on cost (without reference to any rate-based proceeding) and be nondiscriminatory, and may include a reasonable profit. *[252 (d)]*

Checklist item 6. The act requires "... local switching unbundled from transport, local loop transmission, or other services." This means that:

- BOCs must unbundle local switching. *[271(c)(2)(B)(vi)]*
- This is to be provided at any technically feasible point and in a way that is nondiscriminatory, including rates, terms, and conditions that are just, reasonable, and nondiscriminatory. Unbundled network elements will be provided in a manner that allows carriers to combine the elements in order to provide the telecommunications service. *[251(c)(3)]*
- In determining which network elements will be made available, the FCC shall consider, at a minimum, whether (a) access to network elements that are proprietary is necessary, and (b) whether failure to provide access to these network elements would impair the ability of a carrier to provide the services it wishes. *[251(d)(2)]*
- Prices shall be based on cost (without reference to any rate-based proceeding) and be nondiscriminatory, and may include a reasonable profit. *[252 (d)]*

Checklist item 7. The act requires "... nondiscriminatory access to: (I) 911 and E911 services; (II) directory assistance services to allow the other carrier's customers to obtain telephone numbers; and (III) operator call completion services."

Checklist item 8. The act requires "... white pages directory listings for customers of the other carrier's telephone exchange service." This means that access or interconnection provided or generally offered by a BOC to other telecommunication carriers must include white pages directory listings for customers of the other carrier's telephone exchange service. *[271(c)(2)(B)(viii)]*

Checklist item 9. The act requires that "... until the date by which telecommunications numbering administration guidelines, plan, or rules are established, nondiscriminatory access to telephone numbers for assignment to the other carrier's telephone exchange service customers. After that date, compliance with such guidelines, plan, or rules." This means that:

- ➢ The FCC must create or designate one or more impartial entities to administer telecommunications numbering and to make numbers available on an equitable basis. The FCC has exclusive jurisdiction over the U.S. portion of the North American Number Plan, but may delegate any or all jurisdiction to state commissions or other entities. *[251(e)(1)]*
- ➢ BOCs are required to provide nondiscriminatory access to telephone numbers for assignment by other carriers until telecommunications numbering administration guidelines, plan, or rules are established. Once these guidelines, plans, or rules are established, BOCs must comply with them. *[271 (c)(2)(B)(ix)]*

Checklist item 10. The act requires "... nondiscriminatory access to databases and associated signaling necessary for call routing and completion." This means that:

- ➢ Access or interconnection provided or generally offered by a BOC to other telecommunication carriers shall include nondiscriminatory access to databases and associated signaling necessary for call routing and completion. *[271(c)(2)(B)(x)]*
- ➢ In determining which of these network elements will be made available, the FCC shall consider, at a minimum, whether (a) access to network elements that are proprietary is necessary, and (B) whether failure to provide access to these network elements would impair the ability of a carrier to provide the services it wishes. *[251(d)(2)]*
- ➢ Prices of network elements shall be based on cost (without reference to any rate-based proceeding) and be nondiscriminatory, and may include a reasonable profit. *[252(d)(1)]*

Checklist item 11. The act requires that "... until the date by which the Commission issues regulations pursuant to section 251 to require number portability, interim telecommunications number portability through remote call forwarding, direct inward dialing trunks, or other comparable arrangements, with as little impairment of functioning, quality, reliability, and convenience as possible. After that date, full compliance with such regulations." This means that:

- ➢ All local exchange carriers must provide number portability, to the extent feasible, and in accordance with the FCC's requirements. *[251(b)(2)]*
- ➢ Number portability allows customers to retain, at the same location, their existing telecommunications numbers without impairment of quality, reliability, or convenience when switching from one telecommunications carrier to another. *[3(a)(46)]*
- ➢ Until the date that the FCC establishes for number portability, BOCs are required to provide interim number portability through remote call forwarding, direct inward dialing trunks, or other comparable arrangements, with

as little impairment of functioning, quality, reliability, and convenience as possible. BOCs must fully comply with all FCC number portability regulations. *[271(c)(2)(B)(xi)]*

Checklist item 12. The act requires "... nondiscriminatory access to such services or information as are necessary to allow the requesting carrier to implement local dialing parity in accordance with the requirements of section 251(b)(3)." This means that:

- ➢ Access or interconnection provided or generally offered by a BOC to other telecommunication carriers shall include nondiscriminatory access to such services or information as are necessary to allow the requesting carrier to implement local dialing parity in accordance with 251(b)(3). *[271(c)(2)(B)(xii)]*
- ➢ All local exchange carriers have the duty to provide dialing parity to competing providers of telephone exchange service and telephone toll service, and the duty to permit all such providers to have nondiscriminatory access to telephone numbers, operator services, directory assistance, and directory listing, with no unreasonable dialing delays. *[251(b)(3)]*

Checklist item 13. The act requires "... reciprocal compensation arrangements in accordance with the requirements of section 252(d)(2)." This means that:

- ➢ All local exchange carriers must establish reciprocal compensation arrangements for transport and termination of telecommunications traffic. *[251(b)(5)]*
- ➢ The terms and conditions shall allow each carrier to cover its additional costs of terminating the traffic, including offsetting of reciprocal obligations such as bill-and-keep. Commissions may not engage in any rate proceedings nor require record keeping to determine the additional costs of the calls. *[252(d)(2)]*

Checklist item 14. The act requires that "... telecommunications services [be] available for resale in accordance with the requirements of sections 251(c)(4) and 252(d)(3)." This means that:

- ➢ All local exchange carriers must not prohibit and not impose unreasonable or discriminatory restrictions on resale. *[251(b)(1)]*
- ➢ Incumbent local exchange carriers must offer wholesale rates for any telecommunications service that is provided at retail to customers who are not telecommunications carriers. *[251(c)(4)(A)]*
- ➢ Wholesale prices shall be based on retail prices less the marketing, billing, collection, and other costs that will be avoided by selling the service at wholesale. *[252(d)(3)]*
- ➢ State commissions may, to the extent permitted by the FCC, prohibit a reseller from buying a service available only to one category of customers and reselling it to different category of customers. *[251(c)(4)(B)]*

Summary. Whether the act results in actual local telephone competition will depend in large part upon whether this checklist is followed and enforced. If the BOCs comply with all of the requirements to open local telephone markets to competition, the promise of competition will be realized. Initial applications to state and federal regulators in 1996 for permission to offer interLATA long-distance telephone service failed. Although the BOCs believed they had meet the requirements of the 14-point checklist, on further examination the regulators found that the BOCs only articulated a "plan" to meet the checklist requirements and had not actually implemented many of them.

Telecommunications carriers

Telecommunications carriers are companies that provide voice and data transmission services. There are several kinds of carriers, each offering a different combination of services and value-added features, performance guarantees, geographic coverage, volume discounts, and failure-protection mechanisms.

Incumbent Local Exchange Carriers. Incumbent Local Exchange Carriers (ILECs), which are commonly called *telephone companies* or *telcos,* include the 22 former Bell Operating Companies (BOCs) divested from AT&T in 1984, as well as Cincinnati Bell, Southern New England Telephone (SNET), and the telephone companies of GTE and United Telecommunications. In addition, some 1500 smaller telephone companies also are in operation, serving mostly rural areas. The former BOCs are owned by regional holding companies: Ameritech, Bell Atlantic/NYNEX, BellSouth, Pacific Telesis, Southwestern Bell Communications (SBC), and US West.

Telephone companies traditionally operated on a monopoly basis within specified service areas called *Local Access and Transport Areas* (LATAs). The monopoly status of telephone companies officially ended with passage of the Telecommunications Act of 1996. Not only can other types of carriers enter the market for local services in competition with the telcos, but the seven regional parent companies of the telcos can compete in each other's territories. US West, for example, can offer telecommunications services in the region served by Bell South.

Competitive Local Exchange Carriers. Competitive Local Exchange Carriers (CLECs) provide business and residential users with the means to bypass the local exchange, allowing them to save money on local access lines and usage charges. Regional teleports, metropolitan fiber carriers, and CATV operators are among the types of companies that are now involved in providing competitive local exchange services. Typically these alternative access carriers offer service in major cities, where traffic volumes are greatest and, consequently, users are hardest hit with high local exchange charges. Even PCS, the emerging new wireless digital service, is being run over CATV networks, saving on the cost of installing expensive cell towers in metropolitan areas.

With the Telecommunications Act of 1996, the IXCs are allowed to compete in the offering of local exchange services and must be able to obtain the same service and feature connections as the ILECs—and on an unbundled basis. If the ILEC

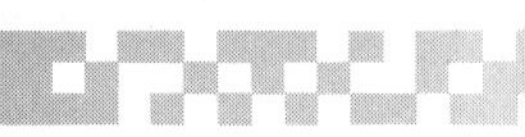

does not meet the requirements of a 14-point checklist to open up its network in this way, it cannot compete with the IXCs in the provision of interLATA long-distance services.

Interexchange Carriers. Interexchange Carriers (IXCs), otherwise known as *long-distance* carriers, include AT&T, MCI, and Sprint. In addition to residential long-distance service, these three carriers have competing offerings in such business areas as WATS, virtual private networks, T-carrier, and a variety of other digital services. Several other IXCs also offer regional interLATA service, but they generally do not match the "Big Three" in terms of the depth and breadth of their service offerings. Typically, these smaller IXCs, also known as *resellers* lease bulk facilities from larger carriers, discount services to users, and reap a profit on call volume.

The Telecommunications Act of 1996 allows IXCs, traditionally limited to providing service between LATAs, to offer local exchange services in competition with the ILECs. Among the methods IXCs are using bypass the local exchange are CATV networks and wireless technology. For example, in the former case AT&T offers a wireless access unit for attachment to homes and office buildings. Long-distance calls go right to AT&T, which does not have to pay the ILEC's local access charge for the call. The savings are passed on to the customer.

Record carriers. Record carriers provide business services such as Telex (teletypewriter exchange service). Telex is a low-speed, character-oriented service that operates at approximately 100 bps and is accessible through the dial-up telephone network. In the United States, Telex is rapidly being displaced by electronic mail and facsimile services. Western Union is probably the most well-known record carrier. In addition to domestic services, Western Union also provides service to overseas locations, qualifying it as an international record carrier. Other international record carriers include TRT in Europe and KDD in Japan.

Specialized mobile radio carriers. Specialized mobile radio carriers provide mobile communications services to business and individual users, including dispatch, paging, and data services. The traditional radio frequency (RF) technology is analog in nature and employs a single-site, high-power transmitter configuration that precludes the use of any given radio frequency by more than one caller at a time within a given service area. The move to digital technology in recent years has created more channel capacity through the subdivision of cells and frequency reuse, increased network reliability and call quality, and made possible a number of advanced call handling features. Among the largest mobile communications providers in the U.S. are ARDIS and RAM Mobile Data.

Satellite carriers. Satellite carriers provide communications services via a network of satellites circling the earth in geostationary orbit. Companies such as GTE Spacenet, Contel/ASC, and Hughes sell satellite capacity to provide a variety of telecommunications services, including PCS and direct broadcast television to consumers. Telecommunications carriers lease transponders from the satellite carriers to back up terrestrial links and to extend their long-distance networks to international locations. Two such satellite networks are Motorola's Iridium and Microsoft/AT&T's Teledesic, which will start coming online in 1998 and 2000, respectively.

Packet data network providers. Packet data network providers offer data services, mostly through X.25 packet data networks (PDNs), and newer frame relay and ATM services, which may be accessed via private leased lines or dial-up connections. PDNs offer such useful features as protocol conversion and flow control, which allow dissimilar terminals and computers to communicate with one another. X.25 networks also provide error correction to ensure that all data is received as the sender intended. Telenet and Tymnet (a British Telecom company) are two of the best-known PDNs. Frame relay and ATM services are offered by IXCs and ILECs, sometimes bundled with applications support (i.e., SNA) and management services.

Value-added service providers. Value-added service providers offer specialized services, such as voice mail, paging, facsimile, electronic mail, and interactive services over the public telephone network. Other types of value-added services include party lines, recorded messages, information services, and live entertainment. Some of these services are offered as add-ons to local or long-distance services. Others, such as live entertainment and information services, are added to the user's phone bill only when they are accessed.

Some value-added service providers offer network services for business and government; these are called *value-added networks* (VANs). They also provide support services such as design, management, and specialized billing—tasks that can be resource-intensive if done by the customer. There are several categories of VANs. With the right combination of services and features, however, a VAN can span two or more of the following categories:

- *Enhanced Transmission Services* offer such value-added enhancements as protocol conversion, encryption, and speed conversion to support the integration of incompatible data systems.
- *Managed Data Networks* provide large corporations with batch processing, e-mail, order entry and processing, report generation, and data archiving facilities from a third-party vendor.
- *Managed Transaction Networks* serve specific interest groups or industries that, for example, allow the exchange of information among manufacturers, warehouses, and retailers (e.g., Electronic Data Interchange, or EDI).
- *Information Providers* offer access to a wide range of databases via dial-up connections. These online information services include specialized financial, scientific, legal, and medical databases. This category also includes more general databases, such as those provided by America Online and CompuServe.

Summary. With continued deregulation, not only are carriers breaking out of their traditional service roles, they also are entering into a variety of business relationships that permit them to offer new types of services. Some of the large regional holding companies are entering into partnerships with cable television operators and entertainment companies, for example, to bring consumers such services as video on demand. In the long term, the distinction between the various types of carriers may become blurred as each tries to enter the markets traditionally held by the others.

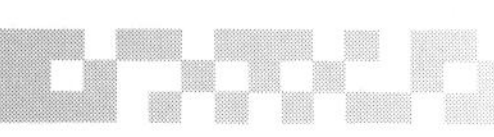

See also

TRANSMISSION FACILITIES

Telecommunications Management Network

Telecommunications carriers today are faced with the task of rapidly introducing and managing new competitive services. To do this successfully, they must quickly integrate communications equipment from multiple vendors. In response to the challenge, the carriers and international standards bodies have defined a solution called the *Telecommunications Management Network* (TMN). TMN consists of a series of interrelated national and international standards and agreements that provide for the surveillance and control of telecommunications service provider networks on a worldwide scale. The result is the ability to achieve higher service quality, reduced costs, and faster product integration. TMN is also applicable in wireless communications, cable television networks, private overlay networks, and other large-scale, high-bandwidth communications networks.

The International Telecommunications Union-Telecommunications (ITU-T) began work on TMN in 1988. The ITU-T (formerly known as CCITT) extended the ISO/OSI standards for systems management by adding its own recommendations for architecture (M.3000 series), modeling (G series), and interfaces (Q series). The standard M.3010 is the high-level specification of the framework that provides for the observation, control, coordination, and maintenance of telecommunications networks. Because the TMN model is based on open interfaces, it is attractive to telecommunications operators and service providers as a way to solve common management problems without being limited by proprietary solutions.

By making all internal network management functions available through standardized interfaces, service providers can achieve more rapid deployment of new services and maximum use of automated functions. Vendors of network elements can offer specialized management systems known as *element managers,* which can integrate readily into a service provider's larger management hierarchy. Groups of service providers can enter into business-level agreements and deploy resource sharing arrangements that can be administered automatically through interoperable interfaces.

Under TMN, management tasks are arranged into Network Element, Element Management, Network Management, Service Management, and Business Management layers. OSI's Common Management Information Protocol (CMIP) is used to communicate between the adjacent layers.

Object-oriented techniques are used for creating data structures and their access methods for the management information base (MIB). Manager/agent concepts are used for the hierarchical exchange of management information between systems. A MIB is located at an agent, which provides an abstraction of network resources for the purpose of management. The agent provides service elements that are a standard set of access methods into the MIB.

Functional architecture. TMN breaks the tasks performed in a management network into function blocks, as follows:

- *Network Element Function* (NEF) contains the telecommunications functions that are the actual subject of management.

- *Operations Systems Function* (OSF) processes information related to management.
- *Workstation Function* (WSF) provides the means to interpret TMN information for the management system operator.
- *QAdaptor Function* (QAF) connects a non-TMN NEF or OSF to a TMN. This is a protocol conversion to a standard TMN interface.
- *Mediation Function* (MF) mediates between a non-OSF and a NEF or a QAF, and presents the information in a different form for the OSF.
- *Data Communication Function* (DCF) transfers telecommunication network management information.

TMN standards define two types of telecommunications resources: *managed systems,* generally known as Network Elements (NE); and *managing systems,* of which the Operation System (OS) is the most prominent. TMN also defines reference points (*f, g, m, q* and *x*) that can exist between these function blocks. Reference points become interfaces when they occur at locations that require data communications between elements. The key interface in the TMN model is the OS-NE (or Q3) interface, which uses a CMIP manager/agent pair to provide access to a standardized MIB.

Physical architecture. For each functional block, a physical block can be implemented, thus leading to a physical architecture. Reference points are a significant part of the TMN functional architecture and are realized within the physical architecture by physical interfaces within systems or equipment. The implementations of the reference points are represented by capital letters (*Q, F, X*) and form the common boundary between associated TMN blocks. The *F* interface is found between workstations and the TMN, the *Q* or *Q3* interface is found between TMN devices, and the *X* interface is found between the devices of one TMN and the devices of another TMN via DCN (Data Communication Network). Figure T8 describes the physical architecture, showing the implementation of reference points (i.e., interfaces) within TMN.

A telecommunications network provides voice and data services to customers; this function is outside of the TMN. However, the network consists of various network elements that can provide TMN functions and services. Central to the TMN is the data communication network (DCN), which typically uses portions of the carrier's telecommunication network for transmission of management information. Telecommunication network NEs, which provide transmission to the DCN, are part of the TMN's DCN.

Most NEs provide services for a facility or a telecommunication network such as environmental alarm notification or telecommunication transmission. Those NEs that provide a TMN management interface as well as non-TMN functions straddle the TMN boundary.

The operation system (OS) provides the mechanism for interaction, observation, and execution of many of the functions within the TMN. All higher-level functionality, however, does not have to be physically located within an OS, NE, or any other single point in the TMN.

TMN elements use object-oriented techniques to represent resources under their control as managed objects. A common understanding of these objects is shared among the element controlling the resources, a mediation device (MD) or NE, and

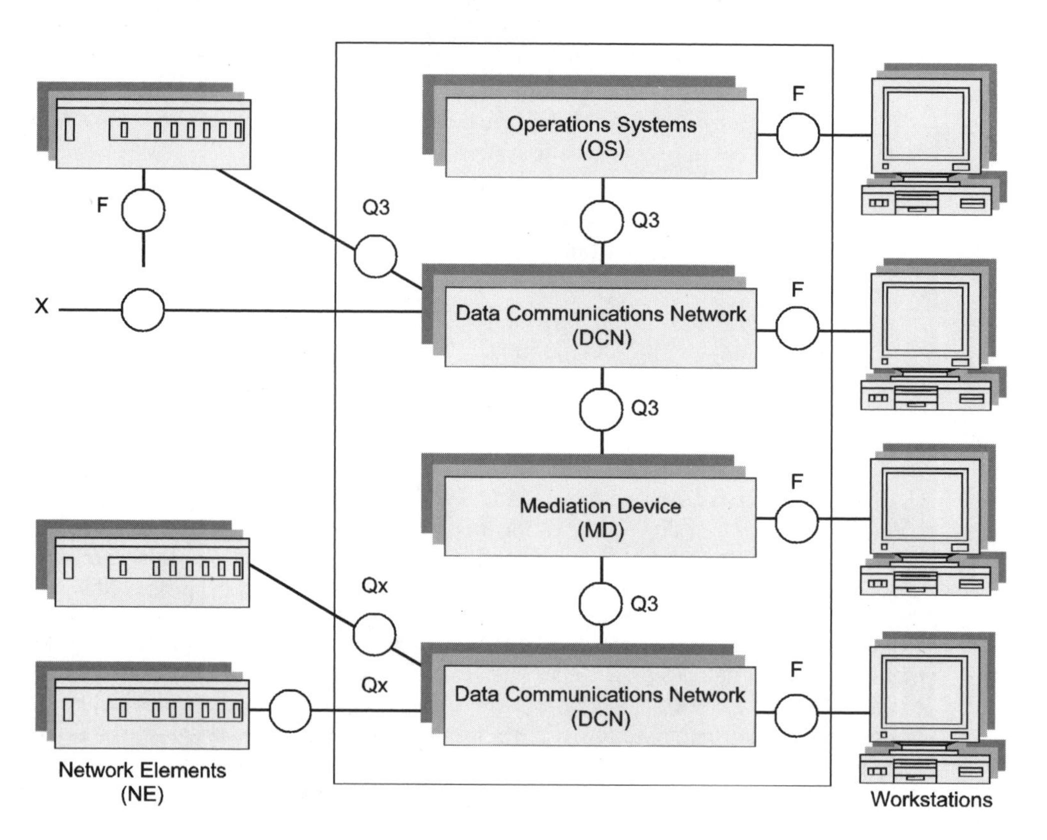

Fig. T8 *TMN physical architecture.*

the manager of the resource, an OS or MD. TMN uses a manager/agent paradigm where the manager issues instructions and the agent carries out the instructions. These operations are accomplished using the Common Management Information Services (CMIS) and Protocol (CMIP) standards. In response to instructions to the agent, methods in the managed object are invoked, resulting in a variety of information storing, retrieving, and processing activities within the NE or MD. A NEF agent can be managed by either an OSF or MD manager, while an MF agent is managed by an OSF manager.

Summary. Architecturally compliant TMNs are growing in popularity and are the path to providing reliable telecommunication network management into the 21st century. The TMN framework holds the promise of a telecommunications network which can be managed easily, even if different technologies from multiple vendors are used. The alternative to a TMN is to continue to create proprietary management networks with inconsistent and incompatible management interfaces. As carrier networks worldwide become more advanced and as carriers continue to engage in mergers and alliances in an attempt to better serve customer needs, adapting the interoperable management solution offered by TMN standards becomes a key ingredient that can determine the success of their efforts.

See also

OPEN SYSTEMS INTERCONNECTION

Telecommunications management systems

Telecommunications management systems capture call records from the PBX or key system for the purpose of identifying and managing costs, keeping the corporate telephone network optimized for maximum savings and availability, and providing decision-making information on the need for more lines and equipment.

The captured call records are stored on a PC or other collection device. The collected data is then processed into a variety of cost and usage reports. With billing reports, for example, call charges can be allocated across the organization. Traffic statistics reports provide a detailed look at trunk usage, call volumes, and calling patterns.

In addition to traditional call-accounting functions, which provide cost allocation and other reports on employee and department telephone usage, numerous other related applications and options are available. For instance, there are software modules that can be added to the system to track equipment and cable assets within the organization.

Some telecommunications management systems can be acquired as individual applications programs. Others are best used as part of a larger, integrated management system that includes call accounting in addition to one or more other application modules. Single-site single-user, single-site multiple user, and multisite packages are available. Some products work completely independently of other computerized systems, while others link under a total network management umbrella.

An alternative to in-house call detail record collection and processing is the *third-party service bureau*. The service bureau collects call records from the call recording device at the client's premises via a dial-up connection or dedicated line (depending on the call record volume) on a daily, weekly, or monthly basis. The call detail records are processed into reports for the client. The client can choose from among a set of standard reports, or have the data processed into custom reports.

Some third-party teleprocessing firms offer billing verification services. Audits verify that carrier billings accurately reflect charges for telephone services (voice and data) and equipment actually contracted for and used. If errors are found, refund requests are prepared and submitted to the appropriate carrier for reimbursement.

System functions. One of the most common telecommunications management applications is call accounting. Such systems produce reports from station message detail recording (SMDR) devices, which capture call detail records generated by PBX, Centrex, hybrid, or key telephone systems, or from tip-and-ring line scanners. Most call accounting systems compute costs for each incoming or outgoing call, whether local or long-distance.

Reporting. Call detail records usually contain the following basic information, which is assembled into summary and detail reports according to such categories as individual station, department, or project:

- Date of the call
- Duration of the call

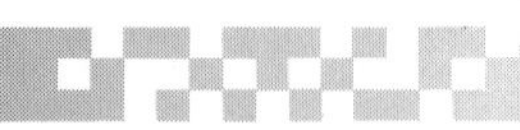

- ➢ Extension number
- ➢ Number dialed
- ➢ Trunk group used

Other types of information might also include the most frequently dialed numbers, longest-duration calls, highest-cost calls, or most calls to a particular area code, exchange, or telephone number.

Call accounting software also can be used to identify possible toll fraud. Standard call accounting reports can be used to locate toll fraud by identifying short frequent calls, long-duration calls, unusual calling patterns, and unusual activity on 800 and 900 numbers. In addition, the reports can be used to identify calls made after hours, on weekends, or on holidays.

Some vendors have introduced special toll fraud detection packages. These programs alert managers to changes in calling patterns or breached overflow thresholds. Several types of alarms can be generated—to a pager, to a local PC, or to a remote PC—so that immediate action can be taken.

Call costing. A key feature of telecommunications management systems is call costing, which can be used for internal charge-back purposes. For each call record, the cost per call can be figured in a variety of ways, including real-time call costing, flat rate or percentage surcharges, and equipment usage. The ability to price calls according to preferred parameters gives organizations flexibility in allocating costs internally and in meeting budgetary targets.

Cost allocation. The cost allocation application distributes calling costs to the appropriate internal departments, projects, clients, workgroups, or subsidiaries, or external customers or users. It also can allocate costs for equipment, trunks, lines, maintenance, or other administrative charges. Some systems also can depreciate equipment according to organizational depreciation schedules.

Like the call accounting application, summary and detail reporting is available for cost allocation by organizational unit or equipment type. Most cost allocation applications also provide an interface to the organization's general ledger.

Equipment inventory. Equipment inventory applications track information on the quantity, type, and location of installed and spare voice equipment (PBX and station), as well as circuits/trunks (e.g., DID, WATS, and private lines). This application usually tracks inventory and indicates if it is committed or available for use. Some systems even provide key sheets or face layouts that include pictures (graphical representations) of each component (e.g., telephone, turret, market data terminal, etc.) within the system, specifying the features and types of circuits associated with each key or button. The emphasis of many equipment inventory systems is to work with call accounting to provide internal billing for equipment usage.

Invoice management. Invoice management systems are designed to compare vendor invoices to equipment inventory databases, and create reports that flag erroneous charges or duplicate billings. Invoice management software also makes billing comparisons using previous-period or previous-year invoices in order to automate approval of vendor invoices, allow telecom managers to analyze usage trends, and facilitate budgeting activities.

Alarm monitoring and reporting. Alarm monitoring and reporting applications monitor voice and data equipment alarm ports, organize alarms by priority, and generate plain-English alarm reports. Some systems automatically dial out to a central reporting center and generate trouble tickets for alarms. The alarm reports generated by alarms monitoring and reporting software provide performance statistics on PBX hardware and software components.

In addition to monitoring and reporting alarms for the telephone system, some telecommunications management systems can act on alarms from other types of equipment, including voice mail systems, environmental control systems, and security systems.

Trouble reporting and tracking. Trouble reporting and tracking applications maintain a log of all system problems, track repair progress, and provide reports on vendor response times and system performance. These applications often provide an interface to help desks or service desks, providing historical problem-resolution information to the operators.

Work order processing and tracking. When a problem occurs, a telecom manager will generate either a paper or electronic work order and give it to a technician. The work order contains information about the work to be performed, the machine or item on which work is to be performed, required completion date, cost of work, and billing information. Work order processing and tracking applications automate the activities involved in processing move/add/change requests. The orders can be categorized as pending, overdue, or completed to help manage vendor and staff resources.

In some work order processing and tracking applications, the software even schedules work orders and automatically updates the equipment and cable inventory databases upon completion of the job. Some applications also commit inventory and schedule vendors, and produce reports that flag overdue orders or scheduling problems associated with large projects.

Network design. Network design applications were originally developed to optimize the use of WATS lines and, for a time, private line services. Now, however, they are in greater demand for intrastate trunk and toll analysis. A tariff database, which is available on a subscription basis, is used to do what-if analysis. Given the data on existing calling patterns, the network design application applies various rate plans from different carriers to find the most economical blend of services.

Summary. Organizations of all types and sizes are turning to telecommunications management systems to optimize their voice networks for efficiency and cost containment. For companies that prefer to outsource this task, there are a number of third-party service firms that remotely capture and process call records into the desired report formats. In addition to offering call record collection equipment and installation services, many offer toll fraud detection tools and billing audits to help clients eliminate wasteful spending.

See also

NETWORK DESIGN

Telecommunications Relay Services

Title IV of the Americans with Disabilities Act of 1990 (ADA) requires the FCC to ensure that Telecommunications Relay Services (TRS) are available, to the extent

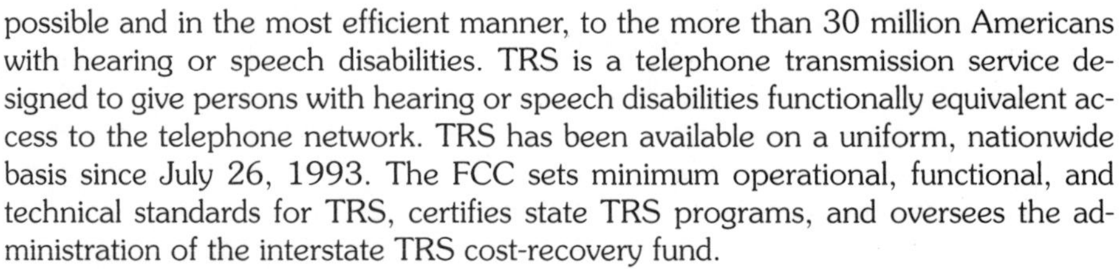

possible and in the most efficient manner, to the more than 30 million Americans with hearing or speech disabilities. TRS is a telephone transmission service designed to give persons with hearing or speech disabilities functionally equivalent access to the telephone network. TRS has been available on a uniform, nationwide basis since July 26, 1993. The FCC sets minimum operational, functional, and technical standards for TRS, certifies state TRS programs, and oversees the administration of the interstate TRS cost-recovery fund.

Currently, TRS provides access to the voice telephone network for over 30 million Americans with hearing and speech disabilities. The service is offered 7 days a week, 24 hours a day in all 50 states, the District of Columbia, Puerto Rico, and the U.S. Virgin Islands.

Operation. TRS relies on communications assistants (CAs) to relay the content of calls between users of text telephones (TTYs) and users of traditional handsets (voice users). For example, a TTY user might telephone a voice user by calling a TRS provider, or *relay center,* where a CA will place the call to the voice user and relay the conversation by transcribing spoken content for the TTY user and reading text aloud for the voice user.

To access TRS, the user dials the local service provider, either through TTY or voice telephone. The call is answered by a CA who takes down the number of the party to contact and completes the call. The CA relays the conversation to the called party, verbatim and in real time.

The CA is not allowed to intentionally alter a relayed conversation and may not limit the length of calls. The CA also may not disrupt the continuity of a TRS call, except when necessary for facilitation of the call (for example, to ask for clarification of unintelligible messages).

The TRS provider number is listed in the local telephone directory, usually in the information section, or may be obtained from directory assistance. Some states have separate TRS provider numbers for TTY and voice callers. The FCC also publishes a TRS directory.

Funding. The cost of interstate TRS is recovered from all providers of interstate telecommunications services, as a percentage of their gross revenues and a contribution factor determined annually by the FCC. Contributions are administered in a TRS fund by the National Exchange Carrier Association (NECA), an association of local telephone companies. TRS providers are compensated for interstate TRS minutes of use based on a payment rate that is also determined annually by the FCC.

The FCC has established an interstate TRS Fund Advisory Council to advise the TRS Fund Administrator on funding issues. The Council comprises consumer representatives, TRS users, state regulatory officials, TRS providers, and state relay administrators. The Advisory Council's meetings are open to the public.

Universal design. *Universal design* is the concept of achieving accessibility of structures, products, and services by planning at the blueprint stage for the fullest range of human function. The dual goals of universal design are accessibility to the widest range of individuals and elimination of the need for retrofitting and reconstruction.

Some examples of universal-designed telecommunications products include televisions with closed-captioning decoder circuitry, telephones with volume control and built-in hearing aid compatibility, and public telephones that are lowered to heights accessible to people who use wheelchairs, and which also feature built-in TTY keyboards.

Closed captioning is the display of audio portions of television and video programming as printed words on the television screen. In addition to displaying spoken dialogue and music lyrics, captions may identify speakers, sound effects, background music, and laughter. *Open captions* always appear directly on the television screen; *closed captions* are hidden as encoded data within the television signal and are displayed only when activated by the viewer. Since 1992, all televisions with 13-inch or larger screens are required to be equipped with the technology to display captioning, and consumers may purchase set-top decoders for older TV models.

Video description is an auditory depiction of a television or video program's visual elements for persons who are blind or visually impaired. Video description is inserted in the natural pauses of a program's dialogue, and may be used to describe visual elements such as body language, settings, and actions. In order to receive video description, an audience member must have a stereo television or VCR that is capable of receiving the Second Audio Program (SAP) channel. The SAP feature is available on most new TVs and VCRs; consumers also may purchase receivers for converting TV sets to stereo with SAP.

Volume control telephones allow the user to amplify the sound output level of the telephone receiver. In addition to benefiting persons with hearing loss and persons losing their residual hearing later in life, volume control also can benefit persons who must use telephones in noisy environments.

Telecommunications Act of 1996. Two provisions of the Telecommunications Act, Sections 255 and 713, focus entirely on access by persons with disabilities.

Section 255 of the act requires all manufacturers of telecommunications equipment and providers of telecommunications services to ensure that such equipment and services are designed and developed to be accessible to and usable by individuals with disabilities, if readily achievable. The FCC will undertake a rulemaking proceeding to implement this provision.

Section 713 aims to ensure that video services are accessible to individuals who have hearing or visual impairments. It requires the FCC to study the level at which video programming is closed-captioned, and then to establish a timetable for closed-captioning requirements. (The FCC is authorized to exempt programming for which the provision of closed captioning would be economically burdensome.) Section 713 also directs the FCC to study the use of video description in order to assure the accessibility of this service to persons with visual impairments.

Other provisions of the act aim to promote access to telecommunications by all Americans, including those with disabilities:

- Section 706 requires the FCC to encourage deployment of advanced telecommunications to all Americans, and to elementary and secondary schools (and classrooms in particular). It requires the FCC to assess the level at which advanced telecommunications are available, and then to take steps, if necessary, to accelerate deployment of such services by removing barriers to infrastructure investment. This provision could significantly benefit children with disabilities as well as children and adults without disabilities.

- Section 254 concerns universal service, and directs the FCC and a Federal-State Joint Board to define what services should be made universally available and to take other actions as needed to further the act's universal service principles. Section 254 also revises the definition of universal service to include schools, libraries, and health care facilities. It says that telecommunications companies must provide services to these public institutions at affordable rates, upon request. The FCC and the states must decide what constitutes "affordable rates," what telecommunications services should be covered, and how discounts should be made available to public institutions.
- Section 256 directs the FCC to establish procedures for oversight of telecommunications network planning and states that the FCC may participate with the industry in developing standards for *interconnectivity* (the ability of telecommunications carriers to connect to each other's networks). Such standards would promote access to telecommunications networks by people with disabilities.
- Section 251 states that telecommunications carriers may not install network features, functions, or capabilities that do not comply with the guidelines and standards established under Sections 255 and 256.

Summary. TRS users pay rates no greater than those paid by other users of functionally equivalent voice communication services with respect to such factors as the duration of the call, the time of day, and the distance from the point of origin to the point of termination. As long as individuals with hearing and speech disabilities are required to purchase specialized customer premises equipment (CPE), including TTYs, however, they might not enjoy the same access to the telephone network as voice telephone users. This is because such specialized equipment can be much more expensive than a regular telephone. The FCC currently permits common carriers to provide, under tariff, appropriate customer premises equipment. Some state legislatures also have passed legislation enacting such equipment distribution programs.

Telecommuting

Telecommuting is a cooperative arrangement between companies and employees that allows employees to work outside of the office, usually in the home or at a satellite office set up for that purpose, on a part-time or full-time basis. Some industry estimates have pegged the number of telecommuters in the United States at 30 million.

Advances in computer and communications technologies, including wireless products and services, have made telecommuting a viable option for many types of jobs. Telecommuting provides participants with a greater sense of job autonomy, which can increase productivity and satisfaction, while providing employees with a means to better manage their work and family lives. A telecommuting program also can help trim corporate overhead expenses by providing savings in office space and other facilities.

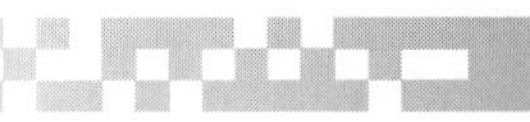

The jobs best suited for telecommuting include those of white collar workers engaged in research, consulting, auditing, illustration, writing, and computer programming. Telecommuting also has proved effective for call center agents, telemarketers, and customer service representatives whose jobs involve a lot of keyboard or telephone interaction. Telecommuting can be useful for people who are highly mobile throughout most of the work day: salespeople, real estate agents, field service personnel, and insurance claims adjusters.

The telecommuting concept reached national prominence with amendments to the Clean Air Act of 1990, which took effect in November, 1992. The amendments require companies with 100 or more employees in certain high-pollution metropolitan areas to develop plans to reduce their number of commuters by 25 percent by 1996. In addition, several states (notably California) have instituted their own laws to reduce commuter traffic. In other states, subsidized public transportation, ride sharing, or van pooling are typical approaches to reducing traffic. These programs have failed to meet expectations, however, but work-at-home programs and satellite offices offer practical alternatives.

Another piece of legislation that encourages companies to set up telecommuting programs is the Family and Medical Leave Act, one of the first laws signed by President Clinton. This law entitles employees to 12 weeks of unpaid, job-protected leave each year for specified family and medical reasons. The act leaves room for companies to develop telecommuting plans that let employees perform some work at home. In this situation, telecommuting can help employees earn some or all of their pay, keep their careers on track, and preserve their seniority—all while attending to prolonged family medical problems.

Management issues. Although telecommuting offers the promise of greater employee productivity and lower overhead costs for the company, it will likely be limited to a few individuals unless companies reorient first-line supervisors to accept it. Many senior first-line supervisors continue to believe that productivity will decrease if workers are not closely watched. Numerous studies have confirmed, however, that employees working at home are generally 20 percent more productive than their counterparts at the office. Some companies and government agencies report productivity gains as high as 40 percent.

Supervising remote employees requires a different set of skills. To ensure that their supervisors have the necessary skills, some companies have issued detailed guidelines, followed by specific supervisory training that emphasizes interpersonal communications and sensitivity to the problems experienced by remote workers and their office-bound colleagues.

Telecommuting does entail some risk, primarily organizational and social. For example, people who spend too much time away from the office are concerned about damaging their relationships, particularly with management. "Out-of-sight and out-of-mind" can stifle or eliminate their advancement opportunities, they believe, or cause them to miss opportunities to cultivate new relationships with face-to-face networking. Over time, telecommuters can feel isolated from the rest of the company and out of step with organizational changes.

In addition, there is the potential for telecommuting to cause resentment among office workers who feel that telecommuters are receiving special treatment.

Only trained managers and supervisors can keep the negative feelings of employees from getting out of hand if they should arise. If left unaddressed, these negative feelings can dampen morale and impede productivity—the very things telecommuting is intended to improve. Not everyone will qualify for the program, so there will be the need to explain to others that the arrangement is work- and job-related, not simply a company perk.

One way to expose potential problems with telecommuting is to implement it as a pilot program. It is much easier for department managers and first-line supervisors to agree to a 3-month project than an indefinite change, and to agree to a pilot if he or she can help define the metrics such as the time frame and what measures constitute a successful test.

Summary. The promises of telecommuting have been touted by advocates for as long as there have been PCs. Due to improvements in computing and communications technologies, plus lower costs for hardware and services, telecommuting is now widely viewed as a practical alternative to working in the office. This is occurring just in time, too: With traffic congestion getting worse in many metropolitan areas and pollution posing health hazards, new work arrangements must be explored to preserve the quality of life.

A noteworthy aspect of telecommuting is that the objectives of reducing travel and associated vehicle pollution need not be achieved at the expense of employee productivity or at great cost to participating companies. With proper planning and sensitivity training for managers and supervisors, telecommuting typically results in lower overhead costs for companies, with the added benefit of increasing the productivity and loyalty of employees. However, the critical factor in the success of any telecommuting program is selecting the right jobs and people for the program.

See also
DISTANCE LEARNING
TELEMEDICINE

Teleconferencing

The term *teleconferencing* refers to the broad category of communications by which three or more people communicate in audio-only mode via telephone lines. The key benefit of teleconferencing is that it eliminates travel while enhancing communications by allowing many people to share information directly and simultaneously. Businesses are implementing teleconferencing strategies to handle large volumes of calls on a daily basis, reducing non-telecommunications costs while improving productivity. Educational institutions are using teleconferencing for cost-effective distance learning programs; government agencies are using teleconferencing for crisis management as well as daily information exchange.

Teleconferencing can be implemented in a variety of ways. For a basic telephone conference involving a limited number of participants, a telephone set with either a three-way calling feature on the line, or a conferencing feature supported by the PBX or key system, is required.

Alternatively, teleconferences can be set up by a system attendant. For a teleconference with more than three parties, the attendant console operator can establish the connections through the PBX and add more participants to the call than can be accommodated from a single set.

As another option, a company can initiate a conference through the telephone company conference operator. The operator sets up a conference by calling each person until all participants are online together. A conference can be established at a prearranged time, or it can be organized so that the participants can phone in to a preassigned number at a designated time.

While individuals use their own telephone sets to participate in a conference, several people at a given location also can participate in the conference as a group. In such cases, specialized equipment is required. The main component is the audio system, which consists of a control unit with an integral keypad, omnidirectional microphone, and a speaker that typically is positioned in the center of a table.

The audio system. At the heart of any teleconferencing system is the audio control unit, since high-quality audio is the key component of teleconferencing. Yet audio quality can be substantially diminished by a condition known as *echo.*

There are two sources of echo, multipath and direct. *Multipath echo* results when sound being emitted from a loudspeaker reflects off surfaces and objects within the room and is passed back into the microphone, creating multiple echoes of the original speech. The microphone transmits the sound to the remote site(s), where it is heard as an echo. *Direct echo* results when sound from the far end is broadcast by the loudspeaker and passed directly into the microphone without reflecting from surfaces in the room. In the latter case the delay is short enough that it can send the system into oscillation, commonly called "feedback."

Most audio systems employ some form of echo control. Nevertheless, audio quality can suffer as a result of an ineffective echo control system. The following conditions therefore may become apparent to teleconference participants:

- *Clipping:* The speaker's voice is broken up.
- *Dropout:* The voice is suddenly cut off when any noise is introduced at the connecting site.
- *Attenuation:* A momentary loss of volume during the conference.
- *Artifacts:* Unintelligible voice remnants that are heard during stuttering pauses.

There are two methods of enhancing audio quality by controlling echo, echo suppression and echo cancellation. *Echo suppression* employs a gating process whereby the loudspeaker and microphones are alternately turned on and off to avoid transmitting the multipath and direct echo. Alternately, if the received level for the microphones is higher than some predefined threshold, the loudspeaker signal is muted. In this way the system attenuates the loudspeaker or the microphones to suppress echo from being picked up and sent.

Echo cancellation employs a high-speed digital signal processor (DSP) or custom application-specific integrated circuit (ASIC) to electronically compare the microphone signal to the transmitted loudspeaker signal. Any similarities in these two signals is recognized as echo. These similarities are then electronically subtracted from the microphone input, allowing only the original speech to be transmitted.

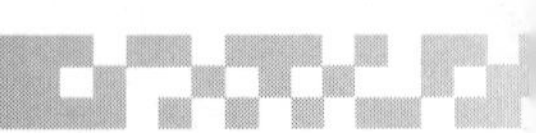

This process eliminates the effects of both direct and multipath echo. Of the two methods for controlling echo, cancellation is the best. It not only permits full-duplex operation, but is more dependable.

The role of bridges. The most economical way to teleconference is to call the various participants on separate telephone lines, then join the phone lines by using a bridge. Bridges can accommodate a number of different types of LAN and WAN interfaces and are capable of linking several hundred participants in a single call. The problem with this method is that the participants cannot always hear one another well. There are teleconferencing devices available, however, that amplify and balance the conversation, allowing everyone to hear everyone else as though the participants were talking one-on-one. The bridge can operate as a stand-alone device or connect through ports of a PBX or Centrex switch, either line-side ports or trunk-side ports.

A bridged conference can be implemented in several ways. In one method, each participant can be called and then transferred to one of the PBX ports assigned to support the conference. Another method entails participants dialing a designated phone number at a preset time and automatically being placed on the bridge. Still another method of access involves participants dialing a main number and being transferred onto the bridge manually by a live operator. In any case, whenever a new person joins the party, a tone indicates his or her presence. The more conversations that are brought into the conference, the more free extensions are required.

An optional moderator phone can be used to set up bridged conferences. The moderator phone allows the operator to initiate conferences, actively participate in conversations, and terminate connections, as well as conduct isolated conversations with one party and then either admit or readmit the call to the conference or disconnect the call.

Conference modes. Depending on the vendor, there are a number of conference modes, implemented by the bridge, from which to choose:

- *Operator Dial-Out:* Participants are brought into the conference by an operator using such methods as manual dialing and abbreviated dialing.
- *Originator Dial-Out:* Additional participants are added to the conference by the conference moderator, who accesses available lines using a tone-dialing telephone.
- *Prearranged:* Conferences are dialed automatically or by a user dialing a predefined code from a tone-dialing phone. In either case, the information needed to set up the conference is stored in a scheduler.
- *Meet-Me:* Participants call into a bridge at a specified time to begin the conference. If teleconferencing is used frequently, a dedicated 800 number can be justified for this purpose.
- *Security Code Access:* Participants enter a conference code and are automatically routed to the appropriate conference. If an invalid password is entered, the call is routed to an attendant station, where an operator screens the caller and offers assistance.

- *Automatic Number Identification:* ANI automatically processes incoming calls based on the phone number of the caller. This includes call branding, call routing, and conference identification. Custom greetings may be designed for each incoming call.

System features. The various features of a teleconferencing system can be grouped into five categories: participant, moderator, conference, administration, and maintenance.

Participant features. Individual participants in any teleconference need only one piece of equipment, a tone-dialing telephone. Some bridges facilitate tone-dialed interaction, increasing the level of participation and end user control dramatically. The conference administrator can configure dual tone multifrequency (DTMF) detection for one or two digits. This allows conference participants to implement various features by pressing one or two buttons on their telephone sets. Among the features a participant can implement are:

- *Help:* If the participant needs help, pressing 0 or *0 signals the conference operator for assistance.
- *Mute:* If a participant wants to mute the line, possibly to talk without having remarks conveyed to the other conferees, pressing 6 or *6 (6=M for "mute") will place the line in listen-only mode.
- *Polling:* Conferees can participate in voting sessions. By pressing one or two digits (to indicate yes or no, for example), each participant's preference can be recorded by the bridge. The results can be read immediately or stored on disk or printed out via the parallel printer port.
- *Question and Answer:* Conference participants can enter a question queue to signal the moderator that a question is waiting. At the appropriate time, the moderator can address each participant's question.

Moderator features. Moderators are given special privileges that provide a higher level of conference control than that afforded to participants. These privileges can be activated using their tone-dial phone:

- *Security:* A moderator can secure a conference by pressing one or two buttons on the tone keypad.
- *Conference Gain:* To level all signals in the conference, the moderator can implement the gain control feature of the bridge, if it is not already set for automatic gain control.
- *Lecture:* The moderator can initiate a lecture in which all lines, except those designated as moderators, are muted. This allows the presenter to convey information without interruption from conference participants.

Conference features. Some bridges provide an extensive array of features, all of which contribute to a productive and successful teleconference. Depending on vendor, these features might include:

- *Polling:* With this feature, the moderator can poll participants using one of the following methods: yes/no, true/false, multiple choice, or assigned

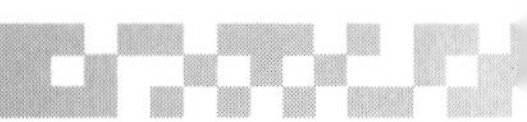

ranges. Participants use their tone pads to make their selections. Results are compiled immediately and can be printed or saved to disk.

- *Question and Answer:* During a lecture, for example, participants can indicate that they have questions by pressing one or two buttons on the tone phones. When the lecture is over, the moderator can take each question randomly or in the order received. Participants also can remove their own lines from the queue.
- *Fast Dial:* Often called *speed dial,* this feature allows an operator to quickly initiate outbound calls from an attendant console. The operator accesses a stored list of phone numbers and highlights the individual to be called into the conference. By merely pressing the Enter key, the number is dialed.
- *Auto Dial:* This feature is similar to fast dial, except that it allows the operator to highlight all of the phone numbers of individuals who must join in a conference. By pressing the Enter key, all of the calls are dialed simultaneously. Each conferee is greeted by a recorded announcement that provides further instructions, such as prompting for a security code.
- *Lecture:* This mode of operation automatically mutes all lines in the conference except that of the moderator, for uninterrupted sessions.
- *Security:* This feature provides confidential conferencing, prohibiting unauthorized individuals from entering. Secured conferences lock out the operator and cannot be recorded. By pressing the same buttons on the tone keypad, the moderator can remove security from the conference.
- *Mute:* This feature places a specific line in listen-only mode. Also, lines are automatically muted by the system when certain features are implemented, such as lecture, polling, and Q&A.
- *Music:* Via an external device attached to the bridge, music-on-hold is provided to entertain participants until the conference begins. Music also provides assurance to waiting participants that the connection is still alive and that they should continue to wait.
- *Record/Playback:* With this feature, conferences can be recorded and played back. Recorded material also can be played into a conference using an external system. All lines are muted during playback. Some systems can be configured to allow the moderator to be heard during the playback.
- *Help:* This feature allows conferees to signal an operator for assistance. Help can be set on a per-line basis, whereby the operator removes the conferee from the conference to provide help. Help also can be set on a conference-wide basis, whereby the operator responds by entering the conference to address the entire group.
- *Conference ID:* This feature provides for the identification of conferences for report generation. The operator can assign the conference ID, or the system can be configured to assign IDs automatically.
- *Conference Note:* The operator can record notations during a conference; notations will appear on the conference report.
- *Conference Scan:* An automatic audio scan of the entire conference can be performed at assigned intervals. Secured conferences are not scanned.

- *Operator Chat:* Two or more operators can send electronic messages to one another without disrupting active conferences.
- *Listen Mode:* For quality-control purposes, operators can listen to individual lines or a range of lines without affecting conference activity.
- *Disconnect Notification:* The system can be configured to notify the operator of a disconnect during the conference.
- *Operator alarms:* The system can be configured to provide the operator with audible and visual signals upon disconnect, help request, or queue activity.

Administration features. From a teleconferencing product's system administration menu, an administrator is able to:

- Modify system configurations,
- Perform supervision of the system during operation,
- Configure the system for auto dial,
- Configure operator functions,
- Configure channels,
- Perform file management,
- Implement disk utilities, and
- Configure the conference scheduler.

Maintenance features. From a system's maintenance menu, a maintenance person can access all conference and administrative functions, plus such special maintenance features as:

- Power-up diagnostics
- Online diagnostics
- Remote diagnostics
- Warm boot
- Maintenance reports
- Alarms

Teleconferencing over the Internet. A relatively new development is the use of IP networks for teleconferencing. Internet telephony software makes it possible for users to engage in long-distance conversation between virtually any location in the world without regard for per-minute usage charges. In most cases, all that is needed is an Internet access connection and a computer equipped with telephony software, sound card, microphone, and speakers or headset.

A teleconference is established by entering one or more e-mail or IP addresses, which alerts the called parties to join at a particular server on the Internet or corporate intranet. The voice of each party is compressed and packetized, sometimes with encryption, for transmission over the IP network. Since the packets may travel a different route to their destination, there may be significant delay as the packets arrive and are put into the proper order. If some packets do not arrive in time, they are discarded. If enough packets are discarded, there can be noticeable gaps in a person's speech.

Companies can fashion their own internal telephone networks by leveraging their existing intranets and conduct teleconferences (and videoconferences) at virtually no extra charge. Alternatively, subscription services are emerging that offer teleconferencing services over private IP networks that can be accessed from the corporate intranet or the Internet at greatly reduced rates.

Summary. For a variety of reasons, many people have lost the daily face-to-face contact that they once had with fellow employees. In addition, the rate of change in the business environment requires that information be distributed more rapidly. These and other factors are making teleconferencing an essential tool for all people in business today, allowing them to stay in contact and be informed without resorting to expensive and time-consuming travel. While the same can be said about videoconferencing, teleconferencing continues to be the more available, economical, and simpler solution for the majority of business needs.

See also

INTERNET TELEPHONY
VIDEOCONFERENCING

Telegraphy

Telegraphy is a form of telecommunication that is based on the use of a signal code. The word comes from Greek, *tele* meaning distant and *graphein* meaning to write. Thus, *telegraphy* is "writing at a distance."

The inventor of the first electric telegraph was Samuel Finley Breese Morse an American inventor and painter. On a trip home from Italy, Morse became acquainted with the many attempts to create usable telegraphs for long-distance telecommunication. He was fascinated by this problem and studied books on physics for 2 years to acquire scientific knowledge.

Early attempts. Morse focused his research on the characteristics of electromagnets, whereby they became magnets only while the current flows. The intermittence of the current produced two states—magnet and no magnet—from which he developed a code for representing characters, which eventually became known as Morse code.

His first attempts at building a telegraph failed, but he eventually succeeded with the help of some friends who were more technically knowledgeable. The signaling device was very simple. It consisted of a transmitter containing a battery and a key, a small buzzer as a receiver, and a pair of wires connecting the two. Morse later improved it by adding a second switch and a second buzzer to enable transmission in the opposite direction as well.

In 1837, Morse succeeded in a public demonstration of his first telegraph. Although he received a patent for the device in 1838, he worked for 6 more years in his studio at New York University to perfect his invention. Finally, on May 24, 1844, Morse unveiled the results of his work. Over a line strung from Washington, D.C. to Baltimore, Morse tapped out the message, "What hath God wrought." The message reached Morse's collaborator in Baltimore, Alfred Lewis Vail, who immediately sent it back to Morse.

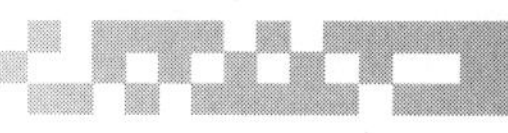

With the success of the telegraph assured, the line was expanded to Philadelphia, New York, Boston, and other major cities and towns. The telegraph lines tended to follow the rights-of-way of railroads and, as the railroads expanded westward, the nation's communications network expanded as well.

Morse code. Morse code uses a system of dots and dashes that are tapped out by an operator using a telegraph key. (It can also be used to communicate via radio and flash lamp.) Various combinations of dots and dashes represent characters, numbers, and symbols separated by spaces. The international Morse code for letters, numbers, and symbols is shown in Table T3.

Table T3.
International Morse Code.

Letters	Numbers
A •–	1 •––––
B –•••	2 ••–––
C –•–•	3 •••––
D –••	4 ••••–
E •	5 •••••
F ••–•	6 –••••
G ––•	7 ––•••
H ••••	8 –––••
I ••	9 ––––•
J •–––	0 –––––
K –•–	
L •–••	**Symbols**
M ––	
N –•	• •–•–•– (period)
O –––	, ••––•• (comma)
P •––•	? ••––••
Q ––•–	; –•–•–• (semicolon)
R •–•	/ –••–•
S •••	: –––••• (colon)
T –	' •––––• (apostrophe)
U ••–	– –••••– (hyphen)
V •••–	= –•••–
W •–	(–•––•
X –••–	) –•––•–
Y –•–	
Z ––••	

Summary. Morse code is the basis of digital communication. Although it has disappeared in the world of professional communication, it is still used in the world of amateur ("ham") radio and is kept alive by history buffs. There are even pages on the World Wide Web that teach telegraphy and perform translations of text into Morse code.

Telemedicine

Telemedicine refers to the delivery of health care services from a distance using a combination of high-speed digital networks, videoconferencing, and medical technologies. In offering the ability to exchange patient records and diagnostic images, while permitting interactive consultations among medical specialists at different locations, telemedicine promises not only to save lives and extend sophisticated medical services to rural areas, but also to cut the overhead costs of hospitals and clinics, saving an estimated US$250 billion of the US$1 trillion Americans now spend each year on health care.

Applications. There are numerous applications of telemedicine. Hospitals, walk-in clinics, mobile medical units, and health care management organizations now routinely exchange information on patient eligibility, claims processing, scheduling, and medical records. What is relatively new is the use of networks for the transmission of diagnostic images and multisite doctor-patient consultations.

For example, radiologists at major medical centers can read magnetic resonance imaging (MRI) and other types of image scans transmitted from other hospitals over telephone wires. Receiving such information in a timely fashion greatly enhances the ability of doctors to offer a quick second opinion or tap into specialized expertise at major medical centers.

Large hospitals that can afford such specializations as nuclear medicine now transmit diagnostic images to satellite locations that are not so elaborately equipped. This arrangement brings economies of scale to otherwise prohibitively expensive medical diagnostic procedures. The central facility can recover capital investments by billing smaller hospitals and clinics for its services. This would not be possible if several hospitals and clinics in the same area competed for patients with their own nuclear medicine facilities.

There now are firms that specialize in providing diagnostic and consulting services to patients and physicians. Online interpretation of x-rays, EKGs, pathology slides, and other medical images can be performed over networks that use compressed interactive video links.

The telemedicine concept also has been applied to the delivery of mental health services. While physicians are in short supply in rural communities, there is an even greater shortage of skilled mental health professionals. Consequently, rural residents must travel even longer distances for psychiatric care.

To improve accessibility to mental health professionals in rural areas, training arrangements based on videoconferencing have been used with great success. Site coordinators at local facilities schedule a room and equipment for training provided by hospital staff via a videoconference. The local coordinators handle equipment setup and operation, as well as registrations for the medical education courses delivered over the network. A project coordinator at the hospital maintains a master schedule to facilitate connection of the local sites to each other and to prevent conflicts in network usage.

Equipment at the rural sites typically includes a video codec, monitor, two video cameras, an audio system, control devices, graphics tablets, and an interface. At the main location, from which the videoconference is moderated, a control console is used to set up the connections to each remote site. If a multipoint control unit (MCU) is used, the participants at each remote site can see one another. The site in the picture is determined by voice-activated switching; that is, when a participant is speaking, that site is displayed on the monitor.

Data from digital medical instruments can be networked for collaborative patient diagnosis. For example, a nurse in the field can use an ENT (ear, nose, throat) videoscope to relay video images of the patient's eardrum to doctors in the nearest city. There, physicians can determine if the problem is a perforated eardrum.

Real-time ultrasound examinations also can be done from the field. Portable ultrasound equipment connected to a computer and phone line relays the image to a hospital's ultrasound department, where physicians can interpret the images.

Another device called a *video dermascope* can be used to examine various skin conditions. A nurse in the field can use the video dermascope to examine the layers of skin in a lesion, for example, while hospital physicians watch on a video screen to determine if the area is malignant.

Likewise, an instrument called an *endoscope* allows images from inside a human digestive tract to be transmitted over a modem or ISDN link for viewing by doctors at major medical centers. Like other such instruments, the endoscope system allows staff at hospitals to spend less time traveling to patients and more time actually diagnosing problems and treating them.

The outputs of such instruments, when coupled to computers and relayed over networks for display on videoconferencing systems, can greatly improve health care in rural areas. In fact, a new medical field called *telediagnosis* is emerging. According to its advocates, telediagnosis is a cost-effective solution to managed care that not only will result in greater access to care, but also improve the quality of care provided in a high-touch, face-to-face application.

Not content with connecting health care facilities to one another, many telemedicine advocates insist that the technology eventually must reach into the home if it is to fulfill its promise of optimal health care access. Already there is a portable defibrillator to restart a heart by remote control over a standard or cellular telephone line. Other devices allow doctors to remotely monitor heart rate, blood pressure, and temperature, and to download data from and reprogram implanted pacemakers.

Eventually, so-called "telehealth" networks will be set up that would not only connect health care providers to patients, but also would tie in any organization connected to the health care process within a given community. Community-wide telehealth networks could combine LANs, MANs, and WANs to provide networking within health care facilities and to link those facilities to one another and to insurance carriers, pharmacies, and academic research centers.

Summary. Telemedicine has the potential to revolutionize health care delivery and access. The integration of various telecommunications technologies, video

conferencing, and sophisticated medical instrumentation brings interactive visual communications into operating rooms and rural clinics alike, providing access to the best doctors while reducing costs and increasing the availability and quality of health care.

See also

DISTANCE LEARNING
VIDEOCONFERENCING

Telemetry

Telemetry is the monitoring and control of remote devices from a central location via wireline or wireless links. Applications include utility meter reading, load management, environmental monitoring, vending machine inventory, and security alarms. Companies are deploying telemetry systems to reduce costs by eliminating the need to read and check remote devices manually. For example, vending machines need not be visited daily to check for proper operation or out-of-stock conditions. Instead, this information can be reported via modem to a central control station so a repair technician or supply person can be dispatched as appropriate. Such telemetry systems greatly reduce service costs.

When telemetry applications use wireless technology, additional benefits accrue. The use of wireless technology allows systems to be located virtually anywhere without depending on the telephone company for line installation. For instance, a kiosk equipped with a wireless modem can be located anywhere in a shopping mall without incurring line installation costs. Via wireless modems, data is collected from all the area kiosks at the end of the day for batch processing by a mainframe. The kiosk also runs continuous diagnostics to ensure proper operation. If a malfunction occurs, problems are reported via the wireless link to a central control facility, which can diagnose and fix the problem remotely, or dispatch a technician if necessary.

Various wireless security systems are available for commercial and residential use (Fig. T9). The use of wireless technology provides installation flexibility because the sensor components can be placed anywhere, without regard for proximity to electrical outlets or phone wiring. A variety of sensors are available to detect such things as temperature, frequency, or motion. The system can be programmed to automatically call monitoring station personnel, police, or designated friends and neighbors when an alert is triggered. Such systems can be set to randomly turn lights on and off at designated times to give the appearance of occupancy. Depending on vendor, the system might even perform continuous diagnostics to report low battery power or tampering.

In a typical implementation, the security system console monitors the sensors placed at various potential points of entry. The console expects the sensor to send a confirmation signal at preset intervals, say every 90 seconds. If the console does not get the signal, it knows that something is wrong. For example, a sensor attached to a corner of the window or other glass panel is specially tuned to vibrations caused by breaking glass. When it detects the glass breaking, the sensor opens its contact, and sends a wireless signal to an audio alarm located on the premises, police station, or private security firm.

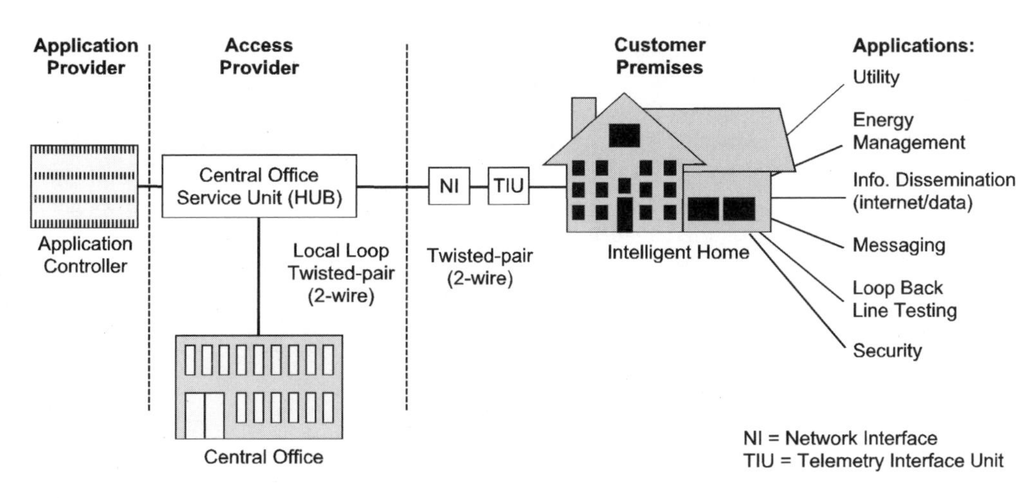

Fig. T9 *Telemetry applications for residential users.*

The use of wireless technology for security applications actually can improve service reliability. A security service that does not require a dedicated phone line is not susceptible to intentional or accidental outages when phone lines are down or there is bad weather. Wireless links offer more immunity to such things.

An emerging application of wireless technology is its use in traffic monitoring. For example, throughout the traffic signal control industry there has been a serious effort to find a substitute for the underground, hard-wired inductive loop that is in common use today to detect the presence of vehicles. Although the vehicle detection loop is inherently simple, it has many disadvantages:

- Slot cutting for loop and lead-in wire is time-consuming and expensive.
- Traffic is disrupted during installation.
- Reliability depends on geographical conditions.
- Maintenance costs are high, especially in cold climates.

These problems are overcome by a wireless presence detector, the signals from which activate traffic lights in the prescribed sequence. The detector, usually mounted on a nearby pole, focuses the wireless signal very narrowly on the road to represent a standard loop. The microprocessor-based detector provides real-time information while screening out such environmental variations as temperature, humidity, and barometric pressure. By tuning out environmental variations, the detectors provide consistent output. This increases the reliability of traffic control systems. Using a laptop computer with a Windows-compatible setup package, information can be exchanged with the detector via an infrared link. From the laptop, the pole-mounted detector can be set up remotely, calibrated, and put through various diagnostic routines to verify proper operation.

Role of cellular carriers. Cellular carriers are well positioned to offer wireless telemetry services. The cellular telephone system has a total of 832 channels, half of which are assigned to each of the two competing cellular carriers in each mar-

ket. Each cellular carrier uses 21 of its 416 channels as control channels. Each control channel set consists of a Forward Control Channel (FOCC) and a Reverse Control Channel (RECC).

The FOCC is used to send general information from the cellular base station to the cellular telephone. The RECC is used to send information from the cellular telephone to the base station. The control channels are used to initiate a cellular telephone call. Once the call is initiated, the cellular system directs the cellular telephone to a voice channel. Once the cellular telephone has established service on a voice channel, it never goes back to a control channel. All information concerning hand-off to other voice channels and termination of the telephone call is handled via communication over the voice channels.

This leaves the control channels free to provide other services, such as telemetry. This can be achieved by connecting a gateway to a port at the local Mobile Switching Center (MSC) or regional facility. The gateway can process the telemetry messages according to the specific needs of each individual application.

For instance, if telemetry is used to convey a message from a alarm panel, the gateway will process the message on a real-time, immediate basis and pass the message to the Central Alarm Monitoring Service. On the other hand, if a soft drink vending company uses telemetry to poll its machines each night for their stock status, the gateway will accumulate all of the responses from the individual vending machines each night and provide them in batch form when requested from the vending company the next morning.

Individual applications can have different responses from the same telemetry radio. While the vending machine uses batch processing for its stock status, it could have an alarm message conveyed to the vending company on an immediate basis, indicating a malfunction. A similar scenario is applicable for utility meter reading. Normal meter readings can be obtained on a batch basis during the night and delivered to the utility company the following morning. However, real-time meter readings can be made any time during the day for customers who wish to close out or open service and require an immediate, current meter reading. Telemetry even can be used to turn on or turn off utility service remotely by the utility customer service representative.

Summary. Telemetry services, once implemented by large companies over private networks, are becoming more widely available for a variety of mainstream business and consumer applications. Wireless technology permits more flexibility in the implementation of telemetry systems and can save on line installation costs. Such systems are inherently more reliable when wireless links are used to convey status and control information, since they are less susceptible to outages due to tampering and severe weather.

See also

CELLULAR COMMUNICATIONS

Telephone

The telephone, invented by Alexander Graham Bell, has undergone dramatic changes in packaging, capabilities, and features since its introduction in 1876. Bell was born in 1847 in Edinburgh, Scotland. He moved to the United States, settling

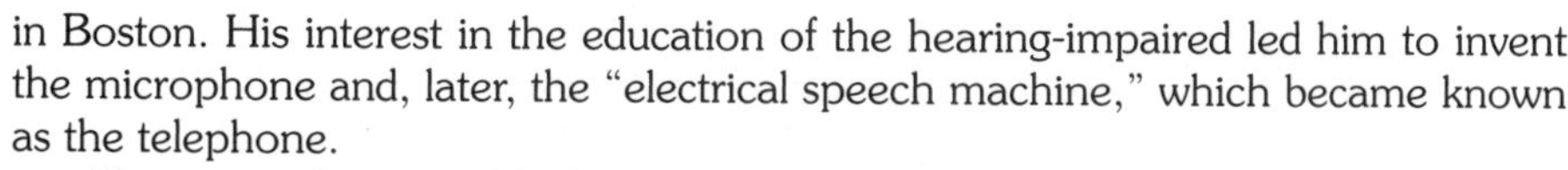

in Boston. His interest in the education of the hearing-impaired led him to invent the microphone and, later, the "electrical speech machine," which became known as the telephone.

Two years later, in 1878, Bell set up the first telephone exchange in New Haven, Connecticut. In essence, this manually operated patch panel was the first central office switch; it also provided the basis for the private branch exchange (PBX). By 1884, lines were strung up on poles between Boston and New York City, marking the start of the long-distance telecommunications industry.

AT&T was incorporated in New York on March 3, 1885, as a wholly owned subsidiary of the American Bell Telephone Company. Its original purpose was to manage and expand the burgeoning toll, or long-distance, business of American Bell and its licensees. It continued as "the long-distance company" until December 30, 1899, when, in a corporate reorganization, it assumed the business and property of American Bell and became the parent company of the Bell System.

Telephone components. Regardless of model or manufacturer, the telephone includes three key components: a transmitter, a receiver, and a dial mechanism. The transmitter converts sound waves emitted by human vocal cords into a fluctuating electric current, while the receiver converts the electric current back into sound waves that can be picked up by the human ear. The dial mechanism, whether pulse or tone, provides the means to reach a telephone with an assigned number at the other end of the circuit. The call itself is switched through a series of telephone company central offices until a dedicated path is created between the calling and called parties. This path, or *circuit,* stays in place for the duration of the conversation. Only when the parties hang up are the lines and central office equipment free to handle other calls.

Call processing. Calls are circuit-switched to their destinations via a system of dialed digits. All telephones are connected to a local central office, each of which is assigned a 3-digit exchange number. Each subscriber's telephone is assigned a 4-digit identification number. Together, the exchange number and subscriber number are what we commonly refer to as a "telephone number," which takes the following format:

XXX-XXXX

When trying to reach a telephone that is located outside of the local service region, a 3-digit area code might have to be dialed as well. The use of an area code tells the local exchange switch to contact an interexchange switch, which might be operated by a different carrier. Eventually a path is established through multiple local and interexchange switches and, when the circuit is complete, the parties at each end can communicate. The area code is entered before the 7-digit phone number. The entire number takes the following format:

XXX-XXX-XXXX

If the call is to a location outside of the country, two additional numbers come into play: a 3-digit international access code and a 2- or 3-digit country code. The 3-digit international access code sends the call from the local exchange switch to one or more interexchange switches, and then to a special gateway switch that just handles calls between other gateway switches at international locations. For example, an international call from the U.S. to Sydney, Australia, takes the following format:

011 61 2 *local_number*

where 011 is the international access code, 61 is the country code for Australia, and 2 is the city code for Sydney. This is followed by the called party's local number.

The dedicated pair of wires running from the subscriber's telephone to the local central office switch is called a *line circuit* or *loop*. The central office places a potential of about 48 V (dc) across the local loop to power the telephones and monitor call activity for billing purposes.

Address signaling. There are two types of address signaling, pulse and tone. Pulse dialing is used when the telephone is connected to an older analog electromechanical central office switch, whereas tone dialing is used when the telephone is connected to a newer digital central office switch. Most telephones sold today for residential use support both types of address signaling.

With either method of address signaling, placing a call starts when the user lifts the handset off the cradle. This closes a contact relay in the telephone and permits current to flow through the loop. This signals the central office that the user would like to place a call. When the user hears a continuous tone from the central office switch (i.e., dial tone), this indicates that an idle line has been secured to handle the call. On the other hand, if a fast-busy tone is sent to the telephone, this indicates that no lines are currently available to handle the call. A recorded announcement is activated, indicating that "all circuits are busy, please try again...."

Pulse dialing. Upon detecting loop current, the central office searches for an unused dial pulse register to store the dialed digits as they are received. The register is connected and dial tone is sent down the line. Upon hearing the dial tone, the user can proceed to dial the desired telephone number.

With rotary dial, or pulse dial, the telephone set rapidly opens and closes the loop at a rate of about 10 pulses per second. The number of pulses corresponds to the digit dialed (Fig. T10). This continues until all of the digits of the telephone number have been dialed. If the number is that of a local subscriber, the connection is made immediately. If the number is that of a subscriber outside of the local calling area, the central office looks for a trunk (interexchange or interoffice line) that will connect it with the appropriate central office.

A trunk circuit provides a signal path between two central offices. Unlike a line circuit, or loop, a trunk circuit is shared by multiple users, although only one uses a trunk circuit at any given time. There might be 100 or more trunk circuits between these central offices, and as one telephone call ends, a trunk circuit is released and made available to handle another call.

When a path has been established, the called party is alerted to the presence of the incoming call. This is done by the local central office, which sends an 88-V, 20-Hz signal down the loop to drive a bell inside the telephone. This signal is repeated—2 seconds on and 4 seconds off—to create the familiar telephone ring. When the phone's handset is lifted, current flows in the telephone loop, telling the central office to disconnect the ringing voltage and establish the two-way voice connection.

Tone dialing. While pulse dialing is still used in many areas, a newer method of address signaling called *tone dialing* is replacing it and has become more common.

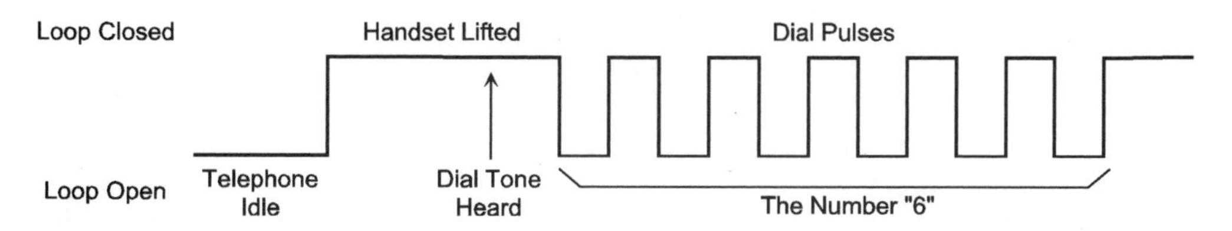

Fig. T10 *In pulse dialing, the number of pulses equates to the number dialed.*

With tone dialing, each digit is formed by selecting two out of seven possible frequencies. These frequencies were selected to minimize the possibility of accidental duplication by voice.

For example, pressing the digit 7 on the phone's keypad sends an 852-Hz tone and a 1209-Hz tone simultaneously (Fig. T11). This is sometimes referred to as dual-tone multifrequency (DTMF) signaling. AT&T has branded the scheme as "Touch-Tone." The central office switch recognizes these different frequencies and associates them with the numbers dialed. The advantage of tone dialing over pulse dialing is faster call processing.

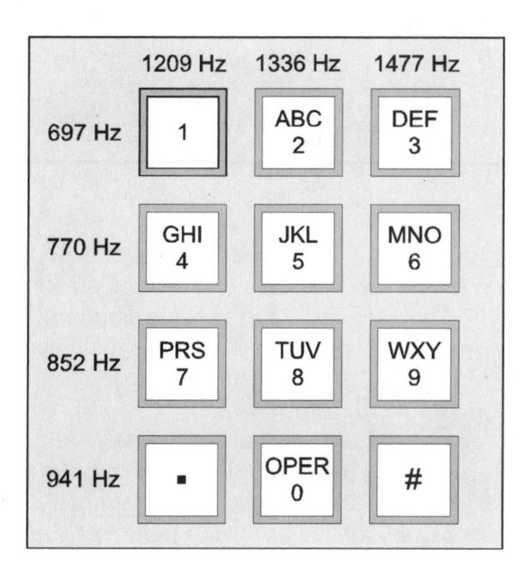

Fig. T11 *DTMF designations on the modern telephone keypad.*

Summary. Bell was not the only person involved in the invention of the telephone. Bell's rival was Elisha Grey, who had filed his intention to invent the telephone with the U.S. Patent Office (now the Patent and Trademark Office). Bell gets credit for the invention, however, because he filed a claim for a workable device only two hours earlier. With support from Western Union, Gray disputed Bell's claim in a lawsuit, but Bell prevailed. Gray made a fortune with other inventions and helped found the Western Electric Company in Chicago. Bell's patent also was contested by Thomas Edison, who claimed he had invented the telephone years earlier. Edi-

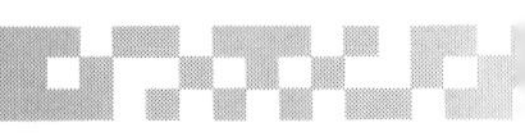

son subsequently made a great improvement in the telephone receiver, which was still in use decades later. Alexander Graham Bell helped form the Bell Telephone Company in 1877, but did not take an active part in the emerging new industry.

This was left to Theodore N. Vail, the first president of AT&T, who believed that the chaotic telephone industry, by the nature of its technology and the need for interoperability, would operate most efficiently and economically as a monopoly providing universal service.

The U.S. government accepted this principle, first informally and then legislatively. As political philosophy evolved, however, federal administrations investigated the telephone monopoly in light of general antitrust law and alleged company abuses. One notable result was an antitrust suit filed in 1949, which led in 1956 to a consent decree signed by AT&T and Department of Justice, whereby AT&T agreed to restrict its activities to the regulated business of the national telephone system and government work.

The changes in telecommunications during these years eventually led to an antitrust suit by the U.S. government against AT&T. The suit began in 1974 and was finally settled in January, 1982, when AT&T agreed to divest itself of the wholly owned Bell operating companies that provided local exchange service. In return, the U.S. Department of Justice agreed to lift the constraints of the 1956 decree. Divestiture took place on January 1, 1984, marking the end of the Bell System. In its place was a new AT&T and seven regional telephone holding companies.

See also

CENTRAL OFFICE SWITCHES
KEY TELEPHONE SYSTEMS
PRIVATE BRANCH EXCHANGE

Telephone fraud

Like many businesses, the telecommunications industry is a target of fraud and abuse. This is especially true of mobile phone services. The underlying wireless technologies make it fairly easy for criminals to intercept signals and obtain electronic identification numbers that can be used to "clone" phones and have their usage billed to legitimate subscribers. In the United States alone, the annual losses from this kind of fraud amount to about US$600 million.

Cellular fraud is an extension of "phone phreaking," the term given to a method of payphone fraud that originated in the 1960s, which employs an electronic box held over the speaker. When a user is asked to insert money, the electronic box plays a sequence of tones that fool the billing computer into thinking that money has been inserted. Since then, criminals and hackers have applied their knowledge and expertise to cellular phones.

Fraud techniques. There are many ways to implement mobile telephone fraud. Some methods of fraud, such as cloning and rechipping, require technical expertise. The method of least risk entails opening a subscriber account and then running up the phone bill as high as possible until the service provider catches on to the scam and terminates service.

Cloning. Dishonest dealers clone large numbers of handsets and sell them to targeted classes of users. Immigrants, both legal and illegal, have long been a favorite target of cloners because they know such people want to stay in touch with friends and relatives back home but usually have no money to call very frequently. Migrant workers, youth gangs, and those engaged in the drug trade represent other potentially lucrative markets for cloned phones. Sometimes the phones are not sold at all. Instead, illegal *phone cells* are set up in apartments or abandoned buildings; people off the street pay a per-minute charge just to use a clone.

For analog systems, each mobile phone carries with it *handshake* information comprising the Electronic Serial Number (ESN) and Mobile Identification Number (MIN). These numbers can fall into the wrong hands in a variety of ways. For instance, lax internal security can allow disgruntled employees to obtain this information from internal computer systems or customer files and sell it to cloners. When a vehicle is left unattended at a public parking lot or repair facility, or driven off by a valet, dishonest individuals can copy the ESN from the car phone and locate the MIN details from phone documents in the glove compartment. Special counterfeiting software is then used to recode the chips of other handsets with the stolen ESN.

Rechipping. This is a technique frequently used to recycle stolen mobile phones that have been reported to network operators and barred from further services. The software gives a stolen mobile phone a new electronic identity that allows it to be reconnected to the network until it is discovered and barred again.

Scanning. A criminal does not have to steal a mobile phone to clone the ESN. With the latest scanning equipment and a laptop computer, a criminal can tune in on a call and steal the ESN from the air. Typically the method used is to sit along a busy highway, intersection, or rest area and wait for unsuspecting business people to make calls from their vehicles. The radio scanner then can pick up a phone's ESN, which is broadcast at the beginning of each call. In only a few hours of scanning, a criminal can walk away with hundreds of legitimate ESNs.

While mobile phones used on analog networks are easy to intercept, phones that are used on such digital networks as the North American Personal Communication Services (PCS), the European Global System for Mobile (GSM) communications, and the pan-Asian Personal Handyphone System (PHS) are not. This is because the signals can be encrypted, making them much more difficult to intercept without expensive equipment and a higher degree of technical expertise.

GSM signal encryption is done via a programmable smart card, the Subscriber Identification Module (SIM), which slips into a slot built into the handset. Each customer has a personal smart card holding personal details (short codes, frequently called numbers, etc.), as well as an international mobile subscriber identity (IMSI)—equivalent to MIN for analog systems—and authentication key on the microprocessor. Plugging the smart card into another phone will allow that phone to be used as if it were the customer's own.

False accounts. By far, the easiest and most risk-free way to engage in mobile phone fraud is to open an account posing as a legitimate subscriber. In this case, false identification, Social Security cards, and addresses are used to open these accounts. Because of competitive pressures, carriers are signing up new customers quickly and are not able to screen all this information to weed out unqualified individuals. New subscribers often are allowed to begin using the service within hours of opening an account. As the carrier closes in on bad accounts when false information does not check out, criminals are already a step ahead, opening new accounts.

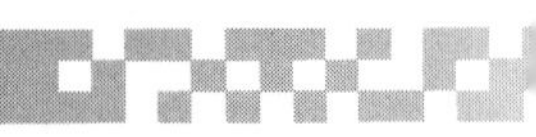

Fraud control technologies. A number of technologies have been developed to discourage attempts at mobile phone fraud; some of them are described in the paragraphs that follow.

Personal identification numbers. Like that of an ATM bank card, a mobile phone PIN is a private number that only the subscriber knows. The PIN is required in order to place a call. Even if the phone's signal is captured, thieves would not be able to use it without the PIN code.

Calling pattern analysis. When a subscriber's phone deviates from its normal call activity, it trips an alarm at the service provider's fraud management system. There it is put into queue, where a fraud analyst ascertains whether the customer has been victimized and then remedies the situation. Newer systems establish and maintain the calling profiles of subscribers. Out-of-profile call requests are identified in real time, whereupon a PIN can be requested to ensure that only legitimate calls are allowed to be placed.

Authentication. This technique uses advanced encryption technology that involves the exchange between the phone and the switch of a secret code based on an intricate algorithm. The cellular network and the authentication-ready phones operating on it carry matching information. When a user initiates a call, the network challenges the phone to verify itself by performing a mathematical equation only that specific phone can solve. An authenticatable phone will match the challenge, confirming that it and the corresponding phone number are being used by the legitimate customer. If it does not match, the network determines that the phone number is being used illegally, and service to that phone is terminated. All this takes place in a fraction of a second.

Radio frequency fingerprinting. Through digital analysis technology that recognizes the unique characteristics of radio signals emitted by cellular phones, a "fingerprint" can be made that can distinguish individual phones within a fraction of a second after a call is made. Once the fraudulent call is detected, it is immediately disconnected. The technology works so well that it has cut down on fraudulent calls by as much as 85 percent in certain high-crime markets, including Los Angeles and New York.

Voice verification. These systems are based on the uniqueness of each person's voice and the reliability of the technology that can distinguish one voice from another by comparing a digitized sample of a person's voice with a stored "voiceprint." The front-end analysis recognizes and normalizes conditions such as background noise, channel differences, and microphone variances. The voice verification system can reside on a public or private network as an intelligent peripheral, or it can be placed as an adjunct serving a PBX or automatic call distributor (ACD). In a cellular environment, the system can be an adjunct to a mobile switching center (MSC).

Summary. While there has been much progress in cracking down on mobile phone fraud in the United States, other countries are experiencing an increase in this kind of criminal activity. According to some experts, the international arena looms as the next frontier for mobile phone fraud, particularly in locations where U.S.-based multinationals are setting up shop and buying this type of service. Foreign governments simply have not been aggressive in finding and prosecuting phone criminals, they note. In some countries, such as China, there are even operations dedicated to building cell phones that get illegally programmed and then are sold on the black market.

Scanning the airwaves for cell phone identification numbers and programming them into clones will be made more difficult with emerging authentication services and digital networks that support encryption. GSM will change the nature of fraud in the future, since authentication facilities are built into the SIM card, forcing many criminals to turn their attention to subscription fraud, which is still time-consuming to track down, even for the largest service providers.

See also

NETWORK SECURITY

Telephone service subscribers

According to a survey conducted by the U.S. Census Bureau in March, 1997, nearly 93.9 percent of all American households have telephone service. The survey breaks down subscribership levels by state, income level, race, age, household size, and employment status. Among the findings:

- 77.1 percent of households with annual income below $5000 and 99 percent of households with annual income above $75,000 had telephone service.
- 86.4 percent of households in New Mexico (lowest) had telephone service, compared to 97.5 percent in Missouri (highest).
- 95 percent of households headed by whites had phones, while 87.3 percent of African American households and 86.3 percent of Hispanic households had service.
- 85 percent of households headed by someone under 25 had service, compared to 96.6 percent of households headed by someone 60 to 64 years old.
- 89 percent of large households (six or more people) had service, while 95 percent of those with 2 to 3 people had service.
- 95.5 percent of employed people had phones, while only 88.2 percent of unemployed people subscribed to phone service.

See also

UNIVERSAL SERVICE

Telex

Telex is the international teleprinter exchange service through which a subscriber can be connected to another telex subscriber, either locally or internationally, for the transmission of text messages. During telex transmission, a copy of the message is produced simultaneously on both the sending and receiving teleprinters. Telex service also is known as teletypewriter exchange or *TWX* (sometimes pronounced "twix").

The transmission standard, character code, and terminal requirement for telex are internationally standardized. This allows a teleprinter of any origin or design to directly exchange messages with any other teleprinters connected to the telex net-

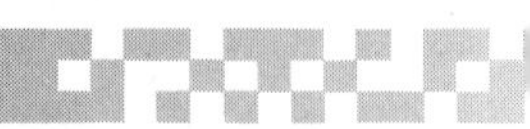

work. The transmission speed is globally standardized at 400 characters per minute (50 baud).

Telex service operates in three modes: conversational, store-and-forward, and operator-assisted. With conversational telex, both sender and receiver can carry on an interactive written telex conversation. With store-and-forward telex, complete text messages are sent out over the telex network to one or more addressees. Operator-assisted telex transmission is used when there is difficulty reaching the destination number.

There are several ways to access the telex network: dial-up, dedicated line, or packet network. Dial-up is used by subscribers who do not have enough message traffic to justify the cost of a dedicated connection. Dedicated lines are used by businesses that use telex frequently, often on a daily basis. Whether dial-up or dedicated, the connection is made to a carrier's switch that provides access to the telex network. Companies that already have packet networks also can access the telex network.

Telex service entails a minimum charge of 1 minute and subsequent increments of 1 minute. Telex subscribers receive a copy of the International Telecommunications Rate Table, a guidebook that contains a complete listing of country codes and pricing.

Applications. There are several applications of telex. For example, a company can send out updated product pricing to hundreds of international locations at once. High-volume store-and-forward telex lets the subscriber store all telex messages on the network so they can be sent them to multiple locations simultaneously.

The telex network is linked to many types of databases, allowing subscribers to receive news, commodity prices, and exchange rates. The network also can be used to make travel reservations.

Depending on the service provider selected, several delivery options are available. Telex messages can be sent to a facsimile machine or e-mail account, or even delivered in hardcopy form like a telegram.

Features. Telex subscribers have access to a full range of convenient, time-saving features, including single-digit dialing, automatic redialing of busy numbers, and multiple calls in a single connection. Other features include the following:

- The International Telex Exchange will automatically print out the chargeable duration if it encounters the string HHHH at the end of the document.
- Mistakes in typing the destination telex number can be corrected immediately by keying the string EEE followed by the correct number.
- When a sender encounters difficulties making a connection (busy line, power failure, etc.) the message can be addressed to the Exchange, to be forwarded later to the intended number. An additional charge is applied to this facility.
- International Telex Exchange can make the transmission of a single message to multiple destination numbers, thereby saving time for the sender. An additional charge is applied to this facility.
- Unique telex answer-backs ensure clear identification of sender and receiver before transmission begins.

Summary. Despite the availability of more advanced, higher-speed telecommunications services, telex remains a valuable source of written communication all over the world. There are approximately 1.8 million telex users in more than 200 countries and locations worldwide.

See also

ELECTRONIC MAIL
FACSIMILE

Time Division Multiple Access

In older FM radio systems, the radio spectrum is divided into 30-kHz channels, with one user assigned to each channel. Frequency Division Multiple Access (FDMA) improves bandwidth utilization and overall system capacity by dividing the 30-kHz channel into three narrower channels of 10 kHz each. Newer digital technologies, such as Time Division Multiple Access (TDMA), allow even more users to be supported by the same channel.

TDMA systems have been providing commercial digital cellular service since mid-1992. TDMA increases the capacity of existing analog cellular systems. (These are called AMPS, for Advanced Mobile Phone Service. Digital cellular is also known as D-AMPS.)

With TDMA, each 30-kHz channel is divided into 3 time slots. Users are assigned their own time slot into which pieces of voice or data are inserted for transmission via synchronized timed bursts. The bursts are reassembled at the receiving end, and appear to provide continuous sound because the process is very fast. The digital bit streams that correspond to the three distinct voice conversations are encoded, interleaved, and transmitted using a digital modulation scheme called differential quadrature phase-shift keying (DQPSK). Together, these manipulations reduce the effects of most common radio transmission impairments.

If one side of the conversation is silent, however, the time slot goes unused. Recent enhancements to TDMA use dynamic time slot allocation to avoid the waste of time slots when one side of the conversation is silent. This technique almost doubles the spectral efficiency of TDMA to about 10:1 over analog. TDMA systems also now use half-rate voice coders, which allow up to 6 users to share one 30-kHz channel. Hierarchical cell topologies further enhance network capacity.

TDMA is a Telecommunications Industry Association (TIA) standard (IS-54) for digital cellular in the United States. IS-136 is the next generation of TDMA, which makes use of the control channel to provide advanced call features and messaging services.

Framing. In a TDMA system, the digitized voice conversations are separated in time, with the bit stream organized into frames, typically on the order of several milliseconds. A 6-ms frame, for example, is divided into six 1-ms time slots, with each time slot assigned to a specific user. Each time slot consists of a header and a packet of user data for the call assigned to it (Fig. T13). The header generally contains synchronization and addressing information for the user data.

If the data in the header becomes corrupted as a result of a transmission problem (signal fade, for example), the entire slot can be wasted, in which case no more

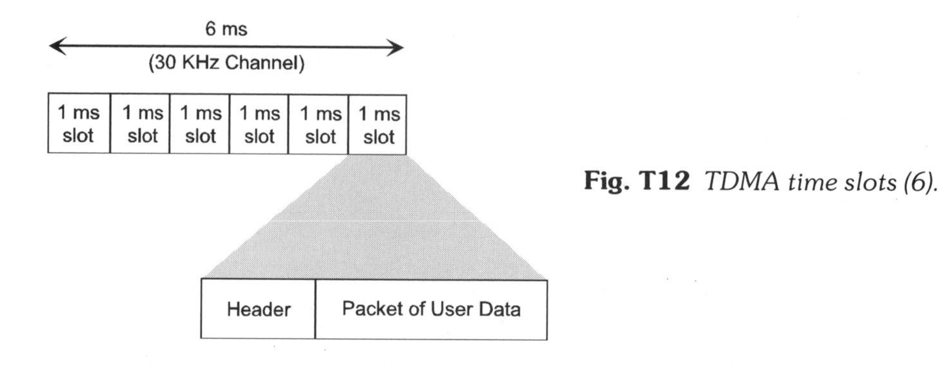

Fig. T12 *TDMA time slots (6).*

data will be transmitted for that call until the next frame. The loss of an entire data packet is called *frame erasure.* If the transmission problem is prolonged (i.e., deep fade), several frames in sequence can be lost, causing clipped speech or forcing the retransmission of data. Most transmission problems, however, will not be severe enough to cause frame erasure. Instead, only a few bits in the header and user data will become corrupted, a condition referred to as *single-bit errors.*

Network functions. The IS-54 standard defines the TDMA radio interface between the mobile station and the cell site radio. The radio downlink (from the cell site to the mobile phone) and the radio uplink (from the mobile phone to the cell site) are functionally similar. The TDMA cell site radio is responsible for speech coding, channel coding, signaling, modulation/demodulation, channel equalization, signal strength measurement, and communication with the cell site controller.

The TDMA system's speech encoder uses a linear predictive coding technique and transfers pulse-code-modulated (PCM) speech at 64 kbps to and from the network. The channel coder performs channel encoding and decoding, error correction, and bit interleaving and de-interleaving. It processes speech and signaling information, builds the time slots for the channels, and communicates with the main controller, modulator, and demodulator.

The modulator receives the coded information and signaling bits for each time slot from the channel coders. It performs DQPSK modulation to produce the necessary digital components of the transmitter waveform. These waveform samples are converted from digital to analog signals. The analog signals are then sent to the transceiver, which transmits and receives digitally modulated RF control and information signals to and from the cellular phones.

The modulator/equalizer receives signals from the transceiver. It performs filtering, automatic gain control, received signal strength estimation, adaptive equalization, and demodulation. The demodulated data for each of the time slots is then sent to the channel coder for decoding.

New voice coding technology is becoming available that produces near-landline speech quality in IS-136 TDMA wireless networks. One technology introduced by Lucent Technologies uses an algebraic code excited linear predictive (ACELP) algorithm, which is an enhanced, internationally established code for dividing waves of sound into binary bits of data. The ACELP coders can be integrated easily into existing wireless base station radios as well as new telephones. As ACELP-capable phones become widely available, they will allow users to take advantage of the improved digital clarity over both North American frequency bands, the 850 MHz cellular and the 1.9 GHz Personal Communications Services (PCS).

Call hand-off. Received signal strength estimation is used in TDMA's mobile-assisted call hand-off process. The traditional hand-off process involves the cell site currently serving the call, the switch, and the neighboring cells that can potentially continue the call. The neighboring cells measure the signal strength of the potential call to be handed off and report that data to the serving cell, which uses it to determine which neighboring cell can best handle the call. TDMA systems, on the other hand, reduce the time needed and the overhead required to complete the hand-off by assigning some of this signal strength data gathering to the cellular phone, relieving the neighboring cells of this task and reducing the hand-off interval.

Digital Control Channel. The capabilities of TDMA are undergoing continual enhancement. For example, the Digital Control Channel (DCCH) described in IS-136 gives TDMA new features that can be added to the existing platform through software updates. Among these new features are:

- *Over-the-air activation* allows new subscribers to activate cellular or PCS service with just a phone call to the service provider's customer service center.
- *Messaging* allows users to receive visual messages up to the maximum length allowed by industry standards (200 alphanumeric characters). Transmission of messages permits a mobile unit to function as a pager.
- *Sleep mode* extends the battery life of mobile phones and allows subscribers to leave their portables powered on throughout the day, ready to receive calls.
- *Fraud prevention* is achieved by the cellular system's ability to identify legitimate mobile phones and block access to invalid ones.

A number of other advanced features also can be supported over the DCCH, such as voice encryption and secure data transmission, caller ID, and voice mail notification.

Summary. Many vendors and service providers have committed to supporting either TDMA or CDMA. Those who have committed to CDMA claim they did so because they consider TDMA to be too limited in meeting the requirements of the next generation of cellular systems. Although TDMA gives service providers a significant increase in capacity over AMPS, the standard was written to fit into the existing AMPS channel structure for easy migration, and it did not appear to meet the requirements for better voice quality and new service requirements.

Proponents of TDMA note that the inherent compatibility between AMPS and TDMA, coupled with the deployment of dual-mode/dual-band terminals, offers full mobility to subscribers, with seamless hand-off between PCS and cellular networks. They also note that the technology is field-proven in many of the world's largest wireless networks and is providing reliable, high-quality service, without additional development or redesign. These TDMA systems can be easily and cost-effectively

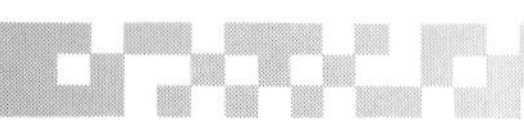

integrated with existing wireless and landline systems, and the technology itself can evolve economically to meet new quality and service requirements.

See also
CODE DIVISION MULTIPLE ACCESS
FREQUENCY DIVISION MULTIPLE ACCESS

Token-Ring

The token ring architecture for local area networks was introduced commercially in 1985 by IBM as their 4 Mbps Token-Ring[1] product. It came in response to the commercial availability of Ethernet, which was developed jointly by Digital Equipment, Intel, and Xerox. When Ethernet was introduced IBM did not endorse it, mainly because its equipment would not work in that environment. Later, in 1989, a 16 Mbps Token-Ring became available.

Token-Ring LANs operate at rates of 4 Mbps or 16 Mbps. The ring is essentially a closed loop, although various wiring configurations that employ a multistation access unit (MAU)[2] and patch panel may cause it to resemble a star topology (Fig. T13). In addition, today's intelligent wiring hubs and Token-Ring switches can be used to create dedicated pipes between rings and provide switched connectivity between users on different rings.

The cable distance of a 4-Mbps Token-Ring installation is limited to 1600 feet between stations, while the cable distance of the 16-Mbps version is 800 feet between. Because each node acts as a repeater in that data packets and the token are regenerated at their original signal strength, Token-Ring networks are not as limited by distance as are bus-type networks. Like its nearest rival, Ethernet, Token-Ring normally uses twisted-pair wiring, shielded or unshielded.

Advantages of Token-Ring. The ring topology offers several advantages:

- Because access to the network is not determined by a contention scheme, a higher throughput rate is possible in heavily loaded situations, limited only by the slowest element, i.e., sender, receiver, or link speed.
- With all messages following the same path, there are no routing problems to contend with. Logical addressing may be accommodated to permit message broadcasting to selected nodes.
- Adding terminals is easily accomplished by unplugging one connector, inserting the new node, and plugging it back into the network. Other nodes are updated with the new address automatically.
- Control is simple, requiring little in the way of additional hardware or software to implement.
- The cost of network expansion is proportional to the number of nodes.

1 Though the term *token ring* describes a particular topology and architecture generically (FDDI also uses it), the IBM-originated implementation is so pervasive that its trade name, Token-Ring, has come to be nearly synonymous with token-passing on twisted-pair rings.

2 A MAU is nonintelligent concentrator that can be used as the basis for implementing Token-Ring LANs.

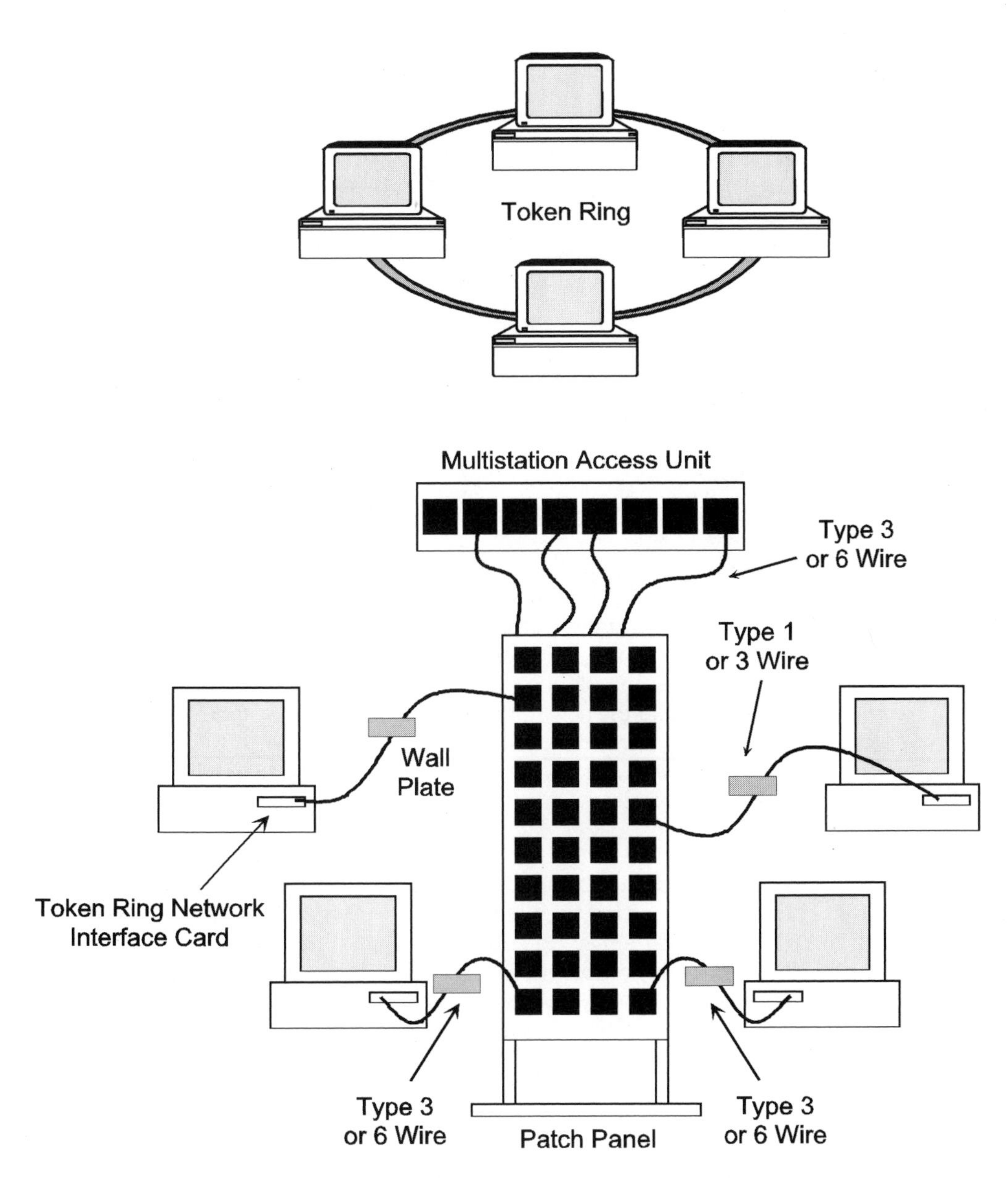

Fig. T13 *Token ring topologies: closed ring (top) and star-wired (bottom).*

Another advantage is that the network can be configured to give high-priority traffic precedence over-lower priority traffic. A station is allowed to transmit only if it has traffic equal to or higher in priority than the priority indicator embedded in the token.

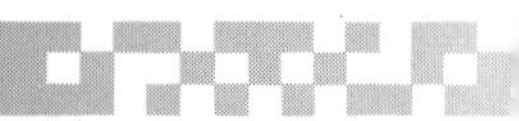

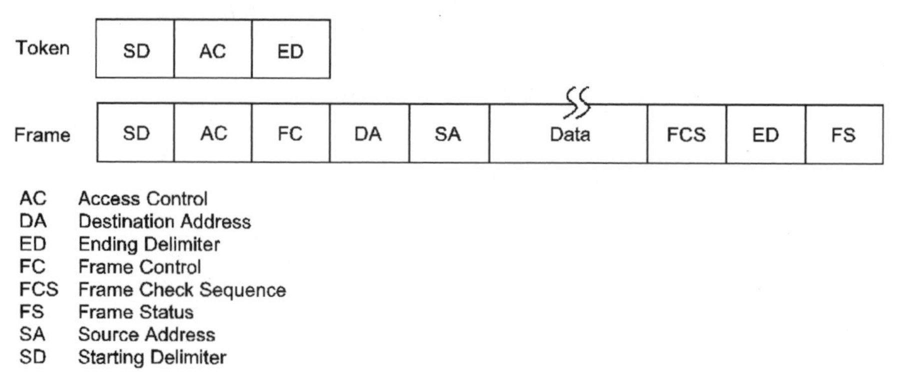

Fig. T14 *Format of IEEE 802.5 token and frame.*

The token ring in its pure, generic configuration is not without liabilities, however. Failed nodes and links can break the ring, preventing all the other terminals from using the network. At extra cost, a dual-ring configuration with redundant hardware and bypass circuitry is effective in isolating faulty nodes from the rest of the network, thereby increasing reliability.

Through the use of bypass circuitry, physically adding or deleting terminals to Token-Ring networks is accomplished without breaking the ring. Specific procedures must be used to ensure that the new station is recognized by the others and is granted a proportionate share of network time. The process for obtaining this identity is referred to as *Neighbor Notification.* This situation is handled quite efficiently, since each station becomes acquainted with the address of its predecessor and successor on the network upon initialization (power-up) or at periodic intervals thereafter.

Frame format. The frame size used on 4-Mbps Token-Ring is 4048 bytes; the frame size used at 16 Mbps is 16,192 bytes. The IEEE 802.5 standard defines two data formats, tokens and frames (Fig. T14). The token, three octets in length, is the means by which the right to access the medium is passed from one station to another. The frame format of IEEE token ring differs only slightly from that of Ethernet. The following fields are specified for IEEE 802.5 token ring frames:

- *Start Delimiter* (SD) indicates the start of the frame.
- *Access Control* (AC) contains information about the priority of the frame and a need to reserve future tokens, which other stations will grant if they have a lower priority.
- *Frame Control* (FC) defines the type of frame, either Media Access Control (MAC) information, or information for an end station. If the frame is a MAC frame, all stations on the ring read the information. If the frame contains information (i.e., user data), it is only read by the destination station.
- *Destination Address* (DA) contains the address of the station that is to receive the frame. The frame can be addressed to all stations on the ring.

- *Source Address* (SA) contains the address of the station that sent the frame.
- *Data* contains the data "payload." If the frame is a MAC frame, this field may contain additional control information.
- *Frame Check Sequence* (FCS) contains error checking information to ensure the integrity of the frame to the recipient.
- *End Delimiter* (ED) indicates the end of the frame.
- *Frame Status* (FS) provides indications of whether one or more stations on the ring recognized the frame, whether the frame was copied, or whether the destination station is not available.

Operation. A token is circulated around the ring, giving each station in sequence a chance to put information on the network. The station seizes the token, replacing it with an information frame. Only the addressee can claim the message. At the completion of the information transfer the station reinserts the token on the ring. A token-holding timer controls the maximum amount of time a station can occupy the network before passing the token to the next station.

A variation of this token-passing scheme allows devices to send data only during specified time intervals. The ability to determine the time interval between messages is a major advantage over the contention-based access method used by Ethernet. This time-slot approach can support voice transmission and video conferencing, since latency is controllable.

To protect the ring from potential disaster, one terminal is typically designated as the control station. This terminal supervises network operations and does important housecleaning chores, such as reinserting lost tokens, taking extra tokens off the network, and disposing of "lost" packets. To guard against the failure of the control station, every station is equipped with control circuitry so that the first station detecting the failure of the control station assumes responsibility for network supervision.

Dedicated token ring. A relatively new technology is dedicated token ring (DTR), which is also known as full-duplex token ring. DTR lets devices directly connected to a token ring switch transmit and receive data simultaneously at 16 Mbps, effectively providing each station with 32 Mbps of throughput.

Under the IEEE 802.5r standard for DTR, which defines the requirements for new end stations and concentrators that operate in full-duplex mode, all new devices will coexist with existing Token-Ring equipment and will adhere to the token-passing access protocol. The DTR concentrator consists of C-Ports and a data transfer unit (DTU). The C-Ports provide the basic connectivity from the device to the ring stations, traditional concentrators, or other DTR concentrators. The DTU is the switching fabric that connects the C-Ports within a DTR concentrator. In addition, DTR concentrators can be linked to each other over a LAN or WAN via data transfer services such as ATM.

Not only does DTR extend existing investments in Token-Ring, it allows expenditures for ATM to be held off for a year or two, during which time the prices for ATM network equipment and interfaces will drop to more affordable levels. Opting for DTR at 32 Mbps also obviates the need for the interim step to 25 Mbps ATM.

Summary. The token-passing ring architecture is a stable technology with a proven capacity for handling today's applications. At the same time, network managers can protect their current investments by understanding application performance and the capacity of the network, and tuning it accordingly. The new DTR standard prolongs the useful life of Token-Ring networks, while meeting the increased bandwidth requirements of emerging applications such as document imaging, desktop video conferencing, and multimedia.

See also
ARCNET
ETHERNET
STARLAN

Transmission facilities

A variety of transmission facilities are available for both voice and data communications, including twisted-pair wiring, coaxial cable, laser, microwave, satellite, optical fiber, and T-carrier systems.

Twisted-pair wiring. Twisted-pair wiring is the most common transmission facility; it is currently installed in most office buildings and residences. Twisted-pair wiring consists of one or more pairs of copper wires. In the local loop, hundreds of insulated wires are bundled into larger cables, with each pair color-coded for easy identification. Bundling facilitates installation and reduces costs. Special sheathing offers protection from natural elements. Most telephone lines between central offices and local subscribers consist of this type of cabling, which is mounted on poles or laid underground.

Twisted-pair also has become the most popular transmission medium for local area networking, where it has been made more reliable by the adoption of the 10Base-T and 100Base-T networking standards. Under these standards, the LAN is arranged in a star topology with an intelligent hub at the center, making it easier to pinpoint faults and other problems. Some hubs support both 10Base-T and 100Base-T, allowing users to protect their investments in 10Base-T infrastructure and providing them with the bandwidth they need to support large applications. Under consideration as a standard is a gigabit-per-second version of Ethernet.

Coaxial cable. Until the 10Base-T standard, coaxial cable was the universal media for local area networks. This type of cable, available in thick (10Base5) and thin (10Base2) versions, contains a conductive cylinder with insulation around a wire in the center. Coaxial cable is typically shielded to reduce interference from external sources of electromagnetic interference (EMI). Coaxial cable can transmit at a much higher frequency than a wire pair, allowing more data to be transmitted in a given period. By providing a "wider" channel, coaxial cable allows multiple LAN users to share the same medium for communicating with host computers, servers, front-end processors (FEPs), peripheral devices, and personal computers. In recent years, standard "thick" coax has fallen out of favor, replaced with the more flexible thin coax. In new installations, less expensive alternatives are used, such as optical fiber for the network backbone and twisted-pair wiring for connecting devices to the backbone.

Coaxial cable is used mostly in CATV networks to bring programming to television sets. A recent innovation allows PCS to be delivered over these cable networks as well. This allows PCS operators to save on the cost and delay inherent in tower installation. It also allows them to extend PCS service very economically to remote locations. There is even a new type of cable to support PCS; called "radiating" cable, it acts as both a transmission medium and an antenna. This type of cable allows PCS to reach into buildings and tunnels where wireless signals normally would be too weak to penetrate.

Radio frequency transmission. Traditionally, radio frequency (RF) transmission has been associated with analog technology used by specialized mobile radio (SMR) carriers to provide communications services to vehicle-mounted and handheld portable telephones and other two-way radio units. In most cases the SMR carrier uses a single-site, high-powered transmitter configuration. Each channel has its own radio frequency and supports one caller at a time within a given service area for the duration of the call.

Many SMR carriers have been upgrading their networks from analog to digital. The advantage of digital is that can be used to integrate a variety of services, allowing the SMR carriers to expand their offerings beyond voice to include text messaging, paging, and facsimile. These digital networks are less crowded and more reliable than traditional analog networks. Channel capacity can be increased by adding cells or subdividing existing cells and through frequency reuse, similar to the way traditional cellular phone networks operate.

Cellular transmission. Cellular transmission is another form of RF communication. It can be either analog or digital, depending on the underlying technology. Although used mostly for voice communication, cellular networks can support data at up to 19.2 kbps. Cellular networks are composed of cells, each with its own low-power transceiver. As the mobile user moves from cell to cell throughout a metropolitan area, the network's central computer arranges a hand-off of the signal to another transceiver in the new cell. The transceiver in the former cell is free to handle another call. When the capacity of a cell reaches its limit, the carrier can keep subdividing cells by adding more transceivers to handle more calls.

Microwave transmission. Microwave systems use the high end of the radio frequency spectrum and require special equipment for transmission and reception. This method of transmission is advantageous in that it does not require stringing wire over long distances, it is immune from the impairments that affect copper wire, and it provides greater bandwidth capacity. However, microwave transmission requires a line-of-sight path between the transmitter and receiver, as well as suitable locations for tower installation. Under ideal conditions, the distance between these stations (or towers) can span up to 30 miles. Each station receives the signals, amplifies them, then passes them to the next relay station along the route.

Microwave technology is used most often for long-haul communications by carriers serving rural areas, and by large companies that wish to bypass the local exchange for intracompany communications. Increasingly, it also is being used in wireless bridges to connect LANs in different buildings without having to string wire over long distances.

Satellite transmission. Satellite transmission is a variation of microwave, but the systems are in geosynchronous orbit above the earth. The coverage area can be as small as a few hundred miles or as large as several thousand miles. To avoid interference between simultaneous transmissions at the same frequency, satellites with adjacent orbits are distanced at least two degrees apart. A satellite network consists of orbiting transponders, ground stations, and gateway switches that provide connections to public and private terrestrial networks. The ground stations are equipped with dish antennas (e.g., very small aperture terminals, or VSATs), satellite control systems, and network management software. Due to orbital decay, each satellite has a 7- to 10-year useful life.

A relatively recent innovation in satellite technology is the low earth orbit (LEO) satellite. Circling the earth only a few hundred miles up, LEO satellites have to keep moving to avoid falling into the atmosphere, so a given satellite passes over a stationary caller rather quickly. A call therefore is handed off from one satellite to the next to keep the session alive. LEO satellites fly low enough that dramatic improvements can be made on the earthbound transmitters: They can be low-powered, handheld devices such as PCS phones. One such network, called Iridium, is being built by a consortium of U.S. and international companies headed by Motorola. The Iridium systems will relay voice, data, facsimile, and paging signals anywhere in the world. Service in the United States is scheduled to begin in 1998. Several competitive LEO networks are being established, including Teledesic, a joint effort by Microsoft and AT&T. Teledesic is being built by Boeing and service is expected to begin in 2002.

Optical fiber. High-capacity fiber-optic transmission systems currently are a popular solution for replacing or supplementing overloaded copper-wire trunks. Among the advantages offered by fiber-optic communications systems over electrical or microwave radio systems, are the following:

- Higher capacity (bandwidth) than copper-based facilities.
- The ability to increase capacity by upgrading the light sources and receivers without changing the cable itself.
- Immunity to electromechanical interference (EMI) and radio frequency interference (RFI).
- Protection from wiretapping, which is extremely difficult over a fiber-optic system.

Laser transmission. Laser-optic transmission systems operate in the near-infrared region of the light spectrum. Utilizing coherent laser light, these wireless line-of-sight links are used in campus environments and urban areas where the installation of cable is impractical and the performance of leased lines is too slow. Unlike microwave transmission, laser transmission does not require an FCC license, and data traveling by laser beam cannot be intercepted.

Laser transmission is not a carrier-provided service, but a method for private network users to bypass the local exchange carrier for certain applications, such as point-to-point LAN interconnection. The lasers at each location are aligned with a simple bar graph and tone-lock procedure. Monitors can be attached to the laser units to provide operational status, such as signal strength, and to implement local and remote diagnostics.

A limitation of laser systems is that transmission can be affected by atmospheric conditions that produce such effects as absorption, scattering, and shimmer. All three can reduce the amount of light energy that is picked up by the receiver and corrupt the data being sent.

T-carrier facilities. T-carrier is not a unique physical entity but rather a technique of using various transmission media. Using multiplexing techniques, a carrier system can accommodate many channels on the same physical line, whether it be wire, coaxial cable, or microwave radio transmission system. This is accomplished by allocating each channel to a different time segment, as in the case of time division multiplexing (TDM).

T-carrier supports the widely available service known as T1, which is a two-way connection offering 24 voice channels operating at Digital Signal Level 1 (DS1), or 1.544 Mbps. In Europe, the United Kingdom, and other countries that adhere to standards issued by the ITU Telecommunications Standardization Sector (ITU-TSS), formerly the CCITT, the analogue of T1 is E1, a 2.048 Mbps service.

T-carrier also is used to support derivatives of T1 service, such as fractional T1 (FT1), which offers bandwidth in 64 kbps increments, and T3, an extension of T1 that is equivalent to 28 T1 lines, operating at the DS3 rate of 44.736 Mbps.

Summary. There are several types of transmission facilities available, each suited for a different purpose. Twisted-pair or coaxial cable is used in the customer premises for local area networking, for program distribution by CATV companies, and in the telco local loop for dial-up communications. Radio frequency (including cellular) is used for wireless data and voice communication serving mobile users. Microwave excels for high-capacity, limited distance communication. Satellite is a viable choice for national or international communications. Optical fiber offers high capacity for terrestrial communications. Laser transmission lends itself to private communications links. Each type of facility can interconnect with the others to form a vast, global communications network.

See also

FIBER OPTICS TECHNOLOGY
HYBRID FIBER/COAX
MICROWAVE COMMUNICATIONS
SATELLITE COMMUNICATIONS
SPECIALIZED MOBILE RADIO
T-CARRIER FACILITIES

Unbundled access

As part of the new rules to ensure competition in the provision of local telecommunications services, the established local exchange carrier (LEC) must provide, to any requesting telecommunications carrier, access to network elements on an unbundled basis at any technically feasible point at rates, terms, and conditions that are just, reasonable, and nondiscriminatory.

Network elements encompass everything, from basic service to call forwarding and caller ID, to loop and/or signalling capabilities. Unbundled access is important because competitive local service providers should not be forced to buy services they really do not need in order to get the services they do need.

According to the competitive guidelines established in the Telecommunications Act of 1996, which are enforced by the FCC, such functions should be available on an à la carte basis and not as part of a package deal, which inflates the operating costs of competitors and, ultimately, the prices customers must pay for service.

Universal messaging

Staying on top of incoming messages from a variety of sources has become a headache for most people. While the ways in which we can communicate has grown and diversified, so to have the number of devices we must use to receive all the messages people are sending us: desk phones, cellular phones, fax machines, alphanumeric pagers, and e-mail systems, to name a few.

Universal messaging brings order to this communications chaos by consolidating the reception, notification, presentation, and management of what have until now been stand-alone messaging systems. This capability is provided through message servers that are connected to carrier switches. These servers can have a distributed architecture, allowing carriers to add universal messaging services incrementally throughout their networks as market demand warrants, or to have a hub architecture with enough capacity to serve a large number of subscribers. The benefits of message service consolidation can be extended to residential subscribers, as well as small and large businesses.

The universal inbox. The service deposits each subscriber's e-mail, fax, and voice messages into a universal messaging inbox, so the subscriber can find all messages in a single place and through a single interface, such as a telephone handset or Web browser. The system notifies subscribers whenever a new message

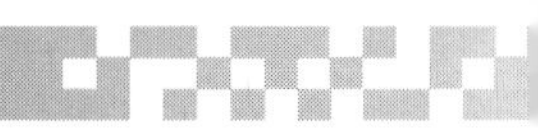

arrives and notifies them via any one of the notification methods the carrier chooses to support, including e-mail notification message, message waiting indicator light, stutter dial tone, pager, and out-dial.

Notification options. While it has long been common for voice messaging subscribers to hear a stutter dial tone upon the arrival of a new voice message, it has not been as common to hear a stutter dial tone when an e-mail or fax message has arrived. With the advent of universal messaging, however, this changes, and the system notifies the subscriber of an incoming message whenever a message arrives, regardless of whether it is a voice, fax, or e-mail message. A stutter dial tone is only one of several potential notification options. The network service provider may choose which notification methods it offers to subscribers.

If the specified notification method is e-mail, for example, the universal messaging system sends an e-mail notification message to a subscriber-specified e-mail address. In the body of the notification message, the system embeds a Universal Resource Locator (URL) in a hypertext link that points back to the voice or fax message in the system's message store. When the subscriber clicks on the hypertext link, the mail application passes the URL to the subscriber's Web browser, which opens up a window to the subscriber's universal inbox, where the voice or fax message appears.

If the message is a voice message and the subscriber's workstation supports multimedia, the subscriber can listen to the voice message over the workstation's audio system. If the message is a fax, the system presents the fax as a graphic image on the workstation's screen. If the message is itself an e-mail message, the system simply passes it along to the subscriber.

Let's say the subscriber wants to be notified of incoming messages via a telephone's message-waiting indicator. When a new voice, fax, or e-mail message arrives and the subscriber checks for messages, the system indicates how many of each type of message have arrived. The system's synthesized voice might tell the subscriber, "You have two new voice messages, three new fax messages, and five new e-mail messages." Using DTMF keys on the handset, the subscriber can listen to the voice messages, output the fax messages to the nearest fax machine, and save the e-mail messages for viewing at a more convenient time and place.

The role of browsers. Because the Web can present text, graphics, and audio efficiently, it provides an effective medium for the consolidation and presentation of e-mail, fax, and voice messages. Some universal messaging systems can take advantage of Web technology to handle these binary objects for large numbers of subscribers securely.

With a Web browser such as Netscape Navigator or Microsoft Internet Explorer, universal messaging system subscribers can view fax and e-mail messages on their computer screens and listen to voice mail messages over headphones or speakers attached to their workstations. They also can use the message composition and reply features of the browser interface to respond to incoming messages.

For example, a subscriber can send a voice mail message to another recognized universal messaging subscriber in response to a voice, fax, or e-mail message. By clicking the *Compose Voice* button in the browser window and picking the recipient of the message from a directory window, the subscriber records a response and, when finished, clicks the *Send* button to deliver the voice mail.

A subscriber also can send an e-mail message in response to an incoming voice, fax, or e-mail message. In response to an incoming voice or fax message, the subscriber clicks the *Compose E-mail* button in the browser window. If the recipient is a recognized universal messaging subscriber, his or her address can be selected from the directory. If the recipient is not a recognized subscriber, the e-mail address must be manually entered. After composing the e-mail address, the subscriber clicks the *Send* button to deliver the e-mail message. If the original message was an e-mail message instead of a voice or fax message, the return address is automatically entered into the reply, whether or not the addressee is a universal messaging subscriber.

Messages also can be forwarded from the browser interface. A *Forward* button allows subscribers to send an incoming e-mail message to another subscriber. If the subscriber wants to forward a voice or fax message, the subscriber saves the voice message in an audio file format (or the fax in a graphic file) and sends it as an attachment to an e-mail message.

Subscribers can use the print capabilities of their Web browsers to output hard copies of received fax and e-mail messages. There usually is no option for printing voice messages, since most universal messaging systems do not yet support speech-to- text conversion. And other than forwarding a fax, these systems also do not provide a means for responding to a message via fax.

Because no subscriber is in front of a computer all the time, the universal messaging system provides users with another way of interacting with their incoming messages: the telephone handset. Subscribers can listen to incoming voice messages from any telephone, wireline or cellular. The system prompts the subscriber to review, save, or delete incoming messages by pressing various DTMF (dual tone multifrequency) key combinations. Subscribers also can forward voice messages to other universal messaging subscribers and send replies to universal messaging subscribers who have left messages.

While certain levels of access are obviously not available (such as viewing a fax or e-mail message over a standard handset), subscribers can use the handset to manipulate even these types of messages. If the fax or e-mail message has been sent from another universal messaging subscriber, the system reports who sent the fax or e-mail message and what time it arrived. If the fax or e-mail message has been sent by a nonsubscriber, the system only reports the type of message and the time it arrived.

Via the handset, subscribers also can redirect fax messages for output on whatever fax machine they designate. This is done by entering the desired fax phone number with the DTMF keys on the handset. The fax machine could be in a hotel lobby, at a remote office, at a customer site, or the subscriber's own home.

Incoming e-mail messages can be output in a similar manner. Through the use of DTMF keys, the subscriber can use the universal messaging system to fax an image of the e-mail message to a specified fax machine. Most universal messaging systems cannot open and fax e-mail message attachments, however.

By pressing DTMF keys at the prompts, a subscriber can record and send a voice mail response to a voice or e-mail message sent from another subscriber within the universal messaging environment. The universal messaging system addresses the reply automatically, so the subscriber does not have to remember (or

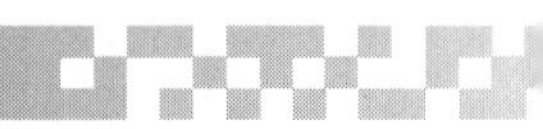

even know) the address of the person to whom they are sending the reply. To reply to an externally originated message, the subscriber must dial the external number directly and rely on the recording capabilities attached to the phone service at the other end of the line.

Although there usually is no support for sending a voice message from the handset in direct response to an incoming fax message, a subscriber can forward voice and fax messages to another recognized universal messaging subscriber by entering that subscriber's mailbox extension number at the prompt. Subscribers also can attach voice annotations to messages they forward from the handset.

Subscribers also can forward fax and e-mail messages to systems outside the universal messaging environment by using the print capabilities of the system. The subscriber forwards the fax by designating a remote recipient's fax machine or fax mailbox as the target output device. This feature also allows handset users to forward e-mail messages to recipients outside the system; the universal messaging system actually faxes an image of the e-mail message to a remote recipient's fax machine. Beyond providing a message to be faxed, the e-mail component of the universal messaging system does not play a role in forwarding messages outside the environment from the handset.

The universal messaging system ensures that actions initiated via the browser and handset interfaces are kept closely synchronized. If a subscriber listens to a new voice message through the browser interface, it is flagged as read and is not announced as a new message when the subscriber later accesses the inbox via the handset interface. And, as noted earlier, if a subscriber deletes a message via one interface, it is deleted from the list of messages accessible via any other interface.

Creating messages. The universal messaging system provides a suite of e-mail and voice messaging capabilities aimed at meeting a wide range of message creation requirements.

To create and send voice messages from a browser window, subscribers typically start by clicking the *Compose* button, which brings up an addressing window that allows the message recipient to be specified. Then the subscriber can speak into the multimedia PC's microphone to record the message. When the message is finished, the system offers to play it back and allows the subscriber to record it again if it is not satisfactory. When the message is ready, the subscriber confirms the instruction to send the message by clicking a button in the browser window.

Subscribers also can click a button in the browser window to create and send e-mail messages. There are no restrictions on the destination of an e-mail message except that the destination be a valid e-mail address. The subscriber can choose the e-mail address of another universal messaging subscriber from the directory, or enter the e-mail address manually if the recipient is not a recognized subscriber.

Once the message has been addressed, the subscriber can type in the body of the e-mail message. If the subscriber's browser supports attachments (for example, a formatted file such as a Microsoft Word document, a graphic, an audio file, or even a executable binary file), the subscriber can attach a file by clicking a button and selecting the appropriate file.

Subscribers can create a voice message by dialing into the universal messaging system and pressing the DTMF keys associated with message creation. The system

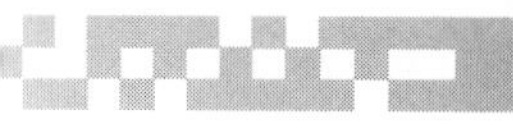

prompts the subscriber to enter an address for the message recipient (a mailbox extension such as 2468) and, upon receiving the addressing information, it prompts the subscriber to record a message. When the subscriber has finished recording the message, the universal messaging system plays back the recorded message, offers the subscriber an opportunity to rerecord the message, then sends the message to the designated recipient upon confirmation by the subscriber.

Voice messages can be sent to single or multiple recipients, and to a list of recipients in a named group. Subscribers can create group lists (if the network service provider allows that privilege) or they can work with group lists that others in their subscriber community have created for public use.

Voice messages composed at the browser can be sent only to other recognized universal messaging subscribers. E-mail messages, however, can be sent to any valid e-mail address. Universal messaging systems usually offer no provisions for creating outgoing fax messages, because subscribers are most likely to send their documents via fax modems or fax machines, or use e-mail instead.

Summary. As we continue to move beyond real-time telephone conversations to rely on voice messaging, fax, and e-mail, the ability to receive and manipulate various types of messages through a single device becomes increasingly important. Universal messaging provides a single view of all voice, fax, and e-mail messages. This means subscribers are no longer forced to delete voice messages through the handset interface and e-mail messages through the workstation interface. Such activities can be performed across all message types from any user interface. The synchronized management of messages is supported through handset or browser commands so that, among other things, messages deleted through one interface become unavailable through other interfaces as well.

See also

ELECTRONIC MAIL
FACSIMILE
PAGING

Universal service

The "universal service" system originally was designed to make local telephone service available to all consumers, including low-income consumers, in all regions of the nation. In many cases, universal service policies have required that rates for certain telecommunications services be set above the cost of providing those services, for the purpose of generating a subsidy to be used to reduce the rates for local service provided to residential customers.

The Telecommunications Act of 1996 updates the traditional universal service system, expanding both the base of companies that contribute to offset communications service rates and the category of customers who benefit from discounts.

Supported services. The specific services that must be given universal service support include:

- Voice-grade access to the public switched network, including, at a minimum, some usage

- ➢ Dual-tone multifrequency (DTMF) signaling or its equivalent
- ➢ Single-party service
- ➢ Access to emergency services, including access to 911, where available
- ➢ Access to operator services
- ➢ Access to interexchange services
- ➢ Access to directory assistance

Whether local telephone service is affordable depends upon several factors apart from local rates. Local calling area size, income levels, cost of living, and other socioeconomic indicators help in assessing affordability. The states, in their rate-setting roles, make the primary determination as to whether rates are affordable and for taking any necessary actions should they determine that the rates are not affordable. The FCC assesses affordability by monitoring subscribership levels.

The statutory criteria for receiving universal service support are as follows: The recipient must be a common carrier and the carrier must offer, throughout a designated service area, all of the services listed above.

Programs for low-income consumers. In 1985 the FCC established two programs that are still available to assist low-income consumers. The Lifeline program reduces qualified low-income consumers' monthly phone charges with matching federal and state funds. A state may choose not to participate in the Lifeline program. Currently 41 states, the District of Columbia, and the U.S. Virgin Islands participate in Lifeline.

The Link Up program provides federal support that reduces qualified low-income consumers' initial local telephone connection charges by up to one half. Link Up currently is funded by contributions from interexchange carriers (IXCs). Support currently is available only to incumbent wireline local exchange carriers.

Under the implementation rules for the Telecommunications Act of 1996, the Lifeline and Link Up programs were revised in the following manner:

- ➢ Both the Lifeline and Link Up programs have been expanded so that eligible low-income consumers in every state and territory can receive support. Every carrier deemed eligible for universal service support is required to participate.
- ➢ For each eligible consumer, federal support amounts to US$5.25. The federal fund also contributes an additional US$1.00 for every US$2.00 a state contributes to Lifeline support. The maximum amount of these federal matching funds is US$1.75, for a total federal support cap of US$7.
- ➢ Eligible low-income consumers receive access to the same designated services identified for support in rural, insular, and high-cost areas. In addition, these consumers pay no charge for access to toll blocking and toll limitation, but only to the extent that the carrier has the technical capability to provide these services.
- ➢ Reduced service deposits if a low-income consumer accepts toll blocking.
- ➢ Carriers cannot disconnect a Lifeline customer's local service for nonpayment of toll charges; a limited waiver of this requirement is available for some carriers, however.

Schools and libraries. Under the Telecommunications Act of 1996, schools and libraries are eligible for the first time to purchase at a discount any telecommunications services, internal connections among classrooms, and access to the Internet. Higher discounts are possible for economically disadvantaged schools and libraries, and those entities located in high-cost areas. Discounts are a minimum of 20 percent and range from 40–90 percent for all but the least disadvantaged schools and libraries. Total expenditures for universal service support for schools and libraries is capped at US$2.25 billion per year, although any funds not disbursed in a given year may be carried forward and also disbursed.

Health care providers. The approximately 9600 health care providers in rural areas in the United States are also eligible to receive telecommunications services supported by the universal service mechanism. Health care providers include teaching hospitals, medical schools, community health centers, migrant health centers, mental health centers, not-for-profit hospitals, local health departments, rural health clinics and consortia, or associations of any of these providers.

Administration of support mechanisms. At this writing, the FCC's Federal-State Joint Board has recommended that the Commission appoint a universal service advisory board, to include state and FCC representatives, to select a neutral, third-party administrator to administer the collection and distribution of the support mechanisms. Because support for schools and libraries will be implemented sooner than support for high-cost areas or low-income consumers, the Joint Board also has recommended the appointment of a temporary administrator of support for schools and libraries and rural health care providers.

In addition, the Joint Board has recommended that all telecommunications carriers that provide interstate telecommunications services be obligated to contribute to universal service. Internet and online service providers would not have to contribute to universal support mechanisms unless they also provide telecommunications services, in which case they would contribute an amount proportional to the revenues they receive from telecommunication services.

Summary. The Telecommunications Act of 1996 directs the FCC to accomplish the following tasks related to improving universal service: Promote the availability of quality services at just, reasonable, and affordable rates; increase access to advanced telecommunications services throughout the nation; and advance the availability of such services to all consumers, including those in low-income, rural, insular, and high-cost areas at rates that are reasonably comparable to those charged in urban areas. In addition, the act states that all providers of telecommunications services should contribute to federal universal service in some equitable and nondiscriminatory manner; there should be specific, predictable, and sufficient federal and state mechanisms to preserve and advance universal service; and all schools, classrooms, health care providers, and libraries should, generally, have access to advanced telecommunications services. The rules for implementing these goals were formalized in mid-1997.

See also

FEDERAL COMMUNICATIONS COMMISSION

Value-added networks

Value-added network (VAN) service providers offer voice and data communications services for large companies, particularly those with international locations. The "value added" comes in the form of in-country local support, network design, integration, and management assistance, as well as economical access to a variety of feature-rich services and customized billing—tasks that can be resource-intensive if done by the subscriber.

VAN providers offer a variety of value-added services across their networks. These include network management, e-mail, EDI, electronic funds transfer (EFT), X.400 global messaging, LAN interconnection, and virtual private data networks (VPDN). These services typically include protocol conversion capabilities between dissimilar network devices, temporary and archival data storage, broadcast services to pre-established distribution lists, timed message delivery, message logging and acknowledgment, usage reports, security, and terminal handling (e.g., polled and scheduled calling). Clients also can set up closed user groups to customize their network requirements.

In addition to providing economical access over a variety of transmission services (such as X.25, frame relay, and TCP/IP), the principal advantages of VANs include:

- *Error control* that ensures a high degree of accuracy for critical data during transmission.
- *Program conversion* utilities that allow a variety of X.25-equipped terminals and computers to intercommunicate without requiring significant software changes.
- *Protocol conversion* that eliminates the need for dedicated protocol conversion devices and frees front-end processors and mainframes from this processing burden.
- *Wide area network support* that allows organizations to configure virtual private networks without incurring the associated setup costs, and ties small remote locations or branches to the main network in a cost-effective manner.
- A high degree of *data security,* including network authorization coding, message recipient authentication, and encryption.

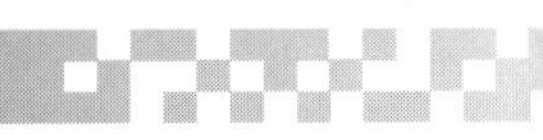

- *Bandwidth-on-demand* that allows users to call up additional bandwidth on an as-needed basis to support specific applications, handle peak-period traffic requirements, or provide convenient backup for private network facilities.
- *Service management,* including equipment, software, and lines.

X.400 global messaging. Among the specific value-added services offered by VAN service providers is X.400 global messaging. For many reasons, large organizations often find it impractical to support only one messaging system. VANs offer store-and-forward mail services that allow users to send and receive messages between diverse e-mail systems. This is accomplished utilizing X.400 gateways. X.400 is the global messaging standard recommended by the ITU (formerly CCITT) international standards committee. Specifically, it is an envelope, routing, and data format standard for electronic messaging that is especially useful for connecting dissimilar e-mail systems. This spares organizations the difficulty of installing and maintaining their own X.400 gateways.

Other services. Another value-added service is SNA interconnectivity, which provides a cost-effective alternative for companies that currently use multidrop private lines to connect remote sites to IBM host systems or to interconnect IBM peer nodes. The VAN service provider supplies the connections between nodes and proactively manages the subscriber's portion of the value-added network from end to end. Users need not purchase additional equipment to use the service because the VAN supports such native IBM protocols as Synchronous Data Link Control (SDLC) and Qualified Logical Link Control (QLLC). The service is available at data rates ranging from 9.6 to 56 kbps.

Some service providers offer secure Internet access through their value-added networks. The aim is to provide fluency of access without exposing businesses to unnecessary security risks. Security is provided with firewalls that act as both buffers and access filters between the corporate portion of the VAN and the greater Internet. Services offered include Domain Name Services, World Wide Web (WWW), Wide Area Information Services (WAIS), Gopher, Telnet, and FTP.

With the number of mobile users growing by leaps and bounds, the need for wireless access to VAN services is becoming a virtual necessity. A user typically connects to a modem pool offered by the VAN service provider. From there, users can gain remote access to corporate LANs or host-based systems as if they were locally attached. The VAN service provider might offer different access plans based on the number of hours of usage.

Summary. As VANs become more economical and feature-rich, there will be less incentive for organizations to set up and maintain their own networks, particularly when a significant portion of these networks spans international boundaries. Via on-premises management terminals, the VAN service providers are even giving subscribers the ability to configure network routing paths and schedule their availability based on time of day. Users also can fine-tune the network to increase data rates and choose alternate routing based on trunk speed. Additionally, VANs offer users increasingly sophisticated diagnostic capabilities that are comparable to those available in the private network environment. This gives organizations the control they need, plus all the value-added features, but without the expense of operating their own networks.

See also

MANAGED SNA SERVICES
VIRTUAL PRIVATE NETWORKS

Videoconferencing

Videoconferencing is the process whereby individuals or groups at different locations meet online to share information through audio and visual communications. The video conferences can be augmented with text and graphics for display on separate screens. In addition, image, text, and graphics from a variety of sources can be multiplexed over the same video circuit to permit interactivity among conference participants without the need to establish additional communications links.

Videoconferencing in recent years has become an increasingly accepted form of communication among businesses and government agencies that want to save on travel costs, encourage collaborative efforts among staff at far-flung locations, and enhance overall productivity among employees. Originally, videoconferencing was seen as a method to link people at remote locations over wide area networks (WANs). More recently, videoconferencing has been used to link desktop computer users over local area networks (LANs) in an effort to obtain the same benefits within a building or campus environment. Today, the technology has progressed to the point where, with relatively inexpensive hardware and software, virtually anyone can participate in videoconferences over the Internet.

Applications. Businesses benefit from videoconferencing in a variety of ways. Cutting travel costs, although important, is not the only motivation for implementing videoconferencing systems. Instead of consuming valuable time with the logistics of travel for face-to-face meetings at individual locations, executives can take advantage of videoconferencing systems to conduct general meetings and individual sessions with appropriate personnel. In the process, the quality and the timeliness of decision-making can be greatly improved.

Because videoconferencing adds the ability to exchange information in a visually compelling way, it can be applied to almost any situation to enhance the quality and effectiveness of the communications process. In the medical profession, videoconferencing often is referred to as *telemedicine.* In academia, videoconferencing is a key component of distance learning. For corporate employees who work full- or part-time out of their homes, videoconferencing makes telecommuting easier.

Whether used for product introductions, sales promotions, employee training, management messages, or collaborative projects among widely dispersed corporate locations, videoconferencing is increasingly viewed not as a prestige technology, but as a practical and immediately useful tool that can yield competitive advantage.

Types of systems. Videoconferencing systems fall into four categories: room-based systems, midrange or roll-about systems, desktop systems, and videophones.

Room-based systems. Room-based systems usually entail the use of one or more large screens in a dedicated meeting room equipped with environmental con-

trols. The system components (screens, cameras, microphones, and auxiliary equipment) can be permanently installed because they will not be moved to another room or building. These systems provide high-quality video and synchronized audio. Prices for room-based systems start at around US$100,000, but can go much higher as more sophisticated equipment and features are added.

Midrange systems. If an organization does not rely extensively on videoconferencing but considers it an important ability to have when the occasion arises, a midrange or roll-about videoconferencing system is a viable solution. Typically these portable systems use one screen and no more than two cameras and three microphones. Prices range from US$20,000 to US$50,000, depending on options.

Desktop systems. Desktop videoconferencing is becoming popular because it allows organizations to leverage existing assets; video conferencing becomes just another application running on the desktop. When equipped for videoconferencing, the desktop machine can be used for video mail over LANs as well as videoconferencing over WANs. The ability to use widely available Ethernet networks makes videoconferencing technology more accessible, less costly, and easy to deploy. Data sharing can be accomplished through an optional whiteboard capability.

Some desktop videoconferencing systems support TCP/IP, the most commonly used wide area network protocol. This support makes the system particularly useful in campus-style settings and for corporations with remote offices. TCP/IP support ensures that the system can be used in conjunction with standard bridges, routers, and dial-up lines, giving users convenient, cost-effective access to videoconferencing capabilities, particularly in areas where ISDN service is not available. Support for TCP/IP also eliminates the need for expensive on-premises switching systems and potentially costly telephone company surcharges for the routing services that are required of ISDN networks.

Fully equipped desktop videoconferencing systems are available in the US$1200 to US$5000 per unit range, depending on options. At the low end of the price range, the videoconferencing capability is added to an existing PC with appropriate hardware and software; at the high end, the vendor provides the PC already configured for videoconferencing. The difficulty in installing and configuring add-in components makes the latter solution very appealing.

Videophones. Videophones are used for one-on-one communication. This type of equipment satisfies the desire for impulse videoconferencing. The videophone unit includes a small screen, built-in camera, video coder/decoder (codec), audio system, and keypad. The handset lets the unit work as an ordinary phone as well as a videoconferencing system. Prices start at about US$1000 for models that work over ordinary phone lines.

Multipoint control units. Videoconferencing among more than two locations requires a multipoint control unit (MCU), which is a switch that connects video signals among all locations, enabling participants to see one another, converse, and work simultaneously on the same document or view the same graphic. The multipoint conference is set up and controlled from a management console connected to the MCU.

The MCU makes it relatively easy to set up and manage conference calls among multiple sites. The following are some of the features that facilitate multipoint conferencing:

- *Meet-Me* allows participants to enter a conference by dialing an assigned number at a prearranged time.
- *Dial-Out* is used to automatically dial out to other locations and add them to the conference at prearranged times and dates.
- *Audio Add-On* allows participants to hear or speak to others who do not have video equipment (or compatible video equipment) at their locations.
- *Tone Notification* provides special tones to alert participants when a person is joining or leaving the conference, and when the conference is about to end.
- *Dynamic Resizing and Tone Extension* allows locations to be added or deleted, and the duration of a bandwidth reservation to be extended, during a conference without the session having to be restarted.
- *Integrated Scheduling* permits videoconferences to be set up and scheduled days, months, or a year in advance using an integral calendar or scheduler application. The MCU automatically reserves the required bandwidth, configures itself at the designated time, and dials out to participating sites to establish the conference.

The MCU also provides the means to precisely control the video conference in terms of who is seeing what at any given time. Some of the advanced conference control features of MCUs include the following:

- *Voice-Activated Switching* allows all participants to see the person speaking, while the speaker sees the last person who spoke.
- *Contributor Mode* works with voice-activated switching to allow a single presenter to be shown exclusively on a conference.
- *Chair Control* allows a person to request or relinquish control, choose the broadcaster, and drop a site or the conference.
- *Presentation or Lecture Mode* allows a speaker to make a presentation and question participants in several locations. Participants can see the presenter at all times, but the presenter sees whoever is speaking.
- *Moderator Control* allows a moderator to select which person or site appears on the screen at any given time.
- *Roll Call* allows a conference moderator to switch to each participant for the purpose of introducing them to others, or to screen the conference for security purposes.
- *Subconferencing* allows the conference operator to transfer participants into and out of separate, private conferences associated with the meeting, without having them disconnect and reconnect.
- *Broadcast with Automatic Scan* allows the participants to see the presenter at all times. To gauge audience reaction, the speaker sees participants in each location on a timed, predetermined basis.

Depending on the choice of MCU, there are a number of network connectivity options available. Generally the MCU can be connected to the network either directly via T1 leased lines or ISDN PRI trunks, or indirectly through a digital PBX. Some MCUs can be connected to the network using dual 56 kbps lines or ISDN BRI. Others connect videoconferencing systems over the WAN using any mix of private and carrier-provided facilities, or any mix of switched services regardless of the carrier (Fig. V1). This capability offers the most flexibility in setting up multipoint conferences.

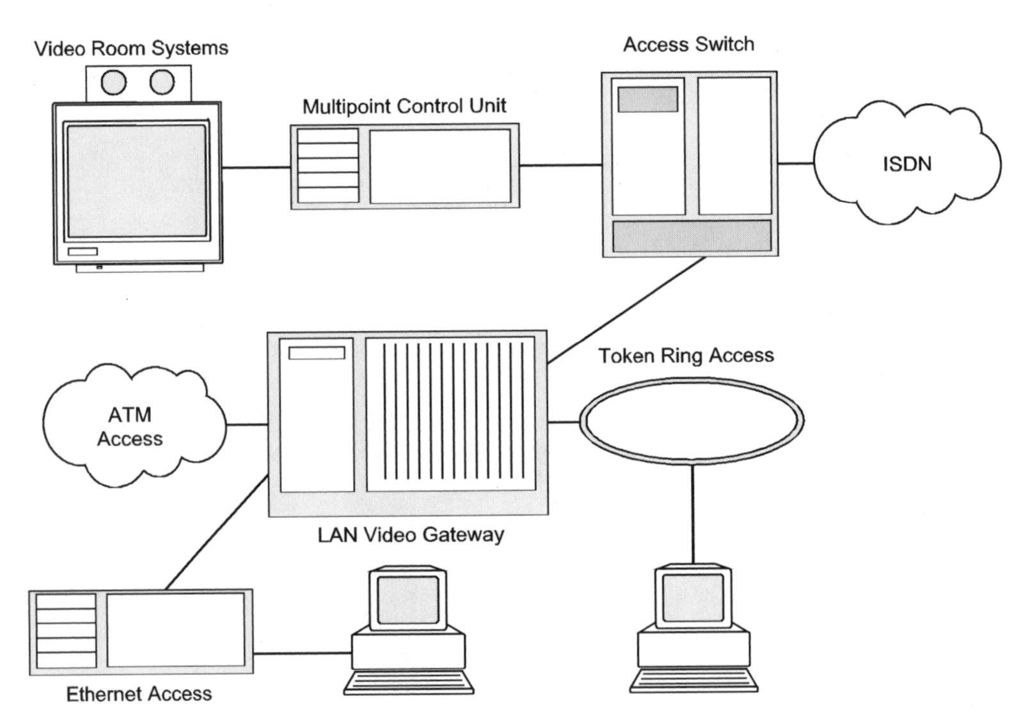

Fig. V1 *A typical videoconferencing arrangement implemented by a multipoint control unit (MCU).*

For users who do not have high-speed links, some MCUs include an inverse multiplexing capability that combines multiple 56/64 kbps channels into a single, higher-speed 384 kbps channel on a demand basis, thus improving video quality. Most MCUs use the inverse multiplexing method that has been standardized by the Bandwidth on Demand Interoperability Group (BONDING).

While most MCUs are designed for use on the WAN, there are some MCUs available for the LAN. These devices are useful for providing multipoint desktop videoconferencing over local networks within a campus environment or among many floors in a high-rise building.

Standards. Standards for videoconferencing and other transmission technologies are established by the International Telecommunications Union (ITU), formerly the

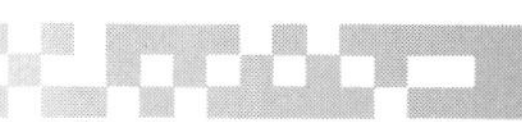

CCITT. By establishing worldwide videoconferencing standards, the ITU-TSS helps ensure that videoconferencing systems from diverse manufacturers will be able to communicate with one another.

Video transmission standards. Recommendation H.320 is a set of videoconferencing standards developed by the ITU's Telecommunications Study Group 15. Recommendation H.320 is the umbrella standard that defines the operating modes and transmission speeds for videoconferencing system codecs, including the procedures for call setup, call teardown, and conference control. The codecs that comply with H.320 are interoperable with those of different manufacturers, delivering a common level of performance.

The H.320 videoconferencing standard includes associated specifications that define how the videoconferencing products from different vendors interoperate. Among the key H.320 standards are:

- *H.322:* A standard for LAN-based videoconferencing with guaranteed bandwidth.
- *H.323:* A standard for LAN-based videoconferencing with nonguaranteed bandwidth (nonisochronous), such as Ethernet or Token-Ring. H.323 specifies G.711 as the mandatory speech codec standard.
- *H.324:* A standard set for videoconferencing over high-speed modem connections using standard telephone lines. H.324 specifies G.723 as the mandatory speech codec standard.

While the H.324 specification facilitates videoconferencing over ordinary phone lines, support for the ITU's V.80 standard also is needed to attain the maximum performance promised by H.324. The V.80 standard is implemented in V.34 modems to enhance the H.324 software applications running on hosts. It is a relatively simple computer-to-modem controller protocol that allows an asynchronous PC interface to talk to a synchronous V.34 modem. The V.80 interface allows the host to define the frame boundaries for the multiplexed video and audio signal and allows the modem to complete the bit-oriented framing. Without V.80, videoconferencers would have to use a special synchronous modem protocol not supported by most modem software, put up with start-stop-start videoconferencing (losing about 20 percent of the available bandwidth in the process), or use a proprietary synchronous controller protocol.

Another important component of H.320 is the H.261 video compression specification, which defines how digital information is coded and decoded. H.261 also permits the signals to be transmitted at a variety of data rates, from 64 kbps to 2.048 Mbps in increments of 64 kbps. H.261 also defines two resolutions. One is the Common Intermediate Format (CIF), a format usually used in high-end room systems, which provides the highest resolution at 352×288 pixels. The other is the Quarter Common Intermediate Format (QCIF), a format used by most desktop videoconferencing systems and videophones, which provides lower resolution at 176×144 pixels.

Another standard, H.230, describes the signals used by conferencing systems and MCUs to communicate during a conference. These signals allow conferencing systems and the MCU(s) to exchange instructions and status information during the initiation of a conference and while the conference is in progress.

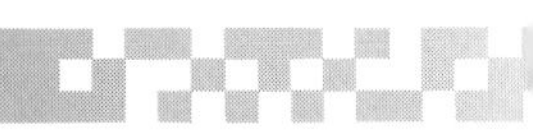

A related standard, Recommendation H.243, defines the basic MCU procedures for establishing and controlling communication between three or more videoconferencing systems using digital channels up to 2 Mbps.

Audio compression standards. There is a set of ITU recommendations that standardizes audio compression for videoconferencing equipment. There are three key standards in this area:

- *G.711* defines the requirements for 64 kbps audio. This is the least compressed and offers the highest audio quality.
- *G.722* defines 2:1 audio compression at 32 kbps.
- *G.728* defines 4:1 audio compression at 16 kbps.

The reason compression is important is that it squeezes the audio component of the videoconference into a smaller increment of bandwidth, freeing more of the available bandwidth for the video component. This results in higher-quality video without appreciably diminishing the audio.

Input/output standards. Many videoconferencing systems have ports for such auxiliary devices as televisions, cameras, and VCRs. Regarding the quality of video input/output for these, the two most pervasive standards are the North American NTSC (National Television Standards Committee), and the European PAL (Phase Alternating by Line). NTSC specifies 320×240 resolution at 27 to 30 frames per second. PAL specifies 384×288 resolution at 22 to 25 frames per second. Most video equipment on the market supports both NTSC and PAL. Other, less-used standards that are country-specific might be supported as options by some vendors.

Summary. Videoconferencing finally is taking its place as a strategically significant corporate communications tool. The ultimate low-cost and ubiquitous method of videoconferencing might well be the Internet. Several vendors offer videoconferencing software that works on desktop computers equipped with a camera, sound card, modem, and Internet connection. Image quality over the Internet currently is not the same as the image quality that can be achieved over high-speed LANs and digital WAN services like ISDN. On the Internet there are the problems of bandwidth and variable delay, which limit video to only 2 to 5 frames per second. As a result, the image tends to be grainy, movement is jerky, and the picture size is small. As more bandwidth is added to the Internet, however, and the use of new resource reservation protocols becomes more widespread, the quality of Internet videoconferencing will improve dramatically.

See also

INVERSE MULTIPLEXING

TELECONFERENCING

Video on demand

Video on demand is a service from which subscribers will be able to select movies for viewing whenever desired, instead of having to rent videotapes from a local store or wait for a scheduled starting time, as in conventional pay-per- view services. To offer video on demand, carriers, TV stations, cable companies, and entertainment conglomerates have entered into highly publicized buyouts, partnerships, and joint ventures. At this writing, on-demand services have been undergoing trials; the results indicate a promising market for this type of service.

Programming is delivered in a variety of ways, including hybrid fiber/Coax (HFC), Asymmetric Digital Subscriber Line (ADSL), and fiber-to-the-curb (FTTC). MPEG1 and MPEG2 encodings are used to compress the program for delivery through the network to any subscriber who requests it. The encoded programs and navigation screens are stored in online transaction computers and video servers.

The service allows subscribers to order programs from an electronic program guide presented as an onscreen menu consisting of hundreds of program choices each month. The offerings are grouped into such categories as entertainment (movies, TV shows); children (movies, TV shows, educational programs), learning and lifestyles (documentaries, how-to programs); and home shopping. The electronic program guide is updated and refreshed each month.

Once a selection is made and the program starts, the subscriber has access to all the features of a VCR: Pause, rewind, and fast forward are controlled with a handheld remote control device. Depending on the program selected, prices range from US$0.49 to US$4.49, making the service very competitive with pay-per-view and video rental.

Trials reveal that the cost for the interactive technology needed to deliver video on demand to a mass market is just pennies per customer per video. This includes the cost for servers, encoding, and the production of navigational interfaces and related processes (but not the cost of transmission or the cost of license fees for programs). Among the trial results are these:

- On average, 73 percent of all participants purchase some programming each month.
- Newly released movies are the most popular content selection, followed by children's programming and library movies.
- On average, more than 70 percent of all program titles were viewed each month, reflecting participants' desire for an offering with both breadth and depth.
- There was high initial satisfaction with the service among trial users: 7.27 on a 10-point scale.

Summary. In contrast to the time-of-day, day-of-week environment of broadcast TV, there is little urgency to view video on demand services. Service providers are experimenting with promotional vehicles to encourage viewership, including a "barker" channel, postcards, and a monthly magazine. Truly robust promotion must wait until video on demand capabilities are integrated with linear TV channels.

See also

DIGITAL SUBSCRIBER LINE TECHNOLOGIES

HYBRID FIBER/COAX

FIBER IN THE LOOP

Virtual private networks

Carrier-provided networks that function like private networks are referred to as *virtual private networks,* or VPNs. With a VPN, corporations can minimize the operating

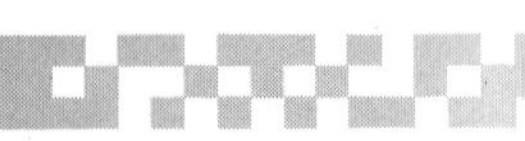

costs and staffing requirements associated with private networks. Additionally, they obtain the advantages of dealing with a single carrier, rather than the multiple carriers and vendors that are normally required to set up and maintain a private network.

AT&T introduced the first VPN service in 1985. Its Software Defined Network (SDN) was a voice-only service offered as an inexpensive alternative to private lines. Since then, VPNs have added more functionality and expanded globally. Today the Big Three carriers—AT&T (Software Defined Network), MCI (Vnet), and Sprint (VPN Service)—each offer virtual private networks. These networks can include high-speed data and cellular calls, all of which may be combined under a single- service umbrella, expanding opportunities for cost savings within a single discount plan.

Advantages of VPNs. An increasing number of companies are finding virtual private networks to be a viable alternative method for obtaining private network functionality without the overhead associated with acquiring and managing dedicated private lines. Additionally, there are several other advantages to opting for a virtual private network, including:

- The ability to assign access codes and corresponding class of service restrictions to users; these codes can be used for internal billing, to limit the potential for misuse of the telecommunications system, and to facilitate overall communications management.
- The ability to consolidate billing, resulting in only one bill for the entire network.
- The ability to tie small remote locations to the corporate network economically, instead of using expensive dial-up facilities.
- The ability to meet a variety of needs (e.g., switched voice and data, travel cards, toll-free service, international and cellular calls, etc.) using a single carrier.
- The availability of a variety of access methods, including switched and dedicated access, 700 and 800 dial access, and remote calling card access.
- The availability of digit translation capabilities that permit corporations to build global networks using a single carrier. Digit translation services can perform 7-to-10, 10-to-7, and 7-to-7 translations, and can convert domestic telephone numbers to International Direct Distance Designator (IDDD) numbers via 10-to-IDDD and 7-to-IDDD translation.
- The ability to have the carrier monitor network performance and reroute around failures and points of congestion.
- The ability to have the carrier control network maintenance and management, reducing the requirement for high-priced, in-house technical personnel, diagnostic tools, and spares inventory.
- The ability to configure the network flexibly, via on-site management terminals that allow users to meet bandwidth application needs and control costs.
- The ability to access enhanced transmission facilities, with speeds ranging from 56 kbps to 384 kbps and 1.536 Mbps, and to plan for emerging broadband services.

- The ability to combine network services pricing typically based on distance and usage with pricing for other services to qualify for further volume discounts.
- The ability to customize dialing plans to streamline corporate operations. A dealership network, for example, can assign a unique four-digit code for the parts department. Then, to call any dealership across the country to find a part, a user would simply dial the telephone number prefix of that location.

VPN architecture. The architecture of the VPN makes use of software-defined intelligence residing in strategic points of the network. AT&T's SDN, for example, consists of a network action point (ACP) connected to the PBX via dedicated or switched lines. The ACPs connect with the carrier's network control point (NCP), where the customer's 7-digit on-net number is converted to the appropriate code for routing through the virtual network (Fig. V2).

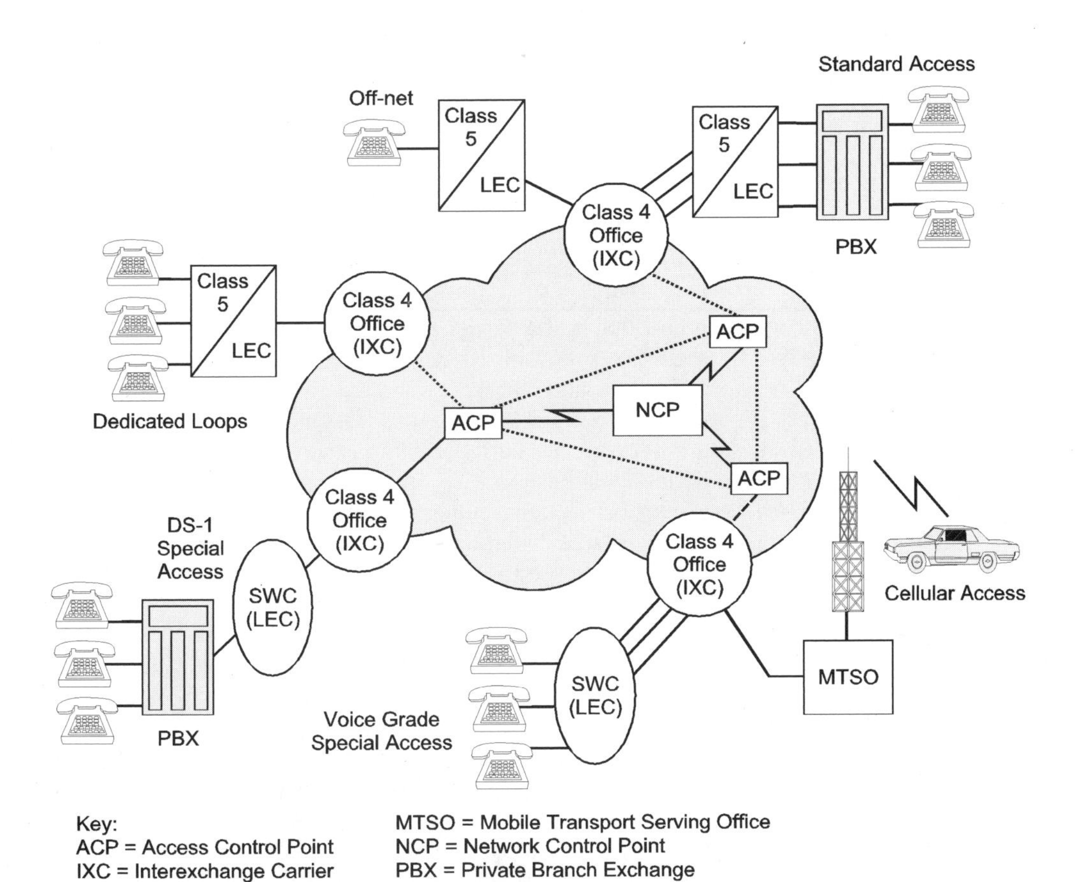

Fig. V2 *The architecture of AT&T's Software Defined Network (SDN).*

Instead of charging for multiple local access lines to support different usage-based services, the carriers allow users to consolidate multiple services over a single T1 access line. A user who needs only 384 kbps for a data application, for example, can fill the unused portion of the access pipe with 18 channels of voice traffic in order to justify the cost of the access line. At the carrier's crossconnect system, the dedicated 384 kbps channel and 18 switched channels are split out from the incoming DS1 signal. The 384 kbps DS0 bundle is then routed to its destination, while the voice channels are handed off to the carrier's Class 4 switch, which distributes the voice channels to the appropriate service.

A variety of access arrangements is available from the VPN service providers, which are targeted for specific levels of traffic, including:

- A single voice-frequency channel,
- 24 voice channels through a DS1 link,
- 44 voice channels through a T1 link equipped with bit-compression multiplexers, and
- A feature that splits a DS1 link into its component 64-kbps DS0s at the VPN serving office for connection to off-net services.

The same DS1 link can be used for a variety of applications, from 800 service to videoconferencing, thereby reducing access costs. Depending on the carrier, there may be optional cellular and messaging links to the VPN as well. Even phone card users can dial into the VPN, with specific calling privileges defined for each card. All of a company's usage can be tied into a single invoicing structure, regardless of access method.

VPN features. VPNs allow users to create their own private networks by drawing on the intelligence embedded in the carrier's network. This "intelligence" actually is derived from software programs that reside in various switch points throughout the network. Services and features are defined in software, giving users greater flexibility in configuring their networks than is possible with hardware-based services. In fact, an entire network can be reconfigured by changing the operating parameters in a network database.

Flexible Routing. Flexible Routing allows the telecom manager to reroute calls to alternate locations when a node experiences an outage or peak-hour traffic congestion. This feature also can be used to extend customer service business hours across multiple time zones.

Location Screening. Location Screening allows the telecom manager to define telephone numbers that cannot be called from a given VPN location. This helps contain call costs by disallowing certain types of outbound calls.

Originating Call Screening. Originating Call Screening gives the telecom manager the means to create caller groups and screening groups. Caller groups identify individual users who have similar call restrictions, while screening groups identify particular telephone numbers that are allowed or blocked for each caller. Time intervals also can be used as a call-screening mechanism, allowing or blocking calls according to time-of-day and day-of-week parameters.

NNX Sharing. With NNX Sharing, VPN customers can reuse NNXs (i.e., exchange numbers) at different network locations to set up 7-digit on-net numbering plans. This provides dialing consistency across multiple corporate locations.

Partitioned Database Management. Partitioned Database Management allows corporations to add subsidiaries to the VPN network while providing for flexible, autonomous management to address local needs when required by the subsidiaries. The VPN even can interface transparently with the company's private network, or the private network of a strategic partner. In the latter case, the VPN caller is not aware that the dialed number is a VPN or private network location because the numbering plan is uniform across both networks.

Automatic Number Identification. With Automatic Number Identification (ANI) data, the telephone numbers of incoming calls can be matched to information in a database, such as the computer and telecommunications assets assigned to each employee. When the call comes through to the corporate help desk, the ANI data is sent to a host, where it is matched with the employee's file. The help desk operator then can have all relevant data available immediately to assist the caller in resolving the problem.

Billing options. One of the most attractive aspects of virtual private network services is customized billing. Users typically can select from among the following billing options:

- The main account can accrue all discounts under the program. In some cases, even the use of wireless voice and data messaging services can qualify for the volume discount.
- Discounts can be assigned to each location according to its prorated share of traffic.
- A portion of the discounts can be assigned to each location based on its prorated share of traffic, with a specified percentage assigned to the headquarters location.
- Usage and access rates can be billed to each location, or subsidiaries can be billed separately from main accounts.
- Billing information and customized reports can be accessed at customer premises terminals, or provided by the carrier on diskette, microfiche, magnetic tape, tape cassette, or CD-ROM, as well as in paper form.
- A name substitution feature allows authorization codes, billing groups, telephone numbers, master account numbers, dialed numbers, originating numbers, and credit card numbers to be replaced by the names of individuals, resulting in a virtually numberless bill for internal distribution. This prevents sensitive information from falling into the wrong hands.

AT&T, MCI, and Sprint all offer rebilling capabilities that can use a percentage or flat-rate formula to mark up or discount internal telephone bills. Billing information even can be summarized in a number of graphical reports, such as bar and pie charts. Carrier-provided software is available that allows users to work with call detail and billing information to generate reports in a variety of formats. Some software even illustrates calling patterns with maps.

Network management. Various management and reporting capabilities are available through a network management database that allows telecom managers to perform tasks without carrier involvement. The network management database contains information about the network configuration, usage, equipment inventory, and call restrictions. On gaining access to the database, the telecom manager can set up, change, and delete authorization codes and approve the use of capabilities such as international dialing by caller, workgroup, or department. The telecom manager also can redirect calls from one VPN site to another to allow, for example, calls to an East Coast sales office to be answered by the West Coast sales office after the East Coast office closes for the day. Once the telecom manager is satisfied with the changes, they can be uploaded to the carrier's network database and will take effect within minutes.

Telecom managers can access call detail and network usage summaries, which can be used to identify network traffic trends and assess network performance. In addition to being able to download traffic statistics about dedicated VPN trunk groups, users can receive 5-, 10-, and 15-minute trunk group usage statistics an hour after they occur; these statistics can then be used to monitor network performance and carry out traffic engineering tasks. Usage can be broken down and summarized in a variety of ways, such as by location, type of service, and time of day. This information can be used to spot exceptional traffic patterns that might indicate either abuse or the need for service reconfiguration.

Via a network management station, the carrier provides network alarms and traffic status alerts for VPN locations using dedicated access facilities. These alarms indicate potential service outages (e.g., conditions that impair traffic and could lead to service disruption). Alert messages are routed to customers in accordance with preprogrammed priority levels, ensuring that critical faults are reviewed first. The system furnishes the customer with data on the specific type of alarm, direction, location, and priority level, along with details about the cause of the alarm (e.g., signal loss, upstream failed signal, or frame slippage). The availability of such detail permits telecom managers to isolate faults immediately.

Additionally, telecom managers can request access-line status information and schedule transmission tests with the carrier. The network management database describes common network problems in detail and offers specific advice on how to resolve them. The telecom manager can submit service orders and trouble reports to the carrier electronically via the management station. Also, telecom managers can test network designs and add new corporate locations to the VPN.

Local VPN service. A new development in the VPN market is the emergence of local service whereby some regional Bell operating companies (RBOCs) allow corporate customers to manage their in-region calls using the public network as if it were their own private network. This allows customers to do such things as access their voice network remotely, make business calls from the road or home at business rates, originate calls from remote locations while billing them to the office, and block calls to certain telephone numbers or regions. Uniform pricing and billing plans also can be arranged for all of the customer's locations to reduce the administrative costs involved with reviewing billing statements, even if each location uses a different carrier.

The service allows large business customers to configure components of the public network like a customized private network without the expense of dedicated lines or equipment. Until now, services of this kind could not be used for local calls because they were offered through long-distance companies. The service also is compatible with Centrex services, PBX systems, or other customer premises equipment.

Summary. VPNs permit the creation of voice and data networks that combine the advantages of both private facilities and public services, drawing on the intelligence embedded in the carrier's network. With services and features defined in software and implemented via out-of-band signaling methods, users have greater flexibility in configuring their networks from on-premises terminals and management systems than is possible with services implemented with manual patch panels and hardwired equipment.

See also

ADVANCED INTELLIGENT NETWORK
SIGNALING SYSTEM 7
VALUE-ADDED NETWORKS

Voice compression

Many compression methods have emerged over the years, employing sophisticated voice encoding schemes to increase the number of channels on the available bandwidth, allowing businesses to achieve substantial savings on leased lines with only a modest cost for additional hardware. Among the most popular compression methods is Adaptive Differential Pulse Code Modulation (ADPCM), which has been a worldwide standard (G.721) since 1984. It is used primarily on private T-carrier networks to increase the channel capacity of the available bandwidth. Although other compression techniques are available for use on wireline networks[1], ADPCM offers several advantages.

Thousands of subjective tests conducted by Bell Labs and independent research firms have confirmed that the compressed voice signals used in ADPCM are virtually indistinguishable in quality from the original, uncompressed voice signals. ADPCM holds up well in the multinode environment, where it may undergo compression and decompression several times before arriving at its final destination. Unlike many other compression methods, ADPCM also does not distort the distinguishing characteristics of a person's voice during transmission.

ADPCM fits very well within the ISDN framework. In fact, the T1 multiplexer equipped with the primary rate interface (PRI) can enhance ISDN by enabling the 64 kbps B channels to carry ADPCM compressed voice at 32 kbps or 16 kbps. With speed selection controlled in software through the multiplexer's transport management system, this capability gives users added configuration flexibility that does not come with ISDN alone. Although the ISDN standards specify the entire B channel as the fundamental unit of circuit switching, logical subchannels may be carried on a single circuit between the same pair of subscribers. Network management information can even be carried within a subchannel to provide users with the kind of end-to-end management and control that is characteristic of the private network.

1 There are numerous compression algorithms for voice and data. This book discusses only the most widely implemented algorithms. For example, there is a worldwide standard for compressing voice over wireless networks, which is discussed under Global System for Mobile (GSM) telecommunications.

Pulse Code Modulation. A voice signal takes the shape of a wave, with the distance from the top to the bottom of the wave defining the signal's power level, or *amplitude.* The voice is converted into digital form by an encoding technique called Pulse Code Modulation (PCM). Under PCM, voice signals are sampled at a minimum rate of twice the highest desired voice frequency; for a top frequency of 4000 Hz, there will be a minimum of 8000 samples per second. The amplitudes of the samples are encoded into binary form using enough bits per sample to maintain a high signal-to-noise ratio. For quality reproduction, the required digital transmission speed for 4 kHz voice signals works out to

8000 samples/sec × 8 bits/sample = 64,000 bps (64 kbps)

The conversion of a voice signal to digital pulses is performed by a coder-decoder, or *codec,* which is a key component of D4 channel banks and multiplexers. The codec translates amplitudes into digital binary values and performs mu-Law quantizing. The mu-Law process (North America only) is an encoding-decoding scheme for improving the signal-to-noise ratio. This is similar in concept to Dolby noise reduction, which ensures quality sound reproduction.

Other components of the channel bank or multiplexer function together to interleave the digital signals representing as many as 24 channels to form a 1.544 Mbps bit stream (including 8 kbps for control) suitable for transmission over a T1 line. PCM exhibits high quality, is robust enough for switching through the public network without suffering noticeable degradation, and is simple to implement. But PCM only allows for 24 voice channels over a T1 line.

Compression basics. The ADPCM device accepts PCM's 8000 sample/sec rate and uses a special algorithm to reduce the 8-bit samples to 4-bit words. These 4-bit words, however, no longer represent sample amplitudes, but only the difference between successive samples. This is all that is necessary for a like device at the other end of the line to reconstruct the amplitudes.

Integral to the ADPCM device is circuitry called the *adaptive predictor,* which predicts the value of the next signal based only on the level of the previously sampled signal. Because the human voice does not usually change significantly from one sampling interval to the next, prediction accuracy can be very high. A feedback loop used by the predictor ensures that voice variations are tracked with minimal deviation. Consequently, the high accuracy of the prediction means that the difference in the predicted and actual signal is very small and can be encoded with only 4 bits, rather than the 8 bits used in PCM. In the event that successive samples vary widely, the algorithm adapts by increasing the range represented by the 4 bits. This adaptation will decrease the signal-to-noise ratio, however, and reduce the accuracy of voice frequency reproduction.

At the other end of the T1 line is another compression device (Fig. V3), where an identical predictor performs the process in reverse to reinsert the predicted signal and restore the original 8-bit code.

By halving the number of bits to encode a voice signal accurately, T1 transmission capacity is doubled from the original 24 channels to 48 channels, providing the user with a 2-for-1 cost savings on monthly charges for leased T1 lines.

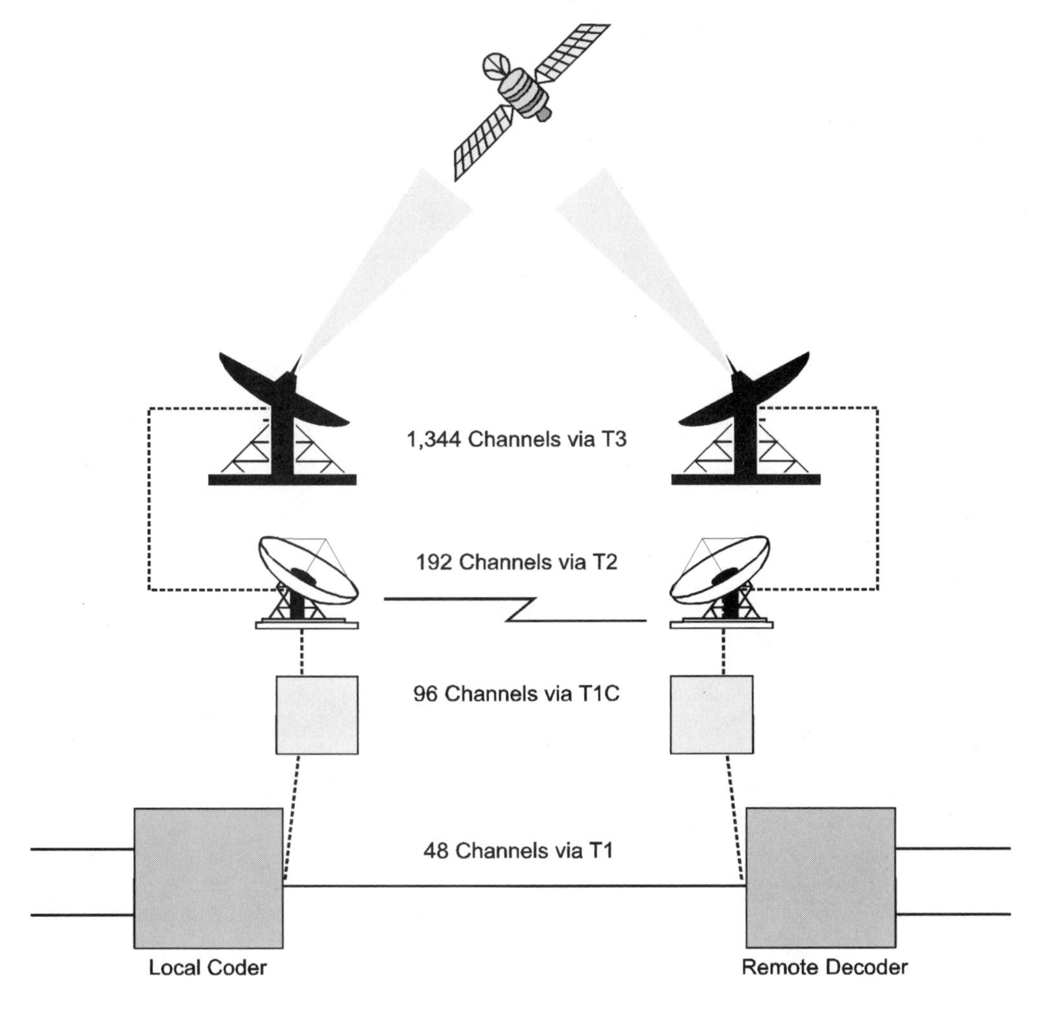

Fig. V3 *Typical network configuration employing ADPCM.*

It also is possible for ADPCM to compress voice to 16 kbps by encoding voice signals with only 2 bits, instead of 4 bits as discussed above. This level of compression provides 96 channels on a T1 line without significantly reducing signal quality.

Variable-rate ADPCM. Some vendors have designed ADPCM processors that not only compress voice, but accommodate 64 kbps pass-through as well. The use of very compact codes allows several different algorithms to be handled by the same ADPCM processor. The selection of algorithm is made by the network manager through software control. Variable-rate ADPCM offers several advantages.

Compressed voice is more susceptible to distortion than uncompressed voice—the 16 kbps rate more so than 32 kbps. When line conditions deteriorate to the point where voice compression is not possible without seriously disrupting com-

munication, a lower compression ratio may be invoked to compensate for the distortion. If line conditions do not permit compression even at 32 kbps, 64 kbps pass-through may be invoked to maintain quality voice communication. Channel availability is greatly reduced, of course, but the ability to communicate with the outside world becomes the overriding concern at this point, rather than the number of channels.

Variable-rate ADPCM provides opportunities to allocate channel quality based on the needs of different classes of user. For example, all intracompany voice links might operate at 16 kbps, while those used to communicate externally may be configured to operate at 32 kbps.

The number of channels may be increased temporarily by compressing voice to 16 kbps instead of 32 kbps until new facilities can be ordered, installed, and put into service. As new links are added to keep up with the demand for channels, others may be returned to 32 kbps operation. Variable-rate ADPCM, then, offers much more channel configuration flexibility than products that offer voice compression at only 32 kbps.

Other compression techniques. Other compression schemes can be used over T-carrier facilities, such as Continuously Variable Slope Delta (CVSD) modulation and Time Assigned Speech Interpolation (TASI).

CVSD. The higher the sampling rate, the smaller the average difference between amplitudes; at a high enough sampling rate (32,000 times per second in the case of 32 kbps voice), the average difference is small enough to be represented by only a single bit. This is the concept behind CVSD, where the bit represents the change in the slope of the analog curve. Successive 1s or 0s indicate that the slope should get steeper and steeper. This technique can result in very good voice quality if the sampling rate is high enough.

Like ADPCM, CVSD will yield 48 voice channels at 32 kbps on a T1 line. But CVSD is more flexible than ADPCM in that it can provide 64 voice channels at 24 kbps, or 96 voice channels at 16 kbps. This is because the single-bit words are sampled at the signaling rate. Thus, to achieve 64 voice channels, the sampling rate is 24,000 times per second, while 96 voice channels takes only 16,000 samples per second. In reducing the sampling rate to obtain more channels, however, the average difference between amplitudes becomes greater. And since the greater difference between amplitudes is still represented by a single bit, there is a noticeable drop in voice quality. Thus the flexibility of CVSD comes at the expense of quality. It is even possible for CVSD to provide 192 voice channels at 8 kbps, but the quality of voice is so poor that this level of compression is rarely used.

TASI. Because people normally are not able to talk and listen simultaneously, network efficiency at best is only 50 percent. And because all human speech contains pauses, which constitute wasted time, network efficiency is further reduced by as much as 10 percent, putting maximum network efficiency at only 40 percent.

Statistical voice compression techniques, such as Time Assignment Speech Interpolation (TASI), take advantage of this quiet time by interleaving various other conversation segments together over the same channel. TASI-based systems actually seek out and detect the active speech on any line and assign only active talkers to the T1 facility. TASI thus makes more efficient utilization of "time" to double T1 capacity. At the distant end, the TASI system sorts out and reassembles the interwoven conversations on the line to which they were originally intended.

The drawback to statistical compression methods is that they have trouble maintaining consistent quality. This is because such techniques require a high number of channels (at least 100) from which a good statistical probability of usable quiet periods may be gleaned. With as few as 72 channels, however, a channel gain ratio of 1.5:1 may be achieved. If the input channels are too few, a condition known as *clipping* may occur, in which speech signals are deformed by the cutting off of initial or final syllables.

A related problem with statistical compression techniques is *freeze out,* which usually occurs when all trunks are in use during periods of heavy traffic. In such cases, a sudden burst in speech energy can completely overwhelm the total available bandwidth, resulting in the loss of entire strings of syllables. Another liability inherent in statistical compression techniques, even for large T1 users, is that they are not suitable for transmissions having too few quiet periods, as when facsimile and music-on-hold is used. Statistical compression techniques, then, work better in large configurations than in small ones.

Summary. Adding lines and equipment is one way organizations can keep pace with increases in traffic. But even when funds are immediately available for such network upgrades, communications managers must contend with the delays inherent in ordering, installing, and putting new facilities into service. To accommodate the demand for bandwidth in a timely manner, communications managers can apply an appropriate level of voice compression to obtain more channels out of the available bandwidth. Depending on the compression technique selected, there need not be a noticeable decrease in voice quality.

See also

DATA COMPRESSION

Wavelength division multiplexing

Wavelength division multiplexing (WDM) technology has been in use by long-distance carriers in recent years to expand the capacity of their trunks by allowing a greater number of signals to be carried on a single fiber. Although the technology has been in existence since the late 1980s, the need among carriers to get more performance and flexibility from their fiber-optic networks only arose in the mid-1990s. AT&T, Sprint, and MCI (among others) have made long-term commitments to WDM technology and will be using it to ramp up their trunk speeds from 2.5 Gbps to over 40 Gbps, without having to install additional fiber.

Applications. WDM will help eliminate capacity constraints in carrier networks brought on by the ever-increasing processing power of computers and the need to link multiple users with multiple sites. WDM will support applications such as the simultaneous distribution of full-motion video and medical images, without forcing carriers to rip out existing fiber backbones and replace them with higher-capacity links. The Department of Defense's Advanced Research Projects Agency (ARPA) also sees a role for WDM-based networks. It intends to deploy WDM to move large amounts of data and images, such as satellite or reconnaissance photos, over long distances in real time.

WDM works with a variety of existing protocols and technologies, such as Synchronous Optical Network (SONET) services, ranging from OC-1 (51.8 Mbps) to OC-192 (9.953 Gbps), and broadband Asynchronous Transfer Mode cell switching.

Operation. WDM allows carriers to divide and condense standard fiber-optic transmissions into separate wavelengths; each wavelength carries different content. Multiple data channels are transmitted over a single optical fiber using distinct colors of light, or optical wavelengths. This is similar to the way radio stations broadcast at different wavelengths without interfering with one another.

Early WDM systems offered only 2 or 4 of these widely spaced channels, which was not particularly cost-effective. But with technological improvements, vendors now offer "dense" WDM solutions that can segment a standard OC-48 (2.488 Gbps) line into as many as 8, 16, or 32 separate channels, each offering a transport rate of up to 2.5 Gbps.

Unidirectional and bidirectional fiber amplifiers impact the way WDM is implemented. In a unidirectional system, two fiber lines with one-way amplifiers are needed for two-way communications. For a bidirectional system, one fiber line with bidirectional amplifiers is needed for two-way communications.

Experimental WDM systems have increased the capacity of fiber to 160 Gbps over short distances. With continual improvements, WDM systems may one day reach the terabits-per-second (Tbps) range.

Summary. WDM technology provides the solution for economically increasing network capacity, without the expense of installing new fiber, to meet the increasing demand for more bandwidth. Adoption of WDM will be an important step toward the goal of all-optical networking, in which optical-to-electrical conversions are minimized by moving more transport and switching duties into the optical domain.

See also

FIBER OPTICS TECHNOLOGY
SYNCHRONOUS OPTICAL NETWORK

Wide Area Telecommunications Service

Wide Area Telecommunications Service, or WATS, was introduced by AT&T in 1961 to allow customers to receive substantial discounts on telephone calls, provided that they stayed with the service for a specified time and adhered to a minimum monthly revenue commitment. The term *WATS* has fallen into disuse in recent years in favor of carrier brand names.

This type of service was basically a bulk-rate toll service priced according to call distance, or rate band. Customers had to commit to the service for a specified duration, usually from 18 to 48 months. Discounts expanded with the length of the plan and the customer's monthly revenue commitment. If a customer did not meet the monthly minimum revenue level, usage charges were adjusted up to that minimum level. Also, if a customer canceled the plan before it expired, the company was billed for the discount accumulated up to the time of cancellation.

The traditional banded WATS facilities introduced 36 years ago have been replaced by more flexible and manageable WATS-like services, in which billing is based on time of day and call duration, as well as distance. Customers not only qualify for additional savings based on call volume, but on the number of corporate locations enrolled in the plan as well.

There is an inbound version of WATS, which allows businesses to offer toll-free calls to their customers and other constituents via 800 and 888 services. Carriers offer either switched-access or dedicated-access 800 and 888 service. Switched service provides businesses with the ability to receive 800 and 888 calls over regular telephone lines, while dedicated service provides a private connection from the carrier's network to the business's network.

Each service provides several options. For example, businesses can geographically screen their calls, or block calls from certain parts of the country; other services automatically route calls to specific locations based on customer-specified requirements.

The larger IXCs now also offer 800 service for international calls. Callers outside the United States use country-specific numbers to route calls to a company's

access line in the United States. There is even personal 800 service for individuals who work out of the home and low call volumes.

Detailed billing reports for 800 services, available on customer request, provide call detail and exception reporting. The carriers also offer call detail information in real time, or on a monthly or daily basis. Such services can be used to measure marketing responses, track lost calls, and gauge the effectiveness of call center operations.

Summary. Over the years, carriers have continually revamped their WATS offerings to take into account changing market conditions. WATS started as an offering for only the largest companies; today a business with only one telephone line can realize significant cost savings with WATS tailored to its specific needs.

Wireless Communications Service

Wireless Communications Service (WCS) is a new category of service that operates at frequencies of 2305 to 2320 MHz and 2345 to 2360 MHz. These frequencies can be used for fixed, mobile, radio-location, or broadcast-satellite (sound) transmission. The FCC completed the auction process for WCS spectrum in April, 1997. There were 17 winning license bidders, eight of whom qualified as small or very small businesses under the FCC's rules.

The FCC granted licensees wide latitude in how WCS spectrum is used. One use for the WCS band of spectrum is the transmission of the personal access communications system (PACS) standard. This would be ideal for consumer-oriented products, such as personal cordless applications, because WCS uses small relay stations and is designed to interface with the existing telephone network.

Where WCS poses interference problems with existing Multipoint Distribution Service (MDS) or Instructional Television Fixed Service (ITFS) operations, the WCS licensees must bear the full financial obligation for the remedy. WCS licensees must notify potentially affected MDS/ITFS licensees, at least 30 days before commencing operations from any new WCS transmission site or increasing power from an existing site, of the technical parameters of the WCS transmission facility. The FCC expects that WCS and MDS/ITFS licensees will coordinate voluntarily and in good faith to avoid interference problems, which will result in the greatest operational flexibility in each of these types of operations.

Summary. The auction for WCS spectrum was expected to bring in about US$2 billion to the U.S. Treasury, but raised only US$13.6 million. Faced with an apparent failure, the FCC claimed that the real issue was not the money raised, but getting spectrum into the marketplace as quickly as possible, and meeting tight congressionally mandated time frames. Because of the short notice of the auction, however, many prospective bidders could not determine how best to use the WCS spectrum and consequently did not bid at all.

See also

MICROWAVE COMMUNICATIONS

Workflow automation

Workflow is the application of computers and networks to automate previously manual business processes. This kind of automation is especially beneficial to or-

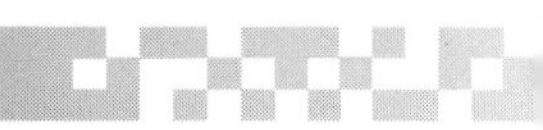

ganizations that rely on standardized forms and multiple steps to complete a transaction, such as fulfilling orders, processing claim forms, and reconciling customer accounts. Instead of just dumping all information into a database, for example, logical queues of documents are established, allowing networked workstation operators to obtain the next available document for processing. This results in more efficient business operations, increased productivity from available staff, faster customer response, and greater accuracy of information.

The workflow process. In an insurance claims processing application (Fig. W1), for example, incoming documents are scanned into the system when received. Then they are indexed by the claim number, name, form type, and scan date. All documents for the same case can be grouped together in an electronic file folder. When a file folder is opened, the whole folder (or individual documents) is routed immediately to the appropriate queue. Queues reside in the image server and are dedicated to particular workstation operators. Other needed information, such as customer account histories, might reside in a mainframe database and can be integrated into the document via terminal emulation. Multiple mainframe sessions may be opened, each displayed in a separate window at the image workstation.

When a document processing step is completed, the file is forwarded to the next step by placing it in the appropriate logical queue, where it can be accessed by the next available operator assigned to that task. If further work is needed on the document, it can be sent back to the person who previously handled it, or it can be held in suspense until additional information is obtained. Supervisors can easily monitor the movement and status of documents and files to ensure that high-prior-

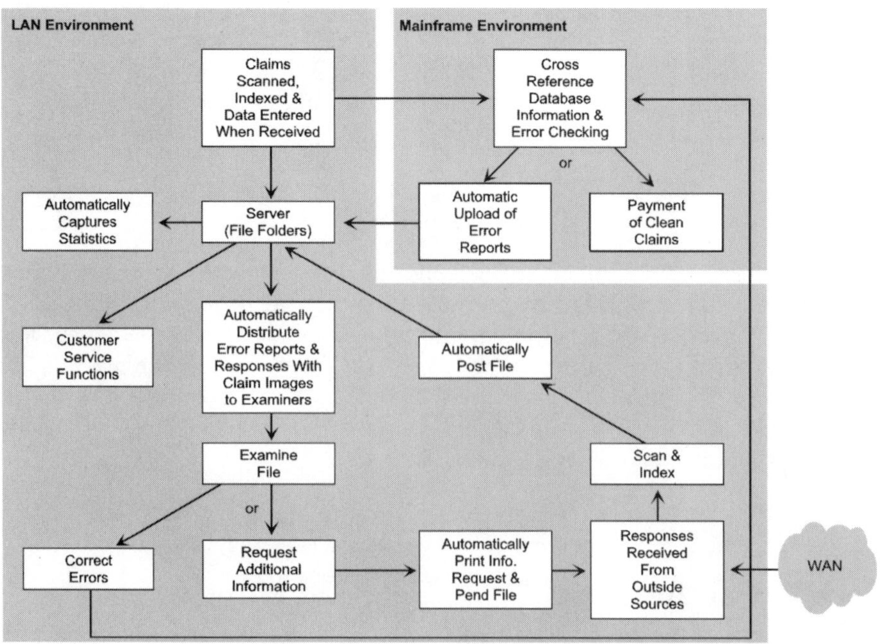

Fig. W1 *The workflow of a typical insurance claims process.*

ity cases are handled expeditiously, exception cases get the attention they deserve, the workload is distributed fairly, and operators meet productivity goals.

A workflow script development kit allows workgroup administrators to automate, manage, and control the queuing and flow of images, data, and text. A script basically is a set of instructions, or calls, that can be activated upon command. The instructions consist of functional statements that automate various tasks such as storing documents on optical disk, retrieving them for processing, moving data from one application to another, and routing information from workstation to workstation. Scripts work together in sets to complete the tasks assigned to the system, with each script performing one or more steps of the task. The advantage of scripts is that the various steps of a task can be put into the order that best fits the application, providing administrators with greater control over the operation.

The role of networks. LANs are particularly well suited for workflow applications because they offer more flexibility and scalability than either minicomputer or mainframe solutions. LANs are able to support the transfer of documents between all nodes—workstations, scanners, storage devices, printers and facsimile machines—thereby mimicking the movement of documents in the paper-based work environment. LANs also allow for greater flexibility in host access, since users typically have a variety of gateways and terminal emulators from which to choose.

LANs also provide more opportunities for performance tuning. Using LAN management systems, analysis tools, and report generators, the administrator can accurately measure performance and take immediate steps to make improvements, such as segmenting the network into subnets, upgrading cache memory, using compression, or adding higher-performance peripherals. Because workflow demands can vary on a daily basis, the ability to respond quickly to changing loads constitutes a key benefit of LAN-based workflow systems over mainframe-based systems.

Network performance can be maintained by putting resource-intensive services, such as scanning, on subnetworks. These subnets can be selectively isolated from the rest of the network using such devices as bridges or routers. This would allow a large accounts payable department, for example, to scan 10,000 invoices a day without bogging down the main network.

Workflow applications are not limited to running over LANs. With the trend toward distributed operations, often LANs must be linked over the wide area network (WAN) via digital facilities. In such cases, the workflow application supports data compression to minimize activity on the network. Compression is especially important when imaged documents must traverse lower-speed WAN links. Table W1 provides some comparisons on the number of images per hour that can be sent over links of various speeds in compressed form.

Summary. Workflow automation relies on LANs and WANs to move documents within a structured transaction processing environment. This kind of automation streamlines the processing of documents, speeds the distribution of information to the right people, and enhances the productivity of corporate staff who use the information on a daily basis. In turn, corporate responsiveness to customers, suppliers, and other constituencies can be greatly improved.

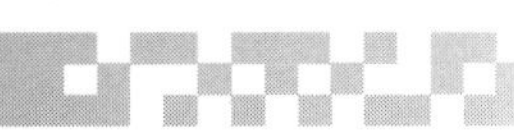

See also
BUSINESS PROCESS REENGINEERING

Table W1. Transmission Times for Compressed Images over the WAN.

Transmission Rate	Image size	Sec/image	Images/hr
56 kbps	50 KB	7.94	168
56 kbps	75 KB	11.90	112
56 kbps	100 KB	15.87	84
384 kbps	50 KB	1.16	1152
384 kbps	75 KB	1.74	768
384 kbps	100 KB	2.31	576
1.536 Mbps	50 KB	.29	4608
1.536 Mbps	75 KB	.43	3072
1.536 Mbps	100 KB	.58	2304

World Trade Organization

Established in January, 1995, the World Trade Organization (WTO) resulted from the Uruguay Round trade negotiations and is the successor to the General Agreement on Tariffs and Trade (GATT). As such, the WTO is the legal and institutional foundation of the multilateral trading system. It provides the principal framework within which governments develop and implement domestic trade legislation and regulations. It is also the platform on which trade relations among countries evolve through collective debate, negotiation, and adjudication. One of the principle objectives of the WTO is the reduction of tariffs and other trade barriers, as well as the elimination of discriminatory treatment in international trade relations.

Structure. The WTO Secretariat is located in Geneva, Switzerland. It has about 450 staff members and is headed by a Director-General. The highest WTO authority is the Ministerial Conference, which meets every other year. The daily work of the WTO, however, falls to a number of subsidiary bodies, principally the General Council, which also convenes as the Dispute Settlement Body and as the Trade Policy Review Body. The General Council delegates responsibility to three other major bodies: the Council for Trade in Goods, the Council for Trade in Services, and the Council for Trade-Related Aspects of Intellectual Property Rights.

Several other bodies have been established by the Ministerial Conference and report to the General Council: the Committee on Trade and Development; the Committee on Balance of Payments; the Committee on Budget, Finance and Administration; and the Committee on Trade and Environment.

The WTO budget is around US$83 million, with individual contributions calculated on the basis of shares in the total trade conducted by members.

Impact on telecommunications. The WTO successfully concluded nearly three years of extended negotiations on market access for basic telecommunications services in February, 1997. A total of 71 governments, accounting for more than 91 percent of global telecommunications revenues in 1995, agreed to set aside national differences in how basic telecommunications might be defined domestically and to negotiate on all telecommunications services, both public and private, that involve end-to-end transmission of customer-supplied information (e.g., simply the transmission of voice or data from sender to receiver).

They also agreed that basic telecommunications services provided over network infrastructure, as well as those provided through resale (over private leased circuits), would fall within the scope of market access commitments. As a result, market access commitments will cover not only cross-border supply of telecommunications, but also services provided through the establishment of foreign firms, or commercial presence, including the ability to own and operate independent telecom network infrastructures.

Examples of the services under negotiation were voice telephony, data transmission, telex, telegraph, facsimile, private leased circuit services (i.e., the sale or lease of transmission capacity), fixed and mobile satellite systems and services, cellular telephony, mobile data services, paging, and personal communications services (PCS).

Value-added services, or telecommunications for which suppliers add value to the customer's information by enhancing its form or content, or by providing for its storage and retrieval, were not formally part of the extended negotiations. Nevertheless, a few participants chose to include them in their offers. Examples include online data processing, online database storage and retrieval, electronic data interchange (EDI), e-mail, and voice mail.

Summary. Under the WTO agreement, which became effective January 1, 1997, each of the signatories agreed to allow resale of current monopoly carrier services, interconnect competitive public networks with existing networks, and let foreign carriers buy or build their own networks. The principal result of the agreement will be competitive pressure to eventually knock down the high cost of international calling, which now averages 99 cents a minute. It will take a number of years, however, before this happens on a wide scale because of the complex system of international pricing and the extra time needed by many countries to implement various provisions of the agreement. Of note is that China and Russia did not sign the agreement.

See also

TELECOMMUNICATIONS ACT OF 1996

World Wide Web

Since its birth in 1990 at the European Particle Physics Laboratory (CERN) in Switzerland, the World Wide Web (WWW) has grown to become one of the most sophisticated and popular services on the Internet. Although no specific organization exercises administrative control of the Web, order is imposed by the languages and protocols that constitute worldwide standards, such as the HyperText Transfer Protocol (HTTP) and HyperText Markup Language (HTML).

HTTP is used to transfer hypertext documents among servers on the Internet and, ultimately, to a client, namely the end user's browser-equipped computer (Fig.

W2). Collectively, the tens of thousands of servers distributed worldwide that support HTTP are known as the World Wide Web. HTML is used to structure information that resides on the servers in a way that can be readily rendered by browser software (such as Internet Explorer, Netscape or Mosaic) installed on the clients (Fig. W3). HTML makes documents portable from one computer platform to another and is intended as a common medium for tying together information from widely different sources.

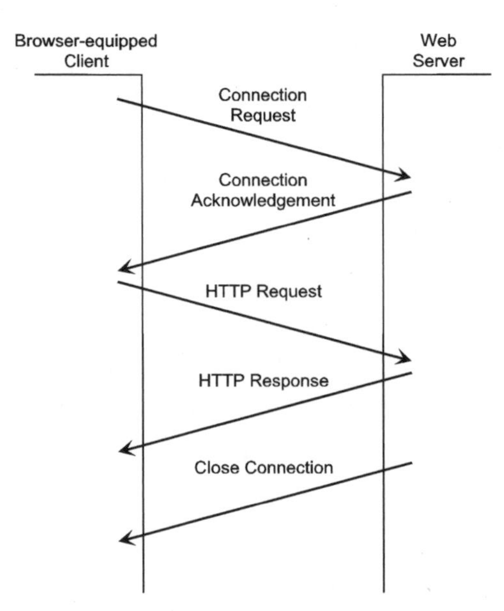

Fig. W2 *The HyperText Transfer Protocol (HTTP) delivers documents from Web servers to browser-equipped clients in response to specific requests, then closes the connection until a new request is made from the client.*

Web characteristics. The Web itself can best be described as a dynamic, interactive, graphically oriented, distributed, platform-independent, hypertext information system.

- The WWW is *dynamic* because it changes daily. New Web servers are continually being added to the Web. New information also is continually being added, as are new hypertext links and innovative services.
- The WWW is *interactive* in that specific information can be requested through various search engines and returned moments later in the form of lists, with each item weighted according to how well it matched the search parameters. Another example of interactivity is online forms for business transactions, whereby users can select items from a catalog and fill out an order form. Credit card numbers are verified and processed through a third-party service. In the case of software purchases, the product can be downloaded from the vendor to the buyer immediately after credit card verification.

- The WWW is *graphically oriented;* in fact, it was designed for the extensive use of graphics. The use of graphics not only makes the Web visually appealing, but easy to navigate as well. Graphical signposts direct users to new sources of information accessed via hypertext links. More recently, sound and video capabilities have been added to the Web.
- The WWW is *distributed,* meaning that information resides on tens of thousands of individual Web servers around the world. If one site goes down, there is no significant impact on the Web as a whole, except that access to that site will be denied. Some servers are mirrored at other sites to keep information available, even if the primary server crashes.
- The WWW is *platform-independent,* which means that virtually any client can access the Web, whether it is based on the Windows, OS/2, Macintosh, or Unix operating environment. This platform independence even applies to the Web servers. Although most Web servers are based on Unix, Windows NT is growing in popularity and may become the platform of choice among developers of new sites.
- The WWW makes extensive use of *hypertext* links. A hypertext link is usually identified by an underlined word or phrase, or a graphic symbol that points the way to other information, which might be found virtually anywhere: the same document, a different document on the same server, or another document on a different server that could be located anywhere in the world. A hypertext link does not necessarily point to text documents; it can point to maps, forms, images, sound and video clips, applications, or other Internet services. Hypertext links can even point to other Internet resources such as FTP and Gopher sites and Usenet newsgroups (Fig. W4).

Business applications. Many companies are finding that the Web is an ideal medium for distributing documents and software, while others are experimenting with electronic commerce. Some companies use the Web to enhance network support. Among the routine tasks that can be implemented over the Web are:

- LAN managers at distributed locations can access an HTML-coded database stored in an internal Web server to troubleshoot system and network problems. The Web server can be a valuable adjunct to the help desk, especially when corporate locations are spread across multiple time zones.
- Service requests can be dispatched electronically to carriers, vendors, and third-party maintenance firms via standardized forms written in HTML.
- For remote sites that are too small to be economically tied into the corporate backbone network, the use of HTML forms can convey move, add, and change information to a central management console to expedite inventory management. Among other things, forms also can be used to report trouble and request technical assistance.

The Web is not only being used for these mundane tasks, but for more sophisticated network support as well. One of the first management applications to make

Fig. W3 *Web browsers, such as Microsoft's Internet Explorer, provide the means to access documents, publications, multimedia services, games, and other content on the World Wide Web.*

use of the Web came from Tribe Computer Works (acquired by Zoom Telephonics in 1996), a provider of routers, switches and remote-access servers. Via firmware called WebManage, which contains an integral ROM-based home page to display and configure network device settings, customers can view and interact with the devices using Netscape Navigator (Fig. W5). Using hypertext links for quick movement between management functions and online resources, including Tribe's Web servers, WebManage allows a network manager to get immediate answers to setup or troubleshooting questions. Different views and access privileges can even be created that vary by user login and password. WebManage complements existing

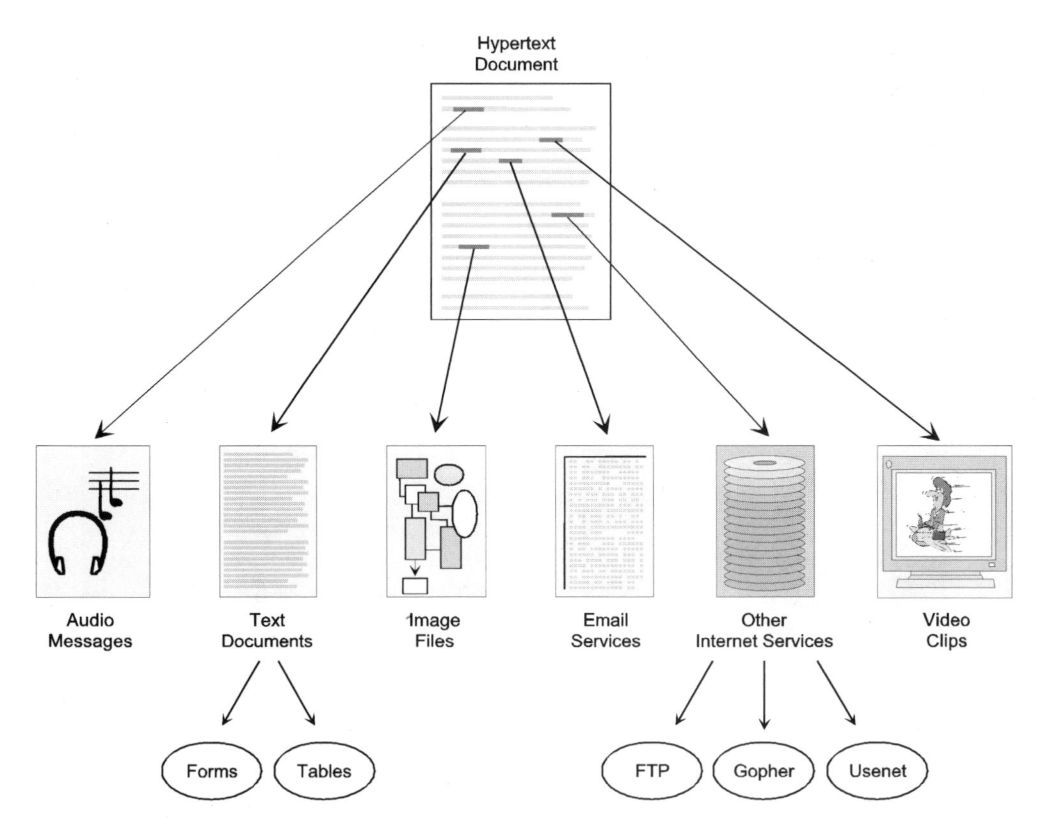

Fig. W4 *Hypertext documents on the World Wide Web can be multimedia in nature, providing links to audio messages, image files, video clips, and other Internet services.*

SNMP installations by providing enhanced device configuration and troubleshooting capabilities, whereas SNMP is most useful for the global monitoring of diverse network equipment.

Summary. The World Wide Web is an ideal medium for information distribution, electronic commerce, and the delivery of support services. The capabilities of the Web are continually being expanded. In addition to text and images, the Web is being used for telephony, videoconferencing, faxing and paging, and collaborative computing.

See also
INTERNET
INTRANETS

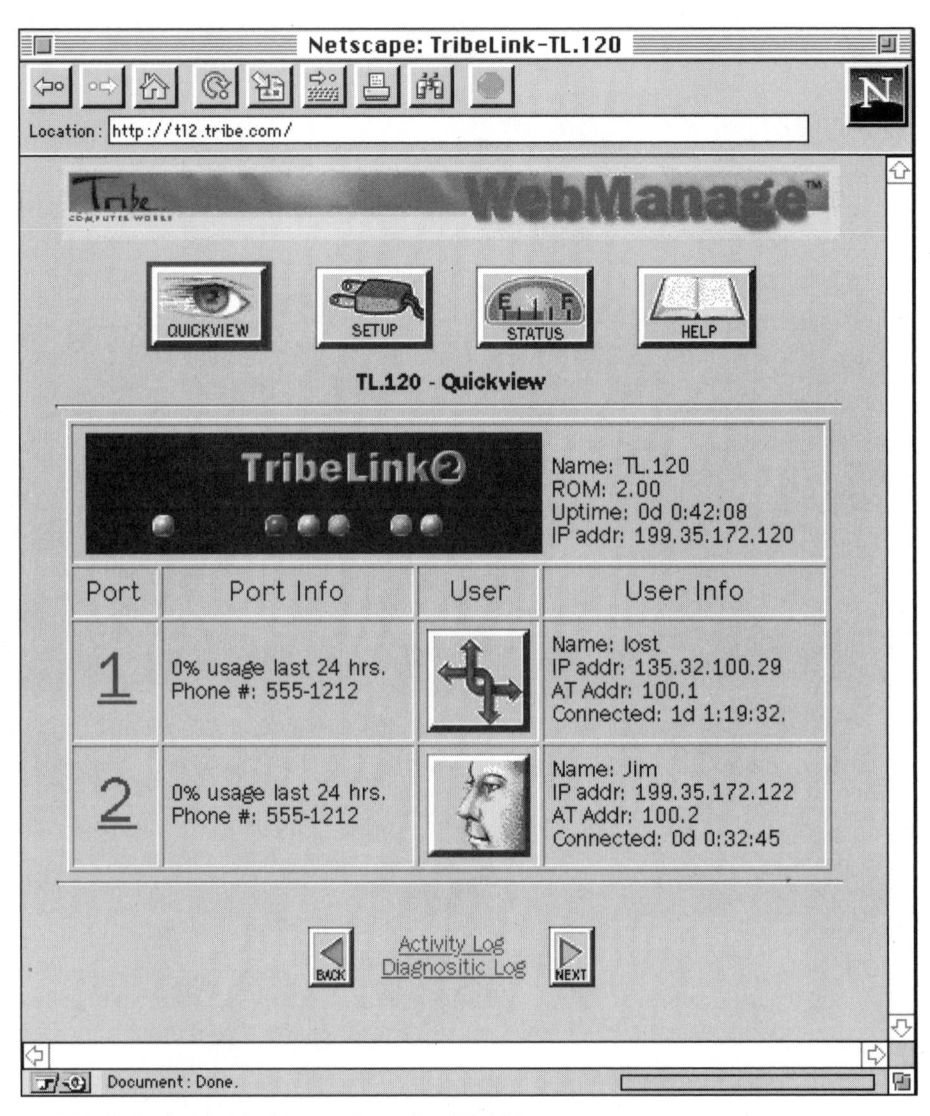

Fig. W5 *TribeLink's Home Page for WebManage.*

Appendix

Abbreviations and Acronyms

AAL	ATM Adaptation Layer
ABATS	Automated Bit Access Test System
ABM	Accunet Bandwidth Manager (AT&T)
ABR	Available Bit Rate
AC	Access Control
AC	Address Copied
ac	alternating current
AC	Authentication Center
ACD	Automatic Call Distributor
ACELP	Algebraic Code Excited Linear Predictive
ACP	Access Control Point
ADA	Americans with Disabilities Act
ADCR	Alternate Destination Call Routing (AT&T)
ADM	Add-Drop Multiplexer
ADN	Advanced Digital Network (Pacific Bell)
ADPCM	Adaptive Differential Pulse Code Modulation
ADSL	Asymmetrical Digital Subscriber Line
AFP	AppleTalk Filing Protocol

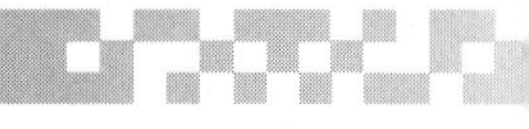

AGRAS	Air-Ground Radiotelephone Automated Service
AIOD	Automatic Identification of Outward Dialed calls
AIN	Advanced Intelligent Network
ALI	Automatic Location Information
AM	Amplitude Modulation
AMPS	Advanced Mobile Phone Service
ANI	Automatic Number Identification
ANR	Automatic Network Routing (IBM Corp.)
ANSI	American National Standards Institute
ANT	ADSL Network Terminator
AOL	America Online
APC	Access Protection Capability (AT&T)
API	Application Programming Interface
APPC	Advanced Program-to-Program Communications (IBM)
APPN	Advanced Peer-to-Peer Network (IBM)
APC	Automatic Protection Switching
ARCnet	Attached Resource Computer Network (Datapoint)
ARB	Adaptive Rate-Based (IBM)
ARP	Address Resolution Protocol
ARPA	Advanced Research Projects Agency
ARQ	Automatic Repeat Request
ARS	Action Request System (Remedy Systems)
AS	Autonomous System
ASCII	American Standard Code for Information Interchange
ASIC	Application-Specific Integrated Circuit
ASN.1	Abstract Syntax Notation 1
ASTN	Alternate Signaling Transport Network (AT&T)
AT&T	American Telephone & Telegraph
ATIS	Alliance for Telecommunications Industry Solutions (formerly ECSA)
ATM	Asynchronous Transfer Mode
ATSC	Advanced Television Systems Committee
AUI	Attachment Unit Interface
AWG	American Wire Gauge
B8ZS	Binary Eight Zero Substitution
BACP	Bandwidth Allocation Control Protocol
BBS	Bulletin Board System
BCCH	Broadcast Control Channel
BDCS	Broadband Digital Crossconnect System
BECN	Backward Explicit Congestion Notification
Bellcore	Bell Communications Research, Inc.
BER	Bit Error Rate
BERT	Bit Error Rate Test (or Tester)
BGP	Border Gateway Protocol
BHCA	Busy Hour Call Attempts
BIB	Backward Indicator Bit
BIOS	Basic Input-Output System

BMC	Block Multiplexer Channel (IBM)
BMS-E	Bandwidth Management Service-Extended (AT&T)
BOC	Bell Operating Company
BONDING	Bandwidth on Demand Interoperability Group
BootP	Boot Protocol
BPDU	Bridge Protocol Data Unit
bps	bits per second
BPV	Biploar Violation
BRI	Basic Rate Interface (ISDN)
BSA	Basis Serving Arrangement
BSC	Base Station Controller
BSC	Binary Synchronous Communications
BSE	Basic Service Element
BSN	Backward Sequence Number
BTS	Base Transceiver Station
CA	Communications Assistant
CAD	Computer-Aided Design
CAM	Computer-Aided Manufacturing
CAN	Campus Area Network
CAP	Carrierless Amplitude/Phase (modulation)
CAP	Competitive Access Provider
CARS	Cable Antenna Relay Services
CASE	Computer-Aided Software Engineering
CATV	Cable Television
CB	Citizens Band
CBR	Constant Bit Rate
CCC	Clear Channel Capability
CCCH	Common Control Channel
CCITT	Consultative Committee for International Telegraphy and Telephony
CCR	Customer Controlled Reconfiguration
CCS	Common Channel Signaling
CCSNC	Common Channel Signaling Network Controller
CCSS 6	Common Channel Signaling System 6
CD	Compact Disk
CDCS	Continuous Dynamic Channel Selection
CD-R	Compact Disk–Recordable
CD-ROM	Compact Disk–Read-Only Memory
CDMA	Code Division Multiple Access
CDO	Community Dial Office
CDPD	Cellular Digital Packet Data
CDR	Call Detail Recording
CEI	Comparably Efficient Interconnection
CENTREX	Central Office Exchange
CGSA	Cellular Geographic Servicing Areas
CHAP	Challenge Handshake Authentication Protocol
CIF	Common Intermediate Format

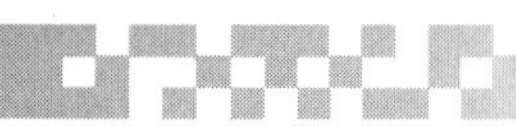

CIR	Committed Information Rate
CLASS	Custom Local Area Signaling Services
CLEC	Competitive Local Exchange Carrier
CLI	Calling Line Identification
CLP	Cell Loss Priority
CMI	Cable Microcell Integrator
CMIS	Common Management Information Services
CMRS	Commercial Mobile Radio Service
CNR	Customer Network Reconfiguration
CNS	Complementary Network Service
CO	Central Office
CON	Concentrator
COT	Central Office Terminal
CP	Coordination Processor
CPE	Customer Premises Equipment
cps	cycles per second (hertz)
CPU	Central Processing Unit
CRC	Cyclic Redundancy Check
CSA	Carrier Serving Area
CSM	Communications Services Management
CSMA/CD	Carrier Sense Multiple Access with Collision Detection
CSU	Channel Service Unit
CT	Cordless Telecommunications
CTI	Computer-Telephony Integration
CVSD	Continuously Variable Slope Delta (modulation)
D-AMPS	Digital Advanced Mobile Phone Service
DA	Destination Address
DACS	Digital Access and Crossconnect System (AT&T)
DAP	Demand Access Protocol
DAS	Dual Attached Station
DASD	Direct Access Storage Device (IBM)
DAT	Digital Audio Tape
dB	decibel
DBMS	Database Management System
DBS	Direct Broadcast Satellite
DBU	Dial Backup Unit
DCCH	Digital Control Channel
DCE	Data Communications Equipment
DCE	Distributed Computing Environment
DCF	Data Communication Function
DCS	Digital Crossconnect System
DDS	Digital Data Services
DDS/SC	Digital Data Service with Secondary Channel
D/E	Debt/Equity (ratio)
DECT	Digital Enhanced (formerly, European) Cordless Telecommunication
DES	Data Encryption Standard
DFSMS	Data Facility Storage Management Subsystem (IBM)

DID	Direct Inward Dialing
DIF	Digital Interface Frame
DDL	Data Link Layer
DLCS	Digital Loop Carrier System
DLL	Dynamic Link Library
DLSw	Data Link Switching (IBM)
DLU	Digital Line Unit
DM	Distributed Management
DME	Distributed Management Environment
DMI	Desktop Management Interface
DMT	Discrete Multitone
DMTF	Desktop Management Task Force
DNS	Domain Name Service
DoD	Department of Defense
DOD	Direct Outward Dialing
DOS	Disk Operating System
DOV	Data Over Voice
DQDB	Distributed Queue Dual Bus
DQPSK	Differential Quadrature Phase-Shift Keying
DS0	Digital Signal Level 0 (64 kbps)
DS1	Digital Signal Level 1 (1.544 Mbps)
DS1C	Digital Signal Level 1C (3.152 Mbps)
DS2	Digital Signal Level 2 (6.312 Mbps)
DS3	Digital Signal Level 3 (44.736 Mbps)
DS4	Digital Signal Level 4 (274.176 Mbps)
DSI	Digital Speech Interpolation
DSL	Digital Subscriber Line
DSN	Defense Switched Network
DSP	Digital Signal Processor
DSS	Decision Support System
DSU	Data Service Unit
DSX1	Digital Systems Crossconnect 1
DTE	Data Terminal Equipment
DTMF	Dual-Tone Multifrequency
DTR	Dedicated Token Ring
DTU	Data Transfer Unit
DTV	Digital Television
DWMT	Discrete Wavelet Multitone
DXI	Data Exchange Interface
e-mail	electronic mail
E-TDMA	Expanded Time Division Multiple Access
ECSA	Exchange Carriers Standards Association
ED	Ending Delimiter
EDI	Electronic Data Interchange
EDRO	Enhanced Diversity Routing Option (AT&T)
EEROM	Electronically Erasable Read-Only Memory
EFRC	Enhanced Full-Rate Codec

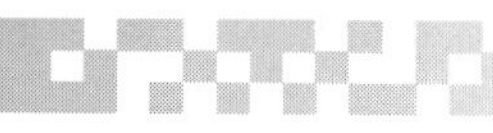

EFT	Electronic Funds Transfer
EGP	External Gateway Protocol
EHF	Extremely High Frequency (more than 30 GHz)
EIA	Electronic Industries Association
EIR	Equipment Identity Register
EISA	Extended Industry Standard Architecture
EMI	Electromechanical Inteference
EMS	Element Management System
EOC	Embedded Overhead Channel
EOT	End of Transmission
ESCON	Enterprise System Connection (IBM)
ESD	Electronic Software Distribution
ESF	Extended Superframe
ESMR	Enhanced Specialized Mobile Radio
ESN	Electronic Serial Number
ETSI	European Telecommunication Standards Institute
4GL	Fourth-Generation Language
FACCH	Fast Associated Control Channel
FASB	Financial Accounting Standards Board
FASC	Fraud Analysis and Surveillance Center (AT&T)
FASTAR	Fast Automatic Restoral (AT&T)
FAT	File Allocation Table
FC	Frame Control
FC	Fibre Channel
FC-0	Fibre Channel Layer 0
FC-1	Fibre Channel Layer 1
FC-2	Fibre Channel Layer 2
FC-3	Fibre Channel Layer 3
FC-4	Fibre Channel Layer 4
FCC	Federal Communications Commission
FCS	Frame Check Sequence
FDDI	Fiber Distributed Data Interface
FDIC	Federal Deposit Insurance Corporation
FDL	Facilities Data Link
FECN	Forward Explicit Congestion Notification
FEP	Front-End Processor
FIB	Forward Indicator Bit
FIB	Forwarding Information Base
FITL	Fiber In The Loop
FM	Frequency Modulation
FOCC	Forward Control Channel
FOD	Fax on Demand
FRAD	Frame Relay Access Device
FRS	Family Radio Service
FS	Frame Status
FSN	Forward Sequence Number
FTAM	File Transfer, Access, and Management

FT1	Fractional T1
FTP	File Transfer Protocol
FTS	Federal Telecommunications System
FTTB	Fiber To The Building
FTTC	Fiber To The Curb
FTTH	Fiber To The Home
FX	Foreign Exchange (line)
GATT	General Agreement on Tariffs and Trade
GDS	Generic Digital Services
GEO	Geostationary-Earth-Orbit
GFC	Generic Flow Control
GHz	gigahertz (billions of cycles per second)
GIS	Geographic Information Systems
GloBanD	Global Bandwidth on Demand
GMRS	General Mobile Radio Service
GPS	Global Positioning System
GSA	General Services Administration
GSM	Global System for Mobile (GSM) telecommunications (formerly Groupe Spéciale Mobile)
GUI	Graphical User Interface
H0	High-capacity ISDN channel operating at 384 kbps
H11	High-capacity ISDN channel operating at 1.536 Mbps
HDSL	High-bit-rate Digital Subscriber Line
HDTV	High-Definition Television
HEC	Header Error Check
HF	High Frequency (3 MHz to 30 MHz)
HFC	Hybrid Fiber/Coax
HIC	Head-end Interface Converter
HLR	Home Location Register
HPPI	High-Performance Parallel Interface
HPR	High-Performance Routing (IBM)
HSCSD	High-Speed Circuit Switched Data
HSM	Hierarchical Storage Management
HST	Helical Scan Tape
HTML	HyperText Markup Language
HTTP	HyperText Transfer Protocol
HVAC	Heating, Ventilation, and Air Conditioning
Hz	Hertz (cycles per second)
I/O	Input/Output
IAB	Internet Architecture Board
IANA	Internet Assigned Numbers Authority
ICI	Interexchange Carrier Interface
ICMP	Internet Control Message Protocol
ICR	Intelligent Call Routing
ICS	Intelligent Calling System
ID	Identification
IDDD	International Direct Dialing Designator

IDPR	Interdomain Policy Routing
IEC	International Electrotechnical Commission
IEEE	Institute of Electrical and Electronics Engineers
IESG	Internet Engineering Steering Group
IETF	Internet Engineering Task Force
IGP	Interior Gateway Protocol
ILEC	Incumbent Local Exchange Carrier
IMAP	Internet Mail Access Protocol
IMEI	International Mobile Equipment Identity
IMS/VS	Information Management System/Virtual Storage (IBM)
IMSI	International Mobile Subscriber Identity
IN	Intelligent Network
INMARSAT	International Maritime Satellite Organization
INMS	Integrated Network Management System
IOC	Interoffice Channel
IP	Internet Protocol
IPH	Integrated Packet Handler
IPI	Intelligent Peripheral Interface
IPN	Intelligent Peripheral Node
IPX	Internetwork Packet Exchange
IrDA	Infrared Data Association
IrLAN	Infrared LAN
IrLAP	Infrared Link Access Protocol
IrLMP	Infrared Link Management Protocol
IrPL	Infrared Physical Layer
IRQ	Interrupt Request
IrTTP	Infrared Transport Protocol
IS	Information System
IS	Industry Standard
IS-IS	Intra-autonomous System to Intra-autonomous System
ISA	Industry Standard Architecture
ISDL	ISDN Subscriber Digital Line
ISDN	Integrated Services Digital Network
ISM	Industrial, Scientific, and Medical (frequency bands)
ISO	International Organization for Standardization
ISOC	Internet Society
ISP	Internet Service Provider
ISSI	Interswitching Systems Interface
IT	Information Technology
ITFS	Instructional Television Fixed Service
ITR	Intelligent Text Retrieval
ITU-TSS	International Telecommunications Union-Telecommunications Standardization Sector (formerly CCITT)
IVR	Interactive Voice Response
IXC	Interexchange Carrier
JIT	Just In Time
JEPI	Joint Electronic Payments Initiative

JPEG	Joint Photographic Experts Group
JTC	Joint Technical Committee
k (kilo)	One thousand (e.g., kbps)
K (kilo)	One thousand and twenty-four (e.g., kilobyte)
KB	Kilobyte
KSU	Key Service Unit
KTS	Key Telephone System
kHz	kilohertz (thousands of cycles per second)
LAN	Local Area Network
LANCES	LAN Resource Extension and Services (IBM)
LAPB	Link Access Procedure-Balanced
LAT	Local Area Transport (Digital Equipment Corporation)
LATA	Local Access and Transport Area
LBO	Line Build-Out
LCD	Liquid Crystal Display
LCN	Local Channel Number
LCP	Link Control Protocol
LD	Laser Diode
LEC	Local Exchange CarrierLED Light-Emitting Diode
LEO	Low-Earth-Orbit
LF	Low Frequency (30 kHz to 300 kHz)
LI	Length Indicator
LLC	Logical Link Control
LMDS	Local Multipoint Distribution System
LMS	Location and Monitoring Service
LSI	Large-Scale Integration
LTG	Line Trunk Group
LU	Logical Unit (IBM)
M (mega)	One million (e.g., Mbps)
MAC	Media Access Control
MAC	Moves, Adds, Changes
MAN	Metropolitan Area Network
MAPI	Messaging Applications Programming Interface (Microsft)
MAU	Multistation Access Unit
MB	Megabyte
MCA	Micro Channel Architecture (IBM)
MCU	Multipoint Control Unit
MD	Mediation Device
MDF	Main Distribution Frame
MDS	Multipoint Distribution Service
MEO	Middle-Earth-Orbit
MES	Master Earth Station
MF	Mediation Function
MF	Medium Frequency (300 kHz to 3 MHz)
MHz	megahertz (millions of cycles per second)
MIB	Management Information Base
MIC	Management Integration Consortium

MIF	Management Information Format
MII	Media-Independent Interface
MIME	Multipurpose Internet Mail Extensions
MIN	Mobile Identification Number
MIPS	Millions of Instructions Per Second
MIS	Management Information Services
MISR	Multiprotocol Integrated Switch-Routing
MJU	Multipoint Junction Unit
MMDS	Multichannel, Multipoint Distribution Service
MO	Magneto-Optical
Modem	Modulator/demodulator
MPEG	Motion Picture Experts Group
MPPP	Multilink Point-to-Point Protocol
MRI	Magnetic Resonance Imaging
ms	millisecond (thousandths of a second)
MS	Mobile Station
MSC	Mobile Switching Center
MSN	Microsoft Network
MSRN	Mobile Station Roaming Number
MSS	Mobile Satellite Service
MTBF	Mean Time Between Failure
MTP	Message Transfer Part
MTSO	Mobile Transport Serving Office
MVC	Multicast Virtual Circuit
MVDS	Microwave Video Distribution System
MVPRP	Multivendor Problem Resolution Process
N-AMPS	Narrowband Advanced Mobile Phone Service
NAM	Numeric Assignment Module
NAP	Network Access Point
NAU	Network Addressable Unit (IBM)
NAUN	Nearest Active Upstream Neighbor
NC	Network Computer
NCP	Network Control Program (IBM)
NCP	Network Control Point
NE	Network Element
NEBS	New Equipment Building Specifications
NECA	National Exchange Carrier Association
NetBIOS	Network Basic Input-Output System
NEF	Network Element Function
NFS	Network File System (or Server)
NIC	Network Interface Card
NiCad	Nickel Cadmium
NiMH	Nickel-Metal Hydride
NIST	National Institute of Standards and Technology
NLM	NetWare Loadable Module (Novell)
nm	nanometer
NM	Network Manager

NMS	NetWare Management System (Novell)
NMS	Network Management System
NNM	Network Node Manager (Hewlett-Packard)
NNTP	Network News Transfer Protocol
NOS	Network Operating System
NPC	Network Protection Capability (AT&T)
NPV	Net Present Value
NSA	National Security Agency
NSF	National Science Foundation
NTSA	Networking Technical Support Alliance
NTSC	National Television Standards Committee
OAM	Operations, Administration, Management
OAM&P	Operations, Administration, Maintenance and Provisioning
OC	Optical Carrier
OC-1	Optical Carrier Signal Level 1 (51.84 Mbps)
OC-3	Optical Carrier Signal Level 3 (155.52 Mbps)
OC-9	Optical Carrier Signal Level 9 (466.56 Mbps)
OC-12	Optical Carrier Signal Level 12 (622.08 Mbps)
OC-18	Optical Carrier Signal Level 18 (933.12 Mbps)
OC-24	Optical Carrier Signal Level 24 (1.244 Gbps)
OC-36	Optical Carrier Signal Level 36 (1.866 Gbps)
OC-48	Optical Carrier Signal Level 48 (2.488 Gbps)
OC-96	Optical Carrier Signal Level 96 (4.976 Gbps)
OC-192	Optical Carrier Signal Level 192 (9.952 Gbps)
OC-256	Optical Carrier Signal Level 256 (13.271 Gbps)
OCR	Optical Character Recognition
OCUDP	Office Channel Unit Data Port
ODBC	Open Database Connectivity (Microsoft)
ODS	Operational Data Store
OEM	Original Equipment Manufacturer
OFX	Open Financial Exchange
OLAP	Online Analytical Processing
OLE	Object Linking and Embedding
OMA	Object Management Architecture
OMAP	Operations, Maintenance, Administration, and Provisioning
OMF	Object Management Framework
OMG	Object Management Group
OOP	Object-Oriented Programming
OPX	Off-Premises Extension
ORB	Object Request Broker
OS	Operating System
OS/2	Operating System/2 (IBM)
OSF	Open Software Foundation
OSF	Operations Systems Function
OSI	Open Systems Interconnection
OSS	Operations Support Systems
OTDR	Optical Time Domain Reflectometry

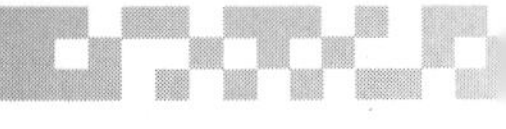

PA	Preamble
PACS	Personal Access Communications System
PAD	Packet Assembler-Disassembler
PAL	Phase Alternating by Line
PAP	Password Authentication Protocol
PBX	Private Branch Exchange
PC	Personal Computer
PCB	Printed Circuit Board
PCH	Paging Channel
PCM	Pulse Code Modulation
PCN	Personal Communications Networks
PCS	Personal Communications Services
PCT	Private Communication Technology
PDA	Personal Digital Assistant
PDN	Packet Data Network
PDU	Payload Data Unit
PEM	Privacy-Enhanced Mail
PGP	Pretty Good Privacy
PHS	Personal Handyphone System
PHY	Physical Layer
PIM	Personal Information Manager
PIN	Personal Identification Number
PIN	Positive-Intrinsic-Negative
PLMRS	Private Land Mobile Radio Services
PMD	Physical Media-Dependent
PnP	Plug-and-Play
POP	Point of Presence
POP	Post Office Protocol
POS	Point of Sale
POTS	Plain Old Telephone Service
PPP	Point-to-Point Protocol
PRI	Primary Rate Interface (ISDN)
PSAP	Public Safety Answering Point
PSN	Packet-Switched Network
PSTN	Public Switched Telephone Network
PT	Payload Type
PTT	Post, Telephone, and Telegraph
PU	Physical Unit (IBM)
PUC	Public Utility Commission
PVC	Permanent Virtual Circuit
QA	Quality Assurance
QAM	Quadrature Amplitude Modulation
QCIF	Quarter Common Intermediate Format
QIC	Quarter-Inch Cartridge
QoS	Quality of Service
QPSK	Quadrature Phase-Shift Keying
RACH	Random Access Channel

RAD	Remote Antenna Driver
RAID	Redundant Array of Inexpensive Disks
RAM	Random Access Memory
RASDL	Rate Adaptive Digital Subscriber Line
RASP	Remote Antenna Signal Processor
RBES	Rule-Based Expert Systems
RCU	Remote Control Unit
RDBMS	Relational Database Management System
RDSS	Radio Determination Satellite Service
RECC	Reverse Control Channel
RFC	Request for Comment
RF	Radio Frequency
RF	Routing Field
RFI	Radio Frequency Interference
RFI	Request for Information
RFP	Request For Proposal
RFQ	Request for Quotation
RIP	Routing Information Protocol
RISC	Reduced Instruction Set Computing
RJE	Remote Job Entry
RMON	Remote Monitoring
ROI	Return on Investment
ROM	Read-Only Memory
RPC	Remote Procedure Call
RSVP	resource ReSerVation Protocol
RT	Remote Terminal
RTNR	Real-Time Network Routing (AT&T)
RTP	Rapid Transfer Protocol (IBM)
RX	Receive
SA	Source Address
SAFER	Split Access Flexible Egress Routing (AT&T)
SAP	Second Audio Program
SAS	Single Attached Station
SBCCS	Single-Byte Command Code Set (IBM)
SCC	Standards Coordinating Committees (IEEE)
SCP	Service Control Point
SCSI	Small Computer Systems Interface
SD	Starting Delimiter
SDCCH	Stand-alone Dedicated Control Channel
SDH	Synchronous Digital Hierarchy
SDLC	Synchronous Data Link Control (IBM)
SDM	Subrate Data Multiplexing
SDN	Software-Defined Network (AT&T)
SDP	Service Delivery Point
SDSL	Symmetric Digital Subscriber Line
SET	Secure Electronic Transaction
SFD	Start Frame Delimiter

SHF	Super High Frequency (3 GHz to 30 GHz)
SHTTP	Secure HyperText Transfer Protocol
SIF	Signaling Information Field
SIM	Subscriber Identity Module
SIP	SMDS Interface Protocol
SLIC	Serial Line Interface Coupler (IBM)
SLIP	Serial Line Internet Protocol
SMDI	Station Message Desk Interface
SMDR	Station Message Detail Recording
SMDS	Switched Multimegabit Data Services
SMR	Specialized Mobile Radio
SMS	Service Management System
SMS	Short Message Service
SMT	Station Management
SMTP	Simple Mail Transfer Protocol
SN	Switching Network
SNA	Systems Network Architecture (IBM)
snagas	SNA Gateway Access Server
SNI	Subscriber Network Interface
SNMP	Simple Network Management Protocol
SONET	Synchronous Optical Network
SPA	Software Publishers Association
SPC	Stored Program Control
SPI	Service Provider Interface
SPX	Sequenced Packet Exchange (Novell)
SQL	Structured Query Language
SS	Switching System
SS7	Signaling System 7
SSCP	System Services Control Point (IBM)
SSCP/PU	System Services Control Point/Physical Unit (IBM)
SSL	Secure Sockets Layer
SSP	Service Switching Point
STDM	Statistical Time Division Multiplexing
STP	Shielded Twisted-Pair (wiring)
STP	Signal Transfer Point
STP	Spanning Tree Protocol
STS	Shared Telecommunications Services
STS	Synchronous Transport Signal
STX	Start of Transmission
SUBT	Subscriber Terminal
SVC	Switched Virtual Circuit
SWC	Serving Wire Center
SYNTRAN	Synchronous Transmission
T1	Transmission service at the DS1 rate of 1.544 Mbps
T3	Transmission service at the DS3 rate of 44.736 Mbps
TA	Technical Advisor
TA	Technical Advisory

TAG	Technical Advisory Group
TAPI	Telephony Application Programming Interface (Microsoft)
TASI	Time Assigned Speech Interpolation
TB	terabyte (trillion bytes)
Tbps	terabits-per-second
TCAP	Transaction Capabilities Applications Part
TCP	Transmission Control Protocol
TDD	Time Division Duplexing
TDM	Time Division Multiplexer
TDMA	Time Division Multiple Access
TDMA/TDD	Time Division Multiple Access with Time Division Duplexing
TDR	Time Domain Reflectometry
TFTP	Trivial File Transfer Protocol
TIA	Telecommunications Industry Association
TIB	Tag Information Base
TIMS	Transmission Impairment Measurement Set
TL1	Transaction Language 1
TMN	Telecommunications Management Network
TRS	Telecommunications Relay Services
TSAPI	Telephony Services Application Programming Interface (Novell)
TSI	Time Slot Interchange
TSR	Terminal Stay Resident
TTRT	Target Token Rotation Time
TTY	Text Telephone
TV	Television
TWX	Teletypewriter Exchange (also known as telex)
TX	Transmit
UART	Universal Asynchronous Receiver/Transmitter
UBR	Unspecified Bit Rate
UDP	User Datagram Protocol
UHF	Ultra High Frequency (300 MHz to 3 GHz)
UI	Unit Intervals
UMS	Universal Messaging System
UN	United Nations
UNI	User-Network Interface
UPS	Uninterruptible Power Supply
USDLA	United States Distance Learning Association
USNC	U.S. National Committee
UTP	Unshielded Twisted-Pair
VAR	Value-Added Reseller
VBNS	Very High-Speed Backbone Network Service
VBR	Variable Bit Rate
VC	Virtual Circuit
VCI	Virtual Channel Identifier
VCR	Videocassette Recorder
VDSL	Very high-speed Digital Subscriber Line
VF	Voice Frequency

VFN	Vendor Feature Node
VG	Voice Grade
VHF	Very High Frequency (30 MHz to 300 MHz)
VLF	Very Low Frequency (less than 30 kHz)
VLR	Visitor Location Register
VLSI	Very Large-Scale Integration
VM	Virtual Machine
VMS	Virtual Machine System (Digital Equipment Corporation)
VOD	Video On Demand
VPI	Virtual Path Identifier
VP	Virtual Path
VPN	Virtual Private Network
VSAT	Very Small-Aperture Terminal
VT	Virtual Terminal
VT	Virtual Tributary
VTAM	Virtual Telecommunications Access Method (IBM)
WACS	Wireless Access Communications System
WAN	Wide Area Network
WATS	Wide Area Telecommunications Service
WCS	Wireless Communications Service
WDCS	Wideband Digital Crossconnect System
WDM	Wave Division Multiplexing
WGS	Worldwide Geodetic System
WLAN	Wireless Local Area Network
WLL	Wireless Local Loop
WORM	Write Once Read Many
WTO	World Trade Organization
WWW	World Wide Web
WWW3	World Wide Web consortium
XNS	Xerox Network System (Xerox Corporation)

Index

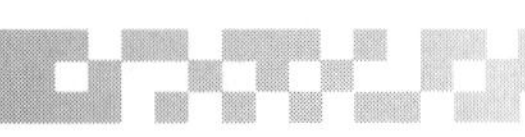

A

B

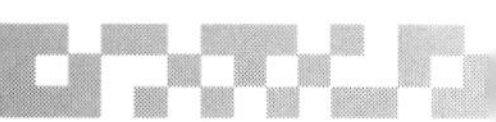

C

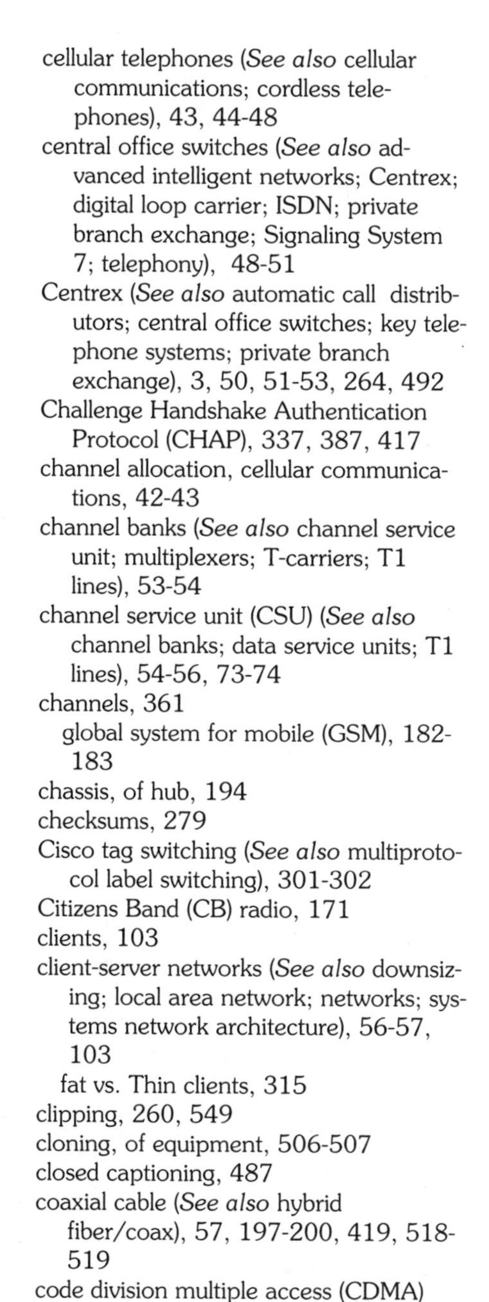

D

E

F

G

H

I

J

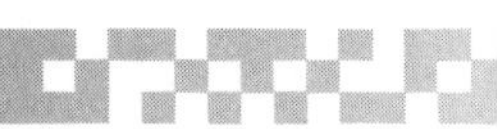

K

L

M

N

O

P

Q

R

 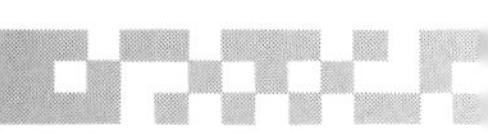

S

T

U

V

W

X

About the Author

Nathan J. Muller (Huntsville, AL) is a consultant who specializes in telecommunications. He has written 1500 articles for nearly 50 magazines, and 14 other books, including *The Totally Wired Web Toolkit*.